D1329864

Solute Processes

LANDSCAPE SYSTEMS

A series in Geomorphology

Editor

M. J. Kirkby, *School of Geography*
University of Leeds

Hillslope Hydrology:
Edited by
M. J. Kirkby, *School of Geography,*
University of Leeds

Soil Erosion:
Edited by
M. J. Kirkby, *School of Geography,*
University of Leeds
and
R. P. C. Morgan, *National College of Agricultural Engineering,*
Bedford

Slope Instability
Edited by
D. Brunsden, *Department of Geography,*
King's College London
and
D. Prior, *Coastal Studies Institute,*
Louisiana State University

Hydrological Forecasting
Edited by
M. G. Anderson, *Department of Geography, University of Bristol*
and
T. P. Burt, *Huddersfield Polytechnic School of Geography,*
University of Oxford

Solute Processes
Edited by
S. T. Trudgill, *Department of Geography,*
University of Sheffield

Solute Processes

Edited by

S. T. TRUDGILL

Department of Geography, University of Sheffield

A Wiley–Interscience Publication

JOHN WILEY & SONS

CHICHESTER · NEW YORK · BRISBANE

TORONTO · SINGAPORE

Library of Congress Cataloging in Publication Data:

Main entry under title:

Solute processes.

(Landscape systems)
'A Wiley–Interscience publication.'
Includes index.
1. Landforms. 2. Solution (Chemistry)
3. Water chemistry. I. Trudgill, Stephen T.
(Stephen Thomas), 1947— II. Series.
GB406.S58 1986 551.3 85–9557
ISBN 0 471 90819 3

British Library Cataloguing in Publication Data:

Solute processes.—(Landscape systems)
1. Geophysics
I. Trudgill, S. T. II. Series
551 QC806

ISBN 0 471 90819 3

Printed and bound in Great Britain

Contents

Contributors

T. P. BURT, *School of Geography, University of Oxford, Mansfield Road, Oxford OX1 3TB, UK*

R. W. CRABTREE, *Water Research Centre, Engineering, P.O. Box 85, Frankland Road, Blagrove, Swindon, Wilts SNJ 8YR*

R. CRYER, *Department of Geography, University of Sheffield, Sheffield S10 2TN, UK*

J. GUNN, *Department of Environmental and Geographical Studies, Manchester Polytechnic, Chester Street, Manchester M1 5GD, UK*

M. J. KIRKBY, *Department of Geography, University of Leeds, Leeds LS2 9JT, UK*

K. R. J. SMETTEM, *CSIRO Division of Soils, Private Bag No. 2, Glen Osmond, S. Australia, 5064*

D. A. SPEARS, *Department of Geology, University of Sheffield, Sheffield S10 2TN, UK*

W. T. SWANK, *Coweeta Hydrologic Laboratory, Route 1, Box 216, Otto, North Carolina 28763, USA*

S. T. TRUDGILL, *Department of Geography, University of Sheffield, Sheffield S10 2TN, UK*

D. E. WALLING, *Department of Geography, University of Exeter, Amory Building, Rennes Drive, Exeter, Devon EX4 4RJ, UK*

B. W. WEBB, *Department of Geography, University of Exeter, Amory Building, Rennes Drive, Exeter, Devon EX4 4RJ, UK*

Series Preface

This is the fifth book in the present series on *Landscape Systems*. Each is intended to present the state of the art in a topic related to the earth's surface, in its natural state and as modified by man. Like other volumes in the series, *Solute Processes* has been written by a number of authors, each of whom is contributing within his own specialist field, but within the context of an overall editorial framework to guarantee proper coverage of the subject at the current research level. This approach provides a comprehensive high level textbook written by acknowledged experts, revealing current controversy and current research directions without needless duplication. The series covers topics in Geomorphology, but is almost equally relevant to Hydrology, Soil Science, Agriculture and Forestry.

Previous books in the series have been on *Hillslope Hydrology*, *Soil Erosion*, *Slope Instability* and *Hydrological Forecasting*. *Solute Processes* covers a topic of considerable interdisciplinary overlap, in which the work of ecologists, hydrologists and geomorphologists is the most central. It also covers a wide range of relevant time spans, often uneasily, from the time spans of kinetic chemical reactions in the soil to the periods over which landscapes effectively evolve. Solutes provide the paradox that while streamwater samples readily provide estimates of long-term catchment denudation, yet our knowledge of the detailed soil chemistry and nutrient exchanges is still far from allowing forecasts of those long-term rates; still less of their spatial distributions. It is thought timely to bring together some of these disparate approaches and data sets to encourage both broader and more detailed studies of solute processes and of the soil and landform assemblages which they help to produce.

Research on our physical environment is still in a phase of rapid development based on incomplete foundations, particularly for the field scale of interest at perhaps a hectare-hour level of spatial and temporal resolution. At these scales our detailed knowledge of physics and chemistry is so complicated by the need to obtain aggregate behaviour for a combination of variable spatial units that it is in many cases largely irrelevant. For many cases we can rely on no relevant and secure foundations beyond statements of mass balance. The problem is no less trivial than the problem of changing from quantum to aggregate descriptions of a solid body, and we are still some way from coming to terms with it in the environmental sciences. It is exacerbated by the practical difficulties of fully representing field variability in material properties, or in water or sediment flows. I believe that the best prospect of major advances in our understanding is through the joint endeavours of workers in many related fields, including both traditionally laboratory and traditionally field scientists. This series is intended to help build bridges between neighbouring disciplies, and to interest a wider scientific community in our common problems, and so stimulate the detailed thought and discussion which lead to scientific advance.

May 1985 Mike Kirkby

Preface

The concern of this book is with solute processes in the environment with special emphasis on the geomorphological context. In the existing literature there is a noticeable gap between the treatments of aqueous chemistry from a chemist's point of view and the work of those who study solute movements and losses in the field. The chemist's work tends to be more rigorous but it is not always applicable to the understanding of field situations. On the other hand, field data on solute mobility at the catchment scale may reflect the operation of natural processes but the data may not always be easy to interpret in terms of theory or in terms of discriminating between the complexities of the sources. In the field, the spatial scales of the variations in chemical processes are often unknown, even though the chemical processes at any one particular point may be accurately described. In addition, there is also a gap between short-term field studies of solute movement and studies of longer-term landform evolution. The aims of this book are to consider the fundamental models of solute movement, to review the work on field observations and to assess the attempts at applying both of these to the study of landforms. The bringing together of these aspects represents a substantial challenge: the book is not able to provide comprehensive, definitive answers in many cases, but it is able to review the problems and the state of the art. It is clear that continued cooperative work is needed by chemists prepared to work in the field environment, field workers trained in chemistry and geomorphological modellers before the gaps can be bridged more satisfactorily.

The focus of the book is a geomorphological one, concerned ultimately with the way in which solute processes interact with and help to produce landforms. The interactions with the wider aspects of environmental and biological systems are also considered. Fundamental theory is considered, and there are assessments of solute processes in rock, soil, vegetation, atmospheric, runoff and fluvial systems. The points of view include assessments of overall budgets, biological regulation of solute flow, soil processes, groundwater processes and models of landform evolution, together with assessments of possibilities and requirements for future work.

Solute Processes
Edited by S. T. Trudgill

CHAPTER 1

Introduction

S. T. Trudgill

Department of Geography,
University of Sheffield

1.1. SOME OF THE CHALLENGES

There exist large numbers of models concerning solute processes under defined experimental conditions, but knowledge of how they may apply at the field scale in relation to landform evolution is limited. It is also often widely known that solutional erosion processes are important in landform evolution, but it is not often widely known how important they are in relation to other landforming processes. In addition, there are few assessments of how solutional denudation is spatially distributed in relation to such features as soil and topography. Until there is a fuller understanding of these topics it is difficult to predict which parts of the landscape are being eroded more than others. This is a crucial consideration since the assessment of differential erosion is a fundamental topic in geomorphology.

There is adequate knowledge to formulate equations concerning dissolutions processes *in vitro*; but, *in vivo*, complex and impure mineral phases may dominate and equilibrium conditions may not prevail because of limited solid-phase:liquid-phase contact. Thus, it can be difficult to apply geochemical theory directly to natural soil–bedrock–water systems.

There is also considerable detailed hydrochemical knowledge concerning solute concentrations in runoff waters (including overland flow, throughflow and streamflow) and in groundwater, but the geomorphological significance of the data gained is, again, often unclear because of the frequently limited knowledge of solute sources and their spatial distribution. Furthermore, it may be known in detail how runoff from a drainage basin responds to rainfall events, but we do not necessarily know what the cumulative effects of such events are on the evolution of landforms.

There are three main reasons for this greater knowledge of hydrochemical processes relative to the study of its application to landform evolution. Firstly, solute concentrations and discharge are relatively easy to measure. The technology of automatic water samples, water discharge measurements and water analysis has been widely adopted and has, indeed, become entrenched in geomorphology. The approach and the data gained have therefore become almost a subject in their own right; but the alliance has often been more with hydrology than with the study of landform evolution. Secondly, there is the related point that this widespread adoption has been encouraged by research council policies and perceived social pressures for the greater applicability of research work: the topic of water quality is clearly a socially useful avenue for research. This is not to be denigrated, but it has meant that the geomorphological aspects of such work may have been relatively under-researched. Thirdly, even when the aim of the study is a geomorphological one, it is often difficult to relate short-term process observations to the longer-term evolution of landforms. This is because landforms frequently show considerable inheritance from former erosional regimes when differing climatic conditions existed. This degree of inheritance is often more evident the larger the size of the landform, with smaller landforms showing greater adjustment to present-day climates. Thus, much of current solutional erosion in Great Britain, for example, is often only a small modification of larger landforms produced under glacial and periglacial regimes.

Geomorphologists have thus tended to focus on the operation of current hydrochemical processes, often in an applied context, and the study of process–form relationships has either been limited to small features, where rapid response is evident, or it has been the province of mathematical modelling. The challenges of process–form studies of the larger landforms which show considerable degrees of inheritance still remain substantially untackled.

The study of the relative importance of solutional and non-solutional processes is an important topic. This is, however, also an under-researched topic especially, again, in an historical dimension where the balance may have been different from that existing at present. It is also clear that some geomorphologists have worked on solutional erosion processes as a topic in itself, neglecting non-solutional processes and viewing the evolution of landforms as a subsidiary consideration. This has often been the case, for example, in limestone solutional studies where the effects of abrasion and frost shattering have not been studied with the same effort as solution processes. Other geomorphologists have tried to provide explanations of landforms which involve a more open approach, assessing the importance of solution processes in relation to other processes and the overall evolution of landforms.

There is thus a great diversity in approach, with substantial challenges remaining in bringing together the different approaches. Most of the chapters in this book are reviews of the state of the art of the various topics and approaches at the present time. They also, however, consider the possible strategies for future research. One fundamental task for the future is the testing and refinement of existing geochemical hypotheses about solute process so that they have application to field phenomena. It is also seen as important that field measurements should have a logical, theoretical basis as well as an empirical one. In addition, there is a need for the further formulation of hypotheses which relate solute processes to landform evolution. Continued work on monitoring solutes in runoff and on *in vitro* experiments are needed, but many of the challenges provided by the study of the relationships between process and form through time need to be tackled more specifically.

In this first chapter, some of the basic considerations and models will be outlined. These provide a basis for discussion in subsequent chapters, especially within a geomorphological context. It should also be stressed, however, that the study of solute processes represents a field of research in its own right, quite apart from its application in a geomorphological (landforming) context (as mentioned, for example, above when discussing applied topics). Thus, some considerations in the book will stand as topics in themselves, irrespective of any geomorphological applications.

1.2 SOLUTE SOURCES

Water running off the land contains chemical elements which have a variety of sources (and which are discussed in a number of subsequent chapters). Some solutes present in runoff are derived from precipitation and dry fallout arriving at the earth's surface (Chapter 2); others are derived during the passage of water through living or dead biomass (Chapter 3), through soils (Chapter 4) or through rock (Chapter 5). It is the latter two which represent a direct denudational component (as discussed further in subsequent chapters). Solutes derived from biological sources will, in part, represent an indirect denudational component as some of them will have been derived initially from the soil and the bedrock. It is the solute losses from the soil and bedrock which are of interest in a geomorphological context, since it is these losses which result in the lowering of land surfaces and the opening up of sub-surface transmission routes for the movement of water and solutes. It is therefore clear that in order to understand solutional denudation by means of a study of solutes in runoff, it is important to be able to partition the solutes into those which have a pedologic, lithologic, atmospheric or biologic origin. Atmospheric inputs should be subtracted from any consideration of the relationships between solutes in runoff and denudation (though this is often less than straightforward, as discussed in Chapter 2). Biological sources are, however, especially difficult to allow for (as discussed in several chapters). Biological solute sources essentially represent leakages of chemical elements from biological cycles where there is uptake by living biomass and release from dead organic material. The uptake may either be directly from the soil or bedrock or from recycled sources from plant litter and animal excreta. The former represents direct denudation. The latter may have been derived from soil and rock, and can thus be seen as a denudational loss, but it is not a loss from the ecosystem; elements in circulation can also have been derived from atmospheric sources. Thus the relationship between nutrient uptake from the soil and rock and solute loss from living and dead biomass to runoff water is not a simple one.

Many studies of denudation assume that there is a steady state between vegetation uptake, cycling and the release of elements into solution from biological sources; however, the validity of this assumption is often unknown. The position is also complicated by the temporary immobilization of elements within biomass, soil and bedorck. Mobilization would have a denudational effect, but storage in the system also has to be allowed for. Such stored solutes would not tend to be found in the runoff, except perhaps as a delayed component; but their mobilization would have had a geomorphological effect: thus output is a nett effect, masking several internal processes. Because of these difficulties, it could be questioned whether such studies are the best way to proceed during geomorphological research and to ask whether more direct studies would not be useful. However, since such studies are already located within solute geomorphology it will be useful to specify the main features below and to return to the discussion in later chapters. The losses, uptakes, immobilizations and circulations can be usefully represented in a summary diagram (Figure 1.1). The notation is as follows:

I_a = inputs from the atmosphere, partitioned into input to plant, I_{ap}, and to the soil, I_{as}
R = recycling
L_c = losses from cycling
L_s = losses from soil
L_r = losses from rock
O_{gw} = output in groundwater
Im_s = immobilization in soil
I_p = Input from plant
S_b = Storage in biomass
O_{ro} = Output in runoff
U_s = Uptake from soil
U_r = Uptake from rock
Im_r = immobilization in rock
M_s = mobilization in soil
M_r = mobilization in rock

Solute output = $O_{ro} + O_{gw}$, where $O_{ro} = L_r + L_s + L_c$. The geomorphic effect may be considered as $O_{ro} + O_{gw} - I_a$, but U_s and U_r also have geomorphic effects since they represent a denudational component. They may or may not be reflected in O_{ro} or O_{gw}, depending on the efficiency of recycling and the

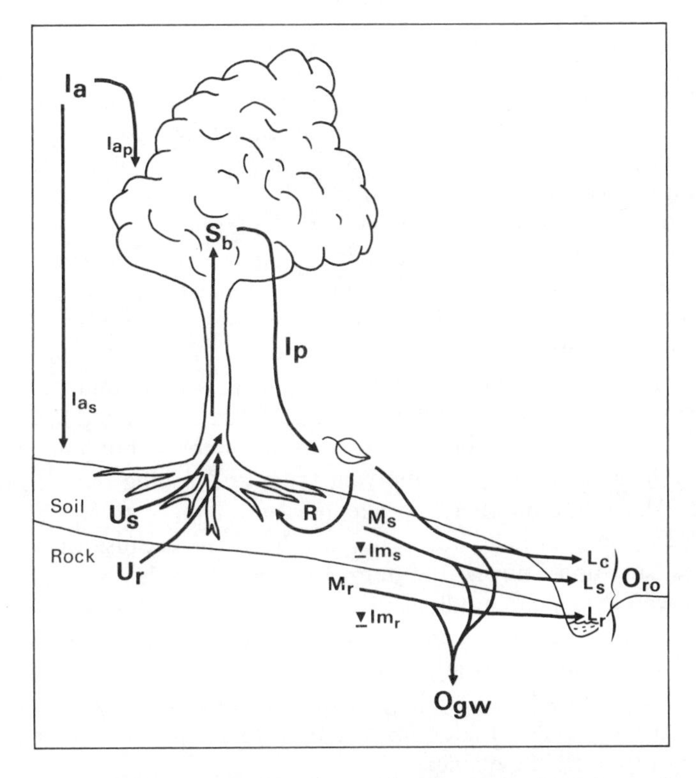

Figure 1.1. Losses, uptakes, immobilizations and circulations of solutes in an ecosystem. For notations see text

partitioning of I_p into R and L_c. Furthermore, I_a, L_c, L_s and L_r can also have Im_s and Im_r subtracted from them at some stage. Thus, the simple study of O_{ro}, even with I_a subtracted, cannot necessarily be expected to yield a true picture of denudation—unless a true steady state exists where R is constant, $U_s + U_r = L_c$, and Im_s, Im_r, O_{gw} and S_b are all zero; then, and only then, will $L_c + L_s + L_r - I_a$ equal the denudation in the soil rock system, as:

$$O_{ro} = L_c + L_s + L_r = (M_s + M_r) - (I_a + (I_p - R) + Im_s + Im_r + O_{gw}) \quad (1.1)$$

where

$$I_p = (U_s + U_r) - (S_b + I_{ap}) \quad (1.2)$$

Specification of solute sources thus remains a major endeavour in solute geomorphology, and the relationship between the presence of solutes in runoff, a variety of sources and landform evolution is a fundamental focus of study.

1.3 MODELS OF SOIL AND WATER FLOW

Accepting the points already made concerning the difficulties of understanding the complexities of natural systems, rather than dwelling within these complexities, science likes to proceed by making simplifying assumptions concerning real-world processes and then it tries to model these simplified systems. If such models are achieved, the intention then is to apply the models to the real world in a predictive manner, calibrating them or modifying them in the light of the evidence available (and assuming that the way in which the models are tested is relevant and appropriate). Often, however, the simplified models have an attraction and rigour in themselves so that this intention is not always fully realized. It is important to remember that the simplified model is not the real thing. Thus, one should have rigorous and defined models to apply to the natural systems, but one must also remember to take the step of applying them and testing them against the evidence. The other pitfall is to take measurements without reference to theory. Field measurements are sometimes justified because they represent the natural situation—this is true but *per se*, they have less value than when they are taken to test and modify existing theories; then they will tend to lead to a more rapid development of new ideas (given also that many ideas were developed when actually testing something else!). Thus, it will be appropriate to outline here some basic existing models of water and solute mobility in order to provide a basis for later discussion, when the models can be reviewed in the light of the field evidence.

1.3.1 Solute models

Models of solute behaviour include those which deal with transfer from the solid phase to the liquid phase and also those which deal with transport of material already in solution. In both these cases, concentration gradients are important driving forces of solute flux but, again in both cases, the motion of the solvent has a crucial role to play in modifying the gradients and the mass transfer of solutes. Models dealing with solid-phase:liquid-phase transfers will be discussed first; these can also be referred to as solubilization or dissolution models.

The simplest situation to model is the dissociation of a soluble salt and its diffusion into static water (Figure 1.2). Here, for example, sodium chloride dissociates into Na^+ and Cl^- which diffuse down the concentration gradient. The overall process can thus be described as a diffusive flux, driven by the steepness of the concentration gradient. In general terms, the rate of mass transfer from the solid phase to the liquid phase is proportional to the difference between the concentration C of the solute in the bulk solution and the maximum concentration at saturation, C_s. Thus, the rate of mass transfer

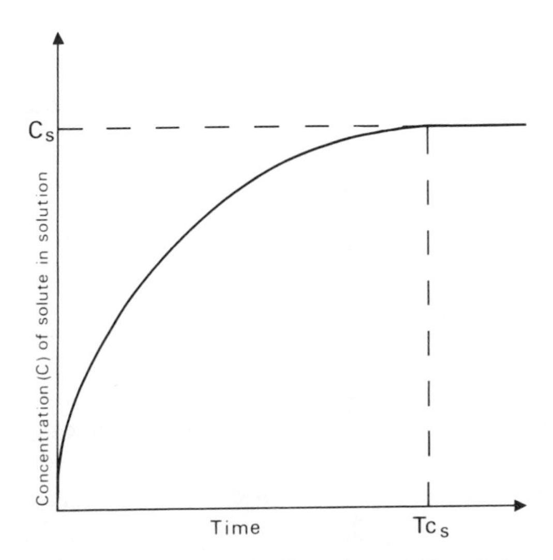

Figure 1.2. Simple dissolution of a soluble salt by dissociation and diffusion. Equilibrium is achieved as C approaches C_s. For notation, see text

decreases as C approaches C_s. This can be described by the Nernst equation (Figure 1.2):

$$\frac{dm}{dt} = K_d(C_s - C) \tag{1.3}$$

Where m is mass transfer, t is time, K_d is a transport rate constant (see Equation (1.4)), C_s is the solute concentration at saturation, C is the solute concentration in the solvent; and:

$$K_d = \frac{D\,A}{\delta\,V} \tag{1.4}$$

where D is the molecular diffusion coefficient, δ is the transport distance, V is the volume of water in contact with A, and A is the solid surface area. For simple solid:liquid configurations over constant values of δ and under defined, comparable situations, K_d reduces to D.

In nature many equilibria are, however, not established so simply, and a reactant is involved (Figure 1.3); and so the equilibria involve other ion species not wholly derived from the solid phase. For example, during hydrolysis of calcite by H^+, the forward reaction R_f involves the movement of H^+ to the solid and the backward reaction R_b, involves the movement of Ca^{2+} and HCO_3^- to the liquid (see §9.2).

During weathering reactions, the principal reactants are H^+ derived from carbon dioxide dissolved in water, H^+ derived from organic acids, O_2 and

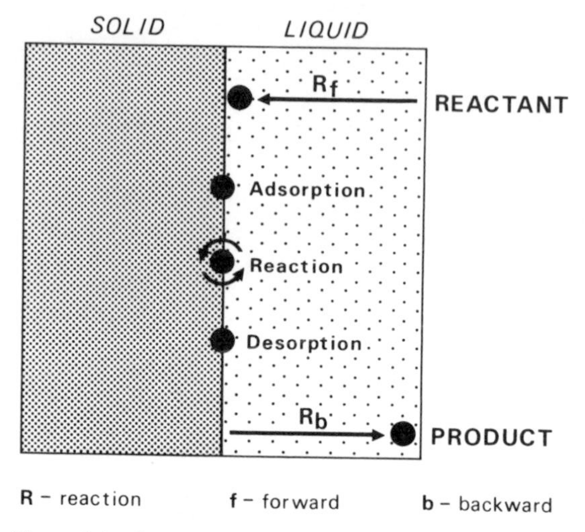

Figure 1.3. Chemical reactants involving a reactant with a forward reaction, R_f, and a backward reaction R_b.

organic chelates. The first of these is derived:

$$CO_2 + H_2O \longrightarrow H_2CO_3 \longrightarrow H^+ + HCO_3^- \qquad (1.5)$$

Thus, measurements of weathering potential—and potential solute provision—focus on pH and Eh measurements. Complex heterogeneous equilibria in soils may also involve OH^- and a number of anions.

Once in solution, in porous media such as soils and rock, solute flux in static water in the pores can be expressed in terms of diffusion down a concentration gradient dC/dx:

$$F = -D\frac{dC}{dx} \qquad (1.6)$$

where F is the flux of solute (mass per distance per time), D is the diffusion coefficient of the solute in question (distance per time), C is the concentration of solute (mass per unit volume of soil), and x is distance. A negative sign is used to imply that movement can only occur down gradient. This equation is similar to Equation (1.3) except that the driving force is seen simply in gradient terms, not equilibrium terms.

If the water itself is moving, mass transfer in moving water—convection—occurs. This has to be added to diffusion for a complete description of solute movement processes. In practice, in soils and other porous media it is difficult to separate the effects of diffusion and convection; and because of a range of pore water velocities, the general term 'dispersion' is used for solute flux in mobile soil water (Schiedegger, 1961; and see further discussion in

Chapter 4):

$$F = -D^* \frac{dC}{dx} + vC_1 \qquad (1.7)$$

where D^* is a dispersion coefficient for the solute in question, v is water flux in direction x, and C_1 is the concentration of solute in mobile soil water.

For further discussions of these topics in soils the reader is referred to Wild (1981) and Nye and Tinker (1977, pp. 9–10, 69–83) and Chapter 4 of this volume.

The mobility of water not only influences mass transfer of solutes already in solution, it also influences the procedure of chemical reactions during dissolution. Rapid movement of water low in solute concentrations will keep the value of $(C_s - C)$ in Equation (1.3) high, encouraging further dissolution; if the mobile water is high in solute concentration, however, the effect on dissolution at any one point will be less.

In an important paper, Berner (1978) considers the role of flushing frequency in a dissolution system. For the purposes of a simple modelling, he considers a finite system where the input water has a concentration of zero. Two situations are considered. Firstly, where detachment (Figure 1.4(a)) is rapid relative to the rate of water flow (i.e. very rapidly soluble material or very slowly flowing water), the products of dissolution readily build up in a saturated layer and the overall dissolution rate is then limited by the rate of flow, i.e. it is transport-controlled (Figure 1.4(b)). Secondly, where detachment is slow relative to the rate of water flow (very slowly soluble material or rapid water movement), the products of dissolution do not build up rapidly, and any produced can be removed. The overall rate is then limited by the rate of detachment, i.e. it is surface-reaction-controlled (Figure 1.4(c)).

Transport control is exhibited with very soluble compounds, such as the simple salts (Table 1.1), whereas surface reaction control is exhibited by the low solubility compounds. Transport controlled dissolution tends to produce smooth surfaces when viewed using a scanning electron microscope. This is because ion detachment over the entire surface is so rapid that no relative etching occurs, giving smooth rounded surface morphologies. By contrast, in surface reaction control, ion detachment is so slow that selective dissolution occurs, with etching out of higher energy positions such as crystal interface intersections and other inhomogeneities. Berner concludes that many weathering processes are governed by surface chemical reactions and not by transport processes. This implies that ion detachment is slow relative to water movement in many earth surface situations. This calls into question the use of equilibrium considerations, since water flow at any one point is liable to be such that soil-phase:liquid-phase equilibrium is not reached at that point. Since the consideration of $C_s - C$ as the driving force for dissolution applies

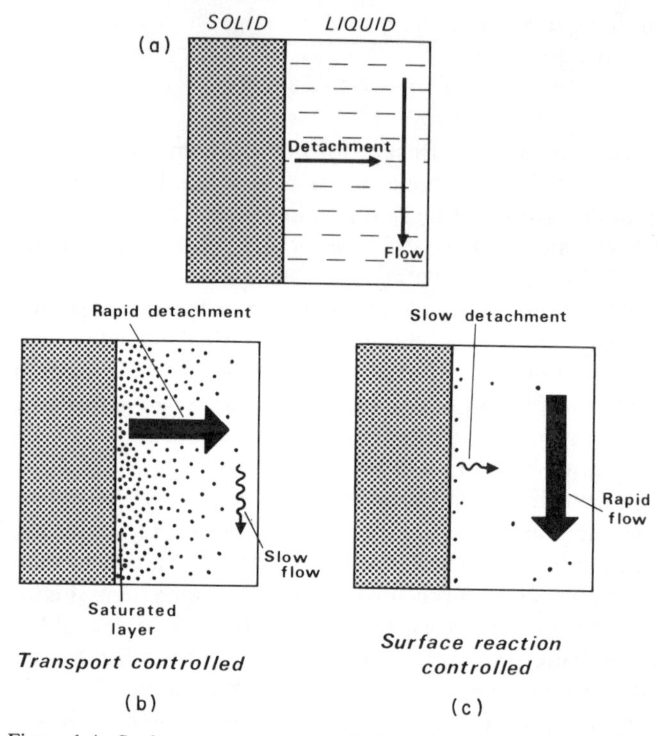

Figure 1.4. Surface reaction controlled and transport controlled dissolution systems. For explanation, see text

Table 1.1. High solubility, transport controlled, and low solubility, surface reaction controlled, solutes (modified from Berner, 1978)

Solute	Solubility ($M\,l^{-1}$)
Highly reactive, transport controlled solutes	
NaCl	5×10^0
$MgCl_2.6H_2O$	5×10^0
KCl	4×10^0
$Na_2CO_3.10H_2O$	3×10^0
$MgSO_4.7H_2O$	3×10^0
$Na_2SO_4.10H_2O$	2×10^{-1}
$CaSO_4.2H_2O$	5×10^{-3}
Low reactive, surface reaction controlled solutes	
Opaline SiO_2	2×10^{-3}
$CaCO_3$	6×10^{-5}
$NaALSi_3O_8$ feldspars	6×10^{-7}
$KAlSi_3O_8$ feldspars	3×10^{-7}

to diffusion down a concentration gradient, and since this gradient will only build up in transport controlled situations, the model is not seen by Berner as an appropriate one for many reaction controlled situations when the rate of reaction provides the limiting factor, rather then the diffusion gradient providing the driving force. However, these considerations apply to solid-phase:liquid-phase equilibrium at a point.

Another important consideration is that the water is making a journey through the porous medium. This means that while equilibrium may not be liable to be established at a point because of slow reaction rates, equilibrium is liable to be established progressively as the water percolates through the porous medium. Thus the progression of C to C_s (Figure 1.2) is liable to occur during percolation through soils and bedrock rather than at any point in the porous medium with static water. This is an important point since the vertical position in the porous medium at which equilibrium occurs during water percolation establishes the position above which denudation will tend to occur and below which it will be minimal. Clearly, this level may vary from solute to solute and with the rate of water flow. Thus, there are two considerations. Firstly, *at a point* the relationship between reaction rate and flow rate is important in governing the overall process. Secondly, there is the time of residence *in the system*. Frequently, although contact time at a point is insufficient to establish equilibrium, equilibrium may be established in the system as a whole. Thus, water output from a porous medium may be in chemical equilibrium with the solid phase. This has led to the assumption that chemical processes can be understood in terms of equilibrium calculations; and indeed they often can when the bulk system is under consideration (Carson and Kirkby, 1972), but the approach does mask the feature that the equilibrium is often obtained progressively during percolation, with greater solutional erosion near the surface. Thus, in terms of surface lowering calculations, the gross calculation may be correct but the precise distribution of the rates will be unknown. Their vertical distribution is liable to be greater at the surface than the mean system rate, and then attenuate with depth.

The solutional denudation model of Carson and Kirkby assumes that the flow of water governs the amount of solute removal because, for the bulk system, the time needed to reach equilibrium is low relative to the residence time of the water in the system as a whole. Their model thus proposes that the progression of weathering is related to discharge and the saturated solubility of the solute:

$$\frac{\partial p}{\partial t} = k \frac{\partial q_z}{\partial x} p \tag{1.8}$$

where p is the amount of solute present in the unweathered mineral phase, expressed as a proportion of that phase (i.e. % weight of solute divided by total weight of 100%), t is the elapsed time, k is the saturated solubility of the solute (assumed to be the same for all silicate minerals), x is the distance travelled by the solvent, and q_z is the solvent volume discharged through unit area of the mineral phase. From this, the amount of mineral remaining unweathered (p) will be:

$$p = p_o e^{-kt_*} \tag{1.9}$$

where p_o is the initial amount of solute present in the unweathered material, and t_* is a measure of relative time.

Although Berner concludes that many processes, at a point, are reaction controlled, the above would appear to be valid for many bulk systems where equilibrium is reached for the solid–water system as a whole. These topics are discussed further in Chapter 10 and by Verstraten (1980). Verstraten considers the relationship between the theoretical water chemistry prediction and actual solute concentrations. There would appear to be limitations on the applications of equilibrium theory (Morgan, 1967), but it is a useful predictive first approximation for hillslope modelling, as discussed further in Chapter 10.

A final weathering model is that of Curtis (1976). He calculated the energy changes involved in weathering reactions from the changes in the energies of the reactants and products involved in the reactions. Chemical reactions involve energy transfer between the reactants and the products, and the energy involved is usually referred to as Gibbs free energy ($G°$). The energy changes for a reaction can be written:

$$\Delta G°_r = \Delta G°_{fp} - \Delta G°_{fr} \tag{1.10}$$

where $\Delta G°$ is the change in energy, and the subscripts are: free energy of the reaction (r), sum of free energies of the products (fp), and sum of free energies of the reactants (fr).

If $\Delta G°r$ is negative, then $\Delta G°_{fr}$ is higher than $\Delta G°_{fp}$ and the reaction is likely to proceed; the greater the negative number, then the less stable is the mineral. For example, calculated values of $\Delta G°_r$ show that methane and pyrite are highly unstable at the earth's surface (with $\Delta G°_r$ values of -184.9 and -583.5 kcal mol^{-1} respectively), compared with some olivines at around -40 and some feldspars at -15 to -20.

The use of such solute models in geomorphology is discussed in Chapter 10 and it will also be mentioned in further chapters. The applicability of many of these models to the field has not yet been fully tested, except for the work of Verstraten and other geochemists mentioned in Chapter 10 and elsewhere; but again the work has often largely focused on water chemistry rather than necessarily applying the work to landform evolution.

1.3.2 Water flow models

As with solute models, water flow models have initially focused on idealized, homogeneous porous media. Thus, under these conditions, water flow can be predicted from a knowledge of hydraulic gradient using Darcy's law (Hillel, 1971):

$$q = K\frac{H}{L} \tag{1.11}$$

q is discharge, K is hydraulic conductivity, and H/L is the hydraulic gradient, changing head H over length L.

Like some of the solute models described above, Darcy's Law adequately predicts macroscopic water flow; but in structured soils or other dual-porosity media, such as fractured aquifers, detailed flow paths may vary considerably. Such models are discussed further in Chapter 4.

Recently, attention has become focused on bypassing flow in soils (Bouma *et al.*, 1981; Trudgill *et al.*, 1983a,b; Smettem *et al.*, 1983; Smettem and Trudgill, 1983). Here, some portion of the infiltrating water flows much faster than others, following preferential pathways such as macropores (Beven, 1983). The occurrence of bypassing flow in strongly structured soil can be predicted in a manner from the overland flow model used by Kirkby (1978). Bypassing flow can be predicted to occur at a rate:

$$Q_f = I - T_p \tag{1.12}$$

where Q_f is bypassing flow, flowing round pedal structures (mm h^{-1}), I is rainfall intensity (mm h^{-1}), and T_p is the transmission rate of soil peds (mm h^{-1}). T_p is not easy to evaluate but it can be approximated for saturated conditions by measurements of ped infiltration using small, ped-sized soil samples (Trudgill *et al.*, 1983b). In terms of solute uptake, such bypassing flow decreases the opportunity for solute uptake, since much of the soil mass is bypassed. The full implications of such flows have not been evaluated in solute geomorphology, though it is possible that when large, bulk systems are modelled, such flow becomes less important to model than when shallow soils or layer models are considered.

Models which partition water flow into contrasting flow rates are relatively well developed in soil science (Chapter 4), but their implications for geomorphology are undeveloped. The relationships between soil water movement and hillslope hydrology are discussed further in Chapter 6.

1.4 CONCLUSION

Two substantial challenges remain in solute geomorphology: firstly, the bringing together of theoretical work and field work and, secondly, the evaluation of the spatial distribution of rates of erosion in order to predict differential landform evolution. In this introductory chapter some of the main challenges have been briefly reviewed. Many of these will be discussed further in succeeding chapters and then reviewed again in the final chapter.

REFERENCES

Berner, R. A. (1978). Rate control of mineral dissolution under Earth surface conditions. *American J. Science*, **278** (9), 1235–1252.

Beven, K. (1983). Macropores in soil. *Water Resources Research.*

Bouma, J., Dekker, L. W. and Huilwijk, C. J. (1981). A field method for measuring short-circuiting in clay soils. *Journal of Hydrology*, **52**, 347–354.

Carson, M. A., and Kirkby, M. J. (1972). *Hillslope Form and Process*, Cambridge University Press, Ch. 9.

Curtis, C. D. (1976). Stability of minerals in surface weathering reactions. *Earth Surface Processes*, **1**, 63–70.

Hillel, D. (1971). *Soil and Water*, Academic Press.

Kirkby, M. J. (1978). Implications for sediment transport. In *Hillslope Hydrology* (M. J. Kirkby, Ed.), John Wiley.

Morgan, J. M. (1967). Applications and limitation of chemical thermodynamics in natural water systems. In *Equilibrium Concepts in Natural Water Systems* (R. F. Gould, Ed.), Am. Chem. Soc. Adv. Chem. Ser. 67, pp. 1–29.

Nye, P. H., and Tinker, P. B. (1977). *Solute Movement in the Soil–Root System*, Blackwell.

Scheidegger, A. E. (1961). General theory of dispersion in porous media. *Journal of Geophysics Research*, **66** (10), 3273–3278.

Smettem, K. R. J., and Trudgill, S. T. (1983). An evaluation of some fluorescent and non-fluorescent dyes in the identification of water transmission routes in soils. *Journal of Soil Science*, **34**, 45–56.

Smettem, K. R. J., Trudgill, S. T., and Pickles, A. M. (1983). Nitrate loss in soil drainage waters in relation to by-passing flow and discharge on an arable site. *Journal of Soil Science*, **34**, 499–509.

Trudgill, S. T., Pickles, A. M., Smettem, K. R. J., and Crabtree, R. W. (1983a). Soil water residence time and solute uptake. 1. Dye tracing and rainfall events. *Journal of Hydrology*, **60**, 257–279.

Trudgill, S. T., Pickles, A. M., and Smettem, K. R. J. (1983b). Soil water residence time and solute uptake. 2. Dye tracing and preferential flow predictions. *Journal of Hydrology*, **62**, 279–285.

Verstraten, J. M. (1980). *Water–Rock Interactions*, BGRG Research Monograph 2, Geo Books, Norwich.

Wild, A. (1981). Mass flow and diffusion. In *The Chemistry of Soil Processes* (D. J. Greenland and M. H. B. Hayes, Eds.), John Wiley.

Solute Processes
Edited by S. T. Trudgill

CHAPTER 2

Atmospheric solute inputs

R. Cryer

Department of Geography,
University of Sheffield

2.1 INTRODUCTION

Geomorphological concern with atmospheric solute inputs is often from the point of view of their contribution to overall solute budgets. Atmospheric inputs need to be subtracted from estimates of outputs in order that assessments may be made of the denudation component of outputs, as was stressed by Goudie (1970). In addition, precipitation acidity is an important factor influencing denudation processes. Much of the work on atmospheric solute inputs has, however, been undertaken from an 'environmental' viewpoint. Thus, it is the current public and scientific preoccupation with 'acid rain' and its potentially damaging interaction with both natural environments and human activities which has focused attention on the atmosphere as a source, not simply of water in various forms, but also of chemical substances.

It has been appreciated for centuries that 'chemicals' are contained in precipitation and that these can be both essential plant nutrients and harmful pollutants, depending on their nature and their concentration. The great expansion of research in the last few decades has, however, demonstrated the complexity of routes by which the atmosphere contributes to the surface and the wide variety of materials contained in this deposition. It has also demonstrated the complexity of the processes involved both before and during deposition, with the resultant spatial and temporal variability in the composition of the deposits and in the rates of deposition.

The term 'atmospheric solute inputs' is used here to include that material, deposited in gas, liquid and solid phases and to vegetation, bare soil, rock and water surfaces, and which may contribute to the solute content of surface waters, soil water and groundwater and thus may figure in the biogeochemical cycling of elements in terrestrial ecosystems. The rate of atmospheric solute input will be of varying significance, depending on the natural background levels of element circulation: in eutrophic systems atmospheric solute input may be negligible, whilst in oligotrophic systems it may provide the sole source of some nutrients or else be a supplement to the nutrient pool which is vital to the maintenance of equilibrium conditions. In addition, some components of the atmospheric input may be chemically reactive and may modify vegetation, enhance chemical denudation and thus affect the chemical quality of drainage waters, while others may be considered essentially inert or 'passive'. Long-term trends in the nature of the atmospheric solute input, both 'active' and 'passive' components, may thus result in changed patterns of nutrient cycling, biological productivity and surface water quality.

The experimental procedures, both in the field and in the laboratory, by which atmospheric solute inputs are monitored and assessed are of paramount importance. Significant uncertainties arise in the interpretation of raw sample quality data derived from 'non-standard', and even from well-designed, instruments. The incorporation of these data into predictive models leads to deficiencies in model forecasts, and the doubtful quality of much historic data makes it difficult to assess long-term trends in the nature of atmospheric solute inputs. Also, reliable input data are required in order to quantify the effects of 'active' atmospheric solute inputs on ecosystems. Accordingly, great pains must be taken to ensure adequate data quality: the procedures described for example by Lindberg *et al*. (1977) show the types of measures necessary to ensure this quality. The following brief review of the processes which are thought to give rise to atmospheric solute input, considered together with the subsequent resumé of the main assessment techniques, should provide a background against which to view the results of experimental assessment of atmospheric solute inputs contained in the last section of this chapter.

2.2 GENERATION, TRANSPORTATION AND DEPOSITION OF ATMOSPHERIC SOLUTES: A REVIEW OF PROCESSES

This chapter is primarily concerned with atmospheric solutes with regard to their observed significance to terrestrial ecosystems. However, it is considered that full awareness of the nature of this input, its magnitude, variability and make-up, will only come from an appreciation of the processes by which atmospheric solutes are generated, transported, transformed and deposited—and the spatial and temporal dimensions of those processes in as much as they affect deposition. Thus any attempt at analysis, explanation and, eventually, prediction of the atmospheric solute input to a point or an area will require early recourse to the physicochemical mechanisms which influence the characteristics of this input. Modelling of the input has therefore progressed from the early empirical approaches of Eriksson (1959) and

Rossby and Egner (1955) to the present application of theoretical principles of atmospheric physics and chemistry by, for example, Slinn (1977). Despite the complexity and sophistication of many current process models the tentative nature of most should, however, be appreciated: 'the present state of knowledge of acid rain is insufficient to permit a quantitative cause and effect analysis' (Durham *et al*., 1981). The ultimate aim of atmospheric solute process models is better prediction of the type and magnitude of both natural and man-influenced effects: as an example, Shaw (1982) expressed the hope that such investigations would help to elucidate the sources and mechanisms of precipitation acidification, thus allowing, in this instance, an assessment of the effects of planned lignite-burning power stations on future atmospheric solute inputs in eastern Canada. Seip (1980) has emphasized the fact that correlations between fish kills in Scandinavian lakes and acid precipitation deposition may in fact be fortuitous and do not necessarily demonstrate any causal relationship: models of the acidification process in the atmosphere and in the lakes are therefore needed to assess the impact of acid precipitation against other possible acidification mechanisms. A better knowledge of the processes of deposition of atmospheric solutes also enables improvements to be made in the design of measurement devices and procedures so that comparable results can be obtained from different periods and from different continents, thus improving the chances of success in process modelling. At present the apparent simplicity of measurement of atmospheric solute inputs in precipitation contrasts sharply with the complexity and variability of the known processes by which gaseous and particulate compounds enter collected precipitation (Skartveit, 1982). Greater understanding of these processes will both result *in* and result *from* improved measurement systems and thus lead to better explanation and prediction of the temporal and spatial patterns of atmospheric solute input.

2.2.1 Sources of atmospheric solutes

According to Chamberlain (1975), in rural areas in Britain and America about $\mu g\ m^{-3}$ of inorganic aerosols is often present in the atmosphere. This is compared with about 14 $\mu g\ m^{-3}$ of organic particles, chiefly pollen grains, at a time of a very high pollen count. The major sources of these inorganic particles are given as primarily 'salt droplets' from sea spray, and blown sand and soil, but also so-called 'anthropic sources'. The latter include the combustion of fuels, reactions of gaseous pollutants in the atmosphere, and sulphates and nitrates either formed by oxidation of SO_2 and NO_2 in the atmosphere or emitted as inorganic particles. As most monitoring and investigation of atmospheric solute inputs has been conducted in coastal or near-coastal areas of North America and Europe it is the sea surface which has most often been referred to as the main source of material which eventually makes up the atmospheric solute input to the land surface. There has been much investigation of the processes of entrainment of material from the sea surface and its 'injection' into the atmosphere, with a view to explaining both the temporal and spatial variability of the magnitude and the composition of atmospheric solute input which is derived from 'natural sources'. Thus the natural background levels of solute input may be established so that 'pollutant' levels can be assessed.

Woodcock (1952) put forward arguments for the generation and entrainment into the atmosphere of small seawater droplets from breaking waves and from jets and film droplets from bursting bubbles in the 'foam mechanism'—a process discussed by Knelman *et al*. (1954) and demonstrated photographically by Koga (1981). Large droplets more than about 80 μm in diameter fall by gravity back to the sea surface immediately, but smaller droplets diffuse upwards, aided by turbulent motions in the atmosphere, and rapidly begin to evaporate to a size determined by the relative humidity. At relative humidity values less than about 40% (Junge, 1963) NaCl forms cubic crystals of around 10 μm diameter (Chamberlain, 1979) which contribute to and dominate the atmospheric aerosol content. Clearly local and regional upwind meteorological conditions and the state of the sea surface influence the entrainment and particle production processes which make material available in the atmosphere for eventual deposition. The natural seasonal and storm-linked temporal variability in the generation mechanisms will be a major factor in determining the spatial and temporal distribution and the nature of deposition, particularly when atmospheric residence times are small. Skartveit (1982) has shown this in southern Norway, where the concentration of sea salt in precipitation at the coast doubles for each 2 ms^{-1} increase in surface (10 m) wind speed in the coastal surf zone.

Given the dominant marine source of material in the atmosphere in coastal regions it might be expected that the detailed composition of this material would be identical with seawater proportions, which are approximately constant. In fact there are frequently significant deviations from the seawater proportions. These deviations are observed to have systematic spatial and temporal dimensions which are normally ascribed to various fractionation and relative enrichment processes, associated with the generation mechanism initially though transportation, transformation and deposition effects have also been noted (e.g. Skartveit, 1982) which will be referred to later. Junge (1963) gave a useful summary of current ideas concerning the fractionation effect, in particular invoking SO_2 and NO_2 reactions in which HCl is liberated. Bloch *et al*. (1966) proposed a mechanism by which bubbles tend to separate out the larger and generally heavier ions, with the greater charge/mass ratio, retain them on their surface, and eject them into the atmosphere on bursting. The resulting droplets thus tend to be enriched in K^+, Br^- and SO_4^{2-} relative to Cl^- and Na^+ compared with seawater in going from the bulk to the disperse phase. Belot *et al.* (1982) have shown that americium-labelled seawater produces bubble droplets which show enrichment in this element by factors of several hundred to 1000 or more. This they ascribe to wide-scale concentration of some elements in the surface microlayer of the water, whereas Blanchard (1982) suggests that a more likely explanation is that it is rising bubbles that selectively scavenge particulate trace metals, a mechanism demonstrated for particulate and dissolved organic matter and for bacteria. Duce and Hoffman (1976) discuss the standard nomenclature and notation concerning enrichment and fractionation indices, and Hunter and Liss (1981) have put forward a surface microlayer sampling device which makes use of the 'bubble-bursting' hypothesis to sample the top 0.5 μm of a water surface. Blanchard also notes that the enrichment factor for any element will vary, if it is due to the bubble-scavenging mechanism, depending on bubble size distribution, bubble-rise distance and the relative production of jet and film drops from the bursting bubbles; it is therefore likely to be very variable in its operation and efficiency as sea surface and meteorological conditions vary. In addition Glass and Matteson's (1973) investigations showed that enrichments in aerosols varied also with aerosol particle size, the relative ion ratios in the bulk solution and its pH. Clearly seawater constituent proportions cannot be assumed to be 'representative' for the movement of material from the sea surface to the atmosphere, nor from the atmosphere back to the surface in deposition.

The existence of processes which modify seawater quality in a variable manner towards precipitation quality was noted some time ago by, among others, Mattson *et al.*: 'It is conceivable that different ions distribute themselves differently under different conditions of dispersion and condensation' (Mattson *et al.*, 1944). Eriksson (1959) found evidence for the existence of sea surface films of organic material which gave rise to an excess of positive over negative inorganic ions, especially divalent ones, which was assumed to exist also in the aerosol droplets produced. Hunter and Liss (1981) and Elzerman (1981) have recently reviewed work on the effect of organic surfactants and other mechanisms and controls of surface enrichment which are particularly effective for heavy metals and organics.

Junge (1972) and Duce and Hoffman (1976) emphasize that the 'marine aerosol' is composed of many substances which have sources other than at the sea surface, which clearly leads to difficulties in interpreting chemical fractionation processes of sea salt particles from atmospheric aerosol composition. Garland (1981) used radioactive sulphate compounds in a laboratory experiment to demonstrate an enrichment of 10–30% in spray relative to seawater, thus showing that not all SO_4^{2-} in precipitation in excess of its seawater proportions relative to Na^+ or Cl^- need be ascribed to anthropic or 'anthropogenic' sources. Boutron (1979), though, found *no* significant enrichment of ocean-derived materials in Antarctic snow (earlier 'significant' enrichments were, instructively, ascribed to sample contamination), but pointed out the difficulties inherent in determining the sources for atmospheric SO_4^{2-} in excess of its seawater proportions when the definition of 'excess' is weak, when production of SO_4^{2-} at the sea surface shows strong variations, and when SO_2 is generated at the surface and undergoes a gas-to-SO_4^{2-} particle transformation in the atmosphere. Garland and Chadwick (1981) found that the sea surface and sea spray may contain substantially enriched concentrations of plutonium relative to seawater below the surface, and suggested limiting discharge to sea of radioactive effluents from nuclear power stations. Clearly much research into the generation of atmospheric 'solutes' has a strong 'applied' flavour. Liss and Slinn (1983) include the results of recent research into the production of 'atmospheric solutes' at the sea surface.

Material which is not produced at the sea surface but which is terrestrially derived is also present in the atmosphere, and this will be an additional source of deviation of atmospheric and precipitation composition from seawater proportions. Junge and Werby (1958) concluded that there was little evidence at that time that the relative enrichment of sea-derived material was due to any factors other than natural terrestrial and industrial inputs. Similarly, Skartveit (1982) has put forward a multiple regression procedure which allows the relative contributions of marine and non-marine sources to be assessed for rainwater samples by making simplifying assumptions concerning element proportions in seawater. Chamberlain (1975) has provided a useful summary of sources of 'non-marine' materials in the atmosphere, mechanisms of entrainment and the conditions which are conducive to their production. Dry, unvegetated, aerodynamically rough surfaces, with wind speeds of around 20 cm s^{-1} and above can disperse soil particles into the atmosphere. The aerosol generated from a soil surface has a median particle size of about 50 μm, though this value will decline to less than 10 μm downwind of the source as larger particles are deposited. Particles more than about 20 μm diameter will be deposited in a manner independent of the nature of the receiving surface, though Junge has noted that the upper limit of particle size tends to be higher over land than over, or downwind of, water because of the rougher land surface. Tenfold increases in inorganic aerosol concentration over 'background' levels have been measured and attributed to the entrainment of soil, dust and sand particles. The existence of a 'background' continental atmospheric aerosol has been assessed by Lawson and Winchester (1979), who measured element concentration in the '>4 μm aerosol' at interior sites in South America and found a stable spatial pattern in aerosol composition when concentrations were expressed relative to an index-element concentration, such as iron, though absolute concentrations still varied considerably. Douglas (1972) had earlier noted that 'the influence of terrestrial dust . . . ranges from virtual dominance of precipitation chemistry in the continental interiors to subordination to marine influences around the coasts', such that marked variations in soil types may be apparent in the spatial pattern of precipitation chemistry as the precipitation acquires the chemical characteristics of the area into which it falls, but at the coast this becomes 'background' to sea-derived materials. Feeley and Liljestrand (1983) have demonstrated a procedure for assessing the relative importance of different source-type contributions to precipitation in Texas.

The common constituents of atmospheric aerosols derived from terrestrial sources include Ca^{2+}, NH_4^+, SO_4^{2-}, HCO_3^- and NO_3^-, while marine sources contribute Na^+, Cl^-, Mg^{2+} and K^+. There are, however, many minor constituents of the atmospheric aerosol which have been recognized but which are in such low concentrations that they are not commonly of interest in geomorphological or hydrological studies except in particular circumstances. Likens *et al*. (1983) have, for example, monitored various components of the total organic carbon content of rainfall at two sites in the north-eastern USA and attributed their sources mainly to the entrainment of soil dust and plant material. Faust and Aly (1981) refer to the minor constituents of precipitation and note that each probably derives from a range of possible sources. Stensland and Semonin (1982) have suggested that a preponderance of dust storms in the mid-1950s in the USA increased Ca^{2+} and Mg^{2+} concentrations in the atmosphere, and eventually in precipitation, such that its acidity was reduced: the subsequent natural reduction in terrestrial input has led to an increase in precipitation H^+ concentrations (i.e. an increase in acidity) which should not therefore be solely attributed to increased acidic emissions from human activities. Volcanic emissions, showing episodic variations, will also complicate the assessment of trends in atmospheric and precipitation chemistry.

All of the constituents of the atmospheric aerosol ascribed to natural marine or terrestrial surface sources may also enter the atmosphere by human agencies: from anthropic or 'anthropogenic' sources. In addition to inorganic aerosol particles, gases may also become a non-natural addition: the question of whether these materials are thus 'pollutants' turns on the question of whether they are present in such high concentrations above 'background levels' (however these may be determined) that they produce 'adverse effects'. SO_2 is emitted from many anthropic sources, particularly those involving the industrial use of fossil fuels and metal smelting, though other gaseous sulphur compounds, including SO_2 and H_2S, are produced by biological decomposition. Nitrogen oxides and ammonia also have considerable anthropic sources largely in vehicle exhausts, though Record *et al*. (1982) note that the ratio of natural to anthropic nitrogen oxides has been inferred to vary from 15 : 1 to 1 : 1. K^+, Ca^{2+} and many other metals are also added to the atmosphere in trace amounts as particles by anthropic emissions, particularly from fossil fuel combustion which also generates considerable amounts of Cl_2 gas. Long-term and 'relative' trends in these emissions may be of significance to the assessment of 'pollution' and precipitation acidity trends.

Last *et al.* (1980a) emphasize that the gaseous and particulate materials emitted from the surface to the atmosphere occur in mixtures and that the nature of these mixtures is constantly changing as input and eventually deposition rates vary, differently for different components. Also chemical reactions will go on in the atmosphere, during the phase of transportation, which will change gaseous to particulate forms and thus alter rates and modes of deposition.

There may be considerable seasonal variations in anthropic emissions in response to summer to winter energy consumption changes, but these variations may be quite different in North America and Europe because of the differences in climate and lifestyle, e.g. with regard to air conditioning (Dovland and Semb, 1980). Such seasonal variations in emission rates and characteristics may or may not be evident in the precipitation quality characteristics at a place because of the great variabilities introduced by the intervening processes of transportation and transformation in the atmosphere and deposition on to the surface. Long-term trends in both sulphur and nitrogen oxide emissions, among other elements, may be considerably affected by technological changes (Galloway and Likens, 1981). Although the trends appear to be for increasing emission of SO_x and, particularly, NO_x gases to the end of the century, Durham *et al.* (1981) note the use, for example, of electrostatic precipitators which means that chemically-basic ash particles are removed while the 'acid' gases are not. Although the ash particles may have low atmospheric residence times because of their large size, the tendency will be towards increasing acidity in the atmosphere and in rainwater. Conversely, the desulphurization of flue gases and changes in process operating techniques to reduce nitrogen oxide emissions may act to reduce acid emissions. The scrubbing of waste gases may remove metallic particles in ash which are known to catalyse SO_2 oxidation and thus also reduce precipitation acidity. Record *et al.* (1982) provide a useful summary of processes of generation of atmospheric aerosols and gases at the surface, particularly referring to the 'acid rain problem', and note that the sources of material and their interactions are so complex that these are 'largely a matter of speculation and hypothesis at present'.

2.2.2 Transport and transformation of atmospheric solutes

It has been noted by Record *et al.* (1982), in referring to acid precipitation generation by anthropic sulphur and nitrogen oxide emissions, that 'no quantitative cause–effect relationship between pollutant (and natural) emissions and the measured acidity of precipitation has yet been determined between individual and regional sources . . . and receptor areas situated some distance downwind'. This situation results from the very complex nature of many chemical and physical processes that are involved in the transformation, transport and deposition of the complex mixture of substances comprising acid precipitation. Indeed, Rosenqvist (1978) has put forward alternative sources and mechanisms for the acidification of river water. Similar comments also apply to the other components of atmospheric solute input, in addition to hydrogen ion input. The complex combination and integration of transport, reaction and deposition processes results in the pattern of deposition in an area over time. The atmosphere is such a sensitive and dynamic system that the investigation, simulation and prediction of the history and fate of introduced material from the surface are extremely complex tasks. Several studies have shown that a variety of surface-derived constituents of air have atmospheric lifetimes of several days, can be transported over many thousands of kilometres, and can be transformed and have considerable effects on precipitation chemistry in areas distant from the areas and points of surface generation. That the distances between surface sources and sinks for atmospheric material may be of the order of many thousands of kilometres is evidenced, for example, by the visible plumes of material from the recent volcanic eruptions of Mount St Helens and El Chichon as viewed from satellites. Sequiera (1982) suggests that arid zone sulphate is a significant component of global precipitation. The implications are that a similar global scale view of atmospheric transportation is appropriate to material from anthropic or 'anthropogenic' sources. Indeed 'acid rain' has been reported from widely separated and isolated areas around the globe from Bermuda to Alaska and the Indian Ocean (Anon, 1982).

Aerosol particles larger than about 1 μm diameter are affected by sedimentation processes under the influence of gravity and are thus readily deposited from an airstream. The size spectrum of the natural atmospheric aerosol is from about 0.1 to 30 μm diameter, dust particles forming the larger end of the spectrum, sea salt particles the middle and material from combustion, natural chemical reactions and from photochemical sources the smaller end. Some particles such as sulphates and nitrates are hygroscopic, and under conditions of high relative humidity will grow into large droplets of about 40 μm diameter and form a mist. The change from a dry aerosol to a water droplet results in an average tenfold increase in particle size and a consequent hundredfold increase in its deposition velocity (Miller and Miller, 1980). Chamberlain (1975) has summarized the way in which the physical forces acting on such particles can be described and resolved. For small particle size viscous forces giving rise to 'drag' predominate over gravitational forces and particle terminal velocities are low—of the order of 10^{-3} cm s^{-1}. As particle size increases terminal velocities increase towards about 10 cm s^{-1} for 60 μm diameter spheres. In a turbulent airstream the effect of this is for small particles to be carried further from their source than larger ones as eddy diffusion near the ground carries the smaller (less than 1 μm diameter) particles to higher altitudes, particularly during daytime. Hogstrom (1973) comments on the factors affecting local fallout and long-distance transport of sulphur compounds. Junge (1963) notes that particles larger than 20 μm radius remain airborne for only a limited time and are therefore restricted to the vicinity of their source, and that upper and lower size limits for aerosol particles in the atmosphere are determined by sedimentation and coagulation respectively. For particles <1 μm radius the effect of gravity becomes negligible and it is these particles that are important in long-range transport. Pollen grains tend to be, like loess particles, around 20–30 μm diameter, which is clearly an important 'evolved' dimension. As the larger aerosol particles, around 1–2 μm radius, tend to be the nuclei for condensation, it should be expected that the cloud droplets will contain most of the aerosol mass in the air column and that therefore material incorporated during the droplets' generation–rainout–is likely to be far in excess of washout in terms of 'solute' input to the surface. The Aitken particles, less than about 0.1 μm radius, produced by gas reactions and industrial processes may be in large concentrations of $>100\ \mathrm{cm}^{-3}$ but Aitken particles do not appear to be important as initiators of growth by accretion of sulphate particles, for example, from oxidation of gaseous SO_2. Thus the absolute concentration of very small aerosol particles, and also their chemical composition (V_2O_5 appears for instance to be a powerful catalyst of gas reactions producing particles) may affect atmospheric chemistry largely independently of the larger particles which affect the generation of precipitation. Recently interest has been directed towards the proteins on certain bacteria which appear to be highly efficient condensation nuclei.

Considering the transportation of solutes in the atmosphere, it is clear that the transportation of gases and the smaller aerosols, and to some extent the transformation of gases to particulates and the eventual deposition of gases and particles, are all a function of meteorological conditions over a period of days or even weeks as well as of the spatial pattern of 'production' (Ragland and Wilkening, 1982). Smith (1983) notes that some 25% of wet-deposited sulphur in Norway originates in the USA and Canada and reaches north-west Europe under particular atmospheric conditions which have prevented deposition during transportation. While it may be difficult to understand and model the 'source–transformation–deposition' mechanisms sufficiently well so as, for example, to justify controls on emissions of acidic 'precursors' by clearly identifying 'emitters' as the pollutors of precipitation, it may be possible to identify sufficiently clearly, in an empirical fashion, the meteorological conditions which affect transport, transformation and deposition. Seasonal changes, for instance, in mean air flow patterns are major simplifications, but they indicate resultant surface and upper air winds

(Record *et al*., 1982) which may provide a surrogate for the processes of advection and turbulent diffusion which affect transportation.

Dovland and Semb (1980) provide a useful summary of material transport processes in the atmosphere. Transport is by both advection and turbulent diffusion, the former due to the mean wind and the latter produced by eddies which disperse surface-derived material in vertical and lateral directions. The degree of turbulence present is most important within tens of kilometres of the source of material, beyond which vertical uniformity of concentration is assumed and transport is taken to be effected by the mean wind, allowing for transformation and deposition.

The atmospheric 'lifetimes' of reactive *gases*, particularly sulphur and nitrogen compounds, will depend on the rapidity of any chemical transformations and on the rate of dry deposition on to the surface. Where these rates are low the occurrence of rain events will control the deposition of these elements and will make prediction very difficult. Inversion layers and very stable non-turbulent conditions, as may most often occur at night, will restrict the effective mixing volume for the material and high concentrations will result, either in an undiluted plume from a point source, or beneath the inversion level. The maximum concentration in the atmosphere may not in fact be at ground level, which will affect the choice of the most appropriate wind data to model transport and trajectories. Dovland and Semb discuss these effects and the importance of selecting the most appropriate trajectory level in the atmosphere to relate to surface deposition concentrations: this level will also depend on the time base of the data.

Munn and Rodhe (1971) have demonstrated a methodology for relating synoptic weather patterns to the chemical composition of monthly precipitation samples which was based on surface geostrophic wind direction during precipitation. Thus it was possible to show an increase over time in sulphur deposition which was not attributable to trends in frequency of wind direction. Skartveit's (1982) analysis of circulation patterns suggested that intensified large scale dispersion in winter was in part, along with enhanced wet scavenging in winter, responsible for the lack of a recognizable winter peak in nitrate and excess sulphate concentrations in coastal Norway to match increased winter-time emission of combustion products. Precipitation frequency is a significant factor in the use of wind direction analysis to assess sources of material, as surface generation just preceding or during a rainfall event may mean a reduction of residence time in the atmosphere from several days to a matter of hours. Cogbill and Likens (1974), Kurtz and Scheider (1981) and Miller *et al*. (1976) have also used 'back-trajectory' analysis of air parcel origins to assess the source of dominant ions in precipitation and to account for variations in precipitation chemistry. Barrie (1982) describes the work of the 'Environment Canada' programme in investigating the long-range transport of pollutants in the atmosphere.

A general conclusion from many investigations into the origins of acidic components in precipitation is that such deposition is highly episodic, mainly because of the variability of climatic factors working in complex combinations. Shaw (1982) used back-trajectory analysis to demonstrate the importance of local (30–50 km radius) sources to precipitation quality with a view to planning the sites of new high-sulphur coal-burning power stations in eastern Canada. It was concluded that 'the relative contributions of each source region to wet and dry deposition, and the relative importance of wet and dry deposition, depend on a complicated interplay of source strengths, distance to the receptor and meteorological factors such as frequency of transport flow to the receptor and precipitation amount'. Durham *et al.* (1981) suggested that the long-term mean trend of storm tracks was responsible in part for the 'acid rain problems' being experienced in north-eastern USA.

Processes of transportation of surface-derived material in the atmosphere are currently receiving much attention as they may make it possible to assess the relative contributions of different sources to eventual precipitation quality. Although progress in this area is said to be at the early exploratory stage (Record *et al.*, 1982) attempts are being made to simulate the regional and long-range transportation processes using mathematical models. This will allow the assessment of the extent of exchange of 'pollutants' across international boundaries and the identification of the effects of isolated events or of point sources on particular receptor areas. Ragland and Wilkening (1982) give an indication of the complexities of one such (sulphur-transport) model:

> '. . . a time-dependent, cell-type model which numerically solves coupled SO_2 and SO_4^{2-} ion diffusion in an Eulerian (fixed) frame of reference in the atmospheric boundary layer. Included are the processes of advection, turbulent diffusion, chemical reactions, source input, and dry and wet deposition. The numerical solution is a first-order fully implicit finite difference scheme.'

Van Dop (1983) provides a comprehensive summary of approaches to modelling pollutant transport in the atmosphere.

Such mathematical transport models as have been developed (Record *et al.* (1982) provide a summary) rely heavily on empirical approximations to describe the chemical transformation processes that take place in the transportation phase between incorporation and deposition of material. The residence time in the atmosphere and the mode, rate and location of deposition of surface-derived materials depend, in addition to meteorological factors, on physical and chemical transformations, often extremely complex, that take place within the atmosphere. Much research into these transformations is linked to 'the acid rain problem' and concerns the transformation of 'acid precursor' gases to particles which form condensation nuclei or are washed out of the atmosphere and largely determine the acidity of precipitation (Cox and Penkett, 1983). The comments of Seip (1980) are instructive here: he notes that in investigations of the processes responsible for production of 'acid rain' many suggested cause–effect relationships are merely correlations in time or space, allowing for lags of time and distance, and do not necessarily give insight into the mechanisms responsible. Rodhe *et al*. (1981) have shown that in-transport chemical changes involve interactions with many other substances which, for example, have the effect that SO_2 emission rates may increase significantly in winter but deposition rates of sulphuric acid, at shorter travel times of less than 10 hours, may in fact not increase. Beilke (1983) provides an up-to-date review of the complex atmospheric evolution of acid precipitation and acidifying substances.

Transformations of 'naturally-derived' components of the atmosphere and the atmospheric aerosol have received less attention than those of 'pollutants'. It has already been mentioned that sea salt particles derived from seawater droplets are hygroscopic and may, under appropriate humidity conditions, grow large and form mist droplets and eventually be affected by sedimentation, or may fragment, thus progressively changing the nature of such material in the atmosphere. Koyama and Sugawara (1953) found that the concentration ratios of Ca^{2+}, Mg^{2+} and SO_4^{2-} to chloride in precipitation increased inland, and invoked the 'evaporite sequence' of salt crystallization in explaining these changes. Eriksson (1959) has discussed the possibility of Cl^- oxidation to Cl_2 in the presence of ozone and the eventual formation of HCl as being a factor in the formation of acid precipitation. Clayton (1972) suggested the idea of a 'base-level' concentration towards which all sea-derived elements tend. Elements originally in high concentrations, such as Na^+, do not reach their base level for greater distances inland than do trace elements, so that element concentration *ratios* in precipitation vary continuously away from the coast. This observation is unexplained but may be linked to the fact that the less hygroscopic, less soluble particles, principally of $CaSO_4$, $CaCO_3$ and $MgCO_3$, are small and so are less affected by sedimentation, and have lower concentrations in the oceanic aerosol.

As suggested above, a great deal of attention has been given to the important chemical transformations in the atmosphere of pollutant emissions which are thought to give rise to an acidic atmospheric deposition which has undesirable effects on natural ecosystems. The two pollutant emissions of greatest concern are of sulphur and nitrogen oxides which are transformed by oxidation during the transportation to more stable particulate sulphate and nitrate aerosols in the sub-micrometre range; this increases their atmospheric lifetime, facilitates long-range transport and thus contributes to the spatial extent of areas affected by 'acid rain' (Record *et al*., 1982). The amount of nitrate and sulphate emitted directly to the atmosphere is relatively small. The chemical transformations involving sulphur and nitrogen oxides are extremely complex and only the sulphur transformations are understood reasonably well. Some modelling of these transformations is necessary in order to estimate deposition rates or potential at sites downwind of major SO_2 sources, but a very large number of factors appears to control the rate of sulphate aerosol production and deposition (Barrie, 1981). Also SO_2 may be directly absorbed at the earth's surface and be 'scavenged' by falling precipitation and add to precipitation acidity. Garland (1978) suggests that about half the SO_2 emitted annually in north-western Europe is 'dry deposited' as SO_2 gas and most of the remainder is oxidized to SO_4^{2-} before being 'wet deposited', though he emphasizes that SO_2 and SO_4^{2-} are probably the oxidation products of more complex sulphur compounds in the atmosphere and therefore it will be very difficult to model their transformation and deposition characteristics, except in a largely empirical fashion. SO_2 is oxidized to SO_4^{2-} by two distinct groups of mechanisms, it is thought. Homogeneous oxidation, in the gas phase, accounts for about 10% of sulphate and is caused by SO_2 reaction with gaseous oxidants such as O_3 and OH and HO_2 radicals produced by photochemical reactions. This possibly involves hydrocarbon and nitrogen oxide emissions and therefore there should be a detectable latitude, season and light intensity-related variation in SO_4^{2-} concentrations as well as a link to the pattern of SO_2 emissions. Skartveit (1982) has noted slower NO_3^- and SO_4^{2-} generation from gaseous precursors in the winter months. Heterogeneous oxidation of SO_2 to SO_4^{2-} is thought to be the more important group of reactions; it involves gas absorption and chemical reaction in the liquid and solid phases in which sulphate is produced in water droplets or on particles. Catalysts in the pollutant mix are of crucial importance to this reaction: these can be soot or heavy-metal particles, particularly transition metals such as vanadium from oil and coal combustion, or can be hydrocarbon–nitrogen oxide mixtures again. In this connection Smith (1983) suggests that a reduction in the emission of these catalysts, which may be more easily and cheaply controlled, could reduce some of the major pollutant depositions, especially close to their source. Careful but difficult investigations will be necessary to assess this effect. Correlation analysis by Sequeira (1982) has shown that SO_4^{2-} is strongly linked with alkaline earth elements in deposition, and he infers that surface-derived particulate matter may be an important sink for SO_2. The oxidation of nitrogen compounds to nitrate is a very much more complex transformation, proceeding via transient species in reactions catalysed by other NO_x species and controlled by other 'synergistic effects' in temporally and spatially complex processes that 'may take hours *or* days' (Record *et al.*, 1982) and are not at present well understood, nor tied firmly in to atmospheric conditions. As an example of the interaction of chemical species, it is thought that increased emissions of nitrogen oxides will decrease the rate of SO_2 transformation because of the common dependence on the OH radical (Rodhe *et al.*, 1981) such that HNO_3/H_2SO_4 ratios in deposition will vary relative to the source. Maclean (1981) has considered the relative proportions of sulphuric and nitric acids in the acidification of natural ecosystems and the implications presented for control strategies. Also, the heterogeneous oxidation rate of SO_2 in droplets decreases rapidly as pH decreases and it has been noted that ammonia is probably the most significant factor in raising the pH of droplets (Barrie, 1981), thus counteracting the acidifying effect of SO_2 oxidation and accounting for the dominance of ammonium sulphate in the atmosphere where it forms about 60% of the total inorganic aerosol in country districts in Europe (Chamberlain, 1975). Dovland and Semb (1980) have reported on monitoring of air quality which has shown that trace metals, introduced to the atmosphere as aerosols, show relatively little chemical transformation and variation in emission rate such that they can be used as references against which climatic and transportation time effects for less stable substances can be assessed. In summary, Smith (1983) has noted that the chemical changes in the atmosphere 'depend on a great number of complex processes within the air which in practice may require basic information totally beyond our capacity to achieve with any kind of precision', but that as far as the determination of long-term deposition fields is concerned this information is unnecessary as these fields may be adequately assessed from air concentrations and long-term rainfall patterns.

2.2.3 Removal from the atmosphere and deposition at the surface

Material in the atmosphere is removed to, and deposited on, the earth's surface by a number of processes which vary in their intensity with meteorological conditions, the nature and the physical and chemical interactions it has undergone and the nature of the surface. As the general descriptive terms which are applied to removal and deposition processes are often not used in a very precise way, it may be useful to characterize these processes generally before looking at more detailed effects which are significant to the assessment of and the characteristics of atmospheric 'solute' input to the surface. It should be noted that the impetus to much of the research in this field has been a desire to model the processes involved in atmospheric input of pollutant gases to a surface (e.g. Garland, 1978) and in the fallout from nuclear explosions, and the spatial and temporal patterns of inputs resulting from these.

As has been made clear in earlier sections, there is material present in the atmosphere, which, on deposition at the surface, becomes what is termed here 'atmospheric solute input'. This input may be in gaseous form, or dissolved in water droplets, or in particulate form, usually in 'dry' suspension but possibly also contained in droplets. This material is deposited at the surface in all three forms and may change between them during the processes of removal and deposition. Atmospheric aerosols may be removed from the atmosphere by processes termed 'rainout' in which small particles act as nuclei for the condensation of water vapour and rainfall results. These particles may, for example, be of sulphate produced by the gas-to-particle oxidation of SO_2. Another removal mechanism is 'washout' in which falling droplets remove or 'scavenge' material from the atmosphere during their descent by exchange of gases, coalescence of other droplets or collision with particles such as those produced by the evaporation of wave-produced droplets or unaltered terrestrial dust. Junge (1963) considered that the uniform vertical distribution of salt aerosol concentrations in the lower atmosphere is indicative of low washout efficiences for these particles. Fowler (1980) emphasizes the point that the 'in-cloud' rainout/'below-cloud' washout distinction is rather arbitrary, in that 'most of the physical processes effecting the transfers (to the surface) do not make this distinction and continue to operate during the lifetime of the droplet'. Rainout and washout are therefore often referred to in combination as 'wet removal'. These processes have proved extremely difficult to model because of their variability and complexity, and more empirical approaches (e.g. Barrie, 1981) have been made using the large amount of available precipitation quality data. Durham *et al.* (1981) demonstrate the very complex chemical interactions which may go on during 'washout'. The combined efficiency of wet removal processes has been expressed as a 'scavenging ratio' (Rodhe and Grandell, 1972) or as a dimensionless washout coefficient (Slinn *et al.*, 1978): this is usually expressed as the concentration of a solute species per unit mass of rain divided by the concentration of that species per unit mass of air. Skartveit (1982) notes that the calculated species washout ratios are not fixed but decrease as precipitation amount increases, especially rapidly for Cl^- and NO_3^- which are scavenged by hydrometors rather than incorporated via the 'cloud condensation nuclei' pathway. The washout ratio for SO_2, for instance, also depends on temperature and rainwater acidity (Barrie, 1981).

Dry removal of material from the atmosphere to a position 1 mm or so above the surface takes place via the processes of sedimentation and turbulent diffusion. Sedimentation is only important for particles larger than about 5 μm diameter and for these large particles is little influenced by the nature of the surface. For smaller particles, turbulent transfer to the surface layer resulting from and enhanced by the degree of 'roughness' of the surface, greatly exceeds simple molecular diffusion, and the turbulent transfer process has been successfully modelled using the physical principles applied by micrometeorologists in assessing the exchange of energy, water vapour and momentum between the atmosphere and the surface (Unsworth, 1980a). While this approach has been successful in approximating rates of particle *removal* to the surface layer, it is difficult to apply physical principles to model actual *deposition* rate. For this a simplification of factors controlling gas and particle fluxes has been derived which incorporates the species concentration in air at a particular level above the surface and a species- and surface-dependent empirical 'deposition velocity':

$$\text{flux (g m}^{-2}\text{ s}^{-1}) = \text{concentration (g m}^{-3}) \times \text{deposition velocity } V_g\text{, (m s}^{-1})$$

This removes the dependence of flux on concentration, the deposition velocity value usually being referred to a height of 1 m above the surface (Chamberlain, 1975). This form of analysis is most successful when applied to small particles rather than gases or restricted to the estimation of long-term gaseous deposition on the regional scale. Short-term deposition on to particular vegetation types has been more successfully modelled by use of the atmospheric and surface resistance approach of Monteith (1973).

In some publications the term 'sedimentation' is used so as to include removal and deposition in precipitation as well as in large (dry) particles. Also deposition via 'turbulent transfer' is applied so as to include very small droplets as well as particles and gases and particularly the 'wet removal' of droplets by the filtering and trapping effect of vegetation (Schlesinger and Reiners, 1974). There are thus overlaps in the terminology which can lead to some imprecision: Fowler (1980) and Miller *et al*. (1980) provide definitions of the main terms involved.

The actual processes of *deposition* and retention on the surface, whether it is open water, bare soil or rock, or vegetation, of material removed from the atmosphere may require separate consideration from the processes of *removal* to just above the surface. Material subject to wet removal in rainfall and snow, together with the larger solid particles, more than about 20 μm in diameter, are affected by sedimentation by gravity and deposited in a way that is largely independent of the characteristics and physical form of the deposition surface. Thus 'wet deposition' to an open collector will inevitably receive some 'contamination' from dry-deposited large particles, even during periods of precipitation; a point given emphasis by Miller *et al*. (1980). Similarly, not all of the material which is wet-deposited will be deposited by a process of sedimentation: small mist or fog droplets, possibly having higher element concentrations than larger droplets, can be captured by impaction on to vegetation, having been delivered to the surface by turbulent transfer processes (Schlesinger and Reiners, 1974). Also, some particulate material may be said to be wet-deposited in suspension in a droplet, and wet and dry deposition processes may interact in that rates of dry deposition of some gases and particles may be enhanced when a surface is wetted (Bache, 1981). Gaseous materials and very small particles ($<c$ 1 μm) are deposited on to surfaces by processes of molecular diffusion and inertial impaction respectively, which take them through the relatively still air of the laminar

boundary layer. Retention on the surface is enhanced by chemical reactions of gases with surface materials and for particles by the size, shape, substance and roughness of the surface (Chamberlain, 1975).

Leachates on leaves and the chemical nature of vegetation surfaces, surface wetness and soil pH can all affect deposition velocities. These effects can be represented as a surface resistance to deposition which controls deposition rate in turbulent conditions (Last *et al.*, 1980a). Particles greater than 1 μm diameter may pass through the 'viscous sub-layer' over a crop or soil surface but, at the exact surface, deposition may be limited by 'bounce-off' saltation and 'blow-off' of particles (Chamberlain, 1975). Deposition velocities for particles smaller than 1.0 μm diameter are, however, usually small, and materials in particles of this size are unlikely to penetrate the boundary layer and thus contribute greatly to dry deposition totals for that material.

With reference to *wet* deposition, Fowler (1980) lists five mechanisms by which material may be removed from the atmosphere in droplets. These are namely: 'diffusiophoresis' in which aerosol particles move with the mean water molecule flux relative to evaporating or condensing droplets; 'Brownian diffusion', particularly efficient for very small particles (<0.1 μm diameter) in which particles in random motion enter water droplets though at relatively small diffusion rates; 'impaction and interception', for larger particles which collide with droplets, both in and below clouds, though the different dominant droplet sizes there are likely to lead to different extents of particle removal by cloud and rain droplets; 'rapid solution of gases', in droplets of all sizes up to a concentration at equilibrium with gas phase concentrations, may be followed by oxidation within the droplet, often catalysed by other solutes present, and the production of aerosol particles; the 'cloud condensation nuclei' pathway, which is by far the most significant contributor to wet removal as sub-micrometre particles, particularly of natural and artificial sulphur and nitrogen compounds, act as condensation nuclei. An important point to appear from this demonstration of the number and complexity of potential contributing processes to wet removal is that their relative importance is likely to change in response to, among other factors, changing droplet size distribution, ambient gas concentrations, atmospheric conditions, droplet lifetimes and precipitation duration and intensity. For these reasons Fowler (1980) has noted that 'the complexity and variability of removal mechanisms have made attempts to model wet deposition from a knowledge of its component processes very difficult, and in a predictive sense not very valuable'. Therefore empirical analysis of collected wet deposition is preferred, especially as it is relatively straightforward to measure amounts of wet deposition and a great number of data are already available. An appreciation of the dominant wet removal mechanisms for a particular solute species will, however, give some insight into the reasons for the variable behaviour of different species in collected precipitation, or vice versa (Gatz and Dingle, 1971): if the cloud condensation pathway is dominant then concentrations in precipitation should remain relatively steady through an event, whereas species removed by other mechanisms may well show dilution effects through a rainstorm or through a series of storms (e.g. Kennedy *et al.*, 1979). Smith (1983) considers that a major source of error in the modelling of wet deposition, in addition to lack of knowledge of wet removal mechanisms, is due to the spatial and temporal smoothing of precipitation fields because of lack of instrumentation, when these data are incorporated in a trajectory analysis. The problem is thought to be particularly significant for the larger, episodic, depositions from convective storms. Rodhe (1980) has subsumed the complexities of the wet removal process by the use of the scavenging coefficient and thus modelled long-term wet deposition around an emitting point source of material: coefficients are derived for wet and dry weather and combined with statistical information on rainfall composition and mean lengths of wet and dry periods to estimate the wet deposition field. Occult precipitation, in which droplets are filtered from an airstream by vegetation, is probably a very important solute input in upland and coastal areas. Skartveit (1982) has noted that the wet deposition of acidic compounds is highly episodic, depending as it does on rainfall occurrence coinciding with aerosol presence in high concentrations, among other factors. An extreme example of this is the apparent concentration of solutes in the first fractions or daily peaks of run-off from snow melt. This is due to the effect of melt–freeze cycles causing or allowing the migration of solutes and their concentration in the lower layers of a snow pack (Colbeck, 1981). This effect may be particularly important in its effects on melt water acidity in sensitive areas (Davies *et al.*, 1982). Johannesen and Henriksen (1978) found such an effect in which the first 33% of snow-melt contained between 41 and 80% of 16 constituents contained in the snowpack.

Referring to dry deposition mechanisms for input of 'solutes' to the surface, Unsworth (1980a) has stated that it has become increasingly apparent in recent years that the dry deposition of gases (and in particular attention has fixed on sulphur dioxide—e.g. Chamberlain (1980)—and other pollutant gases) is a particularly efficient process, especially close to their source. Fowler (1980) suggests that the ratio of dry to wet deposition of sulphur compounds may be 8 to 12 close to sources but 1 in remote areas. Galloway and Whelpdale (1980) assessed the dry deposition of gaseous SO_2 to be about 1.3 times that of particulate SO_4^{2-} in the USA, though only 0.4 times in Canada at a greater distance from the areas of greatest emission. Even at distant sites the deposition of gaseous SO_2 may be significant to studies of plant nutrition (Last *et al.*, 1980a) and to the 'acid precipitation' question. Fowler (1978) determined an SO_2 input of 72 kg $ha^{-1}yr^{-1}$ to an agricultural area in Great Britain. The results of Weeks *et al.* (1982) emphasize the potential importance of dry deposition of gaseous material: man-made fluorocarbons, very stable compounds, were discovered in measurable concentrations at depths of up to 43 m in soil air in unconsolidated sedimentary deposits. Such components of dry deposition are now being more thoroughly assessed in addition to the previously monitored aerosol and mineral dust particles (Unsworth, 1980b). Resistance analogues are being used to model gaseous dry deposition fluxes by analogy with the evapotranspiration process (Monteith, 1973; Fowler, 1980; Chamberlain, 1980) when short-term and smaller-scale flux rates are required. It is considered that in such cases the simplifications in the deposition velocity approach (Smith, 1983) introduce large errors, though they avoid the use of sophisticated instrumentation and modelling procedures.

The dry deposition of aerosol particles and small droplets deposited outside of periods of precipitation (Unsworth, 1980a) is principally by sedimentation close to sources, and also by turbulent impaction at all locations. It has been noted that the low velocities of deposition of small particles, between 0.1 and 1.0 μm diameter, on to soil or vegetation are about two orders of magnitude lower than V_g for some gases and imply very much lower deposition rates compared with wet deposition. However, particulate inputs to vegetated surfaces, particularly to forests, may in fact be more considerable than suggested by this. The rough surface presented by a forest associated with high wind speeds and frequently wet surface conditions increases the capture efficiency of the vegetation for particles and small droplets. Lindberg *et al.* (1982) discuss the interaction of wet and dry deposition with each other and with the forest canopy, and Shuttleworth (1977) has put forward a model for estimating fog and mist input to vegetation which may assist estimation of solute inputs. Azevedo and Morgan (1974) have demonstrated the very high inputs of 'fog precipitation' in the forests of coastal California: 42.5 cm for a summer season and daily inputs of up to 8 cm for a fog event lasting several days. It is likely that the afforestation of areas of natural grassland increases the rates of dry deposition to these areas and that this is a factor in the apparent acidification of lakes and streams in some upland areas (Harriman and Morrison, 1982).

White and Turner (1970) and Chamberlain (1975) have investigated the effects of leaf and foliage characteristics on particle collection or 'trapping' efficiency, and Bache (1981) has assessed the relative importance of sedimentation and impaction for Lycopodium spores as of the order of 1:5, though with considerable foliage density and wetness effects. Dollard and Vitols (1980) used wind-tunnel experiments to assess the effects of tree species' morphology and wind velocity on magnitude of dry deposition and revealed complex species-linked effects among the three tested. It was, for example, found that the deposition velocity for SO_2 decreased by up to three orders of magnitude for dry surfaces in darkness rather than in daylight, indicating the significance of surface resistance to the modelling of these processes. Kellman (1979) has found that some savannah trees and shrubs are capable of acting as effective nutrient sinks through having a particular long and narrow leaf shape conducive to high rates of dry deposition; this is an effect which has been demonstrated in laboratory experiments and reviewed by Chamberlain (1975). Wesely *et al*. (1977) reported experimental results which suggested particle deposition velocities to short grass which are much greater than those previously accepted, and this may indicate the very great significance of the dry deposition pathway in some environmental situations.

Slinn (1977) and Sehmel (1980) have provided useful and comprehensive reviews of progress in modelling the factors influencing dry deposition of particles and gases. In conclusion, Unsworth (1980b), in summarizing the contributions of many workers, has noted that the advances made in assessing and modelling the mechanisms of short-term dry depositional input to forests have not been translated into improved assessment of inputs over longer periods, as is more appropriate to catchment nutrient budgets and to the understanding of the effects of such inputs on ecosystems. Goldsmith *et al*. (1984) and Fowler (1984) have provided up-to-date and comprehensive reviews of current knowledge of atmospheric transport and transformation and subsequent transfer to the surface of 'pollutants', and initial reference should be made to these reviews and to the full symposium proceedings of which they form a part.

2.3 ASSESSMENT OF ATMOSPHERIC SOLUTE INPUTS

The preceding section has provided an indication of the wide range and complexity of processes which are recognized as contributing to the atmospheric solute input to a surface. It must be assumed that there may be other processes, unmentioned here or unrecognized as yet, which become significant contributors or have significant effects from time to time in particular circumstances. The lack of clear understanding of many of these complex processes and their consequent lack of unambiguous definition will mean that major uncertainties and inaccuracies will inevitably attach to estimates of effective total atmospheric solute input to natural ecosystems over a period, and to the components of this total. What are assumed to be complementary and exhaustive estimation procedures may thus overlap or be deficient in their coverage of all components. Also, there will be additional conceptual, operational and instrumental shortcomings attached to the particular estimation procedures themselves owing to the nature of the processes concerned. These estimation deficiencies will 'feed back' into the analysis and modelling of the input processes and thus impede progress towards improving process modelling and operational and instrumental procedures. Thus the very complexity and dynamic characteristics of the processes involved may make such progress comparatively slow, but emphasizes the great importance of careful and painstaking work in assessing atmospheric solute input levels, both in the field and in the laboratory. As an example of this, Goldsmith *et al*. (1984) pointed out that the fact that total pollutant deposition (as measured) does not parallel the increase in emission of sulphur compounds leads to the formulation of hypotheses concerning the transfer of these compounds to the troposphere above the boundary layer, with implications for global background deposition rates. In this case, deficient methods of assessment may be responsible for unnecessary elaboration of hypotheses. Goldsmith *et al*. (1984) then provide an instructive comment:

> 'The results of analyses of data from the European Air Chemistry Network from 1955 to 1975 are briefly described. It is difficult to know to what degree such analyses can be of use other than to reflect the inadequacies of the methods involved.'

The problems of 'atmospheric solute input' assessment therefore form an initial and crucial stage in investigations.

In reviewing the procedures used to evaluate atmospheric solute inputs it is possible to make a first distinction between these procedures as regards whether they attempt to assess either wet deposition only, or dry deposition only, or some combination of these two as 'bulk precipitation'. Secondly, a distinction may be made between techniques of 'direct' assessment by means of actual physical interception of material being deposited from the atmosphere using natural or artificial collecting 'devices' and 'indirect' assessment, such as by analogy with other, better known, atmospheric processes and by using models of these processes, or by estimation 'by difference' between better known or more easily measured parameters in a mass balance calculation.

Considering firstly the assessment of rates and amounts of wet deposition, it is clear that this is mainly accomplished by direct means, rather than indirect means, i.e. by the collection of a sample of precipitation in a container prior to laboratory analysis. This immediately creates the problem of the resolution of the sampling, both in the spatial and temporal senses, necessary to adequately represent areal averages of input and the spatially and temporally varying wet deposition input 'field' over the area of interest. What is considered to be an 'adequate' network of precipitation sampling locations will depend on the particular requirements of the research study or of the users of the information, and can only be assessed, in terms of numbers or density and distribution of precipitation collectors and their operation, by reference to pre-existing information on solute input fields in similar areas or an intensive pilot survey in the study area. Criteria and objective procedures for *rain gauge* network design for precipitation quantity assessment have been thoroughly investigated in recent years (Rainbird, 1965; Nicks, 1965; Nicholas *et al*., 1981). However, the number of rain gauges and their density would generate unmanageable, and probably unnecessary, numbers of water samples for laboratory analysis. A density of the order of 15–20 km^2 per gauge was suggested by O'Connell *et al.* (1978) for a basic rainfall recording network in south-west England designed to satisfy stated error criteria. However, such fine detail in solute input fields is not usually assumed or sought except with reference to known point sources (e.g. Burt and Day, 1977), and point quality data are commonly used to represent spatial patterns and magnitude of solute input over areas of the order of many hundreds of square kilometres. Granat (1976) has provided a review of design requirements for precipitation chemistry networks and procedures by which these requirements might be satisfied. He notes the importance of the 'pioneer', intuitively designed network's results in guiding the improvement of the network as design and operational deficiencies become apparent and new needs arise. The variability of areal mean solute inputs via wet deposition can be considered to contain both a deterministic and a random component. The deterministic variation is that which can be attributed to particular sources or processes of delivery of material, and the remainder of the input is referred to as the randomly varying component. The predominance of one or the other component will affect the required spatial resolution of sampling, as will the size of the area of interest. The presence of a strong deterministic variation in solute input, relative to the coast, to high ground, an urban area or an industrial point source, may require a different spatial pattern and density of gauging compared with an area where a dominantly random variation is apparent, where a form of random spatial sampling may be appropriate. Granat (1976) notes that the characterization of an input field as 'deterministic–or random–dominated' is also a function of the scale of the area of interest, so that in a small catchment investigation the deterministic and random variations are much more significant and require more spatially intensive sampling than over much larger areas, say at a regional, national or continental scale, where the random variability is likely to be less significant and local deterministic source effects subsumed by general regional trends. The timebase for sampling will also be relevant here; greater errors in areal mean input estimation due to random sampling are likely to be associated with event sampling than with weekly or monthly collection of 'temporally-bulked' samples. The extension of point values to areal mean values of solute input may be approached using the established techniques for areal extension of point precipitation values (Hall and Barclay, 1975) either using direct, spatially-weighted, averages of point deposition values (Munn, 1981), or by producing the solute concentration field and combining this with the usually higher-resolution information in the precipitation depth field to arrive at the solute deposition field as the product of the two. In Norway the results of 750 precipitation gauges and 20 precipitation samplers are combined in this way. The attendant assumption of high correlation between precipitation amount and wet deposition of sulphate is based on empirical evidence and the assumed dominance of the cloud condensation nucleation pathway for sulphate in particular (Overrein *et al.*, 1981). The question of the necessary *duration* of sampling to characterize a solute deposition field is a difficult one but is likely to be related to area size and intra-area variability of input values. The separation of deterministic trends in wet deposition concentrations from the considerable 'random' fluctuations at the weekly, monthly or even yearly level is often difficult. This is important, for example, in the 'acid rain problem' where some authors (e.g. Record *et al.*, 1982) have failed to separate a significant long-term trend from the strong, shorter-term variability apparent in long records of bulk precipitation pH and due to changing meteorological conditions. Granat (1976) therefore warns against the 'black box' monitoring of wet deposition composition to assess trends because of the difficulty in controlling for meteorological effects, and instead recommends an emphasis on assessment of areal average inputs in association with evaluation of sources so that input trends can be forecast as sources change in nature and location.

The sampling framework appropriate to relating solute input fields to source area contributions or trajectories may be much less dense than that for assessing areal average inputs and will only be operated on an event-long or daily basis within which meteorological conditions can be characterized. In this case the spacing of sampling points need only be that which is sufficient to demonstrate any input differences which are due to different air masses. When the aim is to assess the solute input contributions from a particular point or line source, however, a greater number of wet deposition collectors is necessary with a good coverage of the area around the source within which it is thought the source is a significant contributor. These comments, though referred to wet-deposition, also apply generally to the sampling of dry deposition and bulk precipitation. Reference should also be made to Rowse (1980) for guidance in the design of data collection networks in general.

The hardware used to obtain samples in order to estimate wet deposition directly is designed to exclude dry-deposited material, though it clearly cannot do this while open to the atmosphere to admit precipitation. Thus 'wet deposition' really means 'wet deposition plus dry deposition during the precipitation event', or during the period that the collector is open to the atmosphere. As the 'wet-only' collector is often a modified bulk precipitation collector, the two types will be discussed together in the following section.

In assessing atmospheric solute input the easiest 'component' to estimate is 'bulk precipitation' quality, since no attempt is made to distinguish between wet- and dry-deposited material as both are collected in the same container. The term 'bulk precipitation' was used by Whitehead and Feth (1964) to refer to the precipitation sample gained from a funnel or other collector which was continuously open to the atmosphere. They considered bulk precipitation to be 'the most significant category in many geochemical considerations' and to be 'the real agent' of rock weathering, soil formation and plant nutrition. Similarly, Pearson and Fisher (1971) referred to bulk precipitation as 'the proper parameter for study'. The types of bulk precipitation collector range from simple plastic sheets laid on the ground and draining to a container, to polyethylene or glass funnels raised above the surface or V-shaped troughs of plate glass (Fish, 1976a). The funnel is the most popular form of bulk precipitation collector, its orifice horizontal and raised at least 2 m above the ground. It should be well away from vegetation and clear of its supports (Slanina *et al*., 1975) and of other obstructions so that it is in a 45° clearing in forested or built-up areas. The material of the funnel is usually acid-washed polyethylene in order to reduce contamination by leaching, though this has a relatively short life, but as some plastics contain fillers which release contaminants, this should be checked for by acid-washing before use (Miller, 1980). Glass funnels, often in Pyrex but sometimes produced from large reagent bottles, may be used, especially when organic materials are to be determined, and heated stainless steel funnels are increasingly used where snowfall is considerable. The collector successfully tested and used in the Hubbard Brook Ecosystem Study (Likens *et al*., 1977) is equipped with a vapour trap to prevent evaporation loss, 'bird discouragers' (spikes fitted to the rim of the funnel), and nylon gauze insect filters in the funnel to keep objects out of the storage container: all these are measures designed to keep the sample uncontaminated and representative of atmospheric solute input at that point, though Lewis and Grant (1978) and Goldsmith *et al*. (1984), for example, reported contamination introduced by the use of glass wool filters, and other sources of unrepresentativeness must be carefully considered (Samant and Vaidya, 1982). On analysis, aberrantly high concentrations of calcium, phosphorus or nitrogen in the collected precipitation may indicate contamination from sources such as bird droppings, and high aluminium concentrations will show up incidences of soil splash. To estimate element input amounts the element concentrations derived from the precipitation collector should always be combined with precipitation amounts from an adjacent, correctly exposed, standard rain gauge. Funnels usually have orifice diameters of at least 20 cm to collect large volumes for analysis from small events; Lewis and Grant (1978) give thorough consideration to the necessary size of collectors to provide a sufficient sample for a full chemical analysis without dilution, and conclude that the collector surface areas theoretically required are far in excess of those currently in use. They include a design for an 'ideal' bulk precipitation collector which has a collecting surface of 2025 cm^2.

It is unlikely that a bulk precipitation collector receives the same solute input that is received by a surrounding vegetated surface. The collecting funnel is certain to have very different characteristics in terms of collection of dry-deposited material when compared with vegetation, particularly for sedimentation into an exposed funnel and impaction on a smooth surface compared with vegetated surfaces. Douglas (1972) notes this, commenting that 'sampling systems cannot always cope with the wide range in aerosol size'. Fowler (1984) notes further sources of error: the rate of dry deposition of SO_2 on a glass collector declines as the glass surface reaches saturation, whereas dry particle deposition continues at a constant rate. In contradiction to this last point Rutter and Edwards (1968) noted that the weekly sum of daily dry particle catches on a vane-type artificial impactor collector was around twice the catch from a single weekly collection, put down to the existence of a stable threshold deposition amount on the collector surface beyond which 'blow off' losses increased rapidly. Such effects can take on considerable significance and data quality control must be stringent. Granat (1976) has a discussion of the interaction of spatial variability of precipitation quality and different 'averaging times', and in a later paper (Granat, 1977) reviews various 'siting criteria' for representative results. A specific investigation of the effect of sampling interval length on the composition of wet deposition samples by Madsen (1982), however, showed no significant sampling period effects. Instruments to sample precipitation sequentially, splitting samples on a precipitation amount or on a time basis, have been developed by many investigators (Gascoyne, 1977; Ronneau *et al.*, 1978; Aichinger, 1980; Coscio *et al.*, 1982). Some of these devices are purely mechanical and others require considerable electronic circuitry.

The European Air Chemistry Network, begun in 1955, had at its peak extent 120 bulk precipitation collectors, each consisting of a permanently open plastic funnel draining to a glass bottle which was collected monthly. Goldsmith *et al.* (1984), analysing the data from British gauges in the Network, showed significant differences in several solute species' concentrations in adjacent check gauges at a site, though there was good serial correlation between the gauges. This illustrates the problem of finding the appropriate balance between precision of individual measurements and numbers of measurements to be made in a network (Miller, 1980). Subsequent investigations of atmospheric solute input, such as the OECD Long-Range Transport of Air Pollutants Project (LRTAP) and the Norwegian SNSF Project, have operated on weekly or daily bases and have been more carefully instrumented. A similar concern with improvement of, and standardization of, sampling and analytical procedures has been evident in North America (Anon., 1978; Pack, 1977; Baker *et al.*, 1981; Wisniewski and Miller, 1977; Wisniewski and Kinsman, 1982; Thompson, 1977).

In order to assess 'wet only' deposition it has been noted that some method of preventing dry deposition into a collector during rain-free periods must be employed. The simplest method is clearly to remove a cover manually from the collector when precipitation starts and replace it at the end of the event. Because of obvious and major problems of convenience and timing, automatic devices have been produced which are activated by precipitation and expose the collector for the duration of the event only. At the 'low technology' end of a series of devices is that used by Torrenueva (1975), in which a weighted cover is retained on the collector by a piece of tissue paper: on wetting, the tissue paper breaks and exposes the funnel but has to be manually reset after the event has finished. The majority of wet-only collectors are electromechanical devices which incorporate a precipitation sensing device which, on wetting, actuates a mechanism to open a cover. The sensor is heated so as to dry quickly at the end of an event and cause the cover to close (e.g. Benham and Mellanby, 1978). Some automatic collectors, such as the Health and Safety Laboratory (USA) type, (HASL), the Aerochem Metrics and the Sangamo have two collecting containers; one is exposed during precipitation by a moveable cover which then covers the other container, thus collecting 'wet deposition' and 'dry deposition' in separate containers. Thompson (1977) has warned against this type of collector on the grounds that what is collected in the 'dry deposition' collector is unlikely to have any meaning or value in calculating atmospheric loadings, it being composed primarily of local ground dust. Lewis and Grant (1978) discuss the effects of collector form, height and surface texture and composition on the collection and retention of dry particles. The situation is so complex that standardization would appear to be a useful first step in resolving these problems. The representativeness of collected dry deposition for vegetated surfaces is a further problem, as is the question of the liquid used to flush the material from the collector surface for analysis: should distilled water or dilute acid be used? Bogen *et al*. (1980) and Lindberg *et al*. (1977) have evaluated the HASL collector, Backlin *et al*. (1977) the collector used in the Swedish precipitation chemistry network, and Slanina *et al*. (1979a) the 'wet only' sampler used in the Netherlands. Zeman and Nyborg (1974) and Raynor and McNeil (1979) describe operating combinations of 'wet–dry' collectors and sequential samplers. Many different 'wet–dry' collectors have been developed and comparative reviews of their performance and analyses of their results have been published. Galloway and Likens (1976a, b) compared 22 collectors of 10 different designs at one site. They found significant differences in performance for certain elements, depending on the collector material, and recommended that dry deposition should be excluded if meaningful and accurate statements are to be made about precipitation quality and if collectors are to be inter-calibrated. They also suggested that one week should be the maximum sampling interval, with event-based sampling preferred. A later paper (Galloway and Likens, 1978) further investigated the detailed behaviour of collectors with regard to collection efficiency, sample storage, length of sampling period and collector location, and concluded by attempting to define an appropriate procedure for collection of representative samples of 'incident precipitation'. The last-named paper should be referred to before embarking on a scheme of precipitation collection for chemical analysis. Galloway and Parker (1980) provide a useful summary of the consensus of reports on bulk and wet deposition collectors. Further intercomparisons of different types of collector have been presented by Weibe (1976) and Slanina *et al.* (1979b). Weibe found significant chemical differences between the collection of wet-only and bulk precipitation collectors, whereas Slanina *et al.* found these differences to be negligible. As noted by Galloway and Likens (1978), 'samples of precipitation are probably the most difficult water samples in nature to collect for chemical analysis'.

The direct measurement of dry deposition is extremely difficult; or rather, the accuracy and interpretation of the results of dry deposition measurements are extremely doubtful. These problems partly arise because, as discussed earlier, the term 'dry deposition' is a rather nebulous one as it includes both particulate and gaseous material. Particulate material may be deposited by processes of impaction and sedimentation, depending on particle size, and gases may be adsorbed and absorbed by the surface and by vegetation. The nature of the surface is very important too: its configuration, roughness, wetness and 'stickiness', and its chemical characteristics have all been noted as affecting dry deposition rates. It is therefore almost impossible for the direct collection technique to duplicate the natural collection mechanism (Galloway and Parker, 1980), as the distinct components of dry deposition require different systems for measurement and the behaviour of each component is itself a function of complex soil and vegetation surface characteristics. Even if the surrogate collecting surface or instrument were standardized, its relative collecting efficiency for the components of dry deposition would still be unknown, so that estimates of input to the earth's surface would be only approximate (Lewis and Grant, 1978).

The dry deposition collectors which have been used include buckets, flat filters and plates, sticky films and moss bags (Ibrahim *et al.*, 1983). Filters are frequently used to sample particulate matter; glass wool, paper, ceramic and fritted glass are among the materials used to remove particles from air (Manaham, 1979). The air may be drawn through the material by a small pump, or natural airflow may be used, more realistically, by locating the filter in a vane-mounted tube (White and Turner, 1970). Membrane filters are available in various pore diameters, 0.45 microns being a popular size. Cascade impactors are also available in which air is drawn through progressively smaller orifices to impinge on to wet or dry plates. Nicholson *et al.* (1980) describe a field set-up in which SO_2, NO_x and O_3 gas concentrations are monitored continuously and directly: more frequently air is bubbled through a solvent or adsorbed on to the surface of a solid for later extraction and analysis. There are significant uncertainties in the results of particulate analysis from filters. Reactions may occur both on the filter and during extraction from the filter. Distilled water (or dilute mineral acid) may be used to leach filters for time durations which are not standard or specified. Lewis and Grant (1978) are particularly concerned about the implications of this last factor for dry deposition assessment. White and Turner (1970) also used natural branches and leaves to assess dry deposition to vegetation: the foliage was eventually washed in distilled water and a procedure developed to express dry deposition input in kg ha^{-1} units.

Dry deposition has most commonly been estimated by exposing open containers or flat plates in the field. Wisniewski and Kinsman (1982), in a review of such collectors, considered that the surface characteristics of collectors represented roughness of vegetation and soils quite well, but the form of the structure itself generated significant discrepancies in collection. Galloway and Likens (1976b) noted that differences in the dry deposition catch of funnels and open-mouthed cylinders could be explained as 'over-catch' in large cylinders combined with 'under-catch' in an open funnel exposed to the scouring action of the wind: it was not clear which collector was most representative of actual dry deposition—if this could be known. They therefore recommended the use of wet-only collectors. Psenner (1984) used trays with a 2 cm depth of distilled water in order to assess bulk precipitation quality, thus reducing the amount of 'blow-off' loss of deposited particles. A comparison of bulk and wet-only catches from a two-bucket automatic sampler can be used to estimate dry deposition by difference and will be referred to in the later discussion of indirect estimates of dry deposition. Unsworth (1980b) has a review of direct measurement techniques to assess dry deposition to vegetation in exposure chambers and in cuvettes under laboratory conditions. Encouraging similarity has been noted between bulk precipitation quality and the sum of the compositions of wet-only and dry-only collectors (Galloway and Likens, 1976b).

It has been found that a forest removes as much as ten times as much particulate material from an airstream as a smooth water surface, but that the forest has a scavenging efficiency which is variable and related to particle size distribution, species, wind speed and humidity. This input is likely to be greatly underestimated by the instruments described above and its chemical composition poorly represented. Assessment of dry-deposited material on a vegetation canopy by analysis of gross precipitation above the canopy and throughfall and stemflow beneath it is an approach which has significant deficiencies. It is unable to measure any gaseous deposition which is adsorbed by the plant surfaces or to identify any particles which are absorbed or re-emitted by, and leached from, these surfaces. Miller (1980) isolates as crucially important that a satisfactory method be devised to separate atmospheric solute input from the crown leaching input, particularly so that environmental factors, such as precipitation acidity affecting crown leaching, may be assessed (Miller, 1984). Miller *et al*. (1976) found that crown leaching accounted for 30–60% of the gain of solutes as precipitation passes through a canopy, the remainder being attributed to wash-down of dry-deposited aerosols. Unsworth (1980a) has noted that:

> 'In humid conditions and in mist and fog, particles probably grow such that their impaction efficiencies to forests are increased by an order of magnitude. In these circumstances they would be trapped by forest trees but not by standard rain gauges which, in this respect, are inefficient.'

Schlesinger and Reiners (1974) demonstrated this effect by comparing open collectors with collectors which had artificial foliage attached: increases in the foliar collectors ranged from 4.9 to 8.3 times for elements in the precipitation and 4.5 times for water collection. While there may be significant edge effects here, clearly both dry and wet deposition of material are considerable. Lindbergh *et al*. (1982) consider the processes involved in some detail and note the positive interaction between wet and dry deposition in a forest canopy, which complicates their measurement. Miller and Miller (1980) employed direct measurement techniques to assess the role of the filtering effect of forests in increasing total element input from the atmosphere: they placed a tube of polyethylene-coated wire mesh over a funnel collector and compared the catch, which should contain more impacted aerosols, with that of a Nipher-shielded gauge, which should retain few aerosol particles in the bulk precipitation collected. Up to sixfold increases in element concentration were noted in the 'filter gauges', and large differences in concentration ratios for several elements between the gauges and compared with seawater were also observed. Miller (1980) has reviewed the direct approaches to assessing atmospheric solute input to forests and noted the difficulties of point assessment using surrogate materials and then extending these values to areas of forest; it appears that indirect approaches are likely to yield better results.

Because of the relative simplicity of direct measurement of bulk precipitation and wet deposition, indirect estimation of these inputs is rarely attempted. If the mean weighted streamwater element concentration is divided by the ratio of precipitation to runoff, then, for a conservative solute species such as chloride, a crude estimate of mean bulk precipitation concentration may be obtained (Juang and Johnson, 1967; Janda, 1971). Baldwin (1971) used a similar data manipulation. Indirect estimation of element inputs from the atmosphere has been accomplished by using seawater proportions between elements where an oceanic source dominates. Cleaves *et al*. (1970) used weathering equations and assumptions about biomass incorporation which could also be used to obtain bulk precipitation element input by difference calculations. Transport and deposition models, such as those discussed by Goldsmith *et al*. (1984) in which meteorological processes are modelled and empirically fitted to collected data to estimate the main features of average annual solute input fields, could also be seen as indirect approaches to determining wet deposition by interpolation to any point within the coverage of the model. Miller (1980) has reviewed the indirect approaches which have been used to separate crown leaching input from the bulk precipitation input when they are combined in the 'bulk net precipitation' collected on the forest floor. Plotting dry deposition on to an automatic collector against 'net-bulk precipitation' gives an intercept which is an estimate of canopy leaching (Miller *et al*., 1976). Mayer and Ulrich (1974, 1977, 1978), by comparison, used the proportional gain of solutes by throughfall in winter under deciduous trees as a baseline with which to partition summer throughfall solutes into atmospheric or canopy leaching sources.

Galloway and Likens (1976b) provide an approach to indirect estimation of the dry deposition input. In this bulk precipitation is compared with wet deposition only and the difference attributed to dry deposition. Allowances must be made for precipitation amount and duration of dry deposition collection, and for the fact that direct measurement of dry deposition is thought to include little input by impaction or gas absorption pathways. The ratio R of mean species concentration in bulk precipitation to that species concentration in wet-only deposition can be used to make inferences about the nature of the solute input. If $R > 1.0$ then the species is present in both wet and dry deposition: if $R = 1.0$ then it is present in wet deposition only, and if $R < 1.0$ then some chemical reaction or sampling effect must be operating.

Because of the practical difficulties with direct measurement of dry deposition input there have been other important developments in its indirect estimation as the significance of this pathway has become more apparent. As noted in an earlier section, these developments are based on either a micrometeorological approach to determining the flux of gas or particles to a surface down a concentration gradient or on a resistance analogue approach. The report edited by Nicholson *et al*. (1980) provides useful summaries of these methods and the often complex instrumentation and analysis required to obtain the necessary calibration data. Fowler (1984) comments that the advantage of the resistance analogy approach is the simple way in which atmospheric and surface resistances, as convenient parameters, may be considered separately to show the relative importance of individual steps in deposition or combined to assess the total input to a surface. At present investigations are proceeding into the way in which a particular deposition velocity (the reciprocal of resistance) varies considerably for different surfaces, through time and with surface conditions and meteorological conditions. These investigations show the approach to be generally promising in the indirect estimation of dry deposition, but that the use of fixed deposition velocities to estimate deposition rates over large areas and for long periods should not be advised, except as a first approximation (Wisniewski and Kinsman, 1982).

The instruments and procedures referred to above are all designed to collect a sample of the total atmospheric solute input, or a component of that input at a point in such a manner that the sample is *representative* of the actual input, both in concentration and in composition; i.e. there is no bias introduced by the sampling process. In addition to the major problems of 'collection' which have been discussed, there are additional problems of storage of the collected samples in the field and in the laboratory before chemical or other analysis, and then further sources of error in the analysis itself. Indeed, deterioration of a sample in storage and poor laboratory procedures may completely negate the care and effort (and expense) which may have gone into the field sampling programme. General comments concerning the need for attention to storage and analysis of natural water samples (Owens *et al*., 1980; Cryer and Trudgill, 1981; Moody, 1983) are particularly important when very dilute and small volume samples, for example of bulk precipitation or wet deposition, are being processed, when minor sources of contamination and environmental changes may profoundly alter the sample's characteristics. In many respects biased data are worse than no data at all in that they may misdirect research effort and provide support for inaccurate hypotheses.

Care must be taken to prevent sample modification in the period of field storage between collection times and during laboratory storage. While most authorities recommend 'immediate' laboratory analysis, practical constraints give rise to typical field storage recommendations of 'preferably less than one, certainly less than two weeks' (Miller, 1980). While in storage, samples may be modified by reactions with the material of the container, by temperature changes and particularly by metabolic activity of organisms present in the container. Galloway and Likens (1978) have produced a list of, and tested, several biocidal chemicals which can be added to containers for field storage, though it should be noted that they did not recommend any one because all had their own peculiar, adverse effects on sample quality. They also list 'manipulative biocides'—techniques to preserve samples, such as refrigeration and dark storage, which Fowler *et al*. (1982) found to be adequate to preserve the integrity of their low solute content samples, which also, helpfully, had low pH values. Allen (1974) has compiled a useful summary of sample treatment procedures which could be used in field storage. The treatments selected to preserve samples should be chosen with the subsequent chemical analysis in mind, as there may be interference by the preservative. While analytically clean containers should be used, Lewis and Grant (1978) note that acid cleaning may in fact 'reactivate' the container surface, thus increasing adsorption and contamination by leaching: they recommend thorough rinsing in distilled water followed by dry sterilization at 150 °C. The choice of sample container material, usually between plastics and glass, will depend on the solute species of interest. Galloway and Likens (1976b) found that plastics generally had a much less active surface than glass, were more robust and were recommended, except where phosphate was to be determined, though long-term storage in plastic is more doubtful because of its adsorptive properties (Bowditch *et al*., 1976). At sites which are left unattended for periods of one month or more at remote sites, it may be that a development of the ion-exchange resin storage technique suggested by Crabtree and Trudgill (1981) may be useful.

Long-term storage in the laboratory is often necessary in order to process large batches of samples at the same time. If preservative chemicals are thought to be necessary and a complete chemical analysis is to be performed, it is common to split samples and apply different treatments as appropriate. There is an increasing tendency to recommend that no chemical treatments are carried out prior to laboratory storage and that cold storage in the dark should be relied on. Galloway and Likens (1978) note that precipitation samples with a pH of below 4.5 are 'essentially self-preserving', though no critical pH value has been accepted. At the low concentrations typical of precipitation samples the formation of precipitates during storage is much less likely than for other natural water samples. Galloway and Likens (1976b) found that a storage temperature of between 4 °C and −4 °C was adequate for a period of 7 months for precipitation samples, though it may be ill-advised to actually freeze samples (McDonald and McLaughlin, 1982; Lewis and Grant, 1978). If a chemical preservative is to be added then high grade reagents should be used to avoid the inadvertent contamination of samples, and control tests should be carried out comparing all combinations of treated and untreated, fresh and stored samples, standards and blank solutions. Slack and Fisher (1965) and Schock and Schock (1982) have presented the results of such tests on the effects of light availability and container type respectively.

Because most sample solutions produced in the assessment of atmospheric solute input are of low ionic strength, particularly if small volume samples have been 'bulked' with distilled water, usually with total dissolved solids concentrations considerably less than 50 mg per litre, then appropriately sensitive analytical procedures should be used. Galloway and Likens (1978) and Lewis and Grant (1978) list the analytical sensitivities usually required and the methods applied for determination of the major constituents of precipitation samples. Thompson (1977) lists the techniques employed by one precipitation quality monitoring network in the USA, and a fuller list, intended for a future US monitoring programme, appears in Anon (1978). Slanina *et al*. (1979a, b) report the analytical procedures used in their comprehensive water analyses. Hunt and Morries (1982) emphasize the significance of the purity of the water used in these analyses and show the bias that may result when the determinand appears at very low concentrations in the blank. There are several publications available which contain descriptions of suitable analytical methods for major ions (Golterman *et al*., 1978; Bauer *et al*., 1978; Allen, 1974; Pagenkopf, 1978; Allen and Kramer, 1972). Particular attention has been paid to the determination of precipitation acidity. Initially this was concerned with the inaccuracies generated by storage and transport from field to laboratory (Roberson *et al*., 1963; Barnes, 1964; Jervis, 1979), but since then there has been a reconsideration and redefinition of the parameter and its measurement and a reappraisal of existing data (Cogbill and Likens, 1974; Hansen and Hidy, 1982). Galloway *et al*. (1979), Granat (1972) and Galloway *et al*. (1976) have all provided significant contributions to a wider knowledge of the deficiencies of some methods in solutions of low ionic strength. Tyree (1981) provides a later review of the alternative procedures for acidity measurement and comments on between-procedure and between-laboratory differences. Trace constituents are of increasing interest in precipitation and demand special analytical procedures: some of these are discussed by Hamilton and Chatt (1982), Slanina *et al*. (1979a) and Strachan and Huneault (1984), the last-named referring particularly to trace organic substances. There are several edited volumes reporting recent conferences on aspects of precipitation quality and 'atmospheric solutes' which contain many references to analytical techniques and procedures which have been selected for and proved in field experiments: among these are Correll (1977), Drablos and Tollan (1980), Nicholson *et al*. (1980) and Eisenreich (1981).

2.4 RESULTS OF ATMOSPHERIC SOLUTE INPUT ASSESSMENT

There has been steadily increasing recognition in recent years of the significance of atmospheric solute inputs to a full evaluation of biogeochemical cycles of elements and the plant nutrient element cycles which are nested within these (Viro, 1953; Gorham, 1961; Kormondy, 1969; Odum, 1971; Duvigneaud, 1971). This recognition has been due in great part to the very long record of precipitation quality produced by the Hubbard Brook Ecosystem Study, carried out in New Hampshire, USA, whose findings are summarized in Likens *et al*. (1977). This project was based on the catchment as the fundamental unit of study and thus took advantage of the intimate interaction between the hydrological cycle and biogeochemical cycles to demonstrate and quantify the element fluxes within the study area (Likens *et al*. 1970). These fluxes may be resolved into three: the meteorological flux, of interest here, into the catchment, the geological flux, and the biological flux. If the net biological flux of elements can be considered to be close to zero in a stable, uniform, natural ecosystem, then estimation of the meteorological flux in, and the geological flux out of a catchment with an impermeable base allows a budget calculation of net gains to, or losses from, the ecosystem. The atmospheric input is thus being designated a 'non-denudational component' of the solute content of streamwater. The use of the drainage basin has in this way allowed element fluxes to be defined and quantitatively determined by establishing functional boundaries in the vertical and horizontal directions for biologic activity and for drainage (Bormann and Likens, 1969, and 1970). The small catchment model has further allowed replication of results and also provides a basis for experimental modification of natural ecosystems, for example, by simulating an increase in the input of acidic components from the atmosphere (e.g. Abrahamsen, 1980; Lee and Weber, 1982).

Atmospheric solute inputs may be considered to have both an 'active' and a 'passive' role in ecosystems. They take on an effectively passive role in high nutrient status and stable ecosystems (Thornton and Eisenreich, 1982) and when the solute species does not act as a plant nutrient. The *net* solute losses from a catchment—that is, streamwater output minus atmospheric input—must come from the chemical weathering of primary and secondary minerals within an ecosystem at or near steady state (Likens *et al*., 1970). In oligotrophic ecosystems, where bedrock is nutrient-poor or resistant to weathering and where precipitation totals are high, atmospheric nutrient inputs may be a major source of plant requirements (Crisp, 1966). Robertson and Davies (1965) noted that 'it may be that the maintenance of the nutrient fund within low-fertility grazing ecosystems depends to a large extent on the input of dissolved nutrients in rainwater'. In this situation, then, atmospheric solute inputs may be considered to have an active role, though this will change seasonally with the activation and relaxation of biological processes and the varying intensity and composition of atmospheric input. Thus in an undisturbed ecosystem, at long-term equilibrium, the atmospheric input of elements should be accounted for in the streamflow output from the catchment. At the shorter term, within one year, biological activity can introduce lags and inequalities between input and output at any time, in the absence of short-term equilibrium conditions within the vegetation cover or the soil. Atmospheric solute input may also be seen as 'chemically active' within the catchment, as 'a factor in the weathering environment' (Douglas, 1972). As an example, Fisher *et al*. (1968) proposed that 'the strong mineral acid from precipitation may provide the principal driving force for mineral weathering as indicated by the solutes contained in drainage waters'. Clearly, 'passive' atmospheric element inputs would need to be allowed for in this context when relating hydrogen ion input to element output in streamwater to avoid over-estimation of the chemical denudation generated. Janda (1971) proposed that this over-estimation could be by a factor of between 1.4 and 2.4. The active role of atmospheric solute input has been quantified in catchment experiments by Bricker *et al.* (1968) and Cleaves *et al.* (1970, and 1974). Huang and Keller (1970) have discussed the weathering reactions in first-stage weathering of silicate minerals which are affected by active solute inputs.

Given the associated difficulties of definition and measurement in atmospheric solute input studies, any published results of such studies should be considered alongside the field and laboratory procedures by which the results were obtained to ensure appropriate interpretation. It is worth repeating that this interpretation is particularly difficult for geomorphological studies, especially in forested areas where there may be large, unmeasured, 'passive' element inputs and also exaggeration of some element inputs by the action of 'active' constituents of the input in canopy leaching etc. Although Walling (1980) states that wet precipitation solute input is more significant to water quality studies than is bulk precipitation input, there is uncertainty here as dry-deposited material may to some extent be dissolved by a subsequent precipitation event and thus be involved in the relatively short-term 'solute dynamics' of streamwater in the same way as wet deposition. Clearly bulk precipitation quality is a more appropriate parameter for the assessment of longer-term biogeochemical budgets, but will be particularly deficient for those elements such as N and S which exist predominantly as gases or aerosols in the atmospheric input. The presence of a significant catch of dry particulate material in bulk precipitation and dry fallout collectors, reported by Finlayson (1977) and Grant and Lewis (1982) respectively, is problematical in that it may originate from within the catchment and thus not be a true atmospheric 'solute' input. Lewis and Grant (1981) could show no significant increase in particulate loadings in the Colorado Rockies following the Mount St Helens eruption, though these inputs would have figured as a true atmospheric input.

A wide variety of units of measurement has been applied to atmospheric solute inputs. For geomorphological studies input amounts are often presented as a concentration in mg per litre (preferred to ppm) together with the precipitation amount and period of measurement. From this an input or load can be expressed in $kg\ ha^{-1}$ per unit time: $kg\ ha^{-1} = (mg\ l^{-1} \times mm)/100$. The element concentrations should be expressed as volume weighted means: $(\Sigma(\text{amount} \times \text{concentration}))/\Sigma\text{amount}$, for any period because of the usual inverse relationship between precipitation amount and concentration. The ratio of the raw mean to the volume-weighted mean element concentration gives a rough indication of its source in the sample: a ratio of above 1.0 suggests a dry-deposited component in the sample. Element loads over a period are then the product of precipitation amount and weighted mean concentration for the period. A purely dry-deposited element should show a generally invariant load with precipitation amount, while a purely wet-deposited element should show an increase in load with precipitation amount. For elements with relatively high concentrations in precipitation, the load–precipitation amount relationship will plot as a steep slope for wet-deposited elements, the line passing through the origin. For dry-deposited elements the load–precipitation amount curve should have a significant intercept value on the load axis. Thus elements with the highest slope/intercept ratios will be basically wet-deposited, and this ratio should be comparable between sites with different total load inputs. Variations around a straight-line plot will result from variations in the rate of wet or dry deposition during the period.

Particularly in the biological literature, input concentrations are preferred in units of equivalents, or micro- or milli-equivalents, per litre. These units are derived by dividing concentrations by equivalent weights for individual species, where equivalent weights are given by the ratio of atomic weight to valency for the ionic species. In calculating acid inputs for element budget studies in which the 'replacing efficiency' (Wiklander and Andersson, 1972) of hydrogen ions is used to estimate chemical weathering rates or to quantify cation exchange reactions, pH units should be converted to equivalents ha^{-1} per unit time in order to estimate the leaching from the system of cations in drainage waters which is due to this external hydrogen ion source (Likens *et al*., 1977). Sutcliffe (1979) has written a useful guide to the appropriate units and data presentation methods for natural water quality data. Many data analyses use the concept of 'excess element concentration' to assess the source of an element (Eriksson, 1960). Excess concentration is the sample concentration remaining after the subtraction of the concentration of that element in standard seawater diluted to the sample's concentration of a particular reference element, usually choride or sodium though magnesium has also been used. An excess element concentration is thus assumed to be due to 'anthropogenic' enrichment of the precipitation, though it may also be due to natural terrestrial sources of the element, or to element enrichment or loss during the processes of generation, transport and deposition which modifies the 'natural' precipitation element ratios away from those present in seawater (Overrein *et al*., 1981). Fisher (1981) has looked at the problem of defining background and excess sulphate concentrations, and Wright and Johannessen (1980) discuss the element ratio methods used in Norway to 'correct' and infill deficient atmospheric solute input data.

Within the last 30 years the increased concern for the quality of the natural environment and the demonstration of the impact that industrial, agricultural, military and even domestic operations may have on this environment have together generated a veritable explosion of environmental research activity. Such research is aimed to both elucidate the natural system at the focus of this activity and to assess the extent of undesirable modifications of the system so as to allow the formulation of 'management' schemes and policies. The recognition of the environmental role of the atmosphere as both a sink and a source for plant nutrients and pollutants has generated a plethora of published material, particularly that concerning 'the acid rain problem'. A spinoff from this activity has been the incorporation of much of the accumulated data and expertise into geomorphological research, where material inputs in precipitation are acknowledged as important to the 'quality dimension' which has of late experienced an upsurge of attention (Walling, 1980). Because of the volume of published work this review cannot hope to be very specific or wholly comprehensive in its coverage of the field of atmospheric solute inputs, but reference will be made particularly to review articles and the more significant and recent research contributions.

As the significant role of atmospheric solute input has become clearer, so this input has been investigated 'in its own right' and its spatial and temporal complexity revealed. The first large-scale maps of major element concentrations in precipitation were produced for western Europe by Eriksson (1955) and Rossby and Egner (1955) and for the USA by Junge and Werby (1958). These demonstrated the importance of proximity to maritime influences, and Rossby and Egner (1955) and Eriksson (1959, 1966) made further inferences concerning the effect of the pattern of atmospheric circulation from the spatial pattern of input observed. Gorham (1958) made early use of maps of precipitation quality to show quantitatively the influence of industrial point sources of pollutants in northern England. Yaalon and Lomas (1970) in Israel, Baldwin (1971) in California, and Stevenson (1968) and Gardiner and McGreal (1978) in the UK also monitored the decline of solute concentrations and systematic variations in ionic ratios away from the coastal zone. Ogden (1980) has shown the significant differences in the relative chemical composition of 'maritime' (Nova Scotia) and 'continental' (north-eastern and eastern USA) precipitation. Such detailed mapping is now more feasible as a result of the national and international precipitation quality monitoring networks which have been set up in North America (Baker *et al.*, 1981) and in western Europe. The inevitably more complex spatial patterns of atmospheric solute input which result from denser monitoring are often difficult to explain and thus lead to a search for improved process explanations (Overrein *et al.*, 1981). Munger and Eisenreich (1983) present maps and discussion of monitored atmospheric solute input in the USA. Data from the remote tropical continental interior sites is now entering the literature (Gaudet and Melack, 1981; Edwards, 1982) and suggests low solute concentration and input loads despite high precipitation totals. Detailed investigations within areas of less than 10 km^2 have also been carried out and have demonstrated significant local spatial variations in solute input magnitude and composition. Martin (1982a) found considerable differences in precipitation quality between adjacent hilltop and valley-bottom sites and the spatial variability discovered by Richter *et al.* (1983) within a 500 h catchment allowed an objective assessment of the number of collectors necessary to maintain a particular level of precision in the precipitation quality data. The spatial pattern of precipitation acidity (e.g. Fowler *et al.*, 1982) has clear significance in geomorphology and in ecology, but is often difficult to explain and to relate to areas where apparent acidification damage to vegetation has occurred.

Superimposed on the spatial pattern of atmospheric solute input, and interacting with it at all scales, is a temporal variation due to variations in the efficiency and intensity of the processes of generation and entrainment of materials and their transportation and deposition. The complexity of these processes, described earlier, is such that they are somewhat crudely represented at present by meteorological variables: by synoptic indices and atmospheric circulation indices at long time scales, by air mass trajectories and wind velocity and direction at the event duration time scale, and by washout and rainout models within the event duration. A very long record of several components of atmospheric solute input, more than 100 years long, has been analysed by Brimblecombe and Pitman (1980) to test for significant long-term trends. The doubtful quality of the data makes this analysis difficult, but on a much shorter time base Barrett and Brodin (1955) identified a winter peak in precipitation solute concentrations and loads in north-western Europe, and Gambell and Fisher (1966) showed the pattern in the eastern USA to be due to changes in the relative dominance of the sea-derived or land-derived solutes in response to seasonally-varying air flow patterns. Allen *et al*. (1968) showed that seasonal variations in the dominant solute source could be represented by changes in concentration ratios. Munn and Rodhe (1971) related monthly precipitation sample quality to contemporary meteorological conditions, and Cryer (1976) used the synoptic indices of Murray and Lewis (1966), as used by Perry (1968), to resolve the temporal variability of solute concentration: in this case improved results were gained when weekly data were aggregated to monthly averages and when the effect of varying precipitation totals on solute concentrations was statistically controlled. Lewis (1981) has demonstrated an extreme situation in a tropical environment where 15% of the annual atmospheric solute input loading came in only one or two weeks of the year.

Significant variations in atmospheric solute input rates from one precipitation event to the next have also been observed. Some of this variation may be due to the length of dry period between events: longer dry periods result in higher concentration inputs because of wash-out and wash-down accumulation and exhaustion effects for aerosols in the atmosphere and on exposed collectors respectively, and a delay in the re-establishment of atmospheric aerosol levels (Walling, 1980; Gascoyne and Patrick, 1981). At coastal sites, Rutter and Edwards (1968) showed the disproportionate effect on salt deposition rates of high and rising wind velocities, and Asman and Slanina (1980) used air mass trajectories to determine precipitation events as 'maritime' or 'continental' and found significant between-event concentration differences for a wide range of solute species. Considerable within-storm variations in element concentrations in precipitation have been monitored, with the aid of automatic sequential samplers. Gascoyne and Patrick (1981) and Kennedy *et al*. (1979) reported results from west coast sites in the UK and USA respectively, the latter finding a strong inverse relationship between concentration and precipitation intensity within the event and both noting a tendency for concentrations to decline to finite, low levels during the event. Hogan (1983), in sampling through frontal precipitation systems in north-eastern USA, found both absolute and relative concentrations also to vary inversely with precipitation intensity, and for loads therefore to remain uniform and proportional to precipitation amount. Cooper (1976) observed the reduction of pH during an event and attributed this to a reduction in chemical buffering effect as calcium, and other neutralizing materials, were removed by washout processes. Variation in trace element concentrations in rainwater during convective storms has been observed and interpreted by Gatz and Dingle (1971).

A major interest in geomorphology is the role of chemical denudation in landform development and landscape evolution, and the rate and nature of solute removal in drainage water is a clear index of this component of landscape development. However, the interpretation of solute load outputs to this end is problematical in that it requires some knowledge of the sources of the solutes—whether these are from within the catchment from rock weathering and biological sources, from outside the catchment as atmospheric solute input, or from 'pollution', including fertilizers and other agricultural applications. There are considerable difficulties in deciding the size of these contributions and the extent to which they represent denudational losses from rock breakdown, or non-denudational losses (Janda, 1971). Careful budget analysis, such as that by Cleaves *et al*. (1970) and Waylen (1979), will allow the estimation of spatially lumped average denudation rates for whole catchments. Further to this, Webb and Walling (1980) suggest that 'scope clearly exists for the coupling of empirical studies of chemical weathering and solute mobilization to models of slope development', thus involving the integration of spatial patterns runoff production and solute production at the hillslope scale to simulate slope form development variations within a catchment—a procedure attempted by Finlayson (1977). The following section here will survey that published research which has had as one of its objectives the estimation of atmospheric solute input as a non-denudational component which can be subtracted from streamflow solute load to arrive at a value for denudation rates. Subsequently reference will be made to those publications which have estimated atmospheric solute inputs within biological and ecological studies as plant nutrients; and finally, some of the major work concerning acidic inputs from the atmosphere will be considered.

Janda (1971) summarized the approach and results of earlier attempts to estimate net denudation rates. He noted that:

> 'Precise computations of rates of chemical denudation must be based on quantitative evaluation of these (terrestrial and atmospheric) potential sources. Such evaluation requires reliable sampling and analytical procedures as well as knowledge of the chemical composition of bulk precipitation (and) sources for the dissolved materials in the precipitation.'

Goudie (1970) similarly reviewed existing research, which underlined the need to measure 'input factors for the proper estimation of solutional denudation'. Carroll (1962) considered precipitation as an active chemical agent in rock breakdown and discussed chemical mass action effects on retention and leaching of soil and rock materials. Miller (1961) and Hembree and Rainwater (1961) constructed element budgets which demonstrated the significance of the input from the atmosphere in the western USA, Miller showing that atmospheric solute inputs provided between 10% and 60% of streamflow output of solutes on sandstones and quartzites respectively. Meade (1969) reviewed the use of streamflow solute load data and warned against simple extension of these data to denudation loss estimates where atmospheric contributions are significant and where areas are affected by human settlement. Van Denburgh and Feth (1965) used chloride content of precipitation and streamflow to show the importance of considering atmospheric inputs in denudation estimation: they suggested that between 2% and 20% of streamflow output across the western USA was, from the chloride balance, derived from the atmosphere, but warned that the results were only first approximations because of the relative paucity and poor spatial cover of atmospheric solute input data at that time. Cleaves *et al*. (1970) and Claridge (1975) provided timely reviews of the experimental procedures necessary to obtain meaningful estimates of atmospheric inputs and their role in denudation. Cleaves *et al*. showed that the figure for the relative effectiveness of solutional to mechanical weathering depended very much on the duration of monitoring: the ratio varied from a value of 5 down to 1 for longer durations which were more likely to include extreme events. Also, they pointed out that the effectiveness of atmospheric input in active weathering would depend on whether precipitation was on floodplain or upslope areas, and thus presumably its effectiveness in catchment denudation is dependent on the relative proportions of floodplain areas within the catchment. Betson (1979) showed the problems in using a 'leaky' limestone-floored catchment in element budgeting: when much of the catchment outflow is not measured at the surface then the basin may be thought to be a sink for atmospheric solutes which may, in fact, be leaving by sub-surface pathways.

In relating precipitation quality to streamwater quality, both Dethier (1979) and Reid *et al*. (1981) showed that only highly mobile ions like sodium directly influence streamwater concentration levels, and that most others reflect biologic and pedogenic processes which 'swamp' or mediate any clear and direct correlation between precipitation quality and streamwater quality, even when between 30% and 50% of streamwater solutes are derived from the atmosphere. Skartveit (1980, 1981) showed high precipitation–streamflow concentration correlations for sulphate and chloride ions at a site in western Norway, especially during periods of high runoff rate. Cryer (1976) demonstrated such a correlation statistically, but in a peat catchment with acid peat soils and rapid hydrological response to precipitation. Bache (1984) reviews the soil–water interactions which are interposed between precipitation solute input and streamwater solute output and which usually blur input–output quality correlations. Walling and Webb (1980) suggested that the variation in atmospheric solute input will complicate simple relationships between discharge and element concentration in streams, and Douglas (1968) demonstrated the interaction of precipitation solute inputs and bedrock lithology in streamflow quality records from eastern Australia. The first comprehensive catchment element budgets were provided by the Hubbard Brook workers (e.g. Fisher *et al*., 1968), and these clearly identified the need to monitor atmospheric solute inputs and influenced many subsequent research projects in other countries. Element budgets produced explicitly for calculation of denudation rates are far less numerous at present than those which result as by-products of nutrient element budgets from biological and ecological research programmes. Edwards (1973), Waylen (1979) and Foster (1980) have, however, drawn up element budgets for denudation rate calculation in catchments in southern England. In North America, Zeman and Slaymaker (1978) produced budgets for catchments in British Columbia. Feller and Kimmins (1979) include a summary of annual chemical budgets for undisturbed forest areas around the world and also provide a useful summary of the difficulties frequently experienced in interpreting the results of these experiments. Foster (1980) has presented alternative procedures which may be used for the computation of chemical denudation rates from element budget data, and notes that poor data quality and different methods of output load calculation produce significant inaccuracies in denudation rate estimates, especially when monitoring continues for only one or two years. Atmospheric solute input data have also proved valuable in assessing recharge rates for shallow unconfined groundwater bodies. Allison and Hughes (1978), Oakes (1979), Kitching *et al*. (1980) and Irving (1982) have used the fact that chloride concentrations in bulk precipitation show a winter peak and that tritium concentrations are declining steadily from a peak in the early 1960s to assess annual and longer-term increments of recharge from the spacing between naturally labelled markers in the interstitial waters from rock cores. Information can also be gained on the nature of flow processes in the unsaturated zone by assessing the attenuation of the atmosphere-derived concentration peaks with vertical progress of the labelled water towards the water table.

Far more information on atmospheric solute inputs has been generated by ecological nutrient budget studies than has come from geomorphological research. These studies are generally concerned with the nutrient cycles of forests and agricultural land in order to discover and quantify the significant sources and pathways for nutrient elements or pollutants, and thus to identify those elements which are in short supply, in excess concentrations, or in a fragile balance which is sensitive to human manipulation. It would seem that much of this nutrient budget information, together with some additional local information on particle density (Foster, 1980), can be used to extend the cover of denudation rate estimates. The data collected by the Hubbard Brook Ecosystem Study (Likens, *et al*., 1977) have been used to produce absolute and relative chemical weathering rates, adjusted for biomass accumulation and state of development of the vegetation cover. The procedures used in these calculations provide useful guidelines and a sound basis for the extraction of maximum information from the results of many catchment-based element budget studies, and indeed a good model for planning such experiments. Overrein *et al*. (1980) point out that 'weathering' may be viewed in slightly differing ways: at Hubbard Brook the term 'cationic denudation' is used to indicate weathering as the rate of cation outflow in streamwater and storage in the biomass, or the rate of hydrogen ion input consumption. Thus, in a degrading ecosystem, a reduction in base saturation in the soil, perhaps due to acidification from the atmosphere, will make an addition to this 'type' of denudation, though the magnitude should be small in undisturbed systems.

Some important early work in measuring atmospheric inputs of macronutrients (Ca, Mg, Na, K, Cl, SO_4–S, NO_3–N) was carried out during studies of moorland nutrient cycles in the north of England (Crisp, 1966; Gore, 1968; White *et al*., 1971), and in the south of Scotland (Boatman *et al*., 1975). In the English Lake District, element budgeting for major cations over a period of one or two years showed considerable temporal variations in element balances and timing differences between different elements (Sutcliffe and Carrick, 1973a, b, c). This research was extended and the results placed in the context of a long-term survey of local lakewater quality in a major publication by Sutcliffe *et al*. (1982) which may be usefully referred to for guidance on field and laboratory methods. There is a wealth of atmospheric solute input data, particularly for Scandinavia, in the 'acid deposition' literature which will be referred to in a following section. Wright (1983) and Calles (1983) present the results of element input–output analysis for catchments in Norway and Sweden respectively, and variable input to a heathland area in Belgium is contained in Van Genechten *et al*. (1981). A comprehensive survey of element budget studies carried out in Norway by the SNSF project is in Wright and Johannessen (1980) and for Scandinavia as a whole in Overrein *et al*. (1981), who have compared weathering rates (in meq m^{-2} yr^{-1}, the accepted units for these estimates) obtained in Scandinavia and in Hubbard Brook. In North America the 1970s saw a great deal of element budget research activity 'in parallel' with the Hubbard Brook study. Swank and Henderson (1976), Swank and Douglas (1975, 1977) and Henderson *et al*. (1977, 1978) presented major nutrient budgets for natural and manipulated, hardwood and softwood, forests in the eastern USA. In addition, Best and Monk (1975), Zeman (1975) and Tabatabai and Laflen (1976) provide useful reference points in this field in the northern American literature. Tomlinson (1978) for Rhodesia, and Westman (1978) for a sub-tropical coastal area, are among the rare atmospheric solute input reports from outside of humid temperate environments. N and P inputs from the atmosphere have been assessed by Fish (1976a, b) in New Zealand and within a full mass balance experiment in the Hubbard Brook catchments and shown here to be small relative to other sources and fluxes, yet not insignificant relative to the net balance of inputs and outputs (Meyer *et al*., 1981). Similarly Schlesinger *et al*. (1982) found that, in a mature chaparral ecosystem in southern California, 'the total atmospheric deposition of nutrients equals or exceeds the annual losses', but that the system would take many years to recover from the nutrient losses during a fire. Nuckols and Moore (1982) found a significantly large N input from the atmosphere in Tennessee and discussed the various input modes and vectors responsible. Schuman and Burwell (1974) and Schreiber *et al*. (1976) measured N input from the atmosphere to be equivalent to streamflow losses. The significance of organic carbon inputs in precipitation and dry fallout has been shown by Likens *et al*. (1983), and Swanson and Johnson's (1980) trace metal budgets in New Jersey forests on acidic soils suggested retention and accumulation of some elements in the soils. Cawse (1977) investigated atmospheric inputs of trace elements in the UK.

The interaction of 'gross' atmospheric solute inputs with forest canopies and other vegetation covers may produce a significantly altered 'net' solute input beneath the vegetation at the soil surface. The solution and cation exchange processes by which this alteration may occur have already been referred to: what follows here is a brief summary of some of the major publications concerned with element budgets for vegetated, particularly forested, catchments which contain atmospheric solute input information. While forested areas may be viewed as relative sinks for precipitation solutes at the regional scale (Characklis *et al.*, 1979) the Hubbard Brook study results showed significant enrichments of gross precipitation passing through forest canopies: 10, 15, 18 and 91-fold increases in concentration between precipitation and throughfall were assessed for Ca, Mg, PO_4–P and K respectively. The greatest absolute increase was for organic carbon concentrations, from 2.4 to 12.0 mg per litre, though it should be recalled that the proportion of all these increases which should be correctly attributed to the meteorological element flux, as true atmospheric solute input, is very difficult to estimate. Some vegetation types on particularly nutrient-poor bedrocks may have developed in response to the nutrients supplied from the atmosphere and thus may be said to 'rely on' these contributions to maintain equilibrium conditions. Robertson and Davies (1965) found that precipitation inputs of calcium and magnesium were sufficient, and of potassium almost sufficient, to replace losses due to burning on a heather moorland. Chapman (1967) similarly found this situation for a dry heathland area in southern England, where precipitation replaced all but the nitrogen and phosphorus of those elements lost by burning. Miller *et al.* (1979), working in a Scottish pine forest, found the low atmospheric input levels of K and Mg were significant to the nutrient cycle because these elements were very efficiently cycled and not immobilized within the soil, while inputs of Ca, N and P were inadequate because of greater demand and immobilization of N in particular. It was suggested that the trees have evolved a mechanism which allows them to exist in K- and Mg-poor soils where there is a sufficient atmospheric input of these elements.

In an early investigation of solutes in net precipitation Madgwick and Ovington (1959) showed the relative magnitudes of atmospheric solute input above and below various forest canopies in an area of south-east England, and they drew attention to the significance of foliar leaching and other mechanisms by which the gross input is enhanced. At the same time, Will (1959) found a twofold increase in P and K under forest cover in New Zealand compared with precipitation outside the forest, and Voigt (1960) recorded comparable results in the USA. A paper by Attiwill (1966) included a comprehensive review of current theories for the increase of concentrations in throughfall and streamflow and presented experimental results for a eucalypt forest. Criticism of the net precipitation sampling schemes used in earlier research came from Carlisle *et al.* (1966), who randomly placed 20 collectors in a 225 m^2 forest plot. They were among the first to note removal rather than addition of nutrients, in this case inorganic nitrogen, from precipitation during interception detention on the canopy. Sokolov (1972) reported on throughfall quality in the USSR, and Turvey (1974) in the tropical rain forest of Papua where it was found that throughfall quality dominated streamflow quality. Reiners (1973) compared net precipitation quality in three forests in Minnesota, USA, and Eaton *et al.* (1973) at

Hubbard Brook showed the significance of solutes in net precipitation as direct and available additions to the forest's nutrient pool. The timing of maximum enrichment of precipitation during the year has been commonly observed to be at times of spring canopy development and autumn leaf fall in deciduous forests (Best and Monk, 1975). The nutrient losses from various row crops in incident precipitation were assessed by White (1981), and Brabec *et al.* (1981) have reported on the magnitude of 'wash down' of particles filtered from an airstream by various forest types. Cole and Johnson (1977) observed exhaustion effects through individual storms in the element concentrations in net precipitation below the canopy.

The effect of high-acidity precipitation on removal of nutrient elements in throughfall has been a major research interest in recent years, presenting as it does a possible mechanism for damage to forests and crops. Cole and Johnson (1977), Eaton *et al.* (1978) and Alcock and Morton (1978) have monitored hydrogen ion and sulphate deposition on various forest canopies and the resulting throughfall qualities. Dollard *et al.* (1983) have investigated the role of occult precipitation and vegetation filtering mechanisms in this process, which may be particularly significant in high-altitude forests (Cronan *et al.*, 1978). Miller (1984) gives an up-to-date summary of current understanding of atmospheric deposition interactions with vegetation.

It is possible that the input of atmospheric solutes directly to lakes may have a significant effect in that the lakes may become eutrophic or hypereutrophic (Brakke, 1977) and may experience damaging acidity increases (Wright *et al.*, 1980), with consequent damage to the lake ecosystem and loss of amenity. Lakes are thought to be particularly sensitive to such direct atmospheric solute inputs, especially when the lake surface forms a large proportion of the catchment area and the buffering effect afforded by a mineral soil cover is small (Scheider *et al.*, 1979). Siegfried (1982) has measured atmospheric P input to a lake as 5% of total inflow and assessed this to be a significant contribution to the nutrient budget which may lead to eutrophic conditions. Psenner (1984) estimated atmospheric input to represent 6% of organic carbon, 15% of nitrogen and 39% of phosphorus loading to an Austrian lake, but with a significant seasonal variability of this input. The N and P additions by dry and wet mechanisms to a large lake have been estimated by Delumyea and Patel (1978), and Kortmann (1980) showed significant P additions, complicated in their effect by water-sediment interchanges of nutrients, to a softwater lake. Concern over the 'ecological health' of Chesapeake Bay and its tributary estuaries, which are experiencing increased incidence of algal blooms which light-stress submerged macrophytic vegetation and reduce fish and shellfish populations, has prompted a series of research investigations into the relative importance of increased or changed atmospheric loadings of nutrients and the effects of land-use practices (Miklas *et al.*, 1977; Correll, 1981; Correll and Ford, 1982). A review of the

Chesapeake Bay Project by Joyce (1983) revealed that the atmospheric input of nitrogen, as nitrate and ammonia primarily, is likely soon to exceed that from surface runoff sources and is probably less amenable to control. A similar problem is emerging for the Venice Lagoon, which, being downwind of a large industrial site, is becoming increasingly affected by atmospheric solute inputs (Guigliano and Cossu, 1983).

The past decade has seen a new scientific, public and political issue gain great momentum—the 'acid rain problem'. At present, the public and political concern which is fuelling the debate and which is centred around fears of irreversible environmental damage and costly countermeasures, is somewhat out of step with the views of the scientific community which is generally agreed that the situation is so complex that simple solutions totally ignore what has been learned from research: this is that, as yet, causes and effects have not been quantitatively and unequivocally linked, the timescales of acidification processes are uncertain and that further research is required to establish the nature and scale of the environmental problem, if indeed we are dealing with real effects and problems at all, before effective and economically acceptable solutions can be devised.

There is already a large number, which is increasing steadily, of published reports of conference proceedings, of government-agency reports and of publications from environmental pressure groups and other private organizations, which address the 'acid rain problem' from one viewpoint or other. These reports contain both papers from individual researchers or research groups and 'keynote' or 'invited survey' contributions from experienced researchers which summarize and critically comment on the report's individual papers and their significance in relation to other and earlier findings. Dochinger and Seliga (1976) have edited the proceedings of the first international symposium on acid precipitation in forest ecosystems, held in the USA, and Hutchinson and Havas (1980) produced a summary of the NATO conference held in Canada and concerned with all aspects of acid deposition into terrestrial ecosystems. The 1980 Sandefjord (Norway) conference, organized by the SNSF-project, is reported by Drablos and Tollan (1980), and the final report of this project, summarizing the results of over 300 research reports which it generated, is edited by Overrein *et al.* (1981). Nicholson *et al.* (1980) produced a summarized report of the 1977 conference concerned with the more practical requirements, in terms of definitions and research methods, of investigations into acidic inputs to forests. The deposition and effects of atmospheric pollution, particularly by acidifying materials, on natural waters and on terrestrial ecosystems have been reported in conference proceedings edited by Shriner *et al.* (1980), Eisenreich (1981), Georgii and Pankrath (1982), D'Itri (1982), Ott and Stangl (1983), Swedish Ministry of Agriculture (1982) and the National Swedish Environment Protection Board (1983). A recent report (UK Review

Group on Acid Rain, 1983) has surveyed the 'acid precipitation' problem in the UK, and the proceedings of an international conference held at the Royal Society of London in 1983, at which summaries of the state of knowledge concerning ecological impacts of acid depositions were presented by acknowledged leading researchers, has now been published (Beament *et al.*, 1984). Record *et al.* (1982) have collated and integrated 'acid rain information' particularly related to North America, and, for Europe, the Watt Committee on Energy (Mellanby, 1984) and Environmental Resources Ltd. (1983) have produced similar publications. Popular summaries of the 'acid deposition problem' have been produced by many environmental-interest groups such as the Nature Conservancy Council (Fry and Cooke, 1984) and the Swedish Ministry of Agriculture (1983).

Historical perspectives on the developing 'acid precipitation problem' have been presented by Cogbill (1976) and Cowling (1982a), reviewing the history of the concept, major publications in the field and the government-funded research programmes which have been instituted. Further status reports on acid precipitation, its causes and biological effects have been compiled by Cowling (1976b, c) and Chester (1983). Clearer definition of the chemical variables involved and the nature of 'acidity' has come from Newman *et al.* (1975), Likens *et al.* (1976) and Van Breemen *et al.* (1984). The extension of areas experiencing acid precipitation in the USA has been observed by Likens and Bormann (1974) and Likens and Butler (1981) with the degree of acidity increasing, and Burns *et al.* (1981) monitored progressive precipitation acidification in the eastern USA. The apparent extent and intensity of precipitation acidity led to a large number of research programmes directed at assessing the possibility of and degree of damage to natural ecosystems, forests in particular, and to surface and groundwater bodies and man-made structures, that acidic depositions could produce.

The idea that low pH precipitation and other acidic atmospheric inputs could leach nutrients from, and ultimately damage, forests, crops and soils gained force through research during the 1970s. Eaton *et al.* (1973) demonstrated the effect of acidic inputs in accelerating nutrient cycling, and Likens and Bormann (1974) drew attention to the environmental problems that could ensue. Cogbill and Likens (1974) demonstrated that sulphate and nitrate components of atmospheric solute input were closely linked to its hydrogen-ion concentration: Cogbill (1976) then attempted to quantify the potential effects over eastern North America but experienced problems in not knowing dates for acid rain commencement and in estimating historic tree growth rates. A flood of research then followed into the effects of acid precipitation on forests other than at Hubbard Brook. The effects of forests themselves in acidifying sensitive environments by efficiently extracting acidic species from the atmosphere have been investigated by Jonsson (1976), Harriman and Morrison (1980, 1982) and Nilsson *et al.* (1982) among others.

Summaries of various acid rain effects on forests have come from Last *et al.* (1980b), Roberts (1983), Johnson *et al.* (1982), Miller (1984), Last (1984) and Jacobson (1984). The mobilization of toxic levels of aluminium in soils at low pH values, suggested by Johnson (1979) and by Cronan and Schofield (1979) as inimicable to tree growth by inhibiting calcium uptake by roots (Hutterman and Ulrich, 1984), is not favoured by Abrahamsen (1984): magnesium loss from leaf tissue is suspected, but from leaves first damaged by very strong leaching, perhaps involving ozone, and more pronounced in older trees, during frosts and at high, exposed sites. Johnson and Reuss (1984) emphasize the very site-specific impacts of acidic deposition, depending on local climate, site nutrient conditions and forest management practices as well as atmospheric inputs. Factors such as the effect of acidic precipitation on soil microbial populations and activity (Killham *et al.*, 1983) have received relatively little attention. Holdgate (1984) identified the interaction between plant roots, soil flora and fauna and soil chemistry as a significant gap in our knowledge of how the effects of atmospheric solute inputs on plant growth are mediated by the soil. The acidification of surface waters in sensitive areas (Hendrey *et al.*, 1980; Clair and Whitfield, 1983) has been attributed to soil acidification and to direct acidic precipitation inputs. Seip and Tollan (1978) and Dillon *et al.* (1978, 1984) have discussed the possible mechanisms involved, and Burton *et al.* (1982), Lewis (1982), Schnoo *et al.* (1983) and Stoner *et al.* (1984) give particular examples of widespread lake acidification and effects on fisheries, which are also discussed by Haines (1981) and Brown *et al.* (1983). The concept of 'base neutralizing capacity' has been applied to small lakes by Driscoll and Schafran (1984) who point to nitrates in atmospheric input as an important contributor to acidic deposition from the atmosphere. A method for dating the onset of acid conditions from the diatom record in lake sediments was applied to small lakes in southern Scotland by Flower and Battarbee (1983). Acidity appeared to have increased before the planting of coniferous forests in the catchment areas. 'Acid surges' in rivers at the onset of annual or daily snowmelt are likely to hinder simple interpretation of lake sediment characteristics, in addition to their damaging functions in providing short-duration but lethal conditions for stream and lake biota (Jeffries *et al.*, 1979; Siegel, 1981; Colbeck, 1981; Bjarnborg, 1983). Johnson (1981) has modelled the potential impact of future acidification of precipitation input to poorly buffered acidic waters in Alaska. Damage by acidic precipitation to drinking-water supplies (Sharpe and Young, 1982; Liebfried *et al.*, 1984) and to buildings and structures (Martin, 1982b) is also receiving attention.

Several researchers have pointed to the fact that low levels of acidic atmospheric input may have various beneficial effects on a forested or agricultural catchment. Voigt (1980) considered that acid precipitation may stimulate plant growth in well-buffered ecosystems. In addition to the potential fertilizer effect of N and S components in atmospheric deposition (Abrahamsen, 1984), weathering production of nutrient elements may be increased (Kilham, 1982) and microbial activity may be stimulated (Killham *et al.*, 1983) by slight acidification. Sulphate addition to a soil may result in sulphate being adsorbed on to iron and aluminium oxides releasing OH^- ions which may then neutralize hydrogen-ion depositions (Johnson and Reuss, 1984). A problem here is that it is at present difficult to establish the early effects of acidic deposition (McFee and Cronan, 1982), whether these are fertilizing or damaging, without more sensitive indicators of relative plant performance (Holdgate, 1984). There is some debate surrounding the sensitivity of different soils and water bodies and the relative significance of internal and external (atmospheric) sources for hydrogen ions (Rosenqvist *et al.* 1980; Driscoll and Likens, 1982; Havas and Hutchinson, 1983; Ulrich, 1983). Extremely acid systems, such as oligotrophic peat bogs, may be little affected by acidic depositions, as may soils with high base status providing significant buffering capacity, whereas soil of 'intermediate' nutrient status may be significantly impoverished by these depositions and thus affect the associated vegetation via various element-deficiency and element-toxicity effects. The slowness of catchment responses to increasing (and currently decreasing in parts of Europe) acidic loadings makes it difficult to clearly link acidic emissions to ecosystem effects (Kish, 1981): there is a major problem of quantification of the relationships and specification of their spatial scale of effectiveness. The models of ecosystem response to acidic deposition currently being developed are a first stage in developing strategies of 'pollution control and acidification mitigation' to counteract damaging effects (Chen *et al.*, 1982; Booty and Kramer, 1984; Goldstein *et al.*, 1984).

Schemes for control of pollutant emissions to the atmosphere and for the management of ecosystems ('mitigation') aimed at reducing any deleterious effects of atmospheric solute input on the environment are fundamentally hampered by a lack of understanding of the mechanisms by which 'acid depositions' come about and the nature and scale of the ecological problem they may be responsible for (Chester, 1983). There is a danger of huge sums of money being allocated to schemes of action which may prove, in the light of future research, to be ineffective and wasteful, and even to exacerbate the present situation. Reductions in sulphur emissions from power stations, though at great cost, may not in fact reduce precipitation acidity if emissions of hydrocarbons, nitrogen oxides, heavy metals and other constituents of the 'pollutant mix' are not also reduced. There is evidence that reducing one component of this mix, nitrogen oxides, could in fact increase the sulphuric acid content of rain. Cowling (1982a) has listed the numerous coordinated research programmes which are currently working in North America and Europe to clarify the processes involved, both in the atmosphere and on the ground, before expensive management plans are formulated. The management options available and the research priorities are discussed by Chadwick (1983), and the political realities which will determine whether they are taken up are fully discussed in Chester (1983) and are apparent in an exchange of views between European politicians in a Council of Europe publication (Muller and Osborn, 1984). The political and legal issues surrounding the 'acid rain' situation in the USA have been fully discussed by Jenkins (1984): the ultimate decisions may be moral ones—whether we prefer to live, or allow some people to live, in an environment which is a product of our times, or whether we should all pay to protect what may be other peoples' environments many miles away. It may be difficult to convince large populations of 'emitters' living in non-sensitive environments to pay for costly equipment such as gas scrubbers and fuel treatment plants to protect the environments of 'receivers' living in distant 'sensitive' environments, particularly as absolute evidence linking cause and effect is not possible and when the scientific establishment is not in full agreement as to these links.

REFERENCES

Abrahamsen, G. (1980). Impact of atmospheric sulphur deposition on forest ecosystems. In: *Atmospheric Sulfur Deposition: Environmental Impact and Health Effects* (D. S. Shriver, C. R. Richmond and S. E. Lindberg, Eds.), Proc. 2nd Life Sciences Symposium, Gatlinburg, Tennessee, 14–18 October 1979, Ann Arbor.

Abrahamsen, G. (1984). Effects of acidic deposition on forest soil and vegetation. *Phil. Trans. R. Soc. Lond.*, **B305**, 369–382.

Aichinger, H. L. (1980). Sampling methods and analysis of wet precipitation. *Science of the Total Environment*, **16** (3), 279–283.

Alcock, M. R., and Morton, A. J. (1981). The sulphur content and pH of rainfall and of throughfalls under pine and birch. *J. Applied Ecology*, **18** (3), 835–840.

Allen, H. E., and Kramer, J. R. (1972). (Eds.) *Nutrients in Natural Waters*, Wiley, New York.

Allen, S. E. (1974). (Ed.) *Chemical Analysis of Ecological Materials*, Blackwell, Oxford.

Allen, S. E., Carlisle, A., White, E. J., and Evans, C. C. (1968). The plant nutrient content of rainwater. *J. Ecology*, **56**, 497–504.

Allison, G. B., and Hughes, M. W. (1976). The use of environmental chloride and tritium to estimate total recharge to an unconfined aquifer. *Australian J. Soil Research*, **16** (2), 181–196.

Anon. (1978). *Research and Monitoring of Precipitation Chemistry in the United States—Present Status and Future Needs*, Federal Interagency Work Group on Precipitation Chemistry, Interagency Advisory Committee on Water Data, US Geol. Survey, Reston, Virginia.

Anon. (1982). Acid rain detected in isolated areas. *Bull. Amer. Met. Soc.*, **63** (12), 1436.

Asman, W. A. H., and Slanina, J. (1980). Meteorological interpretation of the chemical composition of precipitation and some results of sequential rain sampling. In: *Ecological Impact of Acid Precipitation* (D. Drablos and A. Tollan, Eds.), Proceedings of an international conference, Sandefjord, Norway, SNSF project, pp. 140–141.

Attiwill, P. M. (1966). The chemical composition of rainwater in relation to cycling of nutrients in mature forests. *Plant and Soil*, **24**, 309–406.

Azevedo, J., and Morgan, D. L. (1974). Fog precipitation in coastal California forests. *Ecology*, **55**, 1135–1141.

Bache, B. W. (1984). Soil-water interactions. *Phil. Trans. R. Soc. Lond.*, **B305**, 393–407.

Bache, D. H. (1981). Analysing particulate deposition to plant canopies. *Atmos. Env.*, **15** (9), 1759–1761.

Backlin, L., Soderlund, R., and Granat, L. (1977). An improved precipitation collector for subsequent chemical analysis. In: *Air pollution measurement techniques* (W. M. O.), Special Environmental Report No. 10, pp. 58–62.

Baker, M. B., Caniparoli, D., and Harrison, H. (1981). An analysis of the first year of MAP3S rain chemistry measurements. *Atmos. Env.*, **15**, 43–55.

Baldwin, A. D. (1971). Contribution of atmospheric chloride in water from selected coastal streams of central California. *Water Resources Research*, **7** (4), 1007–1012.

Barnes, I. (1964). Field measurement of alkalinity and pH. *U. S. Geol. Survey, Water Supply Paper*, *1535–H*, 18pp.

Barrett, E., and Brodin, G. (1955). The acidity of Scandinavian precipitation. *Tellus*, **7**, 251–264.

Barrie, L. A. (1981). The prediction of rain acidity and SO_2 scavenging in eastern North America. *Atmos. Env.*, **15**, 31–41.

Barrie, L. A. (1982). Environment–Canada's long range of transport of atmospheric pollutants program: atmospheric studies. In: *Acid Precipitation: Effects on Ecological Systems* (F. M. D'Itri, Ed.), Ann Arbor Science, Ann Arbor, pp. 141–164.

Bauer, H. H., Christian, G. D., and O'Reilly, J. E. (1978). *Instrumental Analysis*, Allyn and Bacon Inc., Boston.

Beament, J., Bradshaw, A. D., Chester, P. F., Holdgate, M. W., Sugden, M., and Thrush, B. A. (1984), (Eds.). Ecological effects of deposited sulphur and nitrogen compounds. *Phil. Trans. R. Soc. Lond.*, **B305**, 255–317.

Beilke, S. (1983). Origin, transport, conversion and deposition of air pollutants: report on session. In: *Acid Deposition: a Challenge for Europe* (H. Ott and H. Stangl, Eds.), prelim. edition of symposium proceedings, Karleruhe, 19–20 September, Commission of the European Communities Directorate General for Science, Research and Development, pp. 303–320.

Belot, Y., Caput, C., and Gauthier, C. (1982). Transfer of americium from sea water to atmosphere by bubble bursting. *Atmos. Env.*, **16**, 1463–1466.

Benham, D. G., and Mellanby, K. (1978). A device to exclude dust from rainwater samples. *Weather*, **33** (4), 151–154.

Best, G. R., and Monk, C. D. (1975). Cation flux in hardwood and white pine watersheds. In: *Mineral Cycling in Southeastern Ecosystems* (F. G. Howell, J. B. Gentry and M. H. Smith, Eds.), US Energy Research and Development Administration, Washington, D. C., pp. 847–861.

Betson, R. P. (1978). Bulk precipitation and streamflow quality relationships in an urban area, *Water Resources Research*, **14** (6), 1165–1169.

Bjarnborg, B. (1983). Dilution and acidification effects during the spring flood of four Swedish mountain brooks. *Hydrobiologia*, **101** (1/2), 19–26.

Blanchard, D. C. (1982). Transfer of americium from sea water to atmosphere by bubble bursting. *Atmos. Env.*, **16**, 2273.

Bloch, M. R. D., Kaplan, V., Kertes, V., and Schnerb, J. (1966). Ion separation in bursting air bubbles: an explanation for the irregular ion ratios in atmospheric precipitations. *Nature*, **209**, 802–803.

Boatman, D. J., Hulme, P. D., and Tomlinson, R. W. (1975). Monthly determinations of the concentrations of sodium, potassium, magnesium and calcium in the rain and in pools on the Silver Flowe National Nature Reserve. *J. Ecology*, **63**, 903–912.

Bogen, D. C., Nagourney, S. J., and Torquato, C. (1980). A field evaluation of the HASL wet-dry deposition collector. *Water, Air and Soil Pollution*, **13**, 453–458.

Booty, W. G., and Kramer, J. R. (1984). Sensitivity analysis of a watershed acidification model. *Phil. Trans. R. Soc. Lond.*, **B305**, 441–449.

Bormann, F. H., and Likens, G. E. (1969). The watershed-ecosystem concept and studies of nutrient cycles. In: *The Ecosystem Concept in Natural Resource Management* (G. M. Van Dyne, Ed.), Academic Press, New York, pp. 49–76.

Bormann, F. H., and Likens, G. E. (1970). The nutrient cycles of an ecosystem. *Scientific American*, **223**, 92–101.

Boutron, C. (1979). Alkali and alkaline earth enrichments in aerosol deposited in Antarctic snows. *Atmos. Env.*, **13**, 919–924.

Bowditch, D. C., Edmond, C. R., Dunston, P. J., and McGlynn, J. A. (1976). *Suitability of containers for storage of water samples*. Australian Water Resources Council, Tech. Report No. 16, Australian Government Publishing Service, Canberra.

Brabec, E., Kovar, P., and Drabkova, A. (1981). Particle deposition in three vegetation stands: a seasonal change. *Atmos. Env.*, **15** (4), 583–588.

Brakke, D. F. (1977). Rainwater-nutrient additions to a hypereutrophic lake. *Hydrobiologia*, **52**, 159–164.

Bricker, O. P., Godfrey, A. E., and Cleaves, E. T. (1968). Mineral-water interaction during the chemical weathering of silicates. *Advances in Chemistry Series, American Chemical Society*, **73**, 128–142.

Brimblecombe, P., and Pitman, J. (1980). Long-term deposit at Rothamsted, southern England. *Tellus*, **32** (3), 261–267.

Brown, T. E., Morley, A. W., Sanderson, N. T., and Tait, R. D. (1983). Report on a large fish kill resulting from natural acid water conditions in Australia. *J. Fish Biology*, **22**, 335–350.

Burns, D. A., Galloway, J. N., and Hendrey, G. R. (1981). Acidification of surface waters in two areas of the eastern United States. *Water, Air and Soil Pollution*, **16** (3), 277–285.

Burt, T. P., and Day, M. R. (1977). Spatial variations in rainfall and stream water quality around the Avonmouth industrial complex. *Int. J. Environmental Studies*, **11**, 205–209.

Burton, T. M., Stanford, R. M., and Allan, J. W. (1982). The effects of acidification on stream ecosystems. In: *Acid Precipitation: Effects on Ecological Systems* (F. M. D'Itri, Ed.), Ann Arbor Science, Ann Arbor, pp. 209–236.

Calles, I. M. (1983). Dissolved inorganic substances: a study of mass balance in three small drainage basins. *Hydrobiologia*, **101** (1/2), 13–18.

Carlisle, A., Brown, A. H. F., and White, E. J. (1966). The organic matter and nutrient elements in the precipitation beneath a sessile oak canopy. *J. Ecology*, **54**, 87–98.

Carroll, D. (1962). Rainwater as a chemical agent of geological processes, *U. S. Geol. Surv., Water Supply Paper*, 1535–G.

Cawse, P. A. (1977). *Deposition of trace elements from the atmosphere in the UK*. Agricultural Development and Advisory Service, Conference on Inorganic Pollution and Agriculture, paper SS/OC/77/3.

Chadwick, M. J. (1983). Acid depositions and the environment. *Ambio*, **12** (2), 80–82.

Chamberlain, A. C. (1975). The movement of particles in plant communities. In:

Vegetation and the Atmosphere (J. L. Monteith, Ed.), Academic Press, London. Chap. 5.

Chamberlain, A. C. (1980). Dry deposition of sulfur dioxide. In: *Atmospheric Sulfur Deposition: Environmental Impact and Health Effects* (D. S. Shriner, C. R. Richmond, and S. E. Lindberg, Eds.), Proc. of 2nd Life Sciences Symposium, Gatlinburg, Tenn., 14–18 October 1979, Ann Arbor Science, Ann Arbor, pp. 185–200.

Chapman, S. B. (1967). Nutrient budgets for a dry heath ecosystem in the south of England. *J. Ecology*, **55**, 677–689.

Characklis, W. G., Ward, C. H., King, J. M., and Roe, F. L. (1979). Rainfall quality, land use, and runoff quality. *J. Environmental Engineering Division, Amer. Soc. Civil Engineers*, **105**, EE2, 416–419.

Chen, C. W., Dean, J. D., Gherini, S. A., and Goldstein, R. A. (1982). Acid rain model: hydrologic module, *J. Environmental Engineering Division, Amer. Soc. Civil Engineers*, **108**, EE3, 455–472.

Chester, P. F. (1983). Perspectives on acid rain. *J. Royal Society of Arts*, **131**, 587–603.

Clair, T. A., and Whitfield, P. H. (1983). Trends in pH, calcium and sulphate of rivers in Atlantic Canada. *Limnology and Oceanography*, **28**, (1), 160–165.

Claridge, G. G. C. (1975). Transit time: a factor affecting element balances in small catchments. *N.Z. Journal of Science*, **18**, 297–304.

Clayton, J. L. (1972). Salt spray and mineral cycling in two California coastal ecosystems. *Ecology*, **53**, 74–81.

Cleaves, E. T., Fisher, D. W., and Bricker, O. P. (1974). Chemical weathering of serpentinite in the Eastern Piedmont of Maryland. *Bull. Geol. Soc. Amer.*, **85** (3), 437–444.

Cleaves, E. T., Godfrey, A. E., and Bricker, O.P. (1970). Geochemical balance of a small watershed and its geomorphic implications. *Bull. Geol. Soc. Amer.*, **81** (10), 3015–3022.

Cogbill, C. V. (1976). The history and character of acid precipitation in eastern North America. *Water, Air and Soil Pollution*, **6**, 407–413.

Cogbill, C. V., and Likens, G. E. (1974). Acid precipitation in the north eastern U.S.A. *Water Resources Research*, **10**(6), 1133–1137.

Colbeck, S. C. (1981). A simulation of the enrichment of atmospheric pollutants in snow cover runoff. *Water Resources Research*, **17** (5), 1383–1388.

Cole, D. W., and Johnson, D. W. (1977). Atmospheric sulphate additions and cation leaching in a Douglas Fir ecosystem. *Water Resources Research*, **13**, 313–317.

Cooper, H. B. H. (1976). Chemical composition of acid precipitation in central Texas. In: *Proc. 1st Int. Symposium on Acid Precipitation and the Forest Ecosystem* (L. S. Dochinger and T. A. Seliga, Eds.), U.S.D.A. Forest Service, General Technical Report NE-23, pp. 281–291.

Correll, D. L. (1977). (Ed.). *Watershed Research in Eastern North America: a workshop to compare results*, Smithsonian Institute, Edgewater, Maryland, 2 volumes, 924 pp.

Correll, D. L. (1981). Nutrient mass balance for the watershed headwaters, intertidal zone and basin of the Rhode River estuary. *Limnology and Oceanography*, **26** (6), 1142–1149.

Correll, D. L., and Ford, D. (1982). Comparison of precipitation and land runoff as sources of estuarine nitrogen. *Estuarine, Coastal and Shelf Science*, **15**(1), 45–56.

Coscio, M. R., Pratt, G. C., and Krupa, S. V. (1982). An automatic, refrigerated, sequential precipitation sampler. *Atmos. Env.*, **16** (8), 1939–1944.

Cowling, E. B. (1982a). Acid precipitation in historical perspective. *Environmental Science and Technology*, **16** (2), 110A–123A.
Cowling, E. B. (1982b). A status report on acid precipitation and its biological consequences as of April 1981. In: *Acid Precipitation: Effects on Ecological Systems* (F. M. D'Itri, Ed.), Ann Arbor Science, Ann Arbor, pp. 3–20.
Cowling, E. B. (1982c). An historical resume of progress in scientific and public understanding of acid precipitation and its biological consequences. In: *Acid Precipitation: Effects on Ecological Systems* (F. M. D'Itri, Ed.), Ann Arbor Science, Ann Arbor, pp. 43–86.
Cox, R. A., and Penkett, S. A. (1983). Formation of atmospheric acidity. In: *Acid Deposition: a Challenge for Europe*, (H. Ott and H. Stangl, Eds.), prelim. edition of symposium proceedings, Karlsruhe, 19–21 September, Commission of the European Communities Directorate General for Science, Research and Development, pp. 58–72.
Crabtree, R. W., and Trudgill, S. T. (1981). The use of ion-exchange resin in monitoring the calcium, magnesium, sodium and potassium contents of rainwater. *J. Hydrology*, **53**, 361–365.
Crisp, D. T. (1966). Input and output of minerals for an area of Pennine moorland: the importance of precipitation, drainage, peat erosion and animals. *J. Applied Ecology*, **3**, 314–317.
Cronan, C. S., Reiners, W. A., Reynolds, R. C., and Lang, G. E. (1978). Forest floor leaching: contributions from mineral, organic and carbonic acids in New Hampshire sub-alpine forests. *Science*, **200** (4339), 309–311.
Cronan, C. S., and Schofield, C. L. (1979). Aluminium leaching response to acid precipitation: effects on high-elevation watersheds in the North-east. *Science*, **204** (4390), 304–306.
Cryer, R. (1976). The significance and variation of atmospheric nutrient inputs in a small catchment system. *J. Hydrology*, **28**, 121–137.
Cryer, R., and Trudgill, S. T. (1981). Solutes. In: *Geomorphological Techniques* (A. Goudie, Ed.), Allen and Unwin, London, pp. 181–195.
Davies, T. D., Vincent, C. E., and Brimblecombe, P. (1982). Preferential elution of strong acids from a Norwegian ice cap. *Nature*, **300**, 161–163.
Delumyea, R. G., and Patel, R. L. (1978). Wet and dry deposition of phosphorus into Lake Huron. *Water, Air and Soil Pollution*, **10** (2), 187–198.
Dethier, D. P. (1979). Atmospheric contributions to streamwater chemistry in the north Cascade Range, Washington. *Water Resources Research*, **15** (4), 787–794.
Dillon, P. J., Jeffries, D. S., Snyder, W., Reid, R., Yan, N. D., Evans, D., Moss, J., and Scheider, W. A. (1978). Acidic precipitation in South-Central Ontario: recent observations. *J. Fisheries Research Board of Canada*, **35** (6), 809–815.
Dillon, P. J., Yan, N. D., and Harvey, H. H. (1984). Acidic deposition: effects on aquatic ecosystems. *C. R. C. Critical Reviews in Environmental Control*, **13** (3), 167–194.
D'Itri, F. M. (1982). (Ed.). *Acid Precipitation: Effects on Ecological Systems*, Ann Arbor Science, Ann Arbor, 506 pp.
Dochinger, L. S., and Seliga, T. A. (1976). (Eds.). *Proceedings of the 1st International Symposium on Acid Precipitation and the Forest Ecosystem*, U.S. Dept. Agriculture, Forest Service, General Technical Report NE-23, Northeastern Forest Experiment Station, Upper Darby, PA, 1074 pp.
Dollard, G. J., Unsworth, M. H., and Harve, M. J. (1983). Pollutant transfer in upland regions by occult precipitation. *Nature*, **302**, 241–243.
Dollard, G. J., and Vitols, V., 1980, Wind tunnel studies of dry deposition of SO_2 and H_2SO_4 aerosols. In: *Ecological Impact of Acid Precipitation* (D. Drablos and A.

Tollan, Eds.), Proceedings of an international conference, Sandefjord, Norway, SNSF project, pp. 108–109.
Douglas, I. (1968). The effects of precipitation chemistry and catchment area lithology on the quality of river water in selected catchments in eastern Australia. *Earth Science Journal*, **4**, 125–144.
Douglas, I. (1972). The geographical interpretation of river water quality. *Progress in Geography*, **4**, 1–81.
Dovland, H., and Semb, A. (1980). Atmospheric transport of pollutants. In: *Ecological Impact of Acid Precipitation* (D. Drablos and A. Tollan, Eds.), Proceedings of an international conference, Sandefjord, Norway, SNSF project, pp. 14–21.
Drablos, D., and Tollan, A. (1980). (Eds.). *Ecological Impact of Acid Precipitation.* Proceedings of an international conference, Sandefjord, Norway, SNSF project, 383 pp.
Driscoll, C. T., and Likens, G. E. (1982). Hydrogen ion budget of an aggrading forested ecosystem. *Tellus*, **34** (3), 283–292.
Driscoll, C. T., and Schafran, G. C. (1984). Short-term changes in the base neutralizing capacity of an acid Adirondack lake, New York. *Nature*, **310**, 308–310.
Duce, R. A., and Hoffman, E. J. (1976). Chemical fractionation at the air/sea interface. *Annual Review of Earth and Planetary Sciences*, **4**, 187–228.
Durham, J. L., Overton, J. R., and Aneja, V. P. (1981). Influence of gaseous nitric acid on sulfate production and acidity in rain. *Atmos. Env.*, **15** (6), 1059–1068.
Duvigneaud, P. (1971). Concepts sur la productivite primaire des ecosystemes forestiers. In: *Productivity of Forest Ecosystems* (P. Duvigneaud, Ed.), UNESCO, Paris, pp. 111–140.
Eaton, J. S., Likens, G. E., and Bormann, F. H. (1973). Throughfall and stemflow chemistry in a northern hardwood forest. *J. Ecology*, **61**, 495–508.
Eaton, J. S., Likens, G. E., and Bormann, F. H. (1978). The input of gaseous and particulate sulfur to a forest ecosystem. *Tellus*, **30**, 546–551.
Edwards, A. M. C. (1973). The variation of dissolved constituents with discharge in some Norfolk rivers. *J. Hydrology*, **18**, 219–242.
Edwards, P. J. (1982). Studies of mineral cycling in a montane rain forest in New Guinea: V Rates of cycling in throughfall and litterfall. *J. Ecology*, **70**, 807–827.
Eisenreich, S. J. (1981). (Ed.). *Atmospheric Pollutants in Natural Waters*, Ann Arbor Science, Ann Arbor, 512 pp.
Elzerman, A. H. (1981). Mechanisms of enrichment at the air-water interface. In: *Atmospheric Pollutants in Natural Waters* (S. J. Eisenreich, Ed.), Ann Arbor Science, Ann Arbor, pp. 81–91.
Environmental Resources Ltd. (1983). *Acid Rain; a Review of the Phenomenon in the EEC and Europe*, a report prepared for the Commission of the European Communities, Directorate General for Environment, Consumer Protection and Nuclear Safety, Graham and Trotman, London, 159 pp.
Eriksson, E. (1955). Airborne salts and the chemical composition of river water. *Tellus*, **7** (2), 243–250.
Eriksson, E. (1959). The yearly circulation of Cl^- and S in nature: meteorological, geochemical and pedological implications, part 1. *Tellus*, **11** (4), 375–403.
Eriksson, E. (1960). The yearly circulation of Cl^- and S in nature: meteorological, geochemical and pedological implications, part 2. *Tellus*, **12** (1), 63–109.
Eriksson, E. (1966). Air and precipitation as sources of nutrients. In: *Handbuch der Pflanzenernahrung und Dungung* (K. Scharrer and H. Linser, Eds.), Vienna, pp. 774–792.

Faust, S. D., and Aly, O. M. (1981). *Chemistry of Natural Waters*, Ann Arbor Science, Ann Arbor, Michigan.

Feeley, J. A., and Liljestrand, H. M. (1983). Source contributions to acid precipitation in Texas. *Atmos. Env.*, **17** (4), 807–814.

Feller, M. C., and Kimmins, J. P. (1979). Chemical characteristics of small streams near Haney in southwestern British Columbia. *Water Resources Research*, **15** (2), 247–258.

Finlayson, B. (1977). *Runoff contributing areas and erosion.* School of Geography, Univ. of Oxford, Research Paper No. 18, 41 pp.

Fish, G. R. (1976a). N and P analyses of rainfall at Rotorua. *J. Hydrology, N. Z.*, **15** (1), 17–26.

Fish, G. R. (1976b). The fall out of N and P compounds from the atmosphere. *J. Hydrology, N. Z.*, **15** (1), 27–33.

Fisher, B. E. A. (1981). Acid rain and the long range transport of air pollutants. *Weather*, **36**, 367–371.

Fisher, D. W., Gambell, A. W., Likens, G. E., and Bormann, F. H. (1968). Atmospheric contributions to water quality of streams in the Hubbard Brook Experimental Forest, New Hampshire. *Water Resources Research*, **4**, 1115–1126.

Flower, R. J., and Battarbee, R. W. (1983). Diatom evidence for recent acidification of two Scottish lochs. *Nature*, **305**, 130–133.

Foster, I. D. L. (1980). Chemical yields in runoff, and denudation in a small arable catchment, East Devon, England. *J. Hydrology*, **47**, 349–368.

Fowler, D. (1978). Dry deposition of SO_2 on agricultural crops. *Atmos. Env.*, **12**, 369–373.

Fowler, D. (1980). Removal of sulphur and nitrogen compounds from the atmosphere in rain and by dry deposition. In: *Ecological Impact of Acid Precipitation* (D. Drablos and A. Tollan, Eds.), Proceedings of an international conference, Sandefjord, Norway, SNSF project, pp. 22–32.

Fowler, D. (1984). Transfer to terrestrial surfaces. *Phil. Trans. R. Soc. Lond.*, **B305**, 281–297.

Fowler, D., Cape, J., Leith, I. D., Paterson, I. S., Kinnaird, J. W., and Nicholson, I. A. (1982). Rainfall acidity in northern Britain. *Nature*, **297**, 383–386.

Fry, G. L. A., and Cooke, A. S. (1984). *Acid Deposition and its Implications for Nature Conservation in Britain*. Report No. 7, Nature Conservancy Council, Shrewsbury, 59 pp.

Galloway, J. N., Crosby, B. J., and Likens, G. E. (1979). Acid precipitation: measurement of pH and acidity. *Limnology and Oceanography*, **24**, 1161–1165.

Galloway, J. N., and Likens, G. E. (1976a). Calibration of collection procedures for the determination of precipitation chemistry. *Water, Air and Soil Pollution*, **6**, 241–258.

Galloway, J. N., and Likens, G. E. (1976b). Calibration of collection procedures for the determination of precipitation chemistry. In: *Proc. 1st Int. Symposium on Acid Precipitatiom and the Forest Ecosystem* (L. S. Dochinger and T. A. Seliga, Eds.), U.S.D.A. Forest Service, General Technical Report, NE-23. pp. 137–156.

Galloway, J. N., and Likens, G. E. (1978). The collection of precipitation for chemical analysis. *Tellus*, **30**, 71–82.

Galloway, J. N., and Likens, G. E. (1981). Acid precipitation: the importance of nitric acid. *Atmos. Env.*, **15** (6), 1081–1085.

Galloway, J. N., Likens, G. E., and Edgerton, E. S. (1976). Acid precipitation in the northeastern United States: pH and acidity. *Science*, **194**, 722–724.

Galloway, J. N., and Parker, G. G. (1980). Difficulties in measuring wet and dry deposition on forest canopies and soil surfaces. In: *Effects of Acid Precipitation on*

Terrestrial Ecosystems (T. C. Hutchinson, and M. Havas, Eds.), Plenum Press, New York, pp. 57–68.

Galloway, J. N., and Whelpdale, D. M. (1980). An atmospheric sulfur budget for eastern North America. *Atmos. Env.*, **14**, 409–417.

Gambell, A. W., and Fisher, D. W. (1966). Chemical composition of rainfall, eastern North Carolina and southeastern Virginia. *U.S. Geol. Survey, Water Supply Paper*, 1535–K.

Gardiner, T., and McGreal, W. S. (1978). Variations in the cation and pH content of rainfall from two localities in Co. Down, Northern Ireland. *J. Meteorology*, **3**, 197–203.

Garland, J. A. (1978). Dry and wet removal of sulphur from the atmosphere. *Atmos. Env.*, **12** (1), 349–362.

Garland, J. A. (1981). Enrichment of sulphate in maritime aerosols. *Atmos. Env.*, **15**, 787–792.

Garland, J. A., and Chadwick, R. C. (1981). Plutonium suspension from sea water. *Health Physics*, **41** (2), 279–283.

Gascoyne, M. (1977). Design and operation of a simple, sequential precipitation sampler with some preliminary results. *Atmos. Env.*, **11**, 397–400.

Gascoyne, M., and Patrick, C. K. (1981). Variation in rainwater chemistry, and its relation to synoptic conditions at a site in N.W. England. *International Journal of Environmental Studies*, **17** (3), 209–214.

Gatz, D. F., and Dingle, A. N. (1971). Trace substances in rain water: concentration variations during convective rains, and their interpretation. *Tellus*, **23** (1), 14–27.

Gaudet, J. J., and Melack, J. M. (1981). Major ion chemistry in a tropical African lake basin. *Freshwater Biology*, **11**, 309–333.

Georgii, H.-W., and Pankrath, J. (1982). (Eds.). *Deposition of Atmospheric Pollutants*. Proceedings of a colloquium held at Oberursel/Taunus, W. Germany, 9–11 Nov. 1981, D. Reidel, Hingham, MA, 217 pp.

Glass, S. J., and Matteson, M. J. (1973). Ion enrichment in aerosols dispersed from bursting bubbles in aqueous salt solutions. *Tellus*, **25** (3), 272–280.

Goldsmith, P., Smith, F. B., and Tuck, A. F. (1984). Atmospheric transport and transformation. *Phil. Trans. R. Soc. Lond.*, **B305**, 259–279.

Goldstein, R. A., Gherini, S. A., Chen, C. W., Mok, L., and Hudson, R. J. M. (1984). Integrated acidification study (ILWAS): a mechanistic ecosystem analysis. *Phil. Trans. R. Soc. Lond.*, **B305**, 409–425.

Gore, A. J. P. (1968). The supply of six elements by rain to an upland peat area. *J. Ecology*, **56**, 483–495.

Golterman, H. L., Clymo, R. S., and Ohnstad, M. A. M. (1978). *Methods for Chemical Analysis of Fresh Waters*. I.B.P. Handbook No. 8, 2nd edn., Blackwell Scientific, Oxford, 214 pp.

Gorham, E. (1958). The influence and importance of daily weather conditions in the supply of chloride, sulphate and other ions to fresh waters from atmospheric precipitation. *Royal Society of London, Philos. Trans.* **B241**, 147–178.

Gorham, E. (1961). Factors influencing supply of major ions to inland waters, with special reference to the atmosphere. *Bull. Geol. Soc. Amer.*, **72** (6), 795–840.

Goudie, A. (1970). Input and output considerations in estimating rates of chemical denudation. *Earth Science Journal*, **4**(2), 59–65.

Granat, L. (1972). On the relation between pH and the chemical composition in atmospheric precipitation. *Tellus*, **24**, 550–560.

Granat, L. (1976). Principles in network design for precipitation chemistry. *J. Great Lakes Res.*, **2** (1),42–55.

Granat, L. (1977). Siting criteria for precipitation chemistry measurements—a

method for direct evaluation. In: *Air pollution measurement techniques* (W.M.O.), Special Environment Report No. 10, pp. 78–88.
Grant, M. C., and Lewis, W. M. (1982). Chemical loading rates from precipitation in the Colorado Rockies. *Tellus*, **34**, 74–88.
Guigliano, M., and Cossu, R. (1983). Atmospheric nutrient loadings to the Venetian lagoon. *Science of the Total Environment*, **29** (1/2), 49–63.
Haines, T. A. (1981). Acidic precipitation. *Fisheries*, **5** (6), 2–5.
Hall, A. J., and Barclay, P. A. (1975). Methods of determining areal rainfall from observational data. In: *Prediction in Catchment Hydrology* (T. G. Chapman and F. X. Dunin, Eds.), Australian Academy of Science, pp. 47–58.
Hamilton, E. P., and Chatt, A. (1982). Determination of trace elements in atmospheric wet precipitation by instrumental neutron activation analysis. *J. Radioanalytical Chemistry*, **71** (1/2), 29–45.
Hansen, D. A., and Hidy, G. M. (1982). Review of questions regarding rain acidity data. *Atmos. Env.*, **16** (9), 2107–2126.
Harriman, R., and Morrison, B. (1980). Ecology of streams draining forested and non-forested catchments in Scotland. In: *Ecological Impact of Acid Precipitation* (D. Drablos and A. Tollan, Eds.), Proceedings of an international conference, Sandefjord, Norway, SNSF project, pp. 312–313.
Harriman, R., and Morrison, B. R. S. (1982). Ecology of streams draining forested and non-forested catchments in an area of central Scotland subject to acid precipitation. *Hydrobiologia*, **88**, 251–263.
Havas, M., and Hutchinson, T. C. (1983). The Smoking Hills: natural acidification of an aquatic ecosystem. *Nature*, **301** (5895), 23–27.
Hembree, C. H., and Rainwater, F. H. (1961). Chemical degradation on opposite flanks of the Wind River Range, Wyoming. *U.S. Geol. Survey, Water Supply Paper, 1535-E.*
Henderson, G. S. (1978). Nutrient budgets of Appalachian and Cascade Region watersheds: a comparison. *Forest Science*, **24**, 385–397.
Henderson, G. S., Harris, W. F., Todd, D. E., and Grizzard, T. (1977). Quantity and chemistry of throughfall as influenced by forest-type and season. *J. Ecology*, **65**, 365–374.
Hendrey, G. R., Galloway, J. N., Norton, S. A., Schofield, C. L., Burns, D. A., and Schaffer, P. W. (1980). Sensitivity of the eastern United States to acid precipitation impacts on natural waters. In: *Ecological Impact of Acid Precipitation* (D. Drablos and A. Tollan, Eds.), Proceedings of an international conference, Sandefjord, Norway, SNSF project, pp. 216–217.
Hogan, A. W. (1983). Some characteristics of chemical precipitation. *Tellus*, **35B**, 121–130.
Hogstrom, U. (1973). Comments on the local fallout and long distance transport of sulphur. *Ambio*, **2** (3), 90–91.
Holdgate, M. W. (1984). Concluding remarks. *Phil. Trans. R. Soc. Lond.*, **B305**, 569–577.
Huang, W. H., and Keller, W. D. (1970). Dissolution of rock-forming silicate minerals; simulated first stage weathering of fresh mineral surfaces. *Amer. Mineral*, **55**, 2076–2094.
Hunter K. A., and Liss, P. S. (1981). Principles and problems of modelling cation enrichment at natural air-water interfaces. In: *Atmospheric Pollutants in Natural Waters* (S. J. Eisenreich, Ed.), Ann Arbor Science, Ann Arbor, pp. 99–128.
Hutchinson, T. C., and Havas, M. (1980). (Eds.). *Effects of Acid Precipitation on Terrestrial Ecosystems*. Proceedings of a NATO conference, Toronto, 21–27 May 1978, Plenum Press, New York, 654 pp.

Hutterman, A., and Ulrich, B. (1984). Solid phase-solution-root interactions in soils subjected to acid deposition. *Phil. Trans. R. Soc. Lond.*, **B305**, 353–368.
Ibrahim, M., Barrie, L. A., and Fanaki, F. (1983). An experimental and theoretical investigation of the dry deposition of particles to snow, pine trees and artificial collectors. *Atmos. Env.*, **17** (4), 781–788.
Irving, W. M. (1982). Chloride concentrations as an aid to estimating recharge to the Woburn Sands. *Q.J. Eng. Geol., London*, **15**, 47–52.
Jacobson, J. S. (1984). Effects of acidic aerosol, fog, mist and rain on crops and trees. *Phil. Trans. R. Soc. Lond.*, **B305**, 327–338.
Janda, R. J. (1971). An evaluation of procedures used in computing chemical denudation rates. *Bull. Geol. Soc. Amer.*, **82**, 67–80.
Jeffries, D. S., Cox, C. M., and Dillon, P. J. (1979). Depression of pH in lakes and streams in Central Ontario during snowmelt. *J. Fisheries Research Board of Canada*, **36** (6), 640–646.
Jenkins, T. (1984). The dilemma of an acidic sky. *J. American Water Works Association*, **76** (3), 42–49.
Jervis, T. R. (1979). Rainfall acidity: natural variance and subsequent time dependence of pH. *Atmos. Env.*, **13** (11), 1601.
Johannessen, M., and Henriksen, A. (1978). Chemistry of snow meltwater: changes in concentration during melting. *Water Resources Research*, **14** (4), 615–619.
Johnson, A. H. (1979). Evidence of acidification of headwater streams in the New Jersey Pinelands. *Science*, **206** (4420), 834–836.
Johnson, D. W. (1981). The natural acidity of some unpolluted waters in south-eastern Alaska and potential impacts of acid rain. *Water, Air and Soil Pollution*, **16** (2), 243–252.
Johnson, D. W., and Reuss, J. O. (1984). Soil-mediated effects of atmospherically deposited sulphur and nitrogen. *Phil. Trans. R. Soc. Lond.*, **B305**, 383–392.
Johnson, D. W., Turner, J., and Kelly, J. M. (1982). The effects of acid rain on forest nutrient status. *Water Resources Research*, **18**, 449–461.
Jonsson, B. (1976). Soil acidification by atmospheric pollution and forest growth. In: *Proceedings of the 1st International Symposium on Acid Precipitation and the Forest Ecosystem* (L. S. Dochinger and T. A. Seliga, Eds.), U.S. Dept. Agriculture, Forest Service, General Technical Report NE-23, Northeastern Forest Experiment Station, Upper Darby, PA, pp. 837–842.
Joyce, C. (1983). Can we save the Chesapeake Bay? *New Scientist*, **97** (1350), 797–800.
Juang, F. H. T., and Johnson, N. M. (1967). Cycling of chlorine through a forested watershed in New England. *J. Geophysical Research*, **72** (3), 5641–5647.
Junge, C. E. (1963). *Air Chemistry and Radioactivity*, Academic Press. London.
Junge, C. E. (1972). Our knowledge of the physico-chemistry of aerosols in the undisturbed marine environment. *J. Geophys. Res.*, **77**, 5183–5200.
Junge, C. E., and Werby, R. T. (1958). The concentration of chloride, sodium, potassium, calcium and sulphate in rain water over the United States. *J. Meteorology*, **15** (5), 417–425.
Kellman, M. (1979). Soil enrichment by neotropical savanna trees. *J. Ecology*, **67**, 565–577.
Kennedy, V. C., Kellweger, G. W., and Avanzino, R. J. (1979). Variations in rain chemistry during storms at two sites in Northern California. *Water Resources Research*, **15** (3), 687–702.
Kilham, P. (1982). Acid precipitation: its role in the alkalization of a lake in Michigan. *Limnology and Oceanography*, **27** (5), 856–867.
Killham, K., Firestone, M. K., and McColl, J. G. (1983). Acid rain and soil microbial

activity: effects and their mechanisms. *J. Environmental Quality*, **12** (1), 133–137.
Kish, T. (1981). Acid precipitation: crucial questions still remain unanswered. *J. Water Pollution Control Fed.*, **53** (5), 518–521.
Kitching, R., Edmunds, W. M., Shearer, T. R., Walton, N. R. G., and Jacovides, J. (1980). Assessment of recharge to aquifers. *Hydrological Sciences Bull.*, **25** (3), 217–235.
Knelman, F., Dombrowski, N., and Newitt, D. M. (1954). Mechanism of the bursting of bubbles. *Nature*, **173**, 261.
Koga, M. (1981). Direct production of droplets from breaking wind-waves—its observation by a multi-colored overlapping exposure photographic technique. *Tellus*, **33** (6), 552–563.
Kormondy, E. J. (1969). *Readings in Ecology*, Prentice-Hall, Englewood Cliffs.
Kortmann, R. W. (1980). Benthic and atmospheric contributions to the nutrient budgets of a soft-water lake. *Limnology and Oceanography*, **25** (2), 229–239.
Koyama, T., and Suguwara, K. (1953). Separation of the components of atmospheric salt and their distribution. *Bull. Chem. Soc. Japan*, **26** (3), 123–126.
Kurtz, J., and Scheider, W. A. (1981). An analysis of acidic precipitation in south-central Ontario using air parcel trajectories. *Atmos. Env.*, **15** (7), 1111–1116.
Last, F. T. (1984). Acid Rain: the pollution environment? *N.E.R.C. News Journal*, **3** (3), 4–6.
Last, F. T., Bjor, K., and Nicholson, I. A. (1980a). C. Perspective. In: *Methods for Studying Acid Precipitation in Forest Ecosystems*, Institute of Terrestrial Ecology, Cambridge, pp. 7–8.
Last, F. T., Likens, G. E., Ulrich, B., and Walloe, L. (1980b). Acid precipitation—progress and problems. Conference summary. In: *Ecological Impact of Acid Precipitation* (D. Drablos and A. Tollan, Eds.), Proceedings of an international conference, Sandefjord, Norway, SNSF project, pp. 10–12.
Lawson, D. R., and Winchester, J. W. (1979). A standard crustal aerosol as a reference for elemental enrichment factors. *Atmos. Env.*, **13**, 925–930.
Lee, J. J., and Weber, D. E. (1982). Effects of sulfuric acid rain on major cation and sulfate concentrations of water percolating through two model hardwood forests. *J. Environmental Quality*, **11** (1), 57–64.
Lewis, W. M. (1981). Precipitation chemistry and nutrient loading by precipitation in a tropical watershed. *Water Resources Research*, **17** (1), 169–181.
Lewis, W. M. (1982). Changes in pH and buffering capacity of lakes in the Colorado Rockies. *Limnology and Oceanography*, **27** (1), 167–172.
Lewis, W. M., and Grant, M. C. (1978). Sampling and chemical interpretation of precipitation for mass balance studies. *Water Resources Research*, **14** (6), 1098–1104.
Lewis, W. M., and Grant, M. C. (1981). Effects of the May/June Mt. St. Helens eruptions on precipitation chemistry in central Colorado. *Atmos. Env.*, **15** (9), 1539–1542.
Liebfried, R. T., Sharpe, W. E., and De Walle, D. R. (1984). The effects of acid precipitation runoff on source water quality. *J. American Water Works Association*, **76** (3), 50–53.
Likens, G. E., and Bormann, F. H. (1974). Acid rain: a serious regional environmental problem. *Science*, **184**, 1176–1179.
Likens, G. E., Bormann, F. H., Pierce, R. S., and Fisher, D. W. (1970). Nutrient-hydrological cycle interactions in small, forested watershed ecosystems. In: *Productivity of Forest Ecosystems* (P. Duvigneaud, Ed.), UNESCO, Paris pp. 553–563.

Likens, G. E., Bormann, F. H., Pierce, R. S., Eaton, J. S., and Johnson, N. M. (1977). *Biogeochemistry of a Forested Ecosystem*, Springer Verlag, New York, 146 pp.

Likens, G. E., and Butler, T. J. (1981). Recent acidification of precipitation in North America. *Atmos. Env.*, **15** (7), 1103–1109.

Likens, G. E., Edgerton, E. S., and Galloway, J. N. (1983). The composition and deposition of organic carbon in precipitation. *Tellus*, **35B**, 16–24.

Likens, G. E., Johnson, N. M., Galloway, J. N., and Bormann, F. H. (1976). Acid precipitation: strong and weak acids. *Science*, **194** (4265), 643–645.

Lindberg, S. E., Shriner, D. S., and Hoffman, W. A. (1982). The interaction of wet and dry deposition within the forest canopy. In: *Acid Precipitation: Effects on Ecological Systems* (F. M. D'Itri, Ed.), Ann Arbor Science, Ann Arbor, pp. 385–410.

Lindberg, S. E., Turner, R. R., Ferguson, N. M., and Matt, D. (1977). Walker Branch watershed element cycling studies: collection and analysis of wetfall for trace elements and sulphate. In: *Watershed Research in Eastern North America* (D. L. Correll, Ed.), Smithsonian, Edgewater, pp. 125–150.

Liss, P. S. and Slinn, W. G. N. (1983). (Eds.). *Air-sea exchange of gases and particles*. Proc. NATO Advanced Study Institute, Durham, N.H., 19–30 July. NATO Advanced Science Institutes Series, D. Reidel, Hingham, Maryland.

McDonald, R. W., and McLaughlin, F. A. (1982). The effect of storage by freezing on dissolved inorganic phosphate, nitrate and reactive silicon for samples from coastal and estuarine waters. *Water Research*, **16** (1), 95–104.

McFee, W. W., and Cronan, C. S. (1982). The action of wet and dry deposition components of acid precipitation on litter and soil. In: *Acid Precipitation: Effects on Ecological Systems* (F. M. D'Itri, Ed.), Ann Arbor Science, Ann Arbor, pp. 435–452.

Madgwick, H. A. I., and Ovington, J. D. (1959). The chemical composition of precipitation in adjacent forest and open plots. *Forestry*, **32**, 14–22.

Madsen, B. C. (1982). An evaluation of sampling interval length on the chemical composition of wet-only deposition. *Atmos. Env.*, **16** (10), 2515–2519.

Manahan, S. E. (1979). *Environmental Chemistry* (3rd edn.), Willard Grant Press, Boston.

Martin, A. (1982a). A short study of the influence of a valley on the composition of rainwater. *Atmos. Env.*, **16** (4), 785–793.

Martin, H. C. (1982b). Acid rain: impacts on natural and human environment. *Materials Performance*, **21** (1), 36–39.

Mattson, S., Sandberg, G., and Tersing, P. E. (1944). Electrochemistry of soil formation; 6, Atmospheric salts in relation to soil and peat formation and plant composition. *Ann. Royal Agric. Coll. Sweden*, **12**, 101–118.

Mayer, R., and Ulrich, B. (1974). Conclusions on the filtering action of forests from ecosystem analysis. *Oecol. Plant*, **9**, 157–168.

Mayer, R., and Ulrich, B. (1977). Acidity of precipitation as influenced by the filtering of atmospheric sulphur and nitrogen compounds—its role in the element balance and effect on soil. *Water, Air and Soil Pollution*, **7**, 409–416.

Mayer, R., and Ulrich, B. (1978). Input of atmospheric sulfur by dry and wet deposition to two C. European forest ecosystems. *Atmos. Env.*, **12** (3), 375–377.

Meade, R. H. (1969). Errors in using modern stream load data to estimate natural rates of denudation. *Geol. Soc. Amer. Bull.*, **80**, 1265–1274.

Mellanby, K. (1984). (Ed.). *Acid Rain*, Report No. 14, The Watt Committee on Energy, London, 58 pp.

Meyer, J. L., Likens, G. E., and Sloane, J. (1981). P, N and organic C flux in a headwater stream. *Archiv. fur Hydrobiologie*, **91** (1), 28–44.

Miklas, J., Wu, T. L., Hiatt, A., and Correll, D. L. (1977). Nutrient loading of the Rhode River via land use practices and precipitation. In: *Watershed Research in Eastern North America* (D. L. Correll, Ed.), Smithsonian, Edgewater, pp. 169–191.
Miller, H. G. (1980). E. Throughfall and stemflow: 2. Throughfall, stemflow, crown leaching and wet deposition. In: *Methods for Studying Acid Precipitation in Forest Ecosystems*, Institute of Terrestrial Ecology, Cambridge, pp. 17–20.
Miller, H. G. (1984). Deposistion-plant-soil interactions. *Phil. Trans. R. Soc. Lond.*, **B305**, 339–352.
Miller, H. G., Cooper, J. M., Miller, J. D., and Pauline, O. J. L. (1979). Nutrient cycles in pines and their adaptation to poor soils. *Canadian J. Forest Research*, **9**, 19–26.
Miller, H. G., and Miller, J. D. (1980). Collection and retention of atmospheric pollutants by vegetation. In: *Ecological Impact of Acid Precipitation* (D. Drablos and A. Tollan, Eds.), Proceedings of an international conference, Sandefjord, Norway, SNSF project, pp. 33–40.
Miller, H. G., Unsworth, M. H., and Fowler, D. (1980). B. Definitions and concepts. In: *Methods for Studying Acid Precipitation in Forest Ecosystems*, Institute of Terrestrial Ecology, Cambridge, pp. 3–5.
Miller, J. M., Galloway, J. N., and Likens, G. E. (1976). The use of ARL trajectories for the evaluation of precipitation chemistry data. In: *Proc. 1st. Int. Symposium on Acid Precipitation and the Forest Ecosystem* (L. S. Dochinger and T. A. Seliga, Eds.), U.S.D.A. Forest Serive, General Technical Report NE-23, pp. 131–132.
Miller, J. P. (1961). Solutes in small streams draining single rock types, Sangre de Cristo Range, New Mexico. *U.S. Geol. Survey, Water Supply Paper, 1535—D.*
Monteith, J. L. (1973). *Principles of Environmental Physics*, Arnold, London.
Moody, J. R. (1983). Sampling and storage of materials for trace elemental analysis. *Trends in Analytical Chemistry*, **2** (5), 116–118.
Muller, G., and Osborn, J. (1984). Open Forum: Measures to combat acid rain. *Forum: Council of Europe*, **1/84**, 2–4.
Munger, J. W., and Eisenreich, S. J. (1983). Continental-scale variations in precipitation chemistry. *Environmental Science and Technology*, **17** (1), 32A–42A.
Munn, R. E. (1981). *The design of air quality monitoring networks*, MacMillan, London.
Munn, R. E., and Rodhe, H. (1971). On the meteorological interpretation of the chemical composition of monthly precipitation samples. *Tellus*, **23**, 1–12.
Murray, R. C., and Lewis, R. P. W. (1966). Some aspects of the synoptic climatology of the British Isles as measured by simple indices. *Meteorological Magazine*, **95**, 193–203.
National Swedish Environmental Protection Board (1983). *Ecological Effects of Acid Deposition*, Report and background papers, 1982 Stockholm Conference on the acidification of the environment, Expert Meeting No. 1, PM 1636, 340 pp.
Newman, L., Likens, G. E., and Bormann, F. H. (1975). Acidity in rainwater: has an explanation been presented? *Science*, **188** (4191), 957–958.
Nicholas, C. A., O'Connell, P. E., and Senior, M. R. (1981). Rain gauge network rationalisation and its advantages. *Met. Mag.*, **110**, 92–102.
Nicholson, I. A., Fowler, D., Paterson, I. S., Cape, J. N., and Kinnaird, J. W. (1980). Continuous monitoring of airborne pollutants. In: *Ecological Impact of Acid Precipitation* (D. Drablos and A. Tollan, Eds.), Proceedings of an international conference, Sandefjord, Norway SNSF project, pp. 144–145.
Nicholson, I. A., Paterson, I. S., and Last, F. T. (1980). *Methods for Studying Acid Precipitation in Forest Ecosystems. Definitions and Research Requirements*, I.T.E./N.E.R.C./U.N.E.S.C.O., Cambridge, 36 pp.

Nicks, A. D. (1965). Field evaluation of rain gauge network design principles. *Int. Assoc. Sci. Hydrology, pub. No. 67*, pp. 82–93.

Nilsson, S. I., Miller, H. G., and Miller, J. D. (1982). Forest growth as a possible cause of soil and water acidification: an examination of the concepts. *Oikos*, **39** (1), 40–49.

Nuckols, J. R., and Moore, L. R. (1982). The influence of atmospheric N influx upon the stream N profile of a relatively undisturbed forested watershed. *J. Hydrology*, **57** (1/2), 113–135.

Oakes, D. B. (1979). The movement of water and solutes through the unsaturated zone of the Chalk in the U.K. In: *Surface and Subsurface Hydrology* (H. J. Morel-Seytoux, Ed.), Water Resources Publications, Fort Collins, pp. 447–459.

O'Connell, P. E., Gurney, R. J., Jones, D. A., Miller, J. B., Nicholass, C. A., and Senior, M. R. (1978). Rationalization of the Wessex Water Authority raingauge network. *Institute of Hydrology Report No. 51*.

Odum, E. P. (1971). *Fundamentals of Ecology* (3rd edn.), Saunders, Philadelphia.

Ogden, J. G. (1980). Comparative composition of continental and Nova Scotian precipitation. In: *Ecological Impact of Acid Precipitation* (D. Drablos and A. Tollan, Eds.), Proceedings of an international conference, Sandefjord, Norway, SNSF project, pp. 126–127.

Ott, H. and Stangl, H. (1983). (Eds). *Acid Deposition: a Challenge for Europe.* Preliminary edition of proceedings of a symposium, Karlsruhe, 19–21 September 1983, Commission of the European Communities Directorate General for Science, Research and Development.

Overrein, L. N., Seip, H. M., and Tollan, A. (1981). *Acid precipitation–effects on forest and fish* (2nd edn.), Final Report of the SNSF project, 1972–1980, Oslo.

Owens, J. W., Gladney, E. S., and Purtymun, W. D. (1980). Modification of trace element concentrations in natural waters by various field sampling techniques. *Analytical Letters*, **13** (A4), 253–260.

Pack, D. H. (1977). Global atmospheric monitoring—a status report. In: *Air pollution measurement techniques* (W.M.O.) Special Environmental Report No. 10, pp. 1–8.

Pagenkopf, G. K. (1978). *Introduction to Natural Water Chemistry*, M. Dekker, New York.

Pearson, F. J., and Fisher, D. W. (1971). Chemical composition of atmospheric precipitation in the Northeastern United States. *U.S. Geol. Survey, Water Supply Paper, 1535–P*.

Perry, A. (1968). The regional variation of climatological characteristics with synoptic indices. *Weather*, **23**, 325–330.

Psenner, R. (1984). The proportion of empneuston and total atmospheric inputs of carbon, nitrogen and phosphorus in the nutrient budget of a small mesotrophic lake (Piburger See, Austria). *Int. Revue der Gesamten Hydrobiologie*, **69** (1), 23–39.

Ragland, K. W., and Wilkening, K. E. (1982). Relationship between mesoscale acid precipitation and meteorological factors. In: *Acid Precipitation: Effects on Ecological Systems* (F. M. D'Itri, Ed.), Ann Arbor Science, Ann Arbor, pp. 123–140.

Rainbird, A. F. (1965). Precipitation—basic principles of network design. *Int. Assoc. Sci. Hydrology, pub. No. 67*, pp. 19–30.

Raynor, G. S., and McNeil, J. P. (1979). An automatic sequential precipitation sampler. *Atmos. Env.*, **13**, 149–155.

Record, F. A., Bubenick, D. V., and Kindya, R. J. (1982). *Acid Rain Information Book*, Noyes Data Corporation, Park Ridge, New Jersey.

Reid, J. M., MacLeod, D. A., and Gresser, M. S. (1981). Factors affecting the chemistry of precipitation and river water in an upland catchment. *J. Hydrology*, **50**, 129–145.

Reiners, W. A. (1972). Nutrient content of canopy throughfall in three Minnesota forests. *Oikos*, **23**, 14–22.

Richter, D. D., Ralston, C. W., and Harms, W. R. (1983). Chemical composition and spatial variation of bulk precipitation at a coastal Plain watershed in South Carolina. *Water Resources Research*, **19** (1), 134–140.

Roberson, C. E., Feth, J. H., Seaber, P. R., and Anderson, P. (1963). Differences between field and laboratory determinations of pH, alkalinity and specific conductivity of natural waters, *U.S. Geological Survey Professional Paper 475–C*, 212–215.

Roberts, L. (1983). Is acid deposition killing West German forests? *Bio Science*, **33**(5), 302–305.

Robertson, R. A., and Davies, G. E. (1965). Quantities of plant nutrients in heather. *J. Applied Ecology*, **2**, 211–219.

Rodhe, H. (1980). Estimate of wet deposition of pollutants around a point source. *Atmos. Env.*, **14**, 1197–1199.

Rodhe, H., Crutzen, P., and Vanderpol, A. (1981). Formation of sulphuric and nitric acid in the atmosphere during long-range transport. *Tellus*, **33** (2), 132–141.

Rodhe, H., and Grandell, J. (1972). On the removal time of aerosol particles from the atmosphere by precipitation scavenging. *Tellus*, **24** (5), 442–454.

Rossby, C. G., and Egner, H. (1955). On the chemical climate and its variation with the atmospheric circulation pattern. *Tellus*, **8** (1), 118–133.

Ronneau, C., Cara, J., Navarre, J. L., and Priest, P. (1978). An automatic sequential rain sampler. *Water, Air and Soil Pollution*, **9**, 171–176.

Rosenqvist, I.Th. (1978). Alternative sources for acidification of river water in Norway. *Science of the Total Environment*, **10** (1), 39–49.

Rosenqvist, I.Th., Jorgensen, P., and Rueslatten, H. (1980). The importance of natural H^+ production for acidity in soil and water. In: *Ecological Impact of Acid Precipitation* (D. Drablos and A. Tollan, Eds.), Proceedings of an international conference, Sandefjord, Norway, SNSF project, pp. 240–241.

Rowse, A. A. (1980). Factors affecting environmental surveillance system design. *Science of the Total Environment*, **16** (3), 193–208.

Rutter, N., and Edwards, R. S. (1968). Deposition of airborne marine salt at different sites over the College Farm, Aberystwyth, in relation to wind and weather. *Agricultural Meteorology*, **5**, 235–254.

Samant, H. S., and Vaidya, O. C. (1982). Evaluation of the sampling buckets used in the Sangamo collector, Type A for heavy metals in precipitation. *Atmos. Env.*, **16** (9), 2183–2186.

Scheider, W. A., Snyder, W. R., and Clark, B. (1979). Deposition of nutrients and major ions by precipitation in south central Ontario. *Water, Air and Soil Pollution*, **12**, 171–185.

Schlesinger, W. H., Gray, J. T., and Gilliam, F. S. (1982). Atmospheric deposition processes and their importance as sources of nutrients in a chaparral ecosystem of Southern California. *Water Resources Research*, **18** (3), 623–629.

Schlesinger, W. H., and Reiners, W. A. (1974). Deposition of water and cations on artificial foliar collectors in fir krummholz of New England mountains. *Ecology*, **55**, 378–386.

Schnoor, J. L., Sigg, L., Stumm, W., and Zobrist, J. (1983). Acid precipitation and its influence on Swiss lakes. *EAWAG News*, **14/15**, 6–12.

Schock, M. R., and Schock, S. C. (1982). Effect of container type on pH and alkalinity stability. *Water Research*, **16** (10), 1455–1464.

Schreiber, J. D., Duffy, P. D., and McClurkin, D. C. (1976). Dissolved nutrient losses in storm runoff from five southern pine watersheds. *J. Environmental Quality*, **5** (2), 201–205.

Schuman, G. E., and Burwell, R. E. (1974). Precipitation nitrogen contribution relative to surface runoff discharge. *J. Environmental Quality*, **3** (4), 366–369.
Sehmel, G. A. (1980). Particle and gas dry deposition—a review. *Atmos. Env.*, **14** (9), 981–1011.
Seip, H. M. (1980). Acidification of freshwater-sources and mechanisms. In: *Ecological Impact of Acid Precipitation* (D. Drablos and A. Tollan, Eds.), Proceedings of an international conference, Sandefjord, Norway, SNSF project, pp. 358–366.
Seip, H. M., and Tollan, A. (1978). Acid precipitation and other possible sources for acidification of rivers and lakes. *Science of the Total Environment*, **10** (3), 253–270.
Sequiera, R. (1982). Chemistry of precipitation at high altitudes: inter-relation of acid-base components. *Atmos. Env.*, **16** (2), 329–335.
Sharpe, W. E., and Young, E. S. (1982). The effects of acid precipitation on water quality in roof catchment-cistern water supplies. In: *Acid precipitation: Effects on Ecological Systems*, (F. M. D'Itri, Ed.), Ann Arbor Science, Ann Arbor, pp. 365–384.
Shaw, R. W. (1982). Deposition of atmospheric acid from local and distant sources at a rural site in Nova Scotia. *Atmos. Env.*, **16** (2), 337–348.
Shriner, D. S., Richmond, C. R., and Lindberg, S. E. (1980). (Eds.). *Atmospheric Sulfur Deposition: Environmental Impact and Health Effects*, Ann Arbor Science, Ann Arbor, 568 pp.
Shuttleworth, W. J. (1977). The exchange of wind-driven fog and mist between vegetation and the atmosphere. *Boundary-Layer Meteorology*, **12**, 463–489.
Siegel, D. I. (1981). The effects of snowmelt on the water quality of Filson Creek and Ornaday Lake, northeastern Minnesota. *Water Resources Research*, **17** (1), 238–242.
Siegfried, C. A. (1982). Phosphorus loading to a mountain reservoir in California. *Water Resources Bulletin*, **18** (4), 613–620.
Skartveit, A. (1980). Observed relationships between ionic composition of precipitation and runoff. In: *Ecological Impact of Acid Precipitation* (D. Drablos and A. Tollan, Eds.), Proceedings of an international conference, Sandefjord, Norway, SNSF project, pp. 242–243.
Skartveit, A. (1981). Relationships between precipitation chemistry, hydrology and runoff acidity. *Nordic Hydrology*, **12**, 65–80.
Skartveit, A. (1982). Wet scavenging of sea-salts and acid compounds in a rainy, coastal area. *Atmos. Env.*, **16** (11), 2715–2724.
Slack, K. V., and Fisher, D. W. (1965). Light-dependent quality changes in stored water samples. *U.S. Geological Survey Professional Paper 525-C*, 190–192.
Slanina, J., Van Raaphorst, J. G., and Zijp, W. L. (1979a). An evaluation of the chemical composition of precipitation samples with 21 identical collectors on a limited area. *Int. J. Environ. Anal. Chem.*, **6**, 67–81.
Slanina, J., Mols, J. J., Baard, J. H., Van der Sloot, H. A., and Van Raaphorst, J. G. (1979b). Collection and analysis of rainwater: experimental problems and the analysis of results. *Int. J. Environ. Anal. Chem.*, **7**, 161–176.
Slinn, W. G. N. (1977). Some approximations for the wet and dry removal of particles and gases from the atmosphere. *Wat., Air and Soil Poll.*, **7**, 513–543.
Slinn, W. G. N., Hasse, L., Hicks, B. B., Hogan, A. W., Lal, D., Liss, P. S., Munnich, K. O., Sehmel, G. A., and Vittorri, O. (1978). Some aspects of the transfer of atmospheric trace constituents past the air-sea interface. *Atmos. Env.*, **12**, 2055–2087.
Smith, F. B. (1983). Long range transport of air pollution. *Met. Mag.*, **112**, 237–244.

Sokolov, A. S. (1972). Chemical composition of atmospheric precipitation falling through spruce and birch stands. *Soviet Hydrology*, **4**, 332–336.
Stensland, G. J., and Semonin, R. G. (1982). Another interpretation of the pH trend in the United States. *Bull. Amer. Met. Soc.*, **63** (11), 1277–1284.
Stevenson, C. M. (1968). An analysis of the chemical composition of rainwater and air over the British Isles and Eire for the years 1959–64. *Quarterly J. Met. Soc.*, **94**, 56–70.
Stoner, J. H., Gee, A. S., and Wade, K. R. (1984). The effects of acidification on the ecology of streams in the Upper Tywi catchment in West Wales. *Environmental Pollution*, **35A**, 1–33.
Strachan, W. M. J., and Huneault, H. (1984). Automated rain sampler for trace organic substances. *Environmental Science and Technology*, **18** (2), 127–130.
Sutcliffe, D. W. (1979). Some notes to authors on the presentation of accurate and precise measurements in quantitative studies. *Freshwater Biology*, **9**, 397–402.
Sutcliffe, D. W., and Carrick, T. R. (1973a). Studies on mountain streams in the English Lake District. I. pH, calcium and the distribution of invertebrates in the River Duddon. *Freshwater Biology*, **3**, 437–462.
Sutcliffe, D. W., and Carrick, T. R. (1973b). Studies on mountain streams in the English Lake District. II. Aspects of water chemistry in the River Duddon. *Freshwater Biology*, **3**, 543–560.
Sutcliffe, D. W., and Carrick, T. R. (1973c). Studies on mountain streams in the English Lake District. III. Aspects of water chemistry in Brownrigg Well, Whelpside Ghyll. *Freshwater Biology*, **3**, 561–568.
Sutcliffe, D. W., Carrick, T. R., Heron, J., Rigg, E., Talling, J. F., Woof, C., and Lund, J. W. G. (1982). Long-term and seasonal changes in the chemical composition of precipitation and surface water of lakes and tarns in the English Lake District. *Freshwater Biology*, **12**, 451–506.
Swank, W. T., and Douglass, J. E. (1975). Nutrient flux in undisturbed and manipulated forest ecosystems in the southern Appalachian Mountains. *Proceedings of the Tokyo Symposium on the Hydrological Characteristics of River Basins and the Effects on these Characteristics of Better Water Management, International Association of Hydrological Sciences, Pub. No. 117*, pp. 445–496.
Swank, W. T., and Douglass, J. E. (1977). Nutrient budgets for undisturbed and manipulated hardwood forest ecosystems in the mountains of North Carolina. In: *Watershed Research in Eastern North America* (D. L. Correll, Ed.), Smithsonian, Edgewater, pp. 343–362.
Swank, W. T., and Henderson, G. S. (1976). Atmospheric inputs of some cations and anions to forest ecosystems in North Carolina and Tennessee. *Water Resources Research*, **12** (3), 541–555.
Swanson, K. A., and Johnson, A. H. (1980). Trace metal budgets for a forested watershed in the New Jersey Pine Barrens. *Water Resources Research*, **16** (2), 373–376.
Swedish Ministry of Agriculture (1982). *Acidification Today and Tomorrow*, a report prepared by the Environment '82 committee for the 1982 Stockholm Conference on the acidification of the environment, 28–30 June, 230 pp.
Swedish Ministry of Agriculture (1983). *Acidification: a Boundless Threat to our Environment*, Ministry of Agriculture, Stockholm, 44 pp.
Tabatabai, M. A., and Laflen, J. M. (1976). Nutrient content of precipitation over Iowa. In: *Proc. 1st Int. Symposium on Acid Precipitation and the Forest Ecosystem*, (L. S. Dochinger and T. A. Seliga, Eds.), U.S.D.A. Forest Service, General Technical Report NE-23, 293–308.
Thompson, R. J. (1977). The sampling and analysis techniques in current use in the

EPA/NOAA/WMO precipitation network. In: *Air pollution measurement techniques* (W.M.O.) Special Environmental Report No. 10, 40–49.

Thornton, J. D., and Eisenreich, S. J. (1982). Impact of land-use on the acid and trace element composition of precipitation in the north central U.S. *Atmos. Env.*, **16** (8), 1945–1955.

Tomlinson, R. W. (1978). The supply from rainfall of four elements to an area of N.E. Rhodesia. *Tellus*, **30**, 476.

Torrenueva, A. L. (1975). *Variation in mineral flux to the forest floors of a pine and a hardwood stand in the Georgia piedmont*. Ph.D. dissertation, U. Georgia, Athens.

Turvey, N. D. (1974). Water in the nutrient cycle of a Papuan rain forest. *Nature*, **251** (1574), 414–415.

Tyree, S. Y. (1981). Rainwater acidity measurement problems. *Atmos. Env.*, **5**, 57–60.

U.K. Review Group on Acid Rain (1983). *Acid Deposition in the United Kingdom*, Dept. of Trade and Industry, Warren Spring Laboratory, Stevenage, 72 pp.

Ulrich, B. (1983). Effects on accumulation of air pollutants in forest ecosystems. In: *Acid Deposition: a Challenge for Europe* (H. Ott and H. Stangl, Eds.), Preliminary edition of proceedings of a symposium, Karlsruhe, 19–21 September. Commission of the European Communities Directorate General for Science, Research and Development, pp. 127–134.

Unsworth, M. H. (1980a). D. Evaluation of atmospheric inputs: 1. Dry deposition of gases and particles onto vegetation. In: *Methods for Studying Acid Precipitation in Forest Ecosystems*, Institute of Terrestrial Ecology, Cambridge, pp. 9–15.

Unsworth, M. H. (1980b). D. Evaluation of atmospheric inputs: 2. Evaluation of gaseous and particulate inputs from the atmosphere. In: *Methods for Studying Acid Precipitation in Forest Ecosystems*, Institute of Terrestrial Ecology, Cambridge, pp. 17–20.

Van Breemen, N., Driscoll, C. T., and Mulder, J. (1984). Acidic deposition and internal proton sources in acidification of soils and waters. *Nature*, **307** (5952), 599–604.

Van Denburgh, A. S., and Feth, J. H. (1965). Solute erosion and chloride balance in selected river basins of the western coterminous United States. Water Resources Research, **1**, 537–541.

Van Dop, H. (1983). The residence time and transport of pollutants in the atmosphere—a meteorological problem. In: *Acid Deposition: a Challenge for Europe* (H. Ott and H. Stangl, Eds.), prelim. edition of symposium proceedings, Karlsruhe, 19–21 September, Commission of the European Communities Directorate General for Science, Research and Development, pp. 48–57.

Vangenechten, J. H. D., Bosmans, F., and Deckers, H. (1981). Effects of short-term changes in rainwater supply on the ionic composition of acid moorland pools in the Campine of Antwerp (Belgium). *Hydrobiologia*, **76** (1/2), 149–159.

Viro, P. J. (1953). Loss of nutrients and the natural nutrient balance of the soil in Finland. *Comm. Inst. For. Fenn.*, **42**, 1–45.

Voigt, G. K. (1960). Alteration of the composition of rainwater by trees. *American Midland Naturalist*, **63**, 321–326.

Voigt, G. K. (1980). Acid precipitation and soil buffering capacity. In: *Ecological Impact of Acid Precipitation* (D. Drablos and A. Tollan, Eds.), Proceedings of an international conference, Sandefjord, Norway, SNSF project, pp. 53–57.

Walling, D. E. (1980). Water in the catchment ecosystem. In: *Water Quality in Catchment Ecosystems* (A. M. Gower, Ed.), Wiley, Chichester, pp. 1–47.

Walling, D. E., and Webb, B. W. (1980). The spatial dimension in the interpretation of stream solute behaviour. *J. Hydrology*, **47**, 129–149.

Waylen, M. J. (1979). Chemical weathering in a drainage basin underlain by Old Red Sandstone. *Earth Surface Processes*, **4**, 167–178.
Webb, B. W., and Walling, D. E. (1980). Stream solute studies and geomorphological research: some examples from the Exe basin, Devon, U.K. *Zeitschrift fur Geomorphologie, N. F.*, Supp. Band **36**, 245–263.
Weeks, E. P., Earp, D. E., and Thompson, G. M. (1982). Use of atmospheric fluorocarbons F–11 and F–12 to determine the diffusion parameters of the unsaturated zone in the Southern High Plains of Texas. *Water Resources Research*, **18** (5), 1365–1378.
Weibe, H. A. (1976). The effect of precipitation collector design on the measured acid content of precipitation. In: *Proc. 1st Int. Symposium on Acid Precipitation and the Forest Ecosystem* (L. S. Dochinger and T. A. Seliga, Eds.), U.S.D.A. Forest Service, General Technical Report NE–23, p. 135.
Wesely, M. L., Hicks, B. B., Dannevik, W. P., Frisella, S., and Husar, R. B. (1977). An eddy-correlation measurement of particulate deposition from the atmosphere. *Atmos. Env.*, **11**, 561–563.
Westman, W. E. (1978). Inputs and cycling of mineral nutrients in a coastal sub-tropical eucalypt forest. *J. Ecology*, **66** (2), 513–532.
White, E. J., Starkey, R. S., and Saunders, M. (1971). An assessment of the relative importance of several chemical sources to the waters of a small upland catchment. *J. Applied Ecology*, **8**, 743–750.
White, E. J., and Turner, F. (1970). A method of estimating income of nutrients in a catch of airborne particles by a woodland canopy. *J. Applied Ecology*, **7**, 441–461.
White, E. M. (1981). Nutrient contents of precipitation and canopy throughfall under corn, soybeans and oats. *Water Resources Bulletin*, **17** (4), 708–712.
Whitehead, H. C., and Feth, J. H. (1964). Chemical composition of rain, dry fallout and bulk precipitation, California, 1957–59. *J. Geophys. Res.*, **69** (16), 3319–3333.
Wiklander, L., and Andersson, A. (1972). The replacing efficiency of hydrogen ion in relation to base saturation and pH. *Geoderma*, **7**, 159–165.
Will, G. M. (1959). Nutrient return in litter and rainfall under some exotic-conifer stands in New Zealand. *N.Z. Journal of Agricultural Research*, **2**, 719–754.
Wisniewski, J., and Kinsman, J. D. (1982). An overview of acid rain monitoring activities in North America. *Bull. Amer. Met. Soc.*, **63** (6), 598–618.
Wisniewski, J., and Miller, J. M. (1977). A critical review of precipitation chemistry studies—North America and adjacent areas. In: *Air pollution measurement techniques* (W.M.O.), Special Environmental Report No. 10, pp. 63–69.
Woodcock, A. H. (1952). Atmospheric salt particles and raindrops. *J. Meteorology*, **9**, 202–212.
Wright, R. F. (1983). Input–output budgets at Langtjern, a small acidified lake in S. Norway. Hydrobiologia, **101** (1/2), 1–12.
Wright, R. F., Harriman, R., Henriksen, A., Morrison, B., and Caines, L. A. (1980). Acid lakes and streams in the Galloway area, southwestern Scotland. In: *Ecological Impact of Acid Precipitation* (D. Drablos and A. Tollan, Eds.), Proceedings of an international conference, Sandefjord, Norway, SNSF project, pp. 248–249.
Wright, R. F., and Johannessen, M. (1980). Input–output budgets of major ions at gauged catchments in Norway. In: *Ecological Impact of Acid Precipitation* (D. Drablos and A. Tollan, Eds.), Proceedings of an international conference, Sandefjord, Norway, SNSF project, pp. 250–251.
Yaalon, D. H., and Lomas, J. (1970). Factors controlling the supply and the chemical composition of aerosols in a near shore and coastal environment. *Agricultural Meteorology*, **7**, 443–454.

Zeman, L. J. (1975). Hydrochemical balance of a British Columbian mountain watershed. *Catena*, **2**, 81–94.
Zeman, L. J., and Nyborg, E. O. (1974). Collection of chemical fallout from the atmosphere. *Canadian Agricultural Engineering*, **16** (2), 69–72.
Zeman, L. J., and Slaymaker, O. (1978). Mass balance model for calculation of ionic input loads in atmospheric fallout and discharge from a mountainous basin. *Bull. Int. Ass. Sci. Hydrol.*, **23**, 103–117.

Solute Processes
Edited by S. T. Trudgill

CHAPTER 3

Biological control of solute losses from forest ecosystems

Wayne T. Swank

Coweeta Hydrologic Laboratory,
Otto, North Carolina

3.1 INTRODUCTION

The biological regulation of solute dynamics encompasses numerous interconnected processes distributed throughout a variety of ecosystem compartments. A conceptual framework of processes important in the control of forest nutrient cycling (and hence, solute loss) is shown in Figure 3.1. The relative importance of processes varies considerably between ions and between ecosystems, but most of the processes are basic to all ecosystems. An indication of the influence of forest compartments on dissolved ions as water moves through forest ecosystems is represented in Table 3.1. Data for both sites were derived for mature forests indigenous to the regions of study. The values show substantial alteration of dissolved constituents which reflect the integrated effect of biogeochemical factors.

In this chapter the objective is to identify and demonstrate the influence of selected biotic factors on solute movement in both terrestrial and stream ecosystems. The reader should recognize the complexity of quantitatively separating physical, chemical and biological factors of solute behaviour. The approach taken in this paper uses small catchments (8–60 ha) as the unit of investigation. Furthermore, this discussion will not include all the potential processes regulating solute losses. Emphasis will be placed on biological controls which can be inferred or illustrated by landscape scale experiments and/or recent, original biological process research. Specifically, the topic will be developed by:

1. characterizing biogeochemical behaviour of several different forest ecosystems through budget analysis, sizes of nutrient pools, recycling mechan-

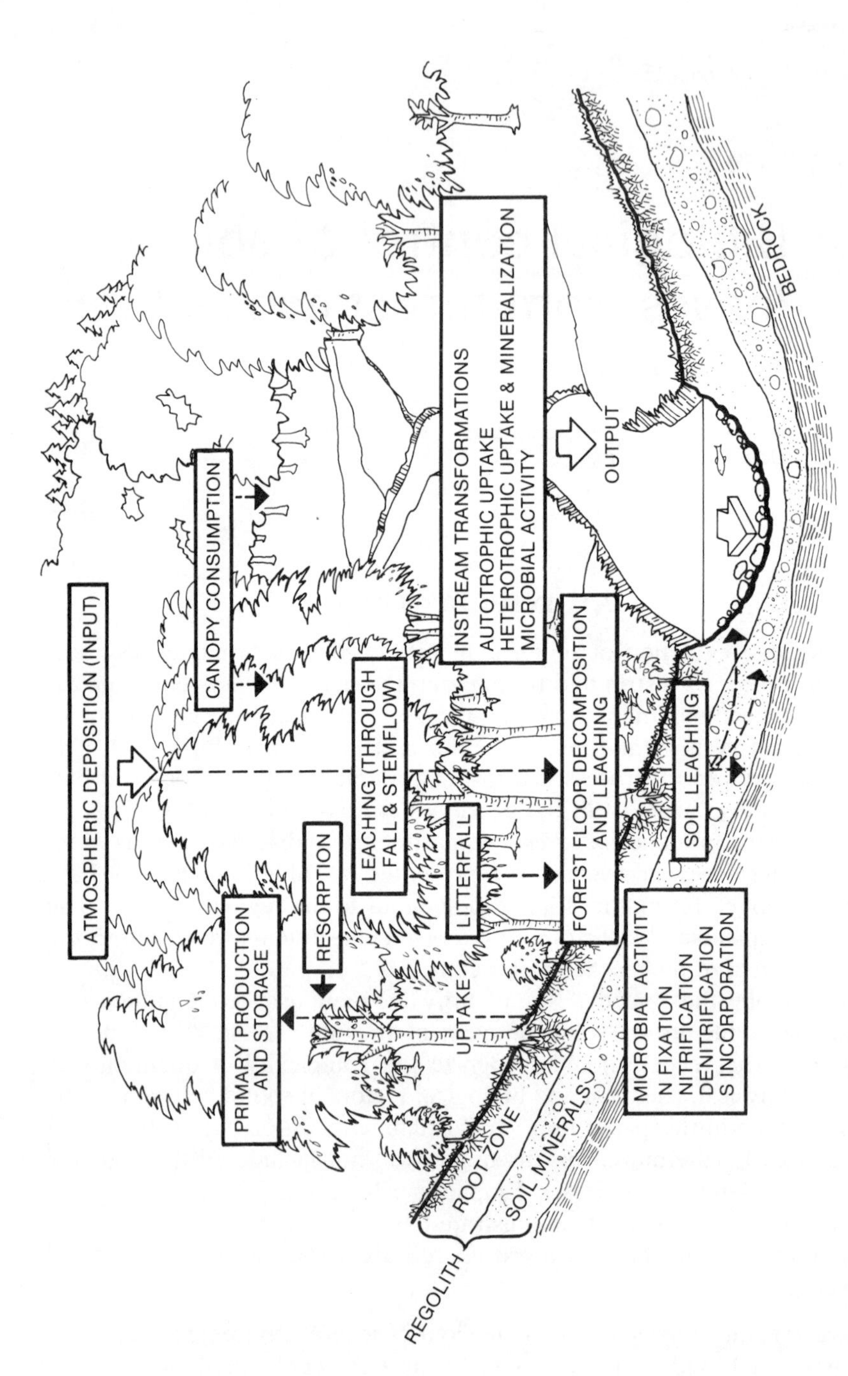

Figure 3.1. Conceptualization of major biological processes important in regulating the cycling, retention and loss of solutes in forest ecosystems

Table 3.1. Vertical profiles of mean annual solute concentrations in hydrologic compartments of two forest ecosystems

	Solute concentration (μeq l^{-1})[d]						
Forest and compartment	H	NO_3–N	K	Na	Ca	Mg	SO_4
1. H. J. Andrews—Old growth Douglas fir forest[a]							
Precipitation	6.3	12	1	10	7	4	10
Throughfall	5.2	1	18	14	18	12	19
Litter solution	1.5	1	33	16	58	23	56
Soil solution (30 cm)	0.1	20	24	292	372	36	518[b]
Streamwater	0.2	19	9	85	160	69	—
2. Coweeta—Mixed deciduous hardwood forest[c]							
Precipitation	17.1	8	2	6	11	3	27
Throughfall	5.1	5	50	7	40	26	77
Litter solution	4.6	1	72	13	65	53	105
Soil solution (30 cm)	1.0	1	33	19	43	53	66
Streamwater	0.2	0.2	13	41	41	30	10

[a]Taken from Sollins, *et al.* (1980).
[b]Concentration measured in soil solution at 100 cm.
[c]After Swank and Swank (1984).
[d]μeq $l^{-1} \rightarrow$ mg l^{-1} = μeq l^{-1}/F_1; divided by 1000 where F = H 0.99209; N =0.07137; K = 0.02557; Na = 0.04350; Ca = 0.04990; Mg = 0.02082.

isms and rates, and examples of the disruption and subsequent recovery of biological controls of solute losses;
2. illustration of how soil microbial populations regulate nitrogen and sulphur transformation, and hence nitrate and sulphate anion fluxes;
3. discussion of the role of terrestrial inputs in controlling solute losses;
4. examination of instream biological transformations that regulate the form and amount of solute losses.

A major portion of the data presented in this paper was derived from the Coweeta Hydrologic Laboratory; therefore, it is appropriate to provide a brief site description. The 2185 ha laboratory is located in the south-west corner of North Carolina in the Southern Appalachians of the USA (Figure 3.2). The area is made up of numerous small catchments that have a 50-year history of continuous hydrologic and climatological monitoring, a well-documented history of vegetation on the catchments, and a variety of experimental manipulations that include species conversions, commercial logging operations, and other land-use treatments. Beginning in 1968, a major collaborative programme between university and Government

Figure 3.2. The Coweeta Hydrologic Laboratory in western North Carolina contains numerous small (8–60 ha) catchments, representing a variety of vegetation types, which are used in biogeochemical cycling research

scientists has emphasized research on biogeochemical processes important in forest ecosystems.

The basin is within the Blue Ridge Province and the bedrock is composed of granodiorite, mica gneiss, and mica schist. The formation is deeply weathered and the regolith depth averages about 7 m. Annual precipitation varies from 180 cm at lower elevations to 250 cm on the upper slopes. Snow contributes less than 2% to total precipitation. Precipitation is rather evenly distributed throughout the year, but stream discharge is usually highest in February and March and lowest in September and October. Quickflow, or direct runoff, usually accounts for less than 10% of the total runoff. The temperate hardwood forests contain a mixture of deciduous species and is dominated by *Quercus*, *Acer* and *Carya* species, along with about 15 other associated species in the overstory.

3.2 CHARACTERIZATION OF SOLUTES IN ECOSYSTEMS

3.2.1 Solute budgets

Input–output solute budgets taken at forest ecosystem boundaries provide an integrated measure of biogeochemical behaviour and are useful in characterizing the net result of physical, chemical and biological processes. Budget data are valuable for characterizing undisturbed ecosystems (Likens, *et al.*, 1967), for generalizing across ecosystems located in diverse physiographic or climatic regions (Henderson *et al.*, 1978), and for documenting changes produced by human activities (Likens *et al.*, 1970; Johnson and Swank, 1973; Swank and Douglass, 1977). Budgets are derived by monitoring quantities of precipitation and streamflow, as well as concentrations of solutes in water entering and leaving catchments. Nitrogen (N), calcium (Ca) and potassium (K) have been selected for discussion here because of their contrasting mobility in biogeochemical cycles, their importance in tree nutrition, and their origin.

Four contrasting baseline ecosystems were selected for budget analysis to represent different physiographic regions of the USA (Figure 3.3). These four systems provide major contrasts in vegetation, bedrock geology, soils and hydrology (Likens *et al.*, 1977; Henderson *et al.*, 1978). The three eastern sites contain deciduous forests (oak–hickory or northern hardwoods), while the western site is characterized by an old-growth coniferous forest. Soils and bedrock underlying these forests vary widely among sites, ranging from typic Paleudults derived from dolomite bedrock at Oak Ridge, soils derived from volcanic tuff overlying andesitic bedrock at H. J. Andrews, typic Hapludults derived from granodiorite, gneiss, and schist at Coweeta, to Spodosols overlying highly metamorphosed mudstones and sandstones at Hubbard Brook. The amount, form and seasonal distribution of annual precipitations are also variable, ranging from 130 to 233 cm with both wet and dry growing seasons and the absence or dominance of snowpack.

Average annual solute inputs and outputs for these systems are shown in Table 3.2. Nitrogen inputs differ by a factor of 10 and outputs by a factor of 2, but net budgets for all systems show an accumulation of N. The N budget is most nearly balanced for the old-growth conifer forest at H. J. Andrews, partly because of small inputs in bulk precipitation and partly because of

Table 3.2. Comparison of nitrogen (organic and inorganic), calcium and potassium inputs in precipitation and outputs in streamflow for four watershed ecosystems in different physiographic regions of the USA (Swank and Waide, 1980)

	Nutrient ($kg\ ha^{-1}\ yr^{-1}$)											
	Organic N			Inorganic N			Ca			K		
Site and vegetation type	Input	Output	Net Budget	Input	Output	Net Budget	Input	Output	Net Budget	Input	Output	Net Budget
Walker Branch, Tennessee[a] (oak–hickory)	3.7	1.6	+2.1	9.3	1.5	+7.8	12.0	148.0	−136.0	3.0	7.0	−4.0
Hubbard Brook, New Hampshire[b] (northern hardwoods)	—	—	—	6.5	3.9	+2.6	2.2	13.7	−11.5	0.9	1.9	−1.0
Coweeta, North Carolina[c] (oak–hickory)	4.3	3.1	+1.2	4.5	0.1	+4.4	4.8	7.7	−2.9	2.1	5.6	−3.5
H. J. Andrews, Oregon[d] (douglas fir)	1.5	1.8	−0.3	0.7	0.1	+0.6	2.3	50.3	−48.0	0.1	2.2	−2.1

[a] Henderson and Harris (1975); Henderson *et al*. (1978).
[b] Likens *et al*. (1977); data not available for organic N.
[c] Swank and Douglass (1975); unpublished data for organic N.
[d] Fredriksen (1975); Grier *et al*. (1974).

Figure 3.3. Location of four experimental sites in different physiographic regions of the USA where long-term research has been conducted on biological processes and solute losses. HJA = H. J. Andrews Experimental Forest, WB = Walker Branch Watershed, HB = Hubbard Brook Experimental Forest, and CHL = Coweeta Hydrologic Laboratory

negative vegetation accretion. This comparison includes only hydrologic constituents of the N budget; large gains and losses which could occur in gaseous form will be discussed later. Budgets for Ca and K show net losses for all ecosystems (Table 3.2). Differences in bedrock mineralogy and solubility are reflected in the wide range of Ca outputs. The dolomitic bedrock at Walker Branch and the andesitic bedrock at Andrews contribute to large stream losses of Ca. Gains and losses of K, a relatively mobile cation, are small. Net budgets reflect the combined effects of weathering rates and biological conservation processes retaining K within ecosystems. Smaller losses of this ion from Andrews and Hubbard Brook may reflect lower rates of K release from bedrock.

3.2.2 Pool sizes and recycling processes

Baseline budgets are useful in evaluating changes in the timing and magnitude of solute losses that accompany forest disturbances or manipulations and in generating hypotheses on processes responsible for such changes. Some examples will be discussed later in this chapter, but first it is appropriate to examine biological processes of nutrient recycling and conservation within forest ecosystems. Analysis of nutrient storage compartments and major pathways of nutrient transfers between compartments for the same four ecosystems included in the budget analysis are summarized in Table 3.3.

Table 3.3. Summary and comparison of compartment sizes and transfer rates for the cycles of nitrogen, calcium and potassium in four forest ecosystems in contrasting physiographic regions of the USA (Swank and Waide, 1980)[a]

	Nutrient and site[b]											
	Nitrogen				Calcium				Potassium			
	WB	HB	CHL	HJA	WB	HB	CHL	HJA	WB	HB	CHL	HJA
Compartment sizes (kg ha^{-1})												
Vegetation (above-ground and below-ground)	470	532	995	560	980	484	830	750	340	218	400	360
Forest floor	310	1,256	140	740	430	372	130	570	20	66	20	90
Mineral soil												
Exchangeable	75	26	117	5	710	510	940	4,450	170	—	510	860
Total[c]	4,700	4,890	6,800	4,500	3,800	9,600	2,500	—	38,000	—	124,000	—
Percentage of system total in vegetation	8.4	8.0	12.4	9.6	16.6	4.4	18.9	—	0.9	—	0.3	—
Transfer rates (kg ha^{-1} yr^{-1})												
Litter fall	39	54	33	21	55	41	44	41	19	18	18	9
Canopy leaching	3	3	4	4	14	4	8	8	19	29	31	15
Woody increment	15	9	13	−2	31	8	23	−4	8	6	13	−1
Uptake[d]	57	66	50	23	100	53	75	45	46	53	62	23

[a]Data from Bormann *et al*. (1977); Likens *et al*. (1977); and Henderson *et al*. (1978).
[b]Site codes as follows: WB, Walker Branch, Tennessee; HB, Hubbard Brook, New Hampshire; CHL, Coweeta Hydrologic Laboratory, North Carolina; HJA, H. J. Andrews, Oregon.
[c]Values to 94.5 cm depth at Hubbard Brook, to 60 cm depth at other sites.
[d]Hubbard Brook values exclude uptake that is recycled in root litter/root exudates.

The major portion of N is found in the soil organic matter fraction for all ecosystems. Comparable values are found at all four sites with a slightly higher compartment size at Coweeta. Vegetation is the second largest N storage pool in the two oak–hickory forests and, again, Coweeta values are highest, partly owing to higher concentrations of N in the roots (McGinty, 1976). The vegetation contains between 8% and 12% of the total system N at all four sites. The forest floor is the second largest N storage compartment at Andrews and Hubbard Brook, and values are two to ten times greater than found in the oak–hickory forests. Slower decomposition rates due to dry summers at Andrews and cold, snowy winters at Hubbard Brook contribute to the large accumulation of N in the forest floor at these sites. The importance of these N storage differences in dissolved N losses will be demonstrated later.

The soil is also the primary storage compartment for Ca, either in bound mineral form for the three deciduous forests or in exchangeable form for the conifer stand. The large Ca pool at Hubbard Brook is probably related to the young, glaciated soils. At Andrews, the high exchangeable Ca value is associated with high cation exchange capacity of soils there (Henderson *et al.*, 1978). Calcium storage in vegetation is similar at three of the sites and is lowest at Hubbard Brook; vegetation pools represent 4–19% of the total system Ca. Litter pools of Ca are again lowest at Coweeta where decomposition is rapid and Ca immobilization in litter is low.

For the K cycle, the dominant storage pool is in a bound form in the soil. Exchangeable K pools vary widely across sites but are generally high in relation to soil cation exchange capacity. The amount of K incorporated into vegetation is comparable across sites and represents less than 1% of the system total; storage in the forest floor is quite low.

Some of the major pathways and transfer rates of nutrients for all four ecosystems are summarized in Table 3.3. Litter fall is the major above-ground transfer pathway in the N and Ca cycles of all ecosystems. Nitrogen values for this transfer are highest at Hubbard Brook, intermediate in the two oak–hickory forests, and lowest in the old-growth Douglas-fir at Andrews. Calcium values of litter transfer are similar across all sites. In contrast to cycles of N and Ca, transfers of K by canopy leaching are large and either equal or greatly exceed K transfer via litter fall. Nitrogen transfer by canopy leaching is small and quite similar across sites. Woody increments and above-ground vegetation uptake values are fairly similar for the three hardwood forests, and are substantially larger than for the old-growth conifer forest at Andrews were mortality exceeds growth.

3.2.3 Root mortality and resorption

Root mortality and nutrient resorption are two potentially important mechanisms of nutrient recycling not considered in Table 3.3. Annual root N turn-

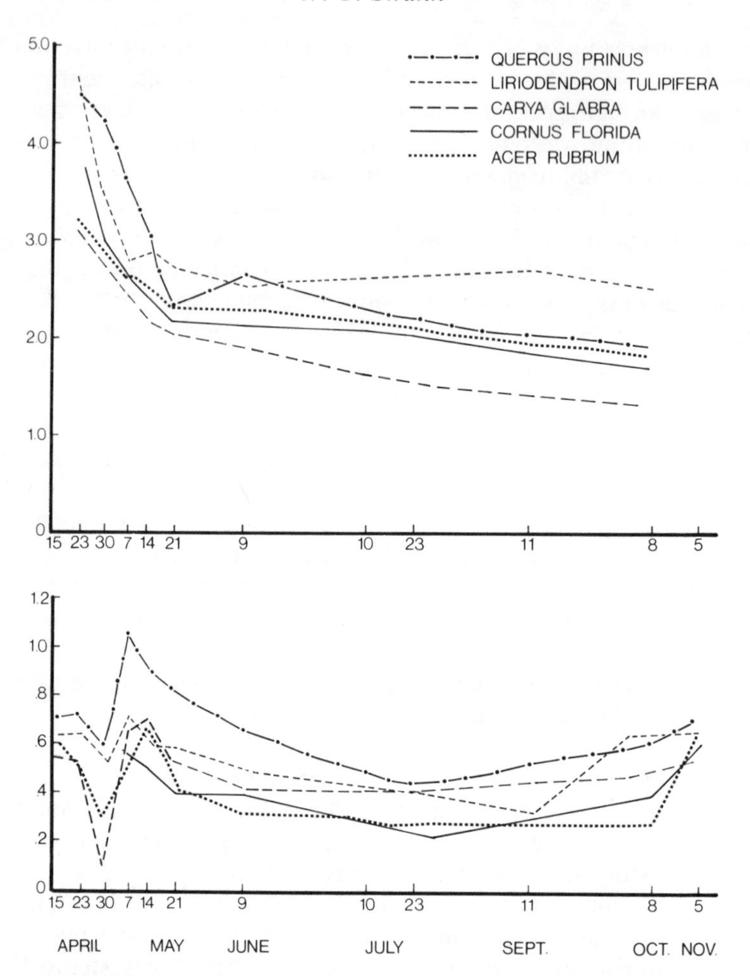

Figure 3.4. Nitrogen concentrations (per cent) in leaves (top) and twigs (bottom) during the growing season for five important hardwood species on Coweeta WS 18 (Day and Monk, 1977). *Reproduced with permission from the* American Journal of Botany

over at Walker Branch was estimated to be 67.5 kg ha^{-1} (Henderson and Harris, 1975), which is about 50% greater than litter fall rates. Root mortality estimates at Coweeta and Andrews (McGinty, 1976; Sollins *et al*., 1980) also suggest that N transfers through root turnover equal or exceed litter fall N transfers. Data on root dynamics are limited and the quality of nutrients going to soil solution or soil organic matter is unknown. Much additional research is needed to more fully quantify and characterize root dynamics as a nutrient recycling process.

The withdrawal of nutrients from senescent leaves to woody tissue (nutrient resorption) is another process that regulates and conserves solute losses.

Resorption occurs in a variety of plant forms, including woody species, and is most important in N, P and K recycling (Chapin, 1980). In the mixed hardwood forest at Hubbard Brook, Ryan and Bormann (1982) estimated resorption of 36, 3.4 and 3 kg ha^{-1} yr^{-1} for N, P and K, respectively, with no resorption of Ca. These investigators calculated similar resorption rates for a vigorous 5-year-old stand dominated by pin cherry (*Prunus pensylvanica* L.). Annual N resorptions in the hardwood forests at Walker Branch and Coweeta were estimated to be about 35 and 51 kg ha^{-1}, respectively (Henderson and Harris, 1975; Mitchell *et al.*, 1975). Significant resorption of K but not Ca was also found at Coweeta. Most of these nutrient resorption fluxes were derived from mass balance calculations. Other studies provide more direct evidence for the resorption process through simultaneous measurement of leaf and twig nutrient concentrations. Day and Monk (1977) followed concentrations of nutrients in leaves and twigs for several hardwood species throughout the growing and senescent periods; the seasonal patterns for N in five species are depicted in Figure 3.4. Nitrogen concentrations in leaves showed substantial reductions in the spring between late April and May. Thereafter, concentrations steadily declined throughout the growing season in most species. Concentrations in twigs also showed a sharp decline during the initial growth period, followed by a rapid rise in very early May with subsequent declining concentrations until early September, when most species exhibited a large increase in concentration (Figure 3.4). When combined with leaf biomass, the absolute amount of leaf N increased in the spring, and Day (1974) attributed concentration decreases to nutrient dilution by rapid leaf growth. Conversely, in the autumn, the absolute amount of N decreased and this was attributed to N withdrawal from leaves to twigs. Similar seasonal patterns were observed in leaf P and K concentrations. Taken collectively, it is apparent that resorption, storage and reuse of some nutrients by forest vegetation provides a buffering mechanism of solute loss.

3.2.4 Disruption of biotic regulation

The foregoing summary of ecosystem budgets and recycling pathways and processes illustrates that much greater quantities of nutrients are stored and recycled within forests than are lost annually in streamwater. The fundamental characteristics of nutrient storage and recycling have been documented for a variety of forest ecosystems (Cole and Rapp, 1980). However, few studies have examined the consequences of biotic deregulation on solute losses from forested ecosystems, and even less experimental evidence is available on alterations of specific biological processes associated with such losses. Long-term ecological research at some of the sites charactertized in Tables 3.2 and 3.3 provide a database for evaluating the effects of ecosystem disturbance and recovery on biogeochemical cycles.

One of the more dramatic experiments was conducted in the northern hardwood forest at Hubbard Brook (Bormann and Likens, 1979). In 1965 all trees on a 15.6 ha watershed were felled in place, no logs were removed, and no roads were constructed on the area. Bromecil and 2,4,5-T were applied during three successive growing seasons to suppress vegetation regrowth and, beginning in the summer of 1969, the watershed was allowed to revegetate. The purpose of this deforestation experiment was to maximize ecosystem nutrient loss by blocking water and nutrient uptake and litter production. An adjacent hardwood forest served as a control to evaluate changes in biological functions and ecosystem solute losses.

Beginning in the first growing season following treatment, concentrations of most dissolved ions in streamwater were substantially higher than the forested watershed. Concentrations of most ions reached a peak during the second year and declined during the third year (Bormann and Likens, 1979). During the 3-year period of deforestation, annual net losses of NO_3–N, Ca and K exceeded the adjacent forested watershed net budgets by 116, 69 and 29 kg ha^{-1}, respectively. A portion of the elevated nutrient export was due to reduced evapotranspiration and, hence, increased streamflow. Annual streamflow during the 3-year deforestation period increased an average of about 28 cm or 31% above values expected if the watershed had not been treated (Pierce *et al.*, 1972). Cutting advanced snowmelt about a week, which increased high flows in the melt period. However, assuming a constant relationship between annual ion concentrations and discharge, it is apparent that increased flow is a minor component of increased solute export.

Bormann and Likens (1979) attributed the elevated solute exports to the following factors:

1. elimination of nutrient uptake and storage by forest vegetation;
2. accelerated decomposition rates of the forest floor early in the treatment period in response to favourable soil moisture and temperature conditions;
3. substantial increases in nitrification rates with accompanying hydrogen ion production and displacement of cations on exchange surfaces.

The release of nutrients from the cut woody biomass on the watershed was judged to be a relatively unimportant source of nutrient export in the 3-year period. Following cessation of herbicide application at the end of the third year, natural revegetation occurred although net primary productivity was initially lower than expected for a commercial clearcut operation. Concurrent with vegetation establishment, there was a sharp decline in dissolved ion concentrations. At the end of 7 years of recovery, hydrologic budgets for N again showed apparent net ecosystem gains; i.e. inputs exceeded outputs. Calcium and K still exhibited net losses compared with exports expected from a forested watershed, but at greatly reduced levels.

Examination of commercial cutting (removal of merchantable saw logs followed by natural regeneration) in other northern hardwood forests in the region also showed increases in dissolved ions of streams draining the cut areas (Pierce *et al*., 1972). Significant increases in Ca and N were observed in the first year after cutting, and even larger increases in export occurred in the second year. Solute losses were not as large as those during the devegetation experiment at Hubbard Brook, partly because of more rapid regrowth.

After commercial clearcutting of mixed hardwood forests at Coweeta, much smaller increases in solute losses were observed compared with Hubbard Brook. The most recent experiment began in 1975 on Watershed (WS) 7, a 59 ha south-facing watershed. Elevations on the watershed range from 720 to 1065 m, slopes average 50%, and the area is drained by a second-order stream. The soil is primarily a Humic Hapludult and the forest cover an oak–hickory (*Quercus-Carya*) community with many associated species. Management of the watershed was separated into three major operations:

1. road construction and stabilization;
2. tree felling and logging;
3. site preparation.

Three logging roads totaling 2.95 km were constructed on the watershed between mid-April and mid-June 1976. Immediately after construction, road cuts and fills were stabilized by seeding grass and applying commercial fertilizer (10-10-10) and lime. Seed, fertilizer and lime were again applied to cuts and fills in July 1977, and to the running surface of the road in June 1978. Logging began in January 1977 and was completed in June. The majority of the logging was conducted from the roads with a mobile cable system and the forest floor generally remained intact. The site was prepared by felling the stems remaining after logging and this operation was completed in October 1977. Prior to treatment, stream chemistry calibration was established with an adjacent control watershed (WS 2). Water samples were collected beginning in 1975 with a flow proportional sampler and weekly grab samples were also taken from each stream. Discharge on WS 2 and WS 7 have been measured continuously since 1935 and provide a firm basis for predicting changes in flow volume due to treatment.

The effects of treatment on concentrations of several selected solutes in stream water of WS 7 are shown in Figure 3.5. During this calibration period, K concentrations on WS 7 were usually below WS 2 (control) concentrations, and this pattern continued through the summer of 1976 following road construction and fertilization. Potassium concentrations on WS 7 increased slightly above control values in late autumn and then remained at higher, elevated levels in 1977 following cutting and logging. During the next four post-treatment years, K concentrations generally remained above the control

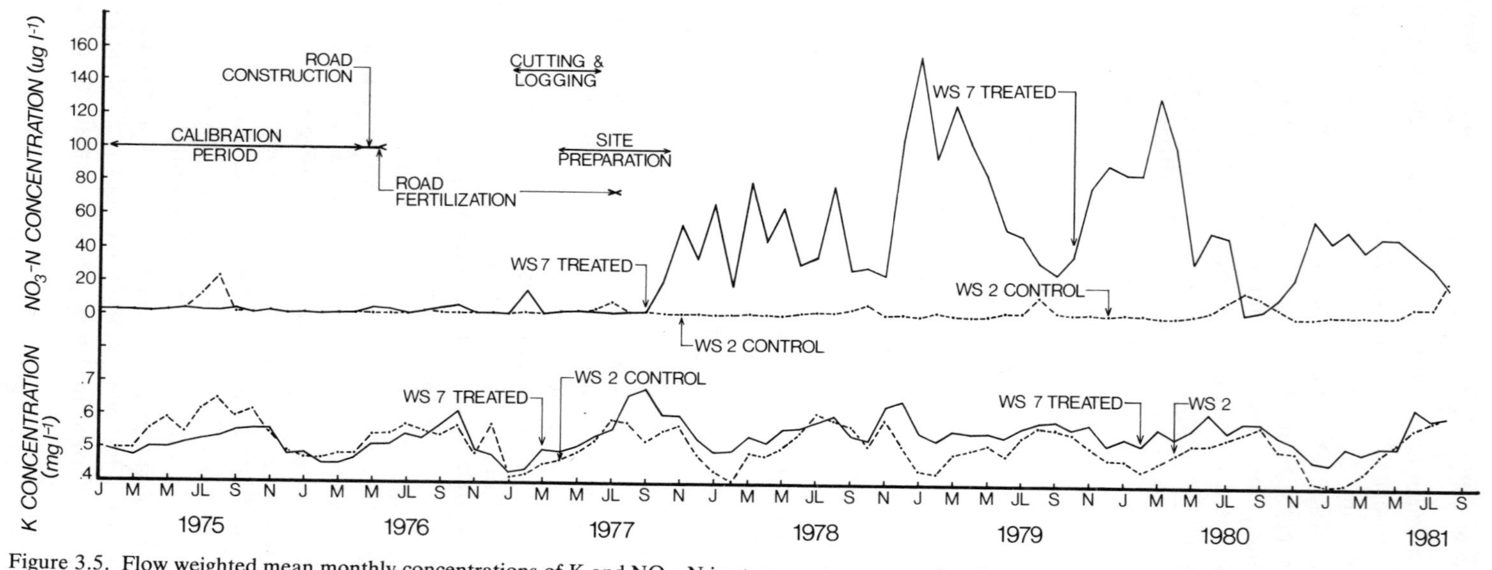

Figure 3.5. Flow weighted mean monthly concentrations of K and NO_3–N in streamwater measured at the weir of Coweeta WS 7 during calibration, road construction, clearcutting and logging, site preparation, and post-harvest recovery periods.

stream, with largest differences (30–100 μg per litre) occurring between November and June. Baseline concentrations on NO_3–N in streams draining undisturbed watersheds at Coweeta are typically near analytical detection limits (2 μg per litre) (Figure 3.5). Increases in NO_3–N on WS 7 were observed about 9 months after initiation of cutting and logging. In the second year after treatment, concentration increases of 100–150 μg per litre were observed between December and April. Thereafter, increased NO_3–N concentrations declined in the summer and autumn months. This annual pattern of concentrations was repeated in 1980 and 1981, but the magnitude of change decreased in succeeding years and values, while elevated, appear to be returning towards pretreatment levels.

Changes in streamwater solute concentrations are indicative of disruption of nutrient recycling processes, but losses must be quantified to evaluate the magnitude of management impacts. Solute concentration data were combined with flow volumes to calculate total flux. Annual changes in NO_3–N, K and Ca were estimated from pretreatment calibration regressions of monthly fluxes between WS 7 and its control, WS 2. Correlations between the two watersheds were very good, with r^2 values of at least 0.97 for Ca and K. Export of NO_3–N was very low on both watersheds, and mean monthly values during the calibration period were nearly identical for the two catchments. Annual increases in streamflow and solute export during the first 5 years after logging are shown in Table 3.4. Increased export was greatest in the second, third and fourth years with annual values for NO_3–N, K and Ca of about 0.9, 2.1 and 2.8 kg ha^{-1}, respectively.

By the fifth year, increases in solute export were substantially diminished and appeared to be approaching prelogging levels. A major factor in increased solute export is the increased flow response that results from reduced evapotranspiration following cutting (Swank *et al.*, 1982). The first

Table 3.4. Annual increases in streamflow and solutes following clearcutting and logging on Coweeta WS 7

	Increase in streamflow and solute export			
Year after treatment (May–April water year)	Flow (cm)	NO_3–N (kg ha^{-1})	K (kg ha^{-1})	Ca (kg ha^{-1})
1[a]	2.8	0.03	0.84	1.85
2	26.5	0.26	1.98	2.60
3	20.5	1.12	1.97	2.53
4	17.3	1.27	2.41	3.17
5	11.9	0.25	0.88	1.66

[a]The first year encompasses only 4 months of the cutting experiment; the second year represents the first full 12-month period of treatment.

full year after cutting, flow increased 26.5 cm and averaged 21.4 cm per year in years 2, 3 and 4 (Table 3.4). Concurrent with the decline in solute export during the fifth year, streamflow increases also declined. Maximum K and Ca increases in export were about 35% greater than values expected had the watershed not been treated, while flow was 20% greater than expected values for the same time period (years 2–4).

These relatively small changes and trends in solute losses clearly illustrate the concepts of ecosystem resistance (the extent of system displacement from a predisturbance functional level) and resilience (the rate at which the ecosystem recovers to predisturbance function). Previous theoretical analyses suggested that mixed hardwood forests at Coweeta exhibit both high resistance and high resilience as related to changes in the N cycle associated with forest harvesting activities (Swank and Waide, 1980). Ecosystem resistance is related to the presence of large storage pools or organic matter and elements which turn over slowly (Webster *et al.*, 1975; O'Neill and Reichle, 1980) as illustrated by Coweeta data in Table 3.3 and relatively small increases in solute losses shown in Table 3.4. Rapid recovery or return of solute losses to baseline levels shown in Table 3.4 are indicative of high ecosystem resilience. Reasons for rapid recovery of biogeochemical cycles in the ecosystems are related to high rates of net primary production (NPP) and incorporation and storage of nutrients in successional vegetation (Boring *et al.*, 1981). By the third year after cutting on WS 7, above-ground NPP was over 80% the NPP of the original mature hardwood forest (Figure 3.6). Early successional species exhibit high nutrient concentrations and in the first year after cutting, nutrient pools in NPP for N, P, K, Mg and Ca were already 29–44% of the NPP of the mature forest (Boring *et al.*, 1981).

Further evidence for control of solute losses by nutrient recycling processes is demonstrated by another long-term watershed experiment at Coweeta. Watershed 6, an 8.8 ha north-west-facing catchment, was converted from hardwoods to Kentucky 31 fescue grass (*Festuca elatior* var. *arundinacea* Schreb.) and maintained in a grass cover from 1960 to 1965 (Hibbert, 1969). Beginning in May 1966 and continuing for a 2-year period, the grass cover on the catchment was killed with herbicides. No further treatment was applied since the spring of 1968, and the catchment has been allowed to revert to successional vegetation (Johnson and Swank, 1973). Weekly stream chemistry determinations were initiated on WS 6 and on WS 18, a nearby control stream, in 1969 and have continued up to the present time. This period of record spans vegetation succession from initial domination by herbaceous species (1969–72) to present day domination by woody species.

The mean monthly weighted (for flow volumes) concentrations of K, Ca and Mg for the periods 1969–72 and 1979–82 for the control catchment (WS 18) are shown in Figure 3.7, and similar plottings for successional WS 6 are depicted in Figure 3.8. The monthly average concentrations for all three

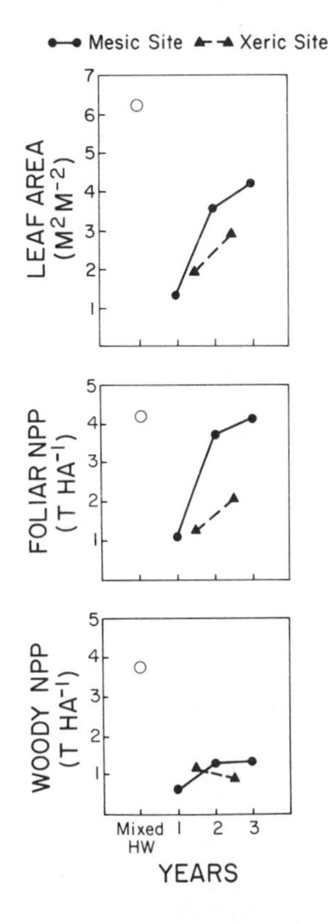

Figure 3.6 Estimates of above-ground net primary production for a mature hardwood forest at Coweeta Hydrologic Laboratory and for regrowth on the same area in the first 3 years after clearcutting

solutes in the stream draining the undisturbed hardwood forest showed a distinct seasonal trend which was repeated between years within a time period and also between time periods. Maximum concentrations occurred during August to October followed by a sharp decline in November with minimum values occurring in the winter months. Thereafter, concentrations increased during the early spring and into the summer months, until a maximum was again reached in the autumn. Similar patterns of solute concentrations have been observed for disturbed Coweeta watersheds with woody regrowth of either white pine (*Pinus strobus* L.) or hardwood coppice (Johnson and Swank, 1973).

In contrast, patterns of solute concentrations were much different in the stream draining the successional watershed during the period 1969–72 (Figure 3.8). That is, minimum concentrations frequently occurred in August or September with a rapid increase in the autumn months, maximum values in early winter, and declining concentrations in the spring. As with other watersheds at Coweeta, K, Ca and Mg usually showed a distinct seasonal trend on the successional watershed which can be described by a sine-wave

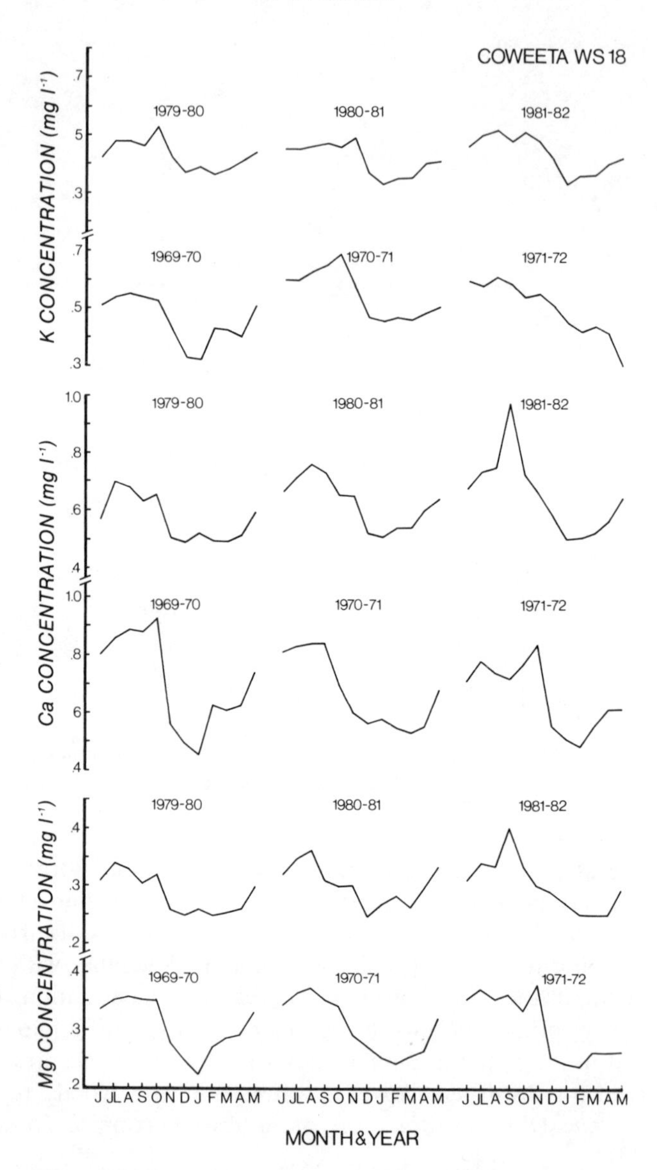

Figure 3.7. Flow weighted monthly concentrations of selected solutes in streamwater of Coweeta WS 18, a control catchment with a mixed hardwood cover, during two separate 3-year periods

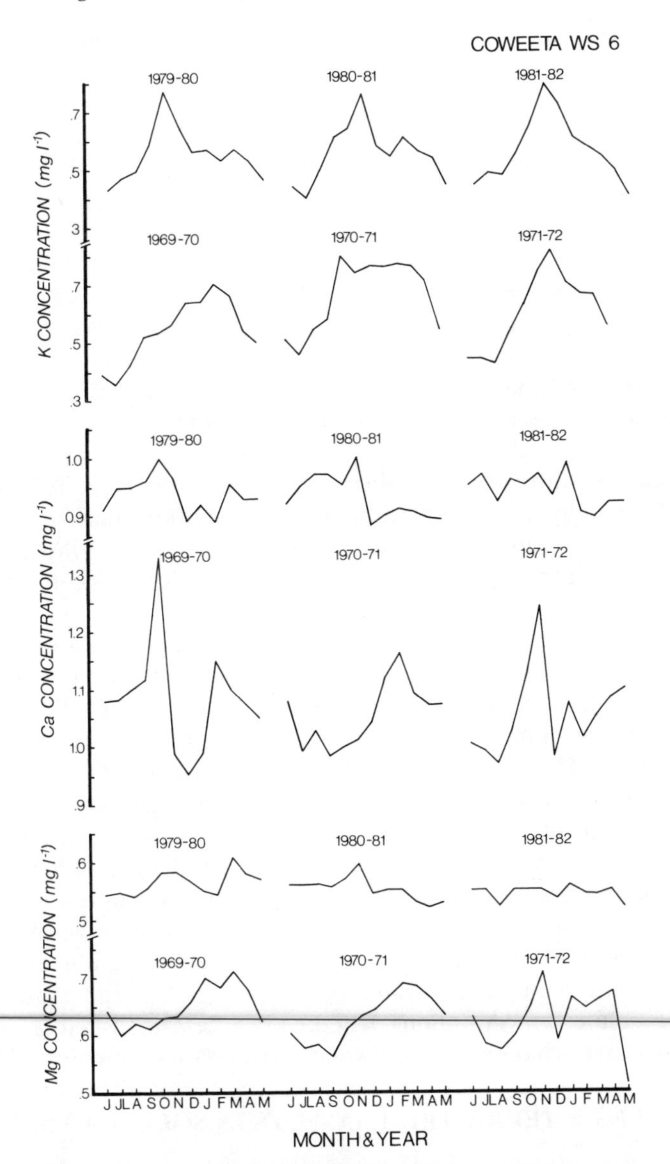

Figure 3.8. Flow weighted mean monthly concentrations of selected solutes in streamwater of disturbed Coweeta WS 6 during two separate 3-year periods. During 1969–72 the vegetation was dominated by herbaceous species, but in the years 1979–82 the vegetation was comprised of woody species

function, but they are of phase; i.e. the timing of minimum and maximum concentrations was shifted. However, 10 years later in succession (the 1979–82 period) the pattern of solute concentrations has changed and may be returning towards trends observed for streams draining forested watersheds. For example, Ca, K and Mg concentrations showed decline during the winter months, minimum concentrations frequently occurred during May or June, and increasing solute concentrations were observed earlier in the autumn months. Comparison of solute values for the early and later successional periods also indicates that the amplitudes of some solute concentrations were damped and overall concentration levels reduced during succession. Monthly concentrations of Ca ranged from about 0.95 to 1.25 mg per litre in 1969–72 and 0.90 to 1.00 mg per litre in 1979–82; Mg ranges for the two periods were 0.55–0.70 and 0.55–0.60 mg per litre, respectively, while K showed little change.

In comparison with control Coweeta watersheds, the treatment on WS 6 did not substantially alter the monthly flow distributions during the periods shown in Figures 3.7 and 3.8, and annual flows were close to those expected for a mature hardwood forest. Thus, changes in solute concentrations are thought to demonstrate the importance of biotic processes in regulating solute losses. In the early years of succession (1969–72), vegetation was dominated by a dense cover of herbaceous species such as horseweed (*Erigeron canadensis* L.) and cottonwood (*Erechtites hieracifolia* (L.) Raf.), but by 1980 the vegetation was dominated by woody species with a density of more than 1400 stems ha^{-1}. The magnitude and timing of nutrient uptake, storage and decomposition are vastly different between herbaceous and woody plants and are postulated to influence both the amount and pattern of solute losses. Increased nitrate export associated with an insect defoliation on WS 6 in 1979 (see next section on terrestrial insects) cannot be discounted as a partial explanation for a shift in solute concentrations. The quantitative contribution of various recycling and other biological processes to solute dynamics is unknown and difficult to ascertain, but it is hypothesized that seasonal patterns of solute concentrations will become re-established as nutrient recycling processes characteristic of woody vegetation continue to develop.

3.3 ROLE OF TERRESTRIAL INSECTS IN SOLUTE DYNAMICS

Biotic regulation of solutes by insect populations is important both in the canopy and forest floor compartments of forest ecosystems. Mattson and Addy (1975) postulated that canopy arthropods act as regulators of forest primary production and nutrient cycling. Although direct evidence was not available, Schowalter (1981) suggested that insect herbivore activity regulates solute transfer rates between vegetation, litter and soil by:

1. stimulating net primary production and hence nutrient uptake by moderately grazed plants;

2. accelerating translocation of nutrients from plant reserves to sites of insect grazing;
3. increased leaching of foliar nutrients from partially consumed leaves;
4. enhancement of litter decomposition, nitrification and nitrogen fixation via nutrient enriched throughfall, litter fall and insect faeces.

Forest floor-soil invertebrates influence solute movement both directly and indirectly by controlling plant and animal litter decomposition concurrent with nutrient release. Direct controls include digestion and assimilation of the litter, although these processes may be less important than indirect controls. For example, using radioactive tracers and field studies, Gist and Crossley (1975) constructed a 10-compartment model of calcium and potassium cycling for an invertebrate food web in a mixed hardwood litter at Coweeta. Their results showed that a significant portion of litter biomass is processed through saprovores, but only 1% of the potassium and 12% of the calcium litter pools are processed by saprovores. More important indirect controls include litter surface area fragmentation through feeding activities which increases leaching and nutrient loss from litter (Crossley, 1970), inoculation of litter with microflora as it passes through the animal gut, dissemination of fungal spores, and physical mixing of soil and litter (Gist and Crossley, 1975).

Quantitative evidence that conclusively illustrates the influence of insect control on solute losses is scarce. In one study at Coweeta, Seastedt *et al.* (1983) examined the effects of low-level consumption by canopy arthropods on foliage nutrient content, throughfall chemistry (canopy leachates), and biomass of 4-year-old black locust (*Robinia pseudoacacia*) and red maple (*Acer rubrum*). A carbaryl insecticide was used to reduce foliage consumption from about 10% to 2% in black locust and from 4% to 1% in red maple; the nutrients examined were N, P, K, Ca, S and Mg. Potassium was the only solute strongly affected by herbivory; the amount of K in throughfall (May–August) for insecticide-protected trees increased from 5.7 kg ha^{-1} to 9.8 kg ha^{-1} for unprotected black locust, and elevated fluxes were also observed for red maple. A small but measurable increase in sulphate–S loss from untreated black locust was noted, but the rates of N and P cycling were unaffected by herbivory. Furthermore, herbivory did not increase foliage and wood production for either tree species during the year of treatment. Seastedt *et al.* (1983) concluded their results suggest that nominal (low-level) herbivory has little role in the biomass and nutrient accretion of successional forests, but the major impact is to increase the rate of cycling for some solutes. The results may not apply to mature forest stands since the research was conducted on young coppice regrowth vegetation which contained large pools of nutrients in the root systems.

Other experiments have been conducted at Coweeta to examine the effects of microarthropods on nutrient dynamics in forest litter. Seastedt and Crossley (1980, 1983) measured the amounts of Ca, K, Mg and P in mixed deciduous leaf litter contained in litter bags. A chemical, naphthalene, was added to half the litter bags which reduced microarthropod densities to about 10% pf the populations found in untreated litter. Losses of P were significantly greater in untreated litter; i.e. microarthropods accelerated P release. After initial losses, the amounts of Ca, K and Mg in untreated litter generally increased over the 12-month period and naphthalene-treated litter showed no seasonal dynamics. Although microarthropods may increase the loss of some nutrients from forest litter by comminution, retention of other nutrients may be increased through microbial stimulation resulting from microarthropod feeding activities (Seastedt and Crossley, 1980). Other studies in forest litter at Coweeta (Seastedt and Tate, 1981) have shown that arthropod remains (exoskeleton) may comprise a significant portion of the total pool of elements such as Ca found within the forest floor.

Results of studies at Coweeta represent one of the clearest demonstrations of functional, ecosystem-level consequences of the feeding activities of forest defoliators (Swank *et al*., 1981). An outbreak population of the fall cankerworm (*Alsophila pometaria*), a spring defoliator of hardwood forests, was first observed adjacent to and on the Coweeta Basin in 1969. Watershed 27, a 38.8 ha control catchment, was the primary site of infestation. Defoliation progressed from the higher elevations (1400 m) on the catchment in 1970 towards lower elevations in ensuing years. Mean monthly concentrations of NO_3–N in streams draining undisturbed hardwood forests at Coweeta are generally below 10 μg per litre; however, levels on WS 27 frequently rose above 30 μg per litre (Figure 3.9). Highest concentrations usually occurred during winter, but large NO_3–N increases were also observed during and immediately following the time period of cankerworm feeding (late April to early June). During 1974 at the peak of infestation, about 33% of the total leaf mass was consumed and NO_3–N concentrations were elevated throughout the year. Subsequently, levels of defoliation were less severe, and in 1978 the population returned to endemic or non-outbreak levels. The decline in the cankerworm population was accompanied by a return of NO_3–N concentrations toward baseline levels (Figure 3.9).

Increased stream export of NO_3–N concomitant with defoliation was also observed on WS 36 (Figure 3.9), another high-elevation catchment at Coweeta. However, cankerworm infestation was not present on the catchment where stream chemistry analyses were initiated; thus, during 1972 and 1973, mean monthly NO_3–N concentrations were representative of other undisturbed forest ecosystems at Coweeta (generally below 10 μg per litre). In late spring of 1974 concentrations began to rise, and egg mass surveys on trees in 1975 confirmed that infestation had occurred on the catchment although defoliation was not as severe as on WS 27. Again, by 1979, concentrations had returned to baseline levels (Figure 3.9). Elevated NO_3–N

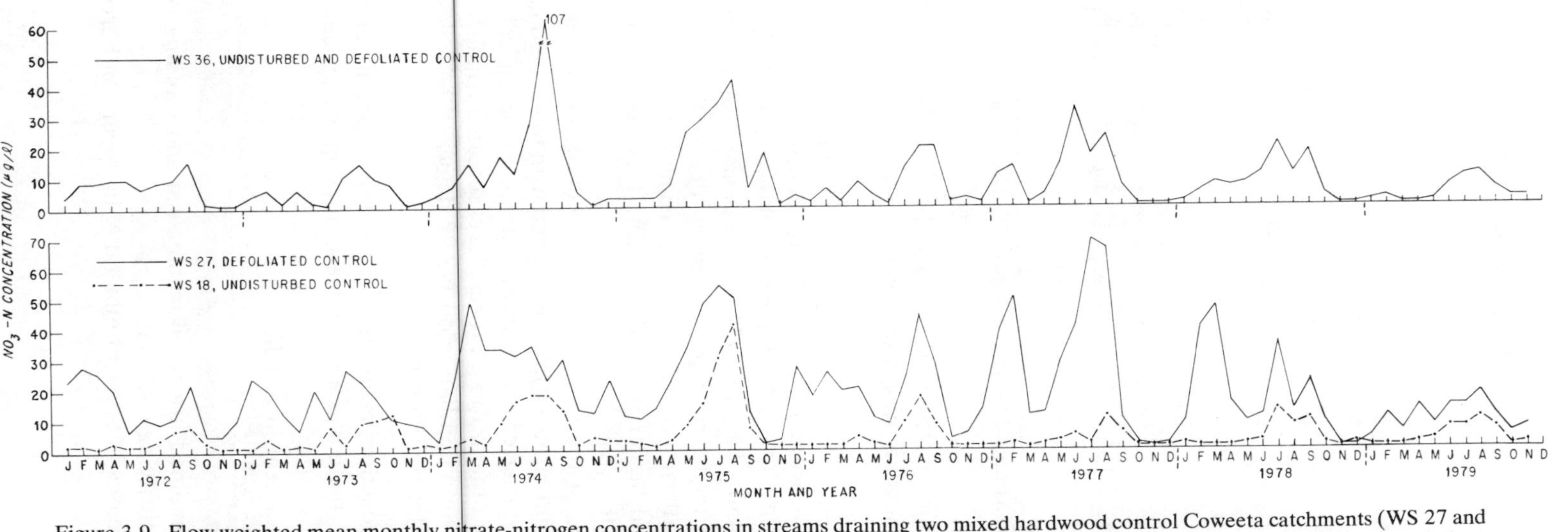

Figure 3.9. Flow weighted mean monthly nitrate-nitrogen concentrations in streams draining two mixed hardwood control Coweeta catchments (WS 27 and WS 36) with outbreak levels of forest defoliators and in one stream (WS 18) draining a control hardwood catchment with endemic defoliator populations (Swank *et al.*, 1981)

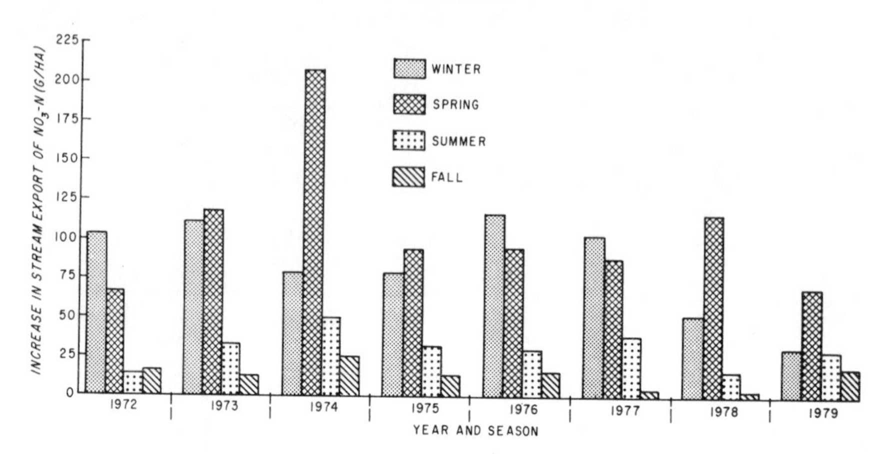

Figure 3.10. Estimated increase in seasonal export of nitrate-nitrogen from the partially defoliated hardwood forest on Coweeta WS 27 over an 8-year period

concentrations for a 'control' stream in association with extensive cankerworm defoliation was also detected for another catchment about 13 km from the Coweeta Basin (Swank *et al*., 1981).

The effects of defoliation on total seasonal export of NO_3–N from the hardwood forest ecosystem on WS 27 are shown in Figure 3.10. Annual increased export of NO_3–N for the period shown ranged from about 200 g ha^{-1} in 1978 to 450 g ha^{-1} in 1974. The seasonal analyses show that more than 80% of the increased export occurred during the winter and spring months. This pattern was due to the combined factors of elevated concentrations and higher streamflow during these seasons compared with the remainder of the year.

The significance of the increased loss of NO_3–N lies in the alteration of nutrient transfer and turnover rates associated with forest defoliation, and the fact that insect grazing can measurably alter biogeochemical cycles at the ecosystem level of organization. Process level research conducted during the period of cankerworm defoliation on WS 27 detected the following changes in the partially defoliated forest:

1. stemwood production decreased but leaf production increased; when taken together, there was an increase in total above-ground net primary production;
2. large increases in leaf litter fall;
3. substantial increases of frass inputs and associated elements to the forest floor;
4. increases in combined litter–soil metabolism and standing crops of total microbes and nitrifying bacteria;
5. significant increases in pools of mineral N in upper soil horizons;

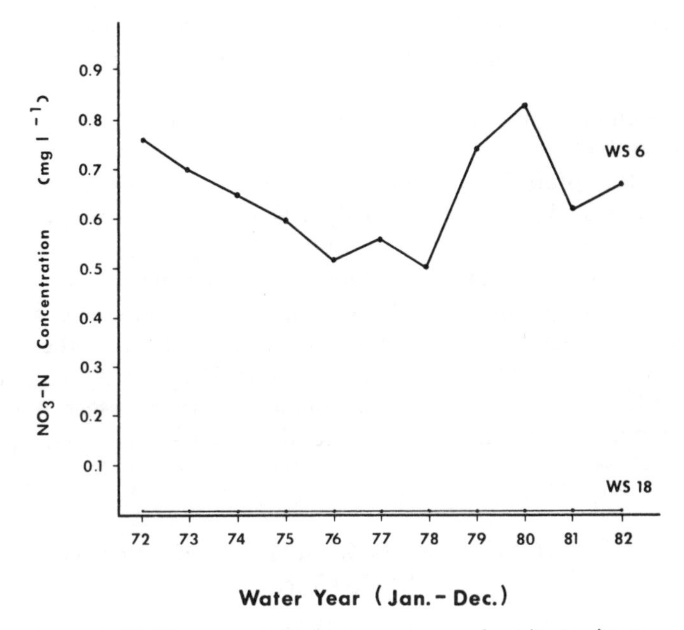

Figure 3.11. Flow weighted mean annual nitrate-nitrogen concentrations in streamwater of Coweeta control WS 18 and disturbed WS 6 over an 11-year period

6. seasonal increases in evapotranspiration, and hence reductions in streamflow owing to increased leaf area index (Swank *et al.*, 1981).

The results suggest a temporary shift from wood to leaf production, increased rates of litter decomposition, nutrient uptake and turnover of nitrogen in litter–soil horizons.

Further evidence for insect regulation of solute losses at the ecosystem level is found on another catchment at Coweeta. Characteristics and the treatment history for WS 6 were given earlier in 3.2.4. During the period of grass and herbaceous-to-forest succession, mean annual NO_3–N concentrations in streamwater gradually declined from about 0.75 mg per litre in 1972 to 0.50 mg per litre in 1978 (Figure 3.11). Then in 1979, NO_3–N concentrations showed an abrupt increase to 0.75 mg per litre concurrent with a heavy infestation of the locust stem borer (*Megacyllene robiniae*). Since black locust was the dominant woody species, the infestation was distributed over the entire catchment. In the subsequent year (1980) following the major infestation by the locust borer, NO_3–N concentrations increased to 0.90 mg per litre and remained above preinfestation levels in the ensuing two years. By 1982, 21% of the black locust trees were dead, 18% were severely injured, and many of the remaining stems showed some evidence of canopy decline

(L. R. Boring, unpublished data). A number of hypotheses are currently under investigation to explain why insect stress produced increased NO_3 loss from the ecosystem; these studies are focused on the effects of insect populations on nutrient uptake, quantity and quality of litter fall, nitrification, and nitrogen fixation.

Both of the preceding experiments represent some of the clearest demonstrations available of functional, ecosystem-level regulation of solute losses by terrestrial insects under outbreak conditions. The catchment responses represent the integrated functional properties of an entire forest ecosystem and changes in biogeochemical processes within the ecosystems. In both cases, changes in mean annual concentrations of other dissolved constituents such as SO_4, K, Ca and other cations are difficult to detect because differences in bedrock mineralogy between insect-infested and control catchments influence ions sufficiently to mask small changes in ionic composition.

3.4 BIOLOGICAL TRANSFORMATIONS OF SULPHUR AND NITROGEN

The majority of the N and S pools reside as organic forms in the forest floor and soil. These organic complexes are subjected to microbial populations which have a dominant influence on anion supply and mobility. Thus, microbial transformations which regulate NO_3^- and SO_4^{2-} dynamics also influence the movement of other solutes as ionic balances of solutions are maintained. Although microbial processes can be regarded to be important in regulating solute movement, quantification of their role at an ecosystem level is not well established, particularly for S transformations. More study has been directed towards the N cycle of forests, but quantification of some processes such as nitrification and denitrification are generally lacking.

3.4.1 Sulphur transformations

Research on sulphur transformations was initiated in 1980 at the Coweeta Hydrologic Laboratory in an effort to assess the importance of the plant sulpholipid as a source of sulphate in forest soils. Results showed not only rapid S-mineralization rates for sulpholipid, but also rapid incorporation of a substantial portion of the released sulphate into organic matter which could only be recovered by acid extraction (Strickland and Fitzgerald, 1983). Subsequent research has attempted to clarify the importance of the sulphate incorporation process and the mineralization of organic forms of S found in the forest floor and soil. Since sulphate is a major constituent of acid precipitation, it is important to understand the fate of exogenous sulphate in forest ecosystems and the impact on movement of other solutes. For example,

Table 3.5. Mean annual (1973–83) sulphate-sulphur budgets for four mixed hardwood watersheds at Coweeta Hydrologic Laboratory

Watershed number	Area (ha)	SO_4–S (kg ha^{-1} yr^{-1})		
		Input	Output	Net difference
2	12.1	9.7	1.4	+8.3
18	12.5	10.6	1.6	+9.0
27	38.8	13.0	7.0	+6.0
36	48.6	11.6	5.5	+6.1

sulphate comprises 68% of precipitation anions at Coweeta, and the average annual hydrologic budget of sulphate for hardwood covered watersheds during the past 11 years shows apparent accumulations of sulphate ranging from 6.0 to 9.0 kg ha^{-1} yr^{-1} of elemental S (Table 3.5). Research at Coweeta has shown that sulphate adsorption to Fe and Al oxides in the soil profile accounts for part of the ecosystem S accumulation (Johnson *et al*., 1980), but other studies show that microbial metabolism of sulphate to organic S forms in soil is also a major process of S immobilization (Fitzgerald *et al*., 1983; Swank *et al*., 1984). The summary for S dynamics presented in the following sections is for Coweeta WS 18, a 12.5 ha mixed hardwood forest. Analyses are for samples of forest floor and soil taken along a transect established at mid-elevation on the catchment which transverses the area from ridge to stream to ridge.

Sulphur forms and *in situ* concentrations for O_1, O_2, and A_1 horizons are characterized in Table 3.6 based on methods and analyses described by Fitzgerald *et al*. (Coweeta files). Carbon-bonded S comprises 87% and 72% of the total S in O_1 and O_2 forest-floor layers, with the remainder in the form of

Table 3.6. Mean concentrations (μg S g^{-1} dry weight) of major forms of S in litter and soil of Coweeta WS 18[a]

Form	O_1 layer	O_2 layer	A_1 horizon
Ester sulphate	185	360	118
Carbon-bonded S[b]	1298	949	161
Amino acid S	270	452	69
Sulphonate S	1028	497	92
Soluble S	0	1.43	1.6
Adsorbed S	0	13.8	17.1
Total S	1482	1325	295

[a]Based on Fitzgerald *et al*., Coweeta files.
[b]Carbon-bonded S = amino acid S + sulphonate S.

ester sulphate. In the A_1 soil horizon, carbon-bonded S comprises only 54% of the total S, followed by 40% as ester sulphate, 6% adsorbed S, and less than 1% as soluble S. Sulphonate-S is a major linkage group present in the carbon-bonded S pool (Table 3.6), particularly in the O_1, while this form slightly exceeds the amino acid linkage in O_2 and A_1 horizons. Other studies at Coweeta have shown that A horizon soils exhibit high rates of sulpholipid and amino acid S mineralization, which suggests that carbon-bonded S is a major source of sulphate in forest soils (Strickland and Fitzgerald, 1983; Fitzgerald and Andrew, 1984). Soluble and adsorbed S comprise a minor portion (6%) of the total S found in the forest floor and soil, which could be expected since these fractions can be highly dynamic.

The metabolic fate of the sulphate anion has been examined using ^{35}S-labelled sulphate, which is incubated with forest-floor and soil samples at concentrations similar to *in situ* concentrations of solutions for these layers. Following incubation of samples for 48 hours at the temperature of interest, the samples are washed with water, extracted with salt and then extracted with a strong acid and a strong base. All extracts were analysed by electrophoresis and radioactive components determined with a scintillation counter. Details of procedures have previously been published and showed that sulphate is the major ^{35}S-labelled component (Fitzgerald *et al*., 1982, 1983).

Total recovery of ^{35}S from forest-floor and soil samples usually exceeded 90% of the added ^{35}S (Table 3.7). Small quantities of adsorbed sulphate (salt extract) were found in O_1 and O_2 layers, but more than 60% of the ^{35}S was adsorbed in the A_1 horizon. Conversely, substantially more sulphur incorporation into the acid and base fractions was observed in the O_1 and O_2 layers compared with the A_1, although this latter horizon typically shows 15–20% of the total ^{35}S incorporated in the acid+base fraction. Several lines of evidence suggest that the incorporation of sulphate into organic matter is microbially mediated and involves the formation of covalent sulphur linkage. For example, incorporation rates have been shown to be temperature- and time-dependent (Fitzgerald *et al*., 1983), sulphate-concentration-dependent (Swank *et al*., Coweeta files), and enhanced by increased energy availability (Strickland and Fitzgerald, 1984). Unequivocal proof for microbial incorporation of sulphate-S into organic matter rather than a physiochemical reaction was obtained by isolating the ^{35}S-labelled organic matter and subjecting it to extraction and dialysis specific for organic S linkage groups (Fitzgerald *et al*., 1985).

Incorporation of $^{35}SO_4$ into organic forms of sulphur for monthly collections of A_1 soil samples from WS 18 showed significant activity in all months (Swank *et al*., 1984). Seasonal differences were observed with highest activities in late summer and lowest rates of formation in winter and late spring months. During one sample period, activity was determined

Table 3.7. Recovery of ^{35}S in various fractions from the O_1 and O_2 forest-floor layers and A_1 soil horizon of WS 18[a]

	Recovery (%)		
Fraction	O_1	O_2	A_1
Soil water	38	49	15
Salt extract	6	15	65
Acid + base extracts	49	26	15
All fractions	93	90	95

[a]After Fitzgerald *et al.* (1983).

throughout the soil profile and for O_1 and O_2 layers (Swank *et al.*, 1984). Sulphate incorporation was about sixfold greater in O_1 and O_2 layers compared with the A_1 horizon (Table 3.8). Measurable activity was observed in all horizons but declined rapidly with soil depth. These activity data were used to place the transformation process in perspective with regard to annual ecosystem flux (Table 3.9) based on several assumptions (Swank *et al.*, 1984). The annual potential flux of sulphate into organic matter was more than 20-fold greater in the A_1 horizon than the flux in the forest floor, owing to differences in quantities of substrate. The B horizons are also major sites of incorporation and the total annual incorporation is estimated as 30 kg ha^{-1} yr^{-1} (Table 3.8). Although additional research is needed to provide more precise estimates of annual incorporation rates, it is apparent that inorganic to

Table 3.8 Annual rates of potential sulphate incorporation into organic sulphur forms by microbial populations in forest-floor and soil horizons of Coweeta WS 18[a]

Location	Incorporation rate (kg S ha^{-1} yr^{-1})
O_1	0.2
O_2	0.2
A_1	10.7
A/B	6.1
B_w	11.1
C_r	1.5
Total	29.8

[a]After Swank *et al.* (1984).

Table 3.9. Incorporation rates of sulphate into organic S forms for forest-floor and soil horizons on Coweeta WS 18 in late August 1982[a]

Forest-floor or soil horizon	Depth (cm)	Incorporation of $^{35}SO_4^{2-}$ (nmol g^{-1} per 48 hours)
O_1	4	11.68
O_2	2	8.73
A_1	0 to 10	1.85
A/B	11 to 25	0.61
B_w	26 to 65	0.44
C_r	66+	0.08

[a]After Swank *et al.* (1984).

organic S metabolism by fungi and bacteria is a major process regulating sulphate mobility and the S cycle. For example, the flux of S via incorporation exceeds the sulphate-S input in bulk precipitation (10 kg ha^{-1} yr^{-1}) at Coweeta and the quantity of S (2 kg ha^{-1} yr^{-1}) sequestered in net primary production of a hardwood forest (Johnson *et al.*, 1982).

Other research at Coweeta has shown that S incorporated into organic matter is mobilized to release sulphate and soluble organic S (Strickland *et al.*, 1984). Unlike the incorporation process, mobilization is not directly linked to microbial metabolism but mediated by depolymerases and sulphohydrolases, enzymes that are present extracellularly in the soil. Mobilization rates for samples of forest floor and soil horizons on Coweeta WS 18 showed a decrease in rates with an increase in soil depth (Table 3.10). In the O_1 layer, less than 20% of the incorporated sulphate was immobilized but increased to about 50% in the O_2 and A_1 layers. It appears that a significant fraction of the incorporated organic S is retained within soil horizons. However, these

Table 3.10. Rates of sulphur incorporation and mobilization in forest-floor and soil of Coweeta WS 18 for samples collected in August 1982[a]

Horizon	Depth (cm)	^{35}S incorporated (nmol g^{-1})	^{35}S mobilized (nmol g^{-1})	^{35}S mobilized (%)
O_1	—	6.85	1.30	19
O_2	—	2.11	0.96	46
A_1	0–5	0.83	0.41	49
A/B	15–30	0.67	0.31	46
B	30–50	0.58	0.26	45
C	70–180	0.43	0.18	43

[a]After Swank *et al.*, Coweeta files.

preliminary calculations are based on a number of assumptions that will be investigated in future studies.

In conclusion, research with Coweeta soils shows that incorporation of sulphate into organic matter by microbial populations is a dominant process in the S cycle of a hardwood forest ecosystem. Preliminary results indicate that mobilization rates of organic S are less than incorporation rates. Clearly, the net balance between these highly dynamic processes is important in regulating the movement of the sulphate anion and losses of other ions. Similar research is needed in other major forest ecosystems to assess the importance of soil biological activity on the S cycle.

3.4.2 Nitrogen transformations

In contrast to S, there is an abundance of scientific literature on the N cycle and the transformations this element undergoes in forest soils (Wollum and Davey, 1975; Keeney, 1980), primarily because N is a nutrient that frequently limits forest productivity. As shown earlier in Table 3.3, soil organic matter is the major pool of N in a variety of forest ecosystems. Although pools of inorganic N forms are small, the release of ammonium from organic matter and subsequent oxidation to nitrate proceed rapidly and are the major sources of biologically available nitrogen. Similar to sulphate, biological regulation of the nitrate anion is important in the transport of other solutes through the soil because of the requirement for electrochemical balances of cations and anions (Johnson *et al*., 1979). The relevance of anion mobility on solute retention and loss has been substantiated by *in situ* forest studies (Johnson and Cole, 1977; Johnson *et al*., 1979; Vitousek, 1984). Physiochemical and biological mechanisms that lead to nitrate retention or loss from disturbed forest ecosystems have been reviewed by Vitousek and Melillo (1979). In the following sections, the objective is to summarize evidence for biologically mediated transformations of N losses and gains in selected forest ecosystems. Specifically, the following processes will be examined: dinitrogen fixation, denitrification, nitrification, and mineralization–immobilization.

Mineralization–immobilization

Rates of N mineralization and immobilization are key factors which regulate the status and fluxes of N in forest ecosystems. The mineralization process entails the decomposition of organic matter in the forest floor and soil in which nitrogenous substances are attacked to release ammonium. Released ammonium can be taken up and immobilized by microflora and fauna associated with decomposing material. Rates of N mineralization in European forests were summarized by Ellenberg (reviewed by Vitousek and Melillo,

1979) and a summary of rates for some North American forest ecosystems was also provided by Vitousek and Melillo (1979). Estimates of N mineralization for ecosystems previously discussed in Table 3.3 are 115, 80, 139 and 19 kg ha^{-1} yr^{-1} for Walker Branch, Hubbard Brook, Coweeta and H. J. Andrews, respectively, with an estimate of 21 kg ha^{-1} yr^{-1} for Thompson Forest, a coniferous forest site in Washington (Vitousek and Melillo, 1979). The much higher rates for deciduous forests than for coniferous forests represented by these sites agrees with findings in other temperate regions of the world. For example, gross annual mineralization for deciduous and coniferous stands in Great Britain were estimated to be between 65–458 kg ha^{-1} and 40–200 kg ha^{-1}, respectively (Heal *et al*., 1982). These same researchers and others (Keeney, 1980; Vitousek and Melillo, 1979) have reviewed microbial immobilization processes with emphasis on the role of C:N ratios during decomposition, N utilization by saprophytic and mycorrhizal fungi, and mechanisms of recycling within the microbial community. The effects of abiotic and biotic factors on mineralization and immobilization were reviewed by Wollum and Davey (1975).

Forest disturbances produce significant changes in rates of mineralization and immobilization. In the first 3 years after clear cutting a hardwood forest at Hubbard Brook, organic matter decreased by 10,800 kg ha^{-1} in the forest floor and by 18,900 kg ha^{-1} in the soil; net N loss from the soil was estimated to be 472 kg ha^{-1} and was accompanied by an increased export of inorganic nitrogen in the stream of 337 kg ha^{-1} during the 3-year period (Bormann and Likens, 1979). These responses were attributed to accelerated rates of decomposition from more favourable temperature–moisture–nutrient relationships in the clearcut and enhancement of nitrification rates. Studies in an age sequence of clearcut and control hardwood forests in Indiana showed greater mineralization rates only in a 4-year-old clearcut (Matson and Vitousek, 1981) but rates for control forests were inherently high. At Coweeta, potential net N mineralization in the soil of the early successional stand on WS 6 greatly exceeded values in the soil of an adjacent mixed deciduous forest (Table 3.11). Carbon:nitrogen ratios were not different in soils of the two ecosystems (an average of 20), so the higher rates on WS 6 may reflect differences in substrate quality and quantity (Sollins *et al*., 1984).

Forest harvesting in loblolly pine increased N mineralization, but N losses from the system were low because microbial populations in residual forest floor and the mineral soil immobilized most of the mineralized N (Vitousek and Matson, in press). Further evidence for the importance of immobilization in delaying or reducing N losses in a variety of ecosystems has been given by Gosz *et al*. (1973), Aber and Melillo (1982) and Heal *et al*. (1982).

The balance between N mineralization and immobilization regulates the magnitude and temporal distribution of various nitrogen forms, their availability, and loss from the system. A more complete understanding of

Table 3.11. Average net N mineralization rates (with standard errors) in soil of a black locust stand on Coweeta WS 6 (n = 18) and control oak–hickory forest on WS 14 (n = 9)[a]

	Net N mineralization rate (mg kg^{-1} soil per 30 days)			
	WS 6, black locust		WS 14, mature oak–hickory	
Soil depth (cm)	March	July	March	July
0–15	34.9 (4.3)	30.9 (1.4)	4.4 (1.4)	2.1 (0.5)
16–30	12.2 (2.9)	7.4 (0.7)	2.9 (0.6)	0.3 (0.3)

[a]After Montagnini *et al*., Coweeta files.

these processes and alterations that accompany disturbance will be advanced through ^{15}N studies combined with well-designed *in situ* measurements and examination of the biochemical quality of substrates.

Nitrification

Nitrification, the microbial oxidation of NH_4–N to NO_3, can be an important N transformation process in forest soils which affects the supply of the highly mobile nitrate anion and movement of other dissolved solutes. The rate of ammonium oxidation varies widely among different ecosystems and is generally considered to be low in the forest floor and soil of most undisturbed forest stands (Keeney, 1980). However, net nitrate production of incubated forest-floor and soil samples from a variety of undisturbed forest ecosystems ranged from 0.1 to more than 800 mg NO_3–N kg^{-1} soil per 8 weeks (Vitousek *et al*., 1982). In this study, no relationship was found between nitrification and soil pH and base saturation, but a positive correlation was observed between nitrate production and mean concentration of soil ammonium. Measurements of *in situ* net nitrification were made during midsummer in the forest floor and throughout the soil profile of four mature forest stands in New England (Federer, 1983). Nitrification did not occur in the forest floor, but rates of 0.1–3.3 kg N ha^{-1} for 28 days were observed for total soil profiles, and this illustrates the importance of mineral soil in the nitrification process. The influence of moisture, temperature, C:N ratio, pH, allelochemicals and soil nutrients on rates of nitrate production were reviewed by Robertson (1982a). In a study of factors regulating nitrification in primary and secondary succession, Robertson (1982b) concluded that ammonium availability frequently controlled nitrification rates. Evidence for allelochemical inhibition of nitrification was found on some sites while changes in soil pH of 0.3 to 1.0 pH units also significantly altered nitrate production.

Disturbed forest ecosystems frequently exhibit increased nitrate losses (Vitousek and Melillo, 1979; Vitousek *et al.*, 1979; Swank and Douglass, 1977; Likens *et al.*, 1977) and elevated nitrification rates (Matson and Vitousek, 1981; Vitousek and Matson, in press). At Coweeta, research efforts have attempted to couple studies in nitrification with whole ecosystem responses to disturbance. In an early study, Todd *et al.* (1975) examined terrestrial nitrifying bacteria populations (most probable number method) throughout the soil profile of three watersheds with contrasting vegetation and a wide range of NO_3 concentrations in streamwater. A positive relationship was evident between the number of nitrifying bacteria and NO_3 concentrations; nitrifying activity appeared to be dependent on vegetation type and successional stage. Subsequent nitrification studies on two of the watersheds using aerobic incubations in the laboratory confirmed substantial differences in potential nitrification activity between early and later successional forests (Montagnini *et al.*, Coweeta files). Soils of a black locust stand on WS 6 (previously described in 3.3) and WS 14 (an adjacent mixed hardwood control catchment) were sampled at depths of 0–15 and 16–30 cm in March and July. Potential nitrification rates decreased with soil depth in both ecosystems, and activities in the black locust on WS 6 were usually 50 to 100 times greater than values for the mixed hardwood soil (Table 3.12). Soil solution and streamwater concentrations were also substantially higher on WS 6 compared with WS 14, which suggests greater leaching losses due to nitrification. Other measurements of soil characterisitics showed that mineralization rates were a major factor controlling nitrification in the early successional forest on WS 6, and there was no evidence of nitrification inhibition in the soil of the oak–hickory forest (Montagnini *et al.*, Coweeta files). It was postulated that low nitrification rates in the mature hardwood forests were due to low soil pH and low levels of cations and phosphorus.

Nitrification studies were also conducted on the commercially clearcut WS 7 at Coweeta (Caskey *et al.*, Coweeta files). After cutting, the numbers of NH_4^+-oxidizing bacteria (Rowe *et al.*, 1976) showed significant increases of about twelvefold in the 0–10 cm soil layer and eightfold in the 11–30 cm depth. Similar increases were observed at Hubbard Brook following clearcutting in a northern hardwood forest (Likens *et al.*, 1968). On WS 7, activity of NH_4^+-oxidizing bacteria appeared to be positively correlated with organic matter and total nitrogen. Numbers of NO_2^--oxidizing bacteria also exhibited substantial increases in the soil on WS 7 after cutting and remained elevated for at least 2 years after treatment. Measurements of nitrification potentials (aerobic incubations) for soil samples taken during a 5-year period after cutting on WS 7 showed about fourfold higher annual rates than the

Table 3.12. Average net nitrification rates (with standard errors) in soil of a black locust stand on Coweeta WS 6 (n = 18) and control, oak–hickory forest of WS 14 (n = 9)[a]

	Net nitrification rate (mg kg^{-1} soil per 30 days)			
	WS 6, black locust		WS 14, mature oak–hickory	
Soil depth (cm)	March	July	March	July
0–15	34.5 (4.3)	34.3 (1.4)	0.3 (0.06)	0.8 (0.5)
16–30	12.0 (2.8)	10.2 (0.9)	0.1 (0.05)	0.3 (0.1)

[a] After Montagnini *et al.*, Coweeta files.

adjacent control mixed hardwood forest. During the same period, NO_3–N concentrations in streamwater on WS 7 showed substantial increases, apparently in partial response to enhanced nitrification rates. However, the magnitude of nitrification potentials and stream NO_3 concentrations were lower after disturbance on WS 7 than values for more severely disturbed WS 6.

Based on existing literature, it is apparent that nitrification rates, and the role played by the process in regulating nitrate losses, varies tremendously across forest ecosystems. Sources of variation for disturbed sites were reviewed by Vitousek *et al.* (1982) and included rates of nitrogen mineralization during pre- and post-disturbance, the influence of other soil processes such as immobilization and delayed nitrate production, and the magnitude of vegetation regrowth and associated nitrogen uptake. Other investigators (Federer, 1983) have pointed out the need to measure nitrogen transformation processes throughout the soil profile using *in situ* techniques. Variable results also reflect many different types of forest disturbances and specific combinations of abiotic variables such as temperature and moisture known to influence nitrification rates. A clearer role of the nitrification process in forest nitrate dynamics will emerge as methods of *in situ* measurement improve and studies of longer duration than typically conducted in the past are implemented.

Denitrification

Denitrification by the dissimilatory pathway entails the reduction of nitrate into nitrogen oxides and dinitrogen by a wide variety of micro-organisms (Wollum and Davey, 1975). Research on this process has been focused primarily on agricultural systems because of interest in accounting for losses of nitrogen fertilizer. There are a few definitive studies of denitrification in forests soils although the process can occur under a variety of conditions. The main requirements for active denitrification are sources of nitrate and energy; in forest soils, carbon compounds can serve as an energy source for heterotrophic denitrifiers (Wollum and Davey, 1975). Moreover, although high oxygen levels inhibit denitrification, moist soils can contain both aerobic and anaerobic microsites.

Methodological limitations have restricted studies in most forest soils until the recent development of the acetylene-inhibition technique (Ryden *et al.*, 1979; Tiedje, 1982). This method was used by Robertson and Tiedje (1984) in short-term incubations of intact soil cores to examine absolute and comparative denitrification rates for soils in 12 non-cultivated vegetation communities, including 9 forested sites. Measurable rates of N_2O production were observed for all soils, but rates showed substantial variation within and among sites. The most active sites of N_2O production were soils of recently clearcut or mature ($\geqq$100 years) hardwood stands with nitrogen losses of 6–30 mg m^{-2} day^{-1}. Extrapolation of rates on the most active sites provided estimates of 2–12 kg Nha^{-1} month^{-1}. This flux of N_2O is of about the same magnitude as combined N inputs in precipitation and fixation. Production of N_2O was much lower in soil of the three pine stands examined. Across all sites, about 65% of the variation in N_2O production was explained by nitrate production, CO_2 production and water content. For some incubations, it was apparent that N_2O production was from sources other than denitrifiers, such as nitrate respiring bacteria, fungi and nitrifiers (Robertson and Tiedje, 1984). The regulating effects of soil characteristics such as acidity, temperature, oxygen, NO_3–N, and available C on denitrification have been demonstrated in other laboratory studies (Firestone *et al.*, 1980; Limmer and Steele, 1982).

Measurable N_2O emissions were also documented for soils in seven natural ecosystems including both deciduous and coniferous forests (Goodroad and Keeney, 1984). Seasonal changes in soil temperature and moisture strongly influenced patterns of N_2O emission. Estimates of N_2O production for the forested sites ranged from 0.2 to 2.1 kg ha^{-1} of N for time increments between 5 and 9 months during 2 years of measurement.

At Coweeta, denitrification rates were measured in soils of control WS 2 and WS 7 over a 2-year period beginning the second year after commercial clearcutting on WS 7. Potential activity was measured as the rate of N_2O production in acetylene-inhibited soil slurries (Swank and Caskey, 1982) during the Phase I of denitrification (Smith and Tiedje, 1979). This measurement represents the maximum activity of denitrifying enzymes present in the soil at the time of sampling. Rates of denitrification on soil samples collected at depths of 0–10 cm and 10–30 cm throughout the year over the watersheds showed substantial spatial and temporal variability

Table 3.13. Average seasonal rates[a] of denitrification in soils from Coweeta control WS 2 and clearcut WS 7 for the period October 1979 to September 1981

	WS 2		WS 7	
Season	0–10 cm	10–30 cm	0–10 cm	10–30 cm
Spring	0.7 ± 0.5	21.6 ± 14.5	8.2 ± 6.0	19.0 ± 10.1
Summer	0.2 ± 0.1	0.2 ± 0.2	3.4 ± 2.4	3.4 ± 3.3
Autumn	0.2 ± 0.1	0.2 ± 0.1	0.3 ± 0.1	0.2 ± 0.1
Winter	0.2 ± 0.1	0.2 ± 0.1	0.4 ± 0.1	0.3 ± 0.1

[a]Rates were multiplied by depth (in cm) and corrected for *in situ* temperature and concentration of NO_3^-. The units are mg Nd^{-1} m^{-2}.

(Table 3.13). Production of N_2O per gram of soil was usually higher in the surface soil (0–10 cm) of both watersheds, but when weighted for depth, the total activity of the two soil layers were similar. Denitrification rates tended to be highest in the spring and summer seasons, but there were no significant differences between watersheds. Phase I rates were corrected for temperature and NO_3 concentrations and related to precipitation events to estimate the amount of denitrification occurring on the two watersheds (Caskey *et al.*, Coweeta files). These procedures yielded estimates of denitrification losses of 4.8 and 1.08 kg Nha^{-1} yr^{-1} for WS 2 and WS 7, respectively. Such estimates must be considered preliminary because of the assumptions involved, but they do serve to place the denitrification process into perspective in relation to other potential N fluxes in forest ecosystems.

From the very limited data available for forest ecosystems, it appears that loss of N via the denitrification pathway could be substantial in comparison with other losses such as dissolved inorganic N. Clearly, initial findings suggest the process is more important in forest soils than originally perceived. Improved methods are needed for *in situ* measurements of denitrification, and research is also needed to assess regulatory parameters contributing to spatial and temporal variability. More precise quantification of denitrification at an ecosystem level will require a combination of modelling and experimental efforts.

Nitrogen fixation

Fixation of dinitrogen to ammonia is a reductive process performed by free-living micro-organisms and through symbiotic association by *Rhizobium* species. Dinitrogen fixation in forest ecosystems can occur on and/or in a variety of terrestrial substrates including foliage, twigs, bark, epiphytic plant parts, mosses, wood, litter, soil and roots (reviewed by Dawson, 1983).

Fixation by free-living bacteria has been detected in a wide range of forest soils, but system level quantification is not well known because of activity variability due to abiotic and biotic factors. Using acetylene reduction methods, Todd *et al.* (1978) derived preliminary estimates of 8.5 kg yr^{-1} for free-living fixation in soils, 1.7 in woody litter, 1.0 on tree boles, 0.6 in leaf litter, and 0.2 in the phyllosphere for a mature hardwood forest at Coweeta. Substantial quantities of total system free-living fixation have also been estimated for other forests (Silvester, 1978; Hicks *et al.*, 1977) and fixation rates are significant in specific forest compartments such as epiphytic lichen (Denison, 1973), forest-floor wood (Cornaby and Waide, 1973), and certain insects (reviewed by Dawson, 1983). Disturbance can increase free-living fixation in the soil as documented on Coweeta WS 7 (Caskey *et al.*, Coweeta files) where, after clearcutting, fixation rates more than doubled in the 0–30 cm soil layer. Significant increases in nitrogen fixation rates were also observed in the soil after repeated burning in loblolly pine (Jorgensen and Wells, 1971). In contrast, very low annual rates of fixation were observed in the soil of other relatively undisturbed forest ecosystems (Vance *et al.*, 1983; Jorgensen, 1975), in an annually burned oak–hickory forest in Missouri (Vance *et al.*, 1983) and harvested loblolly pine stands in South Carolina (Van Lear *et al.*, 1983). A wide range of nitrogen fixation rates in forest soil reflect the influence of variables such as soil temperature, moisture (soil oxygen), pH, nutrients, and the conditions of experimental measurement, including methods.

Dinitrogen fixation by symbiotic nodulated legumes and actinorhizal species (nodulated non-leguminous species) can be a major source of N in some forest ecosystems. Black locust, a woody legume, regenerates rapidly from root and stump sprouts after major forest disturbances and is a dominant species of early forest succession in the southern Appalachians (McGee and Hooper, 1975). Seasonal patterns of *in situ* acetylene reduction activity and nodule biomass were measured in 4-year-old black locust stands on Coweeta WS 7 after commercial clearcutting (Boring and Swank, 1984a). Fixation rates were found to vary with nodule size, time of day, and seasonally with highest rates in the mid-growing season (June) and no detectable activity prior to budbreak or after autumn senescence. Based on several assumptions, black locust was estimated to fix N at the rate of 30 kg ha^{-1} yr^{-1} on mesic sites where it predominates within WS 7, and at 10 kg ha^{-1} yr^{-1} when activity was weighted for black locust density over the entire 59 ha catchment. Another study at Coweeta examined the chronological sequence of N pools and accretion of black locust stands regenerated on clearcut sites (Boring and Swank, 1984b). Standing stocks of N in biomass, forest floor, and soil in 4-, 12- and 38-year-old stands usually exceeded values of unevenaged mixed hardwood stands at Coweeta (Table 3.14). A major portion of the total N pool in young stands was in above-ground biomass; forest-floor N pools in

Table 3.14. Standing stocks of N (kg ha^{-1}) in various compartments of different aged black locust stands and an evenaged mixed hardwood forest at Coweeta[a]

Stand compartment	Age 4	Age 17	Age 38	Unevenaged mixed hardwoods
Above-ground biomass	254	1058	1928	406
Forest floor	537	399	253	137
Soil (0–30 cm)	4209	4466	3378	3960
Roots	129	287	610	435
Total	5129	6210	6169	4938

[a]After Boring and Swank (1984).

young stands also exceeded the control forest. The net average annual rates of N increase (accretion) for the total stand were 48, 75 and 33 kg ha^{-1} yr^{-1} for stand ages 4, 17 and 38 years, respectively. Apparently, peak N fixation occurred in early or mid-stages of forest succession which is similar to results reported for other secondary succession nitrogen-fixing species (Gadgil, 1971; Tarrant *et al*., 1969). Nitrogen fixation by black locust increased the concentration of NO_3 in the soil, which could affect transport of other solutes. Furthermore, the low C:N ratio of black locust litter could result in higher decomposition rates and more rapid mineralization of other elements.

A large number of forest genera form root nodules after infection by the nitrogen-fixing actinomycete *Frankia* (Dawson, 1983). Annual N fixation estimates on a stand basis for *Alnus* species range from 4.7 kg ha^{-1} (Younger and Kapustka, 1983) to more than 362 kg ha^{-1} (Tripp *et al*., 1979; Binkley, 1982), with estimates frequently exceeding 50 kg ha^{-1}. Potential rates of N fixation by wax myrtle (*Myrica cerifera*), a common understory plant in pine flatwoods, was estimated to be 10.6 kg ha^{-1} yr^{-1} (Permar and Fisher, 1983). Other shrub species such as snowbrush (*Ceanothus velutinus*) are also abundant in early successional forests in the Cascade Mountains of the Pacific North-west, and N fixation rates for this species have been estimated to exceed 80 kg ha^{-1} yr^{-1} (Cromack *et al*., 1979; Youngberg and Wollum, 1976; Binkley *et al*., 1982). The fate of nitrogen accretions in forest ecosystems resulting from fixation and the subsequent effects on transport of other elements is not well understood and requires detailed long-term research.

3.5 WITHIN-STREAM TRANSFORMATIONS OF SOLUTES

As solutes enter a stream from the landscape and are transported downstream, they are subjected to additional physical, chemical and biological processes (Figure 3.12). Hydrologic interactions between solutes and the streambed involve physical mechanisms which affect solute transport

(Bencala *et al*., 1984); ion sorption on to stream sediments can also be an important process that influences the transport and behaviour of cations in streams (Kennedy, 1965; Bencala *et al*., 1984). Coupled to such physiochemical mechanisms are a variety of biological processes which affect solute dynamics within a mountain stream. Heterotrophic and autotrophic uptake and release of nutrients, and macroinvertebrate egestion of particles, are among the major processes influencing dissolved and particulate nutrient transport. In the following sections, experimental evidence for biological regulation of solute dynamics in Coweeta streams will be emphasized.

Similar to terrestrial ecosystems, nitrogen fixation and denitrification can be significant processes of N gain and loss in forest streams. Data are unavailable at Coweeta for the former process, but studies on denitrification have been conducted on WS 7 (Swank and Caskey, 1982), a commercially logged 59 ha catchment. The schedule of treatments on WS 7 and monthly weighted NO_3–N concentrations over a 4-year period for the headwater stream and watershed outlet (weir) are illustrated in Figure 3.13. In response to N leached from roadbank fertilization, NO_3–N concentrations increased markedly in the headwater stream, but 650 m downstream at the weir, concentrations were only slightly elevated. A concentration difference of more than 150 μg per litre between headwater and weir sites was maintained for 2 years (1978, 1979) of the treatment period and then declined towards the end of 1980, primarily because of reduced headwater concentrations. Stream discharge and proportional streamwater samples were collected at both headwater and weir sites throughout the period depicted in Figure 3.13, which provided a method for calculating NO_3 depletion from export differences between the two sites (Swank and Caskey, 1982).

Annual calculations showed that in the year of road construction and logging, within-stream NO_3–N depletion was 3 kg, which represented 127% of the NO_3–N exported at the weir (Table 3.15). In each of the next 2 years, N depletion was about 30 kg, but because of increased NO_3 discharge at the weir, the percentage of the total N export decreased. By 1980, stream depletion was about 4 kg, which represented only 5% of the total NO_3–N discharged at the weir. Other studies of stream N mass balances have implicated denitrification in sediments as a significant mechanism of N disappearance in well-oxygenated streams (Kaushik *et al*., 1975; Hill, 1979, 1983). Direct evidence for denitrification in stream sediments of WS 7 was obtained in the fourth year of the treatment. Assays of the quantities of denitrifying enzymes in stream sediment samples were measured during phase I of denitrification as N_2O production in acetylene-inhibited slurries (Swank and Caskey, 1982). Watershed 7 stream sediments were sampled at 10 sites at periodic intervals throughout the year, but only August, October and April results are presented in Table 3.16 since denitrification was not detected during winter months at most sites. Rates of denitrification varied

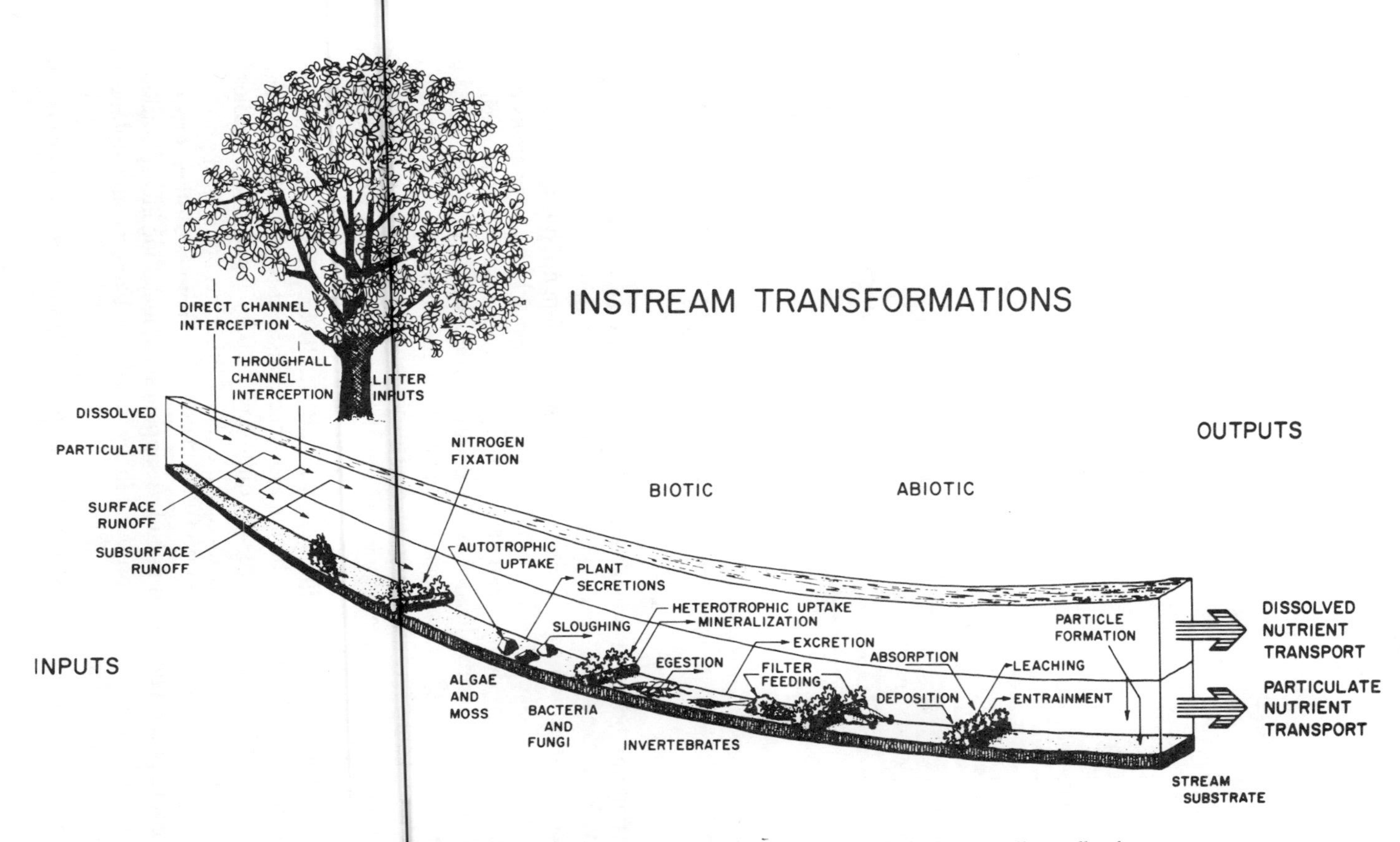

Figure 3.12. A variety of processes can potentially affect solute dynamics within small woodland streams

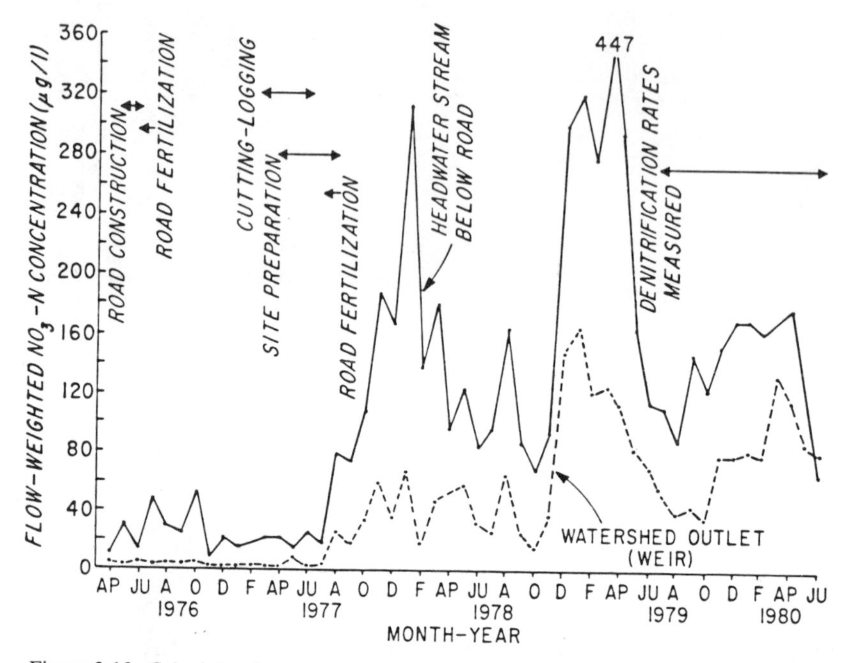

Figure 3.13. Schedule of treatments on commercially logged Coweeta WS 7 and mean monthly weighted NO_3–N concentrations in the headwater stream immediately below a forest road and at the weir of WS 7

substantially between sites (Table 3.16), but analyses showed that the mean activity at each site was positively correlated with total Kjeldahl N and organic matter and negatively correlated with sediment NO_3 concentrations (Swank and Caskey, 1982). Activities measured at each site were projected to monthly rates of denitrification based on volumes of sediment at each site. The potential annual loss of N by denitrification was estimated as 1.7 kg, compared with the mass balance calculation for the same year of 3.9 kg shown in Table 3.15. Concurrent measurements of denitrifying activity and depletion were available only in the fourth year of treatment when depletion was 3.9 kg; however, in the second and third years after harvest, within-stream NO_3–N depletion was about 30 kg. Results suggest that sediment denitrification can be a major mechanism of N loss even in streams where NO_3–N concentrations are quite low (<0.5 μg per litre). Studies of NO_3 transformations in stream sediments of other ecosystems support this interpretation (Wyer and Hill, 1984). However, there are other processes that could potentially result in NO_3 disappearance, including incorporation in primary production, N_2O production by nitrifying bacteria and uptake by heterotrophic bacteria.

The first of these processes (primary production) was also examined in the

Table 3.15. Nitrate-N export and within-stream depletion during the period of system disturbance and recovery[a]

Year (June–May)	NO_3–N exported (kg) Below road[b]	Watershed outlet (weir)	NO_3–N depletion[c] (kg)	Proportion of total NO_3–N discharge at weir (%)
1976–77	5.9	2.6	3.3	127
1977–78	55.9	28.0	27.9	99
1978–79	108.7	76.3	32.4	43
1979–80	83.9	80.0	3.9	5

[a]After Swank and Caskey (1982).
[b]Based on accumulation of N for all drainages below the road.
[c]Calculated from the difference between export from contributing areas below the road and export at the weir.

stream of WS 7. Prior to logging, periphyton primary production was typically low for undisturbed Coweeta streams with estimated rates of 2.9 g dry weight m^{-2} yr^{-1} (Webster *et al.*, 1983). In the first year after treatment, periphyton production significantly increased to about 87 g dry weight m^{-1} yr^{-1} in response to canopy removal and increased solar insolation along with increased dissolved nutrients. In the subsequent 2 years, production approached baseline conditions with rates of 8.7 g dry weight m^{-2} yr^{-1}. Hains (1981) found that increases in moss standing crops the first year after logging were sites of significant phosphorus accumulation.

Table 3.16. Denitrification rates measured with sediment slurries as N_2O evolution in the presence of C_2H_2[a]

Site	N_2O production[b] ($nL\ h^{-1}\ g^{-1}$) August 1979	October 1979	April 1980
1	7.3	18.3	7.6
2	18.8	2.4	20.4
3	9.8	3.7	5.4
4	17.9	16.7	14.8
5	11.9	4.7	9.5
6	27.2	3.0	11.0
7	0.1	11.2	9.0
8	0.1	—	2.5
9	35.0	74.9	58.4
10	34.0	35.0	55.3

[a]After Swank and Caskey (1982).
[b]S.E. of estimates for most values are $\leqslant \pm 10\%$.

3.5.1 Influences of stream fauna

An accumulation of evidence indicates that leaves and small woody material from the surrounding terrestrial environment are the major sources of energy to many headwater streams draining forest ecosystems (Cummings, 1974). Macroinvertebrates and perhaps meiofauna play an important role in the comminution of this particulate organic matter (POM) and the mineralization of nutrients (Webster, 1983) which subsequently influence the form and timing of solute losses. For example, Webster and Patten (1979) estimated that macroinvertebrates annually ingested 80% of the leaf litter input to a small forest stream at Coweeta. Webster (1983) conceptualized and developed a computer model to simulate the functional role of

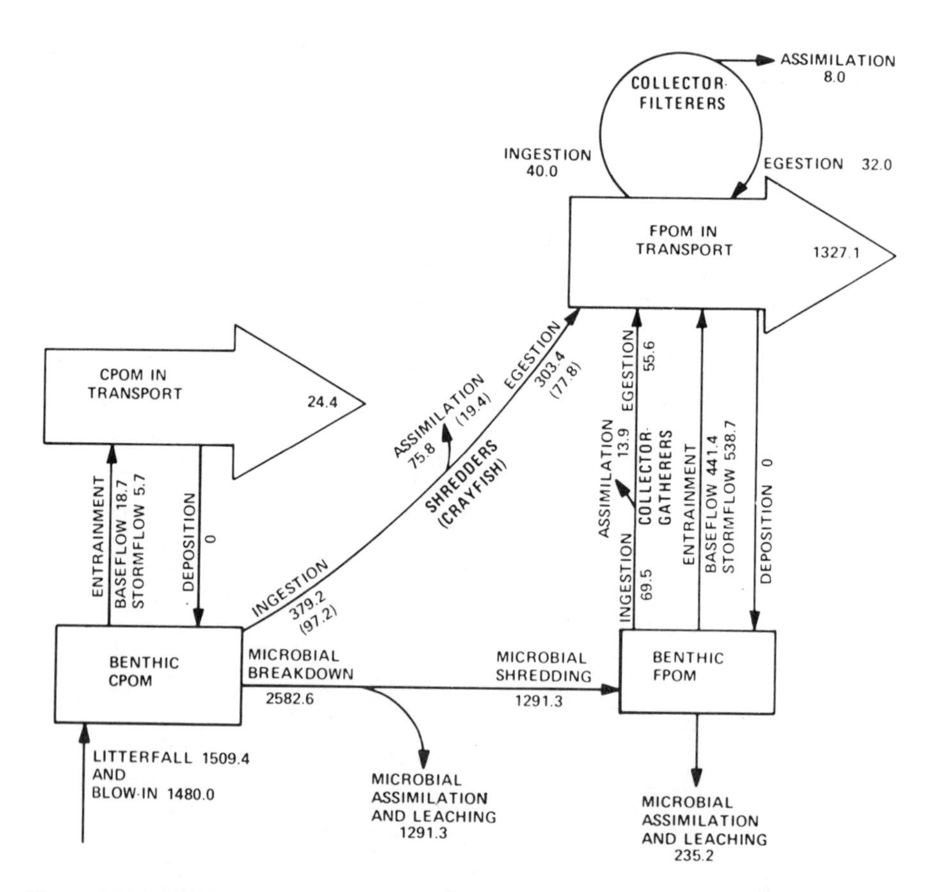

Figure 3.14. Model structure of stream detritus dynamics with annual budgets for major compartments and flows of particulate organic matter (Webster, 1983). *From 'The role of benthic macro invertebrates in detritus dynamics of streams: a computer simulation' by J. R. Webster,* Ecological Monographs, *1983,* ***53****, 385. Copyright © 1983 by the Ecological Society of America*

macroinvertebrates in detritus dynamics and to integrate data collected in a variety of stream studies at Coweeta. A summary of results illustrated in Figure 3.14 (Webster, 1983) showed that only 45% of the nearly 3000 kg of total POM input to the stream was exported from the system in particulate form (coarse [C] POM + fine [F] POM). Furthermore, less than 1% of the input was transported out of the system as CPOM; shredder ingestion and microbial breakdown were primary mechanisms for benthic CPOM losses. Biological processes were less important in regulating losses of benthic FPOM with a combined loss of 24% by macroinvertebrates and microbial populations. Collector–gatherers and shredders activity accounted for 27% of FPOM exported in streamflow. Taken together, this synthesis suggests that the major role of macroinvertebrates is the conversion of benthic POM into

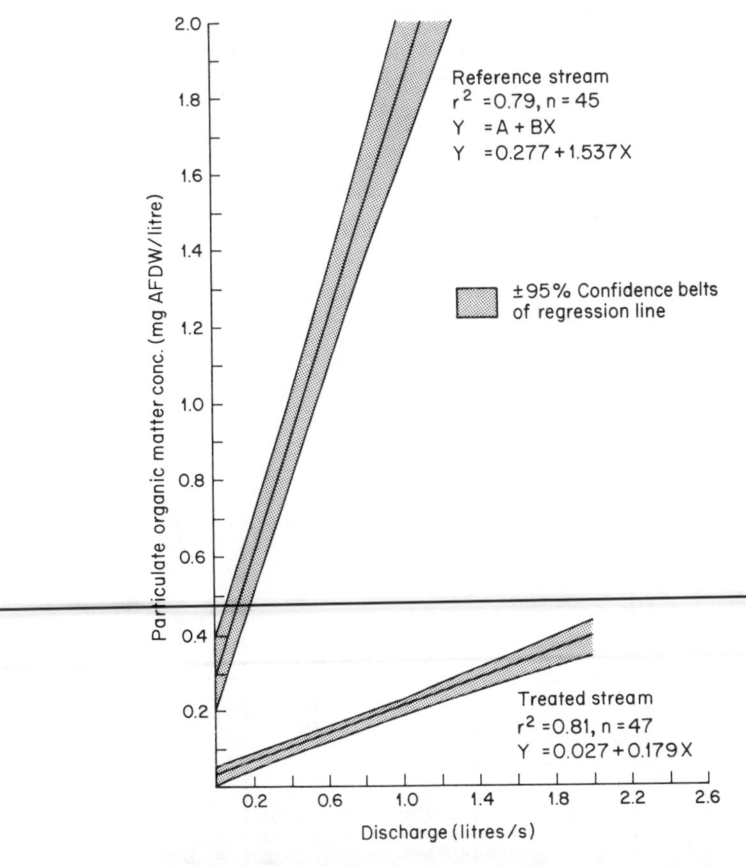

Figure 3.15. Regressions of suspended particulate organic matter versus discharge in two first-order Coweeta streams following application of methoxychor in the treated stream which reduced aquatic insect densities by more than 90%. Data represent low flow conditions, primarily in the autumn months (Wallace *et al.*, 1982). *Reproduced with permission from Springer-Verlag*

Table 3.17. Total annual stream export (kg yr^{-1}) of potassium and calcium in various forms for three Coweeta ecosystems[a]

Form	K			Ca		
	Old field (WS 6)	White pine plantation (WS 17)	Hardwood forest (WS 18)	Old field (WS 6)	White pine plantation (WS 17)	Hardwood forest (WS 18)
Total dissolved	65.26	66.68	86.41	122.83	84.95	121.82
Dissolved excretion (detritivores)[b]	0.14	0.10	0.28	0.71	0.66	0.94
Sediment	0.08	0.04	0.02	1.43	0.96	0.71
Large particulate organic matter	0.006	0.004	0.001	0.16	0.13	0.13
Fine particulate organic matter	—	—	—	0.36	0.24	1.98
Insect emergence and organism drift	0.002	0.002	0.003	0.014	0.007	0.006

[a]After Webster and Patten (1979).
[b]Includes excretion by insects and crayfish.

transported POM which affects the form, availability and timing of detritus and nutrients transported downstream. This suggestion is clearly supported by experimental treatment of a stream at Coweeta, where the introduction of an insecticide reduced aquatic insect densities to below 10% of the untreated reference stream (Wallace *et al.*, 1982). Following stream treatment and the reduction of macroinvertebrates, detrital processing rates were reduced, and this led to an accumulation of benthic organic matter and a significant reduction of particulate organic matter transport (Figure 3.15).

In other investigations at Coweeta, radioisotope and other sampling methods were used to quantify the influence of detritivores on K and Ca dynamics in streams of three different ecosystems (Webster and Patten, 1979). In this assessment, it was assumed that K and Ca excreted as dissolved ions by detritivores are subsequently unavailable to these fauna and that egested material is recycled by filter and deposit feeding detritivores. A summary of the total annual export of K and Ca contained in different forms for the three ecosystems is given in Table 3.17. It is apparent that the dissolved form accounts for more than 95% of the total export for these two cations in each ecosystem. Potassium and Ca excreted by detritivores was only a small fraction of the total dissolved loss, but excreted loads frequently exceeded sediment and other forms of ion export. The effects of macroinvertebrates on leaf breakdown and subsequent release of dissolved organic carbon (DOC) has been studied by Meyer and Tate (1983) and Meyer (1983). Based on this research, it appears that shredder feeding activity could contribute 1–50 $g\ m^{-2}\ yr^{-1}$ of DOC to stream water, which is a significant within-stream source of DOC, particularly during low-flow periods. Similar studies are lacking for dissolved organic nitrogen in Coweeta streams, but macroinvertebrate activity may be more important in regulating dissolved organic solutes than inorganic solutes.

3.6 SUMMARY

This chapter approaches the topic of the numerous biological processes which influence solute retention and loss in forest ecosystems by attempting to illustrate principles derived from landscape scale ecosystem studies. Theoretical considerations are minimized and experimental evidence is emphasized.

A summary of solute inputs–outputs, nutrient pools, and recycling processes for several long-term ecosystem studies in major temperate forests in the USA illustrate the stability of forests. Much greater quantities of nutrients are stored and recycled within forests than are lost annually in streamwater. Solute uptake and incorporation in above- and below-ground biomass are important processes of retention. Solutes are recycled through litter fall, canopy leaching, root mortality, and resorption processes which are mediated by a diverse flora and fauna. The patterns and magnitude of solute fluxes are moderately similar across a range of forest types compared with the variable responses associated with disturbance and alteration of biological processes. Disturbances such as clearcutting and species conversions illustrate the influence of primary production and nitrogen transformations on the magnitude, timing, and recovery of solute losses.

Terrestrial insect populations regulate the transfer of solutes within ecosystems by enhancing litter decomposition and nitrogen transformation processes, through stimulation of net primary production, nutrient uptake and foliar leaching, and by modifying rates of evapotranspiration. Measurable effects of insect herbivory on the nitrogen cycle and nitrate losses have been documented by landscape scale studies.

Microbial transformations of S and N pools in forest floor and soil are among the most important biotic controls of solute losses because they influence the highly mobile sulphate and nitrate anions. Studies show that incorporation of sulphate into organic matter by microbial populations is a significant process in some ecosystems. Mobilization rates of organic S appear to be less than incorporation rates, which leads to sulphate immobilization and contributes to ecosystem S accumulations. Nitrogen mineralization and immobilization are important processes which regulate fluxes of N in forest ecosystems. Mineralization rates are typically higher in deciduous forests than coniferous forests and rates in both systems are increased with disturbance. Microbial immobilization can play an important role in delaying or reducing N losses. Ammonium released by mineralization can also be oxidized to nitrate (nitrification) by micro-organisms. Nitrification is influenced by many abiotic and biotic factors and, hence, rates are highly variable between ecosystems. Disturbed forests frequently exhibit increased nitrification rates which in some ecosystems are associated with substantial nitrate losses. The role of denitrification in N loss from forest soils is not well known, but from recent studies it appears to be more important than originally perceived. A major source of N to forest ecosystems is dinitrogen fixation by free-living micro-organisms and symbiotic associations. Following disturbance, early successional nitrogen fixing woody species frequently become established and can be a significant source of N replenishment. Levels of free-living fixation apparently vary quite widely after disturbance.

Solutes entering a stream are subjected to additional biotic regulation. Major processes which influence solute transport include heterotrophic and autotrophic uptake and release of nutrients and macroinvertebrate processing of particulate organic matter. Several studies indicate that denitrification in stream sediments can result in nitrate depletion in streamwater. Other research has demonstrated that a major role of macroinvertebrates is the conversion of benthic particulate organic matter into transported particulate organic matter. This activity affects the form, availability and timing of detritus and nutrients transported downstream.

ACKNOWLEDGEMENTS

Research findings for the Coweeta Hydrologic Laboratory were supported in part by United States Department of Agriculture, Forest Service, Southeastern Forest Experiment Station, and in part by National Science Foundation grants. The author thanks L. R. Boring, W. H. Caskey, D. A. Crossley, Jr., J. R. Webster, J. B. Wallace, J. W. Fitzgerald, F. Montagnini, M. G. Rollins and S. E. Waldroop for their assistance in several phases of preparing this chapter.

REFERENCES

Aber, J. D., and Melillo, J. M. (1982). Nitrogen immobilization in decaying hardwood leaf litter as a function of initial nitrogen and lignin content. *Can. J. Bot.*, **60**, 2263–2269.

Bencala, K. E., Kennedy, V. C., Zellweger, G. W., Jackman, A. P., and Avanzino, R. J. (1984). Interactions of solutes and streambed sediment.1. An experimental analysis of cation and anion transport in a mountain stream. *Water Resour. Res.*, **20**, 1797–1803.

Binkley, D. (1982). Nitrogen fixation and net primary production in a young Sitka alder stand. *Can. J. Bot.*, **60**, 281–284.

Binkley, D., Cromack, K., and Fredriksen, R. L. (1982). Nitrogen accretion and availability in some snowbrush ecosystems. *For. Sci.*, **28**, 720–724.

Boring, L. R., Monk, C. D., and Swank, W. T. (1981). Early regeneration of a clear-cut Southern Appalachian forest. *Ecology*, **62**, 1244–1253.

Boring, L. R., and Swank, W. T. (1984a). Symbiotic nitrogen fixation in regenerating black locust (*Robinia pseudoacacia* L.) stands. *For. Sci.*, **30**, 528–537.

Boring, L. R., and Swank, W. T. (1984b). The role of black locust (*Robinia pseudoacacia*) in forest succession. *J. Ecol.*, **72**, 749–766.

Bormann, F. H., and Likens, G. E. (1979). *Pattern and Process in a Forested Ecosystem*, Springer-Verlag, New York.

Bormann, F. H., Likens, G. E., and Melillo, J. M. (1977). Nitrogen budget for an aggrading northern hardwood forest ecosystem. *Science*, **196**, 981–983.

Caskey, W. H., Swank, W. T., Todd, R. L., Boring, L. R., and Waide, J. B. (Coweeta files). Nitrogen cycling in a forested watershed in the Southern Appalachians. Paper presented at symposium on Long-Term Research on Forested Watersheds at Coweeta, 15–17 October 1984, Athens, Georgia.

Chapin, F. S. (1980). The mineral nutrition of wild plants, annual review. *Ecol. Syst.*, **II**, 233–260.

Cole, D. W., and Rapp, M. (1980). Elemental cycling in forest ecosystems. In: *Dynamic Properties of Forested Ecosystems* (D. E. Reichle, Ed.), Cambridge Univ. Press, pp. 341–409.

Cornaby, B. W., and Waide, J. B. (1973). Nitrogen fixation in decaying chestnut logs. *Plant Soil*, **39**, 445–448.

Cromack, K., Delwiche, C., and McNabb, D. H. (1979). Prospects and problems of nitrogen management using symbiotic nitrogen fixers. In: *Symbiotic Nitrogen Fixation in the Management of Temperate Forests* (J. C. Gordon, C. T. Wheeler, and D. A. Perry, Eds.), Oregon State Univ., Corvallis, pp. 210–233.

Crossley, D. A. (1970). Roles of microflora and fauna in soil systems. In: *Pesticide in the Soil: Ecology, Degradation and Movement*, Michigan State Univ., East Lansing, pp. 30–35.

Cummings, K. W. (1974). Structure and function of stream ecosystems. *BioScience*, **24**, 631–641.
Dawson, J. O. (1983). Dinitrogen fixation in forest ecosystems. *Can. J. Microbiol.*, **29**, 979–992.
Day, F. P. (1974). *Primary Production and Nutrient Pools in the Vegetation on a Southern Appalachian Watershed*. Dissertation, University of Georgia, Athens.
Day, F. P., and Monk, C. D. (1977). Seasonal nutrient dynamics in the vegetation on a Southern Appalachian watershed. *Am. J. Bot.*, **64**.
Denison, W. C. (1973). Life in tall trees. *Sci. Am.*, **228**, 74–80.
Federer, C. A. (1983). Nitrogen mineralization and nitrification: depth variation in four New England forest soils. *Soil Sci. Soc. Am. J.*, **47**, 1008–1014.
Firestone, M. K., Firestone, R. B., and Tiedje, J. M. (1980). Nitrous oxide from soil denitrification: factors controlling its biological production. *Science*, **208**, 749–751.
Fitzgerald, J. W., and Andrew, T. L. (1984). Mineralization of methionine sulfur in soils and forest floor layers. *Soil Biol. Biochem.*, **16**, 565–570.
Fitzgerald, J. W., Ash, J. T., Strickland, T. C., and Swank, W. T. (1983). Formation of organic sulfur in forest soils: a biologically mediated process. *Can. J. For. Res.*, **13**, 1077–1082.
Fitzgerald, J. W., Strickland, T. C., and Ash, J. T. (1985). Isolation and partial characterization of forest floor and soil organic sulfur. *Biogeochemistry*, **1**, 155–167.
Fitzgerald, J. W., Strickland, T. C., and Swank, W. T. (1982). Metabolic fate of inorganic sulfate in soil samples from undisturbed and managed forest ecosystems. *Soil Biol. Biochem.*, **14**, 529–536.
Fitzgerald, J. W., Swank, W. T., Strickland, T. C., Ash, J. T., Hale, D. D., Andrew, T. L., and Watwood, M. E. (Coweeta files). Sulfur pools and transformations in litter and surface soil of a hardwood forest. Paper presented at symposium on Long-Term Research on Forested Watersheds at Coweeta, 15–17 October 1984, Athens, Georgia.
Fredriksen, R. L. (1975). *Nitrogen, Phosphorous, and Particulate Matter Budgets of Five Coniferous Forest Ecosystems in the Western Cascades Range, Oregon*. Dissertation, Oregon State University, Corvallis.
Gadgil, R. L. (1971). The nutritional role of *Lupinas arboreus* in coastal sand dune forestry. 3: Nitrogen distribution in the ecosystem before tree planting. *Plant Soil*, **35**, 113–126.
Gist, C. S., and Crossley, D. A. (1975). The litter arthropod community in a Southern Appalachian hardwood forest: numbers, biomass and mineral element content. *Am. Midl. Nat.*, **93**, 107–122.
Goodroad, L. L., and Keeney, D. R. (1984). Nitrous oxide emission from forest, marsh, and prairie ecosystems. *J. Environ. Qual.*, **13**, 448–452.
Gosz, J. R., Likens, G. E., and Bormann, F. H. (1973). Nutrient release from decomposing leaf and branch litter in the Hubbard Brook Forest, New Hampshire. *Ecol. Monogr.*, **43**, 173–191.
Grier, C. C., Cole, D. W., Dyrness, C. T., and Fredriksen, R. L. (1974). Nutrient cycling in 37 and 450-year-old Douglas-fir ecosystems. In: *Integrated Research in the Coniferous Forest Biome* (R. H. Waring and R. L. Edmonds, Eds.), University of Washington, Seattle, pp. 21–34.
Hains, J. J. (1981). *The Response of Stream Flora to Watershed Perturbation*. Thesis, Clemson University, Clemson, South Carolina.
Heal, O. W., Swift, M. J., and Anderson, J. M. (1982). Nitrogen cycling in United Kingdom forests: the relevance of basic ecological research. *Phil. Trans. R. Soc. Lond.*, **B296**, 427–444.
Henderson, G. S., and Harris, W. F. (1975). An ecosystem approach to the

characterization of the nitrogen cycle in a deciduous forest watershed. In: *Forest Soils and Forest Land Management* (B. Bernier and C. H. Winget, Eds.), Laval Univ. Press, Quebec, Canada, pp. 179–193.

Henderson, G. S., Swank, W. T., Waide, J. B., and Grier, C. C. (1978). Nutrient budgets of Appalachian and Cascade region watersheds: a comparison. *For. Sci.*, **24**, 385–397.

Hibbert, A. R. (1969). Water yield changes after converting a forested catchment to grass. *Water Resour. Res.*, **5**, 634–640.

Hicks, B. J., Silvester, W. B., and Taylor, F. J. (1977). Nitrogen fixation in mangroves. *Proc. N. Z. Ecol. Soc.*, **24**, 130–131.

Hill, A. R. (1979). Denitrification in the nitrogen budget of a river ecosystem. *Nature (London)*, **281**, 291–292.

Hill, A. R. (1983). Denitrification: its importance in a river draining an intensively cropped watershed. *Agric. Ecosyst. Environ.*, **10**, 47–62.

Johnson, D. W., Breuer, D. W., and Cole, D. W. (1979). The influence of anion mobility of ionic retention in waste water irrigated soils. *J. Environ. Qual.*, **8**, 246–250.

Johnson, D. W., and Cole, D. W. (1977). Sulfate mobility in an outwash soil in western Washington. *Water Air Soil Pollut.*, **7**, 489–495.

Johnson, D. W., Henderson, G. S., Huff, D. D., Lindberg, S. E., Richter, D. D., Shriner, D. S., Todd, D. E., and Turner, J. (1982). Cycling of organic and inorganic sulphur in a chestnut oak forest. *Oecologia (Berl.)*, **54**, 141–148.

Johnson, D. W., Hornbeck, J. W., Kelly, J. M., Swank, W. T., and Todd, D. E. (1980). Regional patterns of soil sulfate accumulation: relevance to ecosystem sulfur budgets. In: *Atmospheric Sulfur Deposition: Environmental Impact and Health Effects* (D. S. Shriner, C. R. Richmond and S. E. Lindberg, Eds.), Ann Arbor Press, pp. 507–520.

Johnson, P. L., and Swank, W. T. (1973). Studies of cation budgets in the Southern Appalachians on four experimental watersheds with contrasting vegetation. *Ecology*, **54**, 70–80.

Jorgensen, J. R. (1975). Nitrogen fixation in forested Coastal Plain soils. USDA For. Serv. Res. Pap. SE-130, Southeast. For. Exp. Sta., Asheville, North Carolina.

Jorgensen, J. R., and Wells, C. G. (1971). Apparent nitrogen fixation in soil influenced by prescribed burning. *Soil Sci. Soc. Am. Proc.*, **35**, 806–810.

Kashik, N. K., Robinson, J. B., Sain, P., Whitely, H. R., and Stammers, W. N. (1975). A quantative study of nitrogen loss from water of a small spring-fed stream. In: *Water Pollution Research in Canada*, University of Toronto, pp. 110–117.

Kennedy, V. C. (1965). Mineralogy and cation-exchange capacity of sediments from selected streams. *U.S. Geol. Survey Prof. Paper 433-D*.

Keeney, D. R. (1980). Prediction of soil nitrogen availability in forest ecosystems: a literature review. *For. Sci.*, **26**, 159–171.

Likens, G. E., Bormann, F. H., and Johnson, N. M. (1968). Nitrification: importance to nutrient losses from a cutover forested ecosystem. *Science*, **163**, 1205–1206.

Likens, G. E., Bormann, F. H., Johnson, N. M., Fisher, D. W., and Pierce, R. S. (1970). Effects of forest cutting and herbicide treatment on nutrient budgets in the Hubbard Brook watershed-ecosystem. *Ecol. Monogr.*, **40**, 23–47.

Likens, G. E., Bormann, F. H., Johnson, N. M., and Pierce, R. S. (1967). The calcium, magnesium, potassium, and sodium budgets in a small forested ecosystem. *Ecology*, **48**, 772–785.

Likens, G. E., Bormann, F. H., Pierce, R. S., Eaton, J. S., and Johnson, N. M. (1977). *Biogeochemistry of a Forested Ecosystem*, Springer-Verlag, New York.

Limmer, A. W., and Steele, K. W. (1982). Denitrification potentials: measurement of seasonal variation using a short-term anaerobic incubation technique. *Soil Biol. Biochem.*, **14**, 179–184.

Matson, P. A., and Vitousek, P. M. (1981). Nitrification potentials following clearcutting in the Hoosier National Forest, Indiana. *For. Sci.*, **27**, 781–791.

Mattson, W. J., and Addy, N. D. (1975). Phytophagous insects as regulators of forest primary production. *Science*, **190**, 515–522.

McGee, C. E., and Hooper, R. M. (1975). Regeneration trends ten years after clearcutting of an Appalachian hardwood stand. USDA For. Serv. Res. Note SE-227, Southeast. For. Exp. Sta., Asheville, North Carolina.

McGinty, D. T. (1976). *Comparative Root and Soil Dynamics on a White Pine Watershed and in the Hardwood Forest in the Coweeta Basin*. Thesis, University of Georgia, Athens.

Meyer, J. L. (1983). Leaf-shredding insects as a source of dissolved organic carbon in headwater streams. *Am. Midl. Nat.*, **109**, 175–183.

Meyer, J. L., and Tate, C. M. (1983). The effects of watershed disturbance on dissolved organic carbon dynamics of a stream. *Ecology*, **64**, 33–44.

Mitchell, J. E., Waide, J. B., and Todd, R. L. (1975). A preliminary compartment model of the nitrogen cycle in a deciduous forest ecosystem. In: *Mineral Cycling in Southeastern Ecosystems* (F. G. Howell, J. B. Gentry and M. H. Smith, Eds.), ERDA Symp. Ser. (Conf-740513), pp. 41–52.

Montagnini, F., Haines, B., Boring, L., and Swank, W. (Coweeta files). Nitrification potentials in early successional black locust and in mixed hardwood forest stands in the Southern Appalachians. Southeast. For. Exp. Sta., Coweeta Hydrologic Laboratory, Otto, North Carolina.

O'Neill, R. V., and Reichle, D. E. (1980). Dimensions of ecosystem theory. In: *Forests: Fresh Perspectives from Ecosystem Analysis* (R. H. Waring, Ed.), Oregon State Univ. Press, Corvallis, pp. 11–26.

Permar, T. A., and Fisher, R. F. (1983). Nitrogen fixation and accretion by wax myrtle (*Myrica cerifera*) in slash pine (*Pinus elliottii*) plantations. *For. Ecol. Manage.*, **5**, 39–46.

Pierce, R. S., Martin, C. W., Reeves, C. C., Likens, G. E., and Bormann, F. H. (1972). Nutrient loss from clear-cutting in New Hampshire. In: *Proceedings Symposium Watersheds in Transition*, Am. Water Resour. Assoc. Proc. Ser. 14, pp. 285–295.

Robertson, G. P. (1982a). Nitrification in forested ecosystems. *Phil. Trans. R. Soc. Lond.*, **B296**, 445–457.

Robertson, G. P. (1982b). Factors regulating nitrification in primary and secondary succession. *Ecology*, **63**, 1561–1573.

Robertson, G. P., and Tiedje, J. M. (1984). Denitrification and nitrous oxide production in successional and old-growth Michigan forests. *Soil Sci. Soc. Am. J.*, **48**, 383–389.

Rowe, R., Todd, R. L., and Waide, J. B. (1976). A micro-technique for MPN analysis. *Appl. Environ. Microbiol.*, **33**, 675–680.

Ryan, D. F., and Bormann, F. H. (1982). Nutrient resorption in northern hardwood forests. *BioScience*, **32**, 29–32.

Ryden, J. C., Lund, L. J., and Focht, D. D. (1979). Direct measurement of denitrification loss from soils: I: Laboratory evaluation of acetylene inhibition of nitrous oxide reduction. *Soil Sci. Soc. Am. J.*, **43**, 104–110.

Schowalter, T. C. (1981). Insect herbivore relationships to the state of the host plant: biotic regulation of ecosystem nutrient cycling through ecological succession. *Oikos*, **37**, 126–130.

Seastedt, T. R., and Crossley, D. A. (1980). Effects of macroarthropods on seasonal dynamics of forest litter. *Soil Biol. Biochem.*, **12**, 337–342.

Seastedt, T. R., and Crossley, D. A. (1983). Nutrients in forest litter treated with naphthalene and simulated throughfall: a field microcosm study. *Soil Biol. Biochem.*, **15**, 159–165.

Seastedt, T. R., Crossley, D. A., and Hargrove, W. W. (1983). The effects of low-level consumption by canopy arthropods on the growth and nutrient dynamics of black locust and red maple trees in the Southern Appalachians. *Ecology*, **64**, 1040–1048.

Seastedt, T. R., and Tate, C. M. (1981). Decomposition rates and nutrient contents of arthropod remains in forest litter. *Ecology*, **62**, 13–19.

Silvester, W. B. (1978). Nitrogen fixation and mineralization in kauri (*Agathis australis*) forest in New Zealand. In: *Microbial Ecology* (M. W. Loutit and J. A. R. Miles, Eds.), Springer-Verlag, New York, pp. 138–143.

Smith, M. S., and Tiedje, J. M. (1979). Phases of denitrification following oxygen depletion in soil. *Soil Biol. Biochem.*, **11**, 261–267.

Sollins, P., Grier, C. C., McCorison, F. M., Cromack, K., Fogel, R., and Fredriksen, R. L. (1980). The internal element cycles of an old-growth Douglas-fir ecosystem in western Oregon. *Ecol. Monogr.*, **50**, 261–285.

Sollins, P., Spycher, G., and Glassman, C. A. (1984). Nitrogen mineralization from light- and heavy-fraction forest soil organic matter. *Soil Biol. Biochem.*, **16**, 31–37.

Strickland, T. C., and Fitzgerald, J. W. (1983). Mineralization of sulphur in sulphoquinovose by forest soils. *Soil Biol. Biochem.*, **15**, 347–349.

Strickland, T. C., and Fitzgerald, J. W. (1984). Formation and mineralization of organic sulfur in forest soils. *Biogeochemistry*, **1**, 79–95.

Strickland, T. C., Fitzgerald, J. W., and Swank, W. T. (1984). Mobilization of recently formed forest soil organic sulfur. *Can. J. For. Res.*, **14**, 63–67.

Swank, W. T., and Caskey, W. H. (1982). Nitrate depletion in a second-order mountain stream. *J. Environ. Qual.*, **11**, 581–584.

Swank, W. T., and Douglass, J. E. (1977). Nutrient budgets for undisturbed and manipulated hardwood forest ecosystems in the mountains of North Carolina. In: *Watershed Research in Eastern North America: A Workshop to Compare Results* (D. L. Correll, Ed.), Smithsonian Institution, Edgewater, Maryland, pp. 343–363.

Swank, W. T., Douglass, J. E., and Cunningham, G. B. (1982). Changes in water yield and storm hydrographs following commercial clearcutting on a Southern Appalachian catchment. In: *Hydrological Research Basins and their Use in Water Resource Planning: Proceedings of the International Symposium*, Berne, Switzerland, pp. 583–594.

Swank, W. T., Fitzgerald, J. W., and Ash, J. T. (1984). Microbial transformation of sulfate in forest soils. *Science*, **233**, 182–184.

Swank, W. T., Fitzgerald, J. W., and Strickland, T. C. (Coweeta files). Transformations of sulfur in forest floor and soil of a forest ecosystem. Paper presented at Acid Rain and Forest Resources Conference, 14–17 June 1983, Quebec City, Canada.

Swank, W. T., and Swank, W. T. S. (1984). Dynamics of water chemistry in hardwood and pine ecosystems. In: *Catchment Experiments in Fluvial Geomorphology* (T. P. Burt and D. E. Walling, Eds.), Geo Books, Norwich, United Kingdom, pp. 335–346.

Swank, W. T., and Waide, J. B. (1980). Interpretation of nutrient cycling research in a management context: evaluating potential effects of alternative management strategies on site productivity. In: *Forests: Fresh Perspective from Ecosystem Analysis* (R. W. Waring, Ed.), Oregon State Univ. Press, Corvallis, pp. 137–158.

Swank, W. T., Waide, J. B., Crossley, D. A., and Todd, R. L. (1981). Insect defoliation enhances nitrate export from forest ecosystems. *Oecologia (Berl.)*, **51**, 297–299.

Tarrant, R. F., Lu, K. C., Bollen, W. B., and Franklin, J. F. (1969). Nitrogen enrichment of two forest ecosystems by red alder. USDA For. Serv. Res. Pap. PNW-76, Pacific Northwest For. Exp. Sta., Portland, Oregon.

Tiedje, J. M. (1982). Denitrification. In: *Methods of Soil Analysis, Part 2* (2nd edn.) (A. L. Page, Ed.), Am. Soc. Agron., pp. 1011–1024.

Todd, R. L., Meyer, R. D., and Waide, J. B. (1978). Nitrogen fixation in a deciduous forest in the southeastern United States. *Ecol. Bull. (Stockholm)*, **36**, 172–177.

Todd, R. L., Swank, W. T., Douglass, J. E., Kerr, P. C., Brockway, D. L., and Monk, C. D. (1975). The relationship between nitrate concentration in the southern Appalachian Mountain streams and terrestrial nitrifiers. *Agro-Ecosystems*, **2**, 127–132.

Tripp, L. N., Bezdicek, D. F., and Heilman, P. E. (1979). Seasonal and diurnal patterns and rates of nitrogen fixation by young red alder. *For. Sci.*, **25**, 371–380.

Vance, E. D., Henderson, G. S. and Blevins, D. G. (1983). Nonsymbiotic nitrogen fixation in an oak–hickory forest following long-term prescribed burning. *Soil Sci. Soc. Am. Proc.*, **47**, 134–137.

Van Lear, D. H., Swank, W. T., Douglass, J. E., and Waide, J. B. (1983). Forest management practices and the nutrient status of a loblolly pine plantation. In: *IUFRO Symposium on Forest Site and Continuous Productivity*, USDA For. Serv. Gen. Tech. Rep. PNW-163, Pacific Northwest For. and Range Exp. Sta., Portland, Oregon, pp. 252–258.

Vitousek, P. M. (1984). Anion fluxes in three Indiana forests. *Oecologia (Berl.)*, **61**, 105–108.

Vitousek, P. M., Gosz, J. R., Grier, C. C., Melillo, J. M., and Reiners, W. A. (1982). A comparative analysis of potential nitrification and nitrate mobility in forest ecosystems. *Ecol. Monogr.*, **52**, 155–177.

Vitousek, P. M., Gosz, J. R., Grier, C. C., Melillo, J. M., Reiners, W. A., and Todd, R. L. (1979). Nitrate losses from disturbed ecosystems. *Science*, **204**, 469–474.

Vitousek, P. M., and Matson, P. A. (in press). Disturbance, nitrogen availability, and nitrogen losses: an experimental study in an intensively managed loblolly pine plantation, *Ecology*.

Vitousek, P. M., and Melillo, J. M. (1979). Nitrate losses from disturbed forests: patterns and mechanisms. *For. Sci.*, **25**, 605–619.

Wallace, J. B., Webster, J. R., and Cuffney, T. F. (1982). Stream detritus dynamics: regulation by invertebrate consumers. *Oecologia (Berl.)*, **53**, 197–200.

Webster, J. R. (1983). The role of benthic macroinvertebrates in detritus dynamics of streams: a computer simulation. *Ecol. Monogr.*, **53**, 383–404.

Webster, J. R., Gurtz, M. E., Hains, J. J., Meyer, J. L., Swank, W. T., Waide, J. B., and Wallace, J. B. (1983). Stability of stream ecosystems. In: *Stream Ecology* (J. B. Barnes and G. W. Minshall, Eds.), Plenum Publishing Corp., pp. 355–395.

Webster, J. R., and Patten, B. C. (1979). Effects of watershed perturbation on stream potassium and calcium dynamics. *Ecol. Monogr.*, **49**, 51–72.

Webster, J. R., Waide, J. B., and Patten, B. C. (1975). Nutrient recycling and the stability of ecosystems. In: *Mineral Cycling in Southeastern Ecosystems* (F. G. Howell, J. B. Gentry and M. H. Smith, Eds.), ERDA Symp. Ser. (Conf-740513), pp. 1–27.

Wollum, A. G., and Davey, C. B. (1975). Nitrogen accumulation, transformation, and transport in forest soils. In: *Forest Soils and Forest Land Management* (B. Bernier and C. H. Winget, Eds.), Laval Univ. Press, Quebec, Canada, pp. 67–106.

Wyer, M. D., and Hill, A. R. (1984). Nitrate transformations in southern Ontario stream sediments. *Water Resour. Bull.*, **20**, 581–584.
Youngberg, C. T., and Wollum, A. G. (1976). Nitrogen accretion in developing *Ceanothus velutinus* stands. *Soil Sci. Soc. Am. J.*, **40**, 109–112.
Younger, P. D., and Kapustka, L. A. (1983). $N_2(C_2H_2)$ASE activity by *Alnus incana* ssp. *rugosa* (*Betulaceae*) in the northern hardwood forest. *Am. J. Bot.*, **70**, 30–39.

Solute Processes
Edited by S. T. Trudgill

CHAPTER 4

Solute movement in soils

K. R. J. Smettem

Department of Soil Science,
University of Sydney

4.1 INTRODUCTION

The fundamental physical processes governing solute transport in porous media have been reasonably well understood for a considerable period of time. However, prediction of solute concentration changes with depth or over time in field soils remains particularly problematic. Difficulties arise owing to the complexity of natural soils, coupled with solute–soil interactions and frequently changing field conditions. For these reasons, quantification of solute movement in soils must involve some degree of approximation, while retaining the important aspects of real phenomena.

In this chapter, the current state of knowledge regarding the prediction of solute transport in soils is reviewed. Initially, however, since not all readers may be familiar with the basic theoretical concepts, these will be outlined. In developing the theory, a macroscopic viewpoint is adopted in which the details of within-pore distributions are replaced by macroscopic variables which are volume averages for the distributions of the corresponding microscopic variables. From this viewpoint, the mass flux density of a non-reactive solute in a rigid porous medium will comprise diffusion, bulk flow, and as a consequence of this flow, dispersion. It is convenient to commence by dealing briefly with each of these in turn.

4.2 DIFFUSION

Diffusion results from the random thermal motion of ions, atoms and molecules (Nye and Tinker, 1977). Although diffusion can occur in all phases, only liquid-phase diffusion will be considered in the present context.

Random thermal motion results in the eventual removal of any irregularities of the solution concentration (Wild, 1981), at a rate given by Fick's first law:

$$J = -D_1(dc/dx) \tag{4.1}$$

which is an empirical relationship between the net amount of solute crossing a unit cross-section in unit time and the driving force exerted by the concentration gradient (dc/dx). The relationship yields D_1, the liquid-phase diffusion coefficient, with the negative sign indicating movement from high to low concentration.

It can be noted that Equation (4.1) is mathematically similar to Darcy's law (see §1.3.2), and the coefficient D_1 is thus subjected to similar constraints in the soil as the hydraulic conductivity coefficient, being reducing as water content decreases (Porter *et al.*, 1960) and as tortuosity increases (Gardiner, 1965). The diffusion coefficient in soils D_s can, therefore, be defined as:

$$D_s = D_1\theta f_1 \tag{4.2}$$

where θ is the volumetric moisture content (cm^3cm^{-3}) and f_1 is an impedance factor (dimensionless).

The impedance factor accounts for tortuosity and has the effect of increasing the traversed path length and reducing the concentration gradient along this path length (Nye and Tinker, 1977). Since tortuosity is related to water content it follows that so too will be the impedance factor.

In soils, therefore, Equation (4.1) is generally expressed as:

$$J = -D_s(\theta)(dc/dx) \tag{4.3}$$

In order to describe transient state processes Equation (4.3) must be combined with the law of mass conservation as expressed in the continuity equation. For a uniform cube of soil (Figure 4.1), with solute entering through one plane and leaving through the opposite plane, the balance between the solute entering and leaving the cube in one direction is given by:

$$A(\partial c/\partial t)\,\Delta x = A[J + (\partial J/\partial x)\,\Delta x] - AJ \tag{4.4}$$

which reduces to:

$$\partial c/\partial t = -\,\partial J/\partial x \tag{4.5}$$

Combining equations (3) and (5) gives:

$$\partial c/\partial t = \partial[D_s(\theta)\,\partial c/\partial x]/\partial x \tag{4.6}$$

If D_s in Equation (4.6) is constant, then the analogy with Fick's second

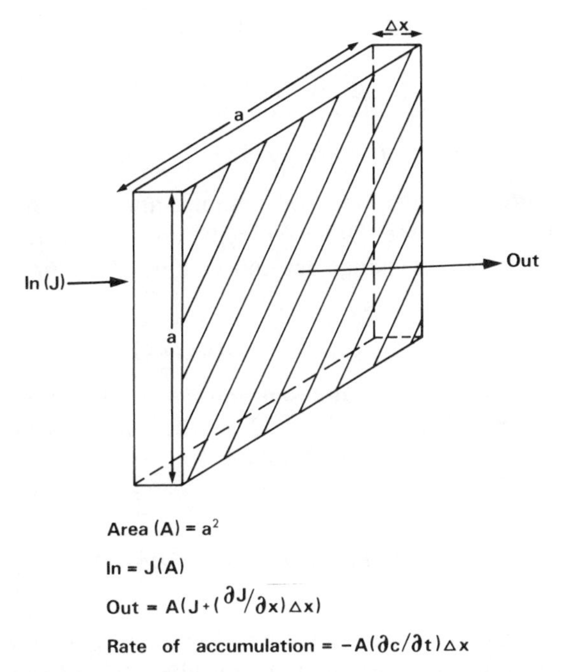

Figure 4.1. Diffusion through a horizontal slab

law can be used and Equation (4.6) reduces to:

$$\partial c/\partial t = D_s(\partial^2 c/\partial x^2) \tag{4.7}$$

Equation (4.7) should be recognized as a special case, since D_s is not only dependent on soil wetness but is also influenced by the proportion of diffusible ion in solution and by adsorption (Wild, 1981). The latter affects the rate of diffusion by desorption buffering of the solution concentration.

A wide range of solutions to Equation (4.7) satisfying different initial and boundary conditions have been given by Crank (1975). A number of useful numerical solutions are also given in Crank *et al*. (1981). Under many rainfall and irrigation regimes, the rate of solute diffusion is considerably slower than the rate of mass flow (convective transport). For example, one of the most diffusively mobile solutes, nitrate, has a diffusion coefficient at 0.01 bar of 10^{-6} cm^2 s^{-1} (Nye and Tinker, 1977). Since movement by diffusion results from random thermal motion, the concentration distribution is Gaussian (normal) with a root-mean square distance diffused of $(2D_st)^{\frac{1}{2}}$. For nitrate, the root-mean square movement is therefore about 0.5 cm in a day. An equivalent movement could be initiated by a rainfall event of 0.2 cm or less (Wild and Cameron, 1981), indicating the dominant effect of mass flow under most climatic conditions.

4.3 MASS FLOW

Following Darcy's law, the flux J of a solute due to mass flow in direction x can be expressed following Gardiner (1965) as:

$$J = qc = U\theta c \tag{4.8}$$

where c is the solute mass per unit soil volume, and U is the pore water velocity given by $(q/A)/\theta$, the Darcy velocity divided by the volumetric moisture content. The rate of change per unit volume of soil under mass flow is thus:

$$\partial c/\partial t = -\partial J/\partial x = -\partial(Uc/\partial x) \tag{4.9}$$

where c is the concentration per unit soil volume. It should be noted that Equation (4.9) assumes free drainage and all the solute to be in the soil solution at a uniform concentration (Wild and Cameron, 1981).

4.4 HYDRODYNAMIC DISPERSION

Spreading and mixing of new input water and solutes with that already in the soil system can be initiated by diffusion, but the dominance of mass flow tends to emphasize interactions between the fluid and the porous medium through which it passes. Dispersion that results primarily from the complexities of the pore system, rather than from the intrinsic motion of the molecules, is referred to as hydrodynamic dispersion (Bear, 1969; Biggar and Nielsen, 1963).

In even the simplified case of a straight capillary tube under constant laminar flow, hydrodynamic dispersion will occur owing to the parabolic velocity distribution that is associated with frictional resistance close to the periphery of the pore, decreasing towards the centre (Figure 4.2). In a volume of soil with a non-uniform pore size distribution this effect will be magnified owing to velocity changes with position within the medium.

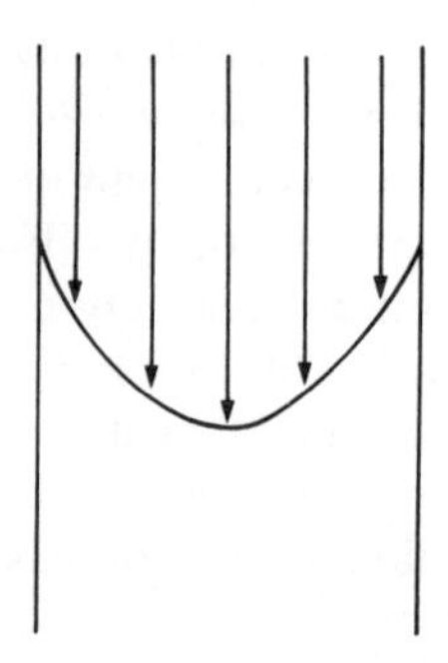

Figure 4.2. Dispersion in a tube

Furthermore, in field soils with tortuous flow paths, hydrodynamic dispersion will be affected to an even greater extent owing to the existence of different flow rates per unit length of soil.

It is clear from the above discussion that solute transfer in soils involves a subtle interplay between a number of variables. With the addition of further variables, such as density differences between incoming and resident solutions, or solute–solid phase interactions (anion exclusion, adsorption and chemical transformation), the enormity of the problem faced by the scientist wishing to quantify the process of solute movement begins to become clear. The reader should appreciate this and carefully examine the assumptions underlying a particular model, since none as yet is 'universal' and all are contingent upon particular specified conditions.

4.5 DESCRIPTION OF MASS TRANSFER

4.5.1 Saturated flow in a uniform porous medium

Miscible displacement experiments have become important tools in the quantitative analysis of solute concentration changes in soils. In order to introduce the basics of miscible displacement it is convenient to start with a simple system.

Consider a vertical column of sand, saturated, with a steady flow of solution from the influent (Figure 4.3). At a given time period the influent solution is step-changed to one containing a solute. (In this case it will be assumed that the solute does not interact with the porous medium and has identical physical properties to the original solution flowing through the column). If there is no dispersion (i.e. 'piston' flow is occurring) then the solute will arrive at any point down the column after the original solution above that point has been displaced.

For the finite system depicted in Figure 4.3, the solute will appear in the effluent after:

$$T = Qt/V_0 = 1 \tag{4.10}$$

where T is the number of pore volumes collected, Q is the volume discharge (cm^3), V_0 is the column volume occupied by fluid (cm^3), and t is time. It can be seen that Equation (4.10) is an expression of dimensionless time. This facilitates comparison between columns of different sizes.

The effluent concentration C_e can also be expressed in dimensionless ratio form as:

$$C_e = (C - C_i)/(C_0 - C_i) \tag{4.11}$$

where C is the solute concentration in the output, C_0 is the input

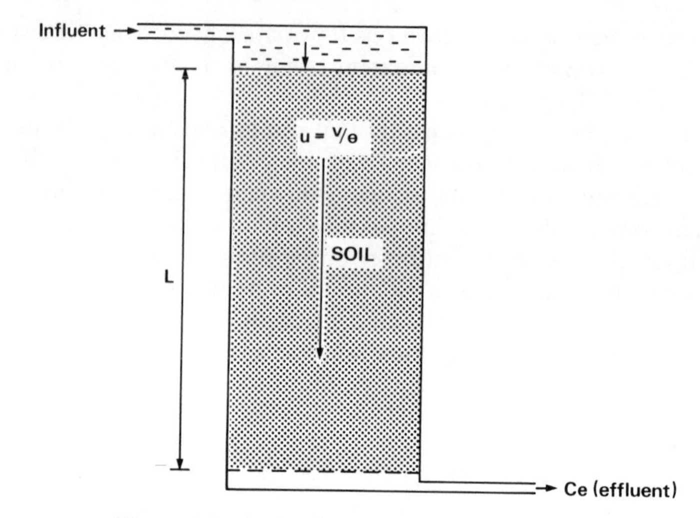

Figure 4.3. A simple experimental flow system

concentration, and C_i is the initial concentration in the system (assumed to be zero in the present discussion, reducing C_e to $= C/C_0$).

The graphical presentation of effluent concentration data in dimensionless form is a common practice in displacement studies and is referred to as the breakthrough curve (BTC). BTC (a) in Figure 4.4 demonstrates how the solute front would appear graphically in the absence of dispersion.

Spreading of the solute front by hydrodynamic dispersion is generally

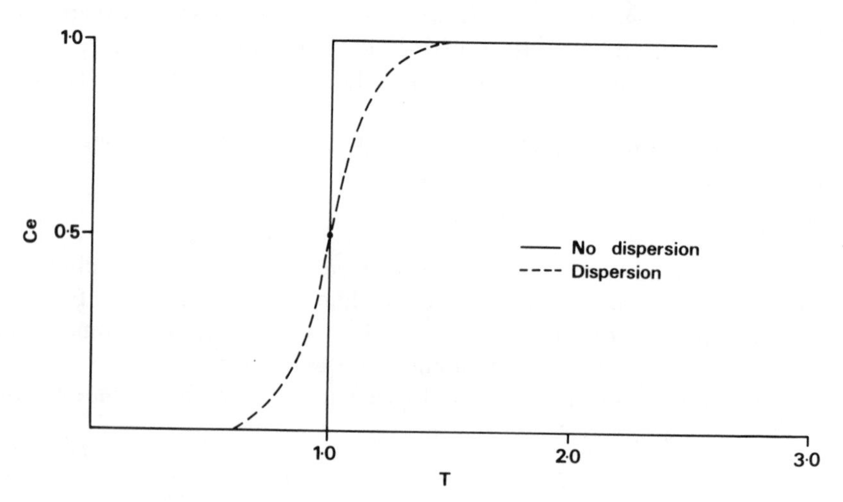

Figure 4.4. Frontal displacement breakthrough curves

described by the equation of Lapidus and Amundson (1952):

$$\partial c/\partial t = D'(\partial^2 c/\partial x^2) - U(\partial c/\partial x) \tag{4.12}$$

This in effect combines Equations (4.7) and (4.9) with the coefficient of diffusion D replaced by one of hydrodynamic dispersion D', the latter being independent of concentration. Equation (4.12) will generally yield sigmoid or symmetrical concentration distributions (Figure 4.4, curve (b)) when solved subject to specified initial and boundary conditions. The solutions, however, become skewed when U/D' is small, and this should not be confused with asymmetry arising out of the influence of secondary processes (see §4.7).

4.5.2 Pulse-type displacement

The type of displacement described in the previous section is referred to as frontal displacement. This involves a single-step concentration change, followed by a continuous application of solute at a particular concentration.

If the concentration of the influent were suddenly stepped back to the original concentration, then the solute would have been applied as a 'pulse' or discrete band. Two pulse-type BTCs are shown in Figure 4.5. Curve (a) displays no dispersion ('piston' flow) and curve (b) shows the influence of dispersion. It should be noted that owing to conservation of mass, the areas under the curves are identical.

For pulse-type displacement and semi-infinite boundary conditions, one possible analytical solution describing the effluent concentration is (Kirkham

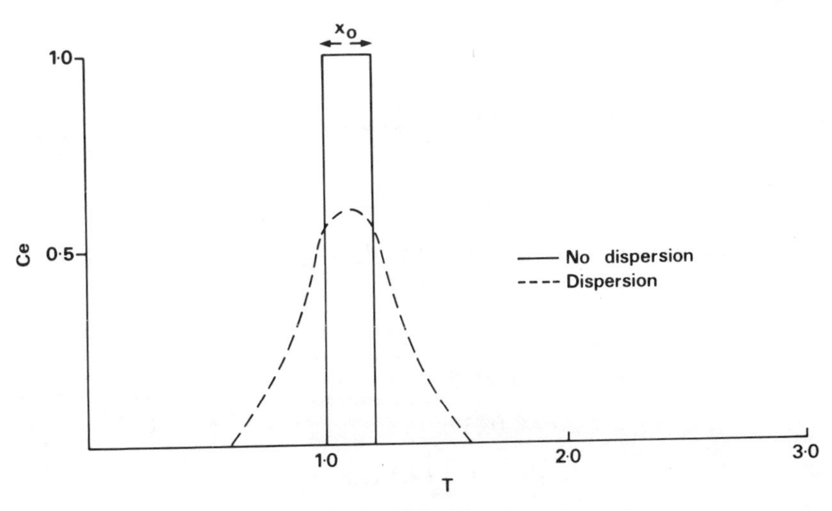

Figure 4.5. Pulse type breakthrough curves

and Powers, 1972):

$$C_e = 0.5\left\{\operatorname{erf}\left[\frac{1 + x_0/L - T}{(2D'T/UL)^{\frac{1}{2}}}\right] - \operatorname{erf}\left[\frac{1 - T}{(2D'T/UL)^{\frac{1}{2}}}\right]\right\} \tag{4.13}$$

where erf is the error function and x_0 is the length of the solute pulse.

4.5.3 Calculation of the dispersion coefficient

In order to solve hydrodynamic dispersion equations such as (4.13), a value must be derived for the only adjustable parameter, the dispersion coefficient D'. A number of methods exist for obtaining this from experimental data, most of which have been reviewed by Wild (1981).

For frontal displacement, the method of Rose and Passioura (1971) is generally regarded as the most accurate procedure for evaluating D' rapidly, and this should be consulted together with Rose (1973) for full details. Briefly, the method involves linearizing the BTC by plotting inverf $(2C_e - 1)$ against ln T on probability paper and obtaining the slope s which is a unique function of the Brenner (or Peclet) number ($P = UL/D'$) which defines the flow system (extreme limits being 'piston' flow; $P = 0$ and diffusion; $P \rightarrow \infty$). Although the exact relationship between s and P is mathematically complex, Rose (1973) has given some empirically derived polynomial expressions which hold with an error of no more than 0.05% in most cases.

Their method is, however, only applicable to frontal displacement experiments and other procedures must be employed to evaluate D' for pulse-type curves. One suitable equation given by Kirkham and Powers (1972) is:

$$D' = \frac{UL(x_0/2L)^2}{2(1 + x_0/2L)Z^2} \tag{4.14}$$

Where Z is the normal probability integral corresponding to $0.5C_{emax}$; i.e.

$$0.5C_{emax} = \frac{1}{\sqrt{(2\pi)}}e^{-x^2/2} \tag{4.15}$$

As a numerical example, consider the data given in Table 4.1. From this it is possible to derive:

$$x_0/2L = 0.4715/(2 \times 23.0) = 0.01025$$

$$0.5C_{emax} = 0.0095$$

$$Z = 0.0237$$

Inserting these values into Equation (4.14) results in:

$$D' = (0.51 \times 23 \times 0.01025)^2/2(1.01025 \times 0.0237^2) = 1.086 \text{ cm h}^{-1}$$

Table 4.1. Soil column data for a displacement experiment

Darcy velocity	0.18 cm h^{-1}
Pore water velocity	0.51 cm h^{-1}
Volumetric moisture content	0.355 cm^3 cm^{-3}
Column length	23.0 cm
Pulse volume	50 cm^3
Column cross-sectional area	798.7 cm^3
Pulse length	0.4715 cm

To calculate C_e at 1 pore volume D' is introduced into Equation (4.13):

$$C_e = 0.5[\text{erf}(0.0205/0.6085) - \text{erf}(0/0.6085)]$$

$$= 0.019$$

Subsequent points would be calculated in a similar manner.

4.5.4 Anion exclusion

Equation (4.13) has been derived on the assumption that all the liquid phase is available for mass flow. However, in many circumstances anions are negatively adsorbed near solid surfaces owing to electrostatic and chemical forces (Elprentice and Day, 1977). A situation can thus arise in which the anion concentration is greater in the fast-moving region of the velocity profile than in the slow-moving region close to the pore peripheries. This results in the anion travelling faster than the fluid as a whole (Bigger and Nielsen, 1962; Thomas and Swoboda, 1970). To account for this, the value of T in Equation (4.13) can be adjusted to allow for an excluded fraction, and Equation (4.10) becomes:

$$T' = Qt/V_s \quad (4.16)$$

where V_s is the 'effective' solution volume (cm^3) (i.e. that portion containing anions). A procedure for fitting Equation (4.13) using two adjustable parameters has been outlined by Elprentice and Day (1977).

4.6 MASS TRANSFER WITH ADSORPTION

The process of adsorption involves the passage of a chemical species from one bulk phase to the surface of another where it accumulates without penetrating the structure of this second phase. (Burchill *et al.*, 1981). In terms of displacement, interest is focused on solute accumulation at the liquid/solid interface which results in attenuation of the BTC.

Most solutes are adsorbed to a greater or lesser extent and it is, therefore, of practical importance to include this process when describing mass transfer. Procedures for obtaining a quantitative description of adsorption are outlined below.

In order to describe adsorption at the macroscopic scale, molecular considerations must be set aside and the process studied empirically in relation to bulk quantities of material.

4.6.1 Equilibrium adsorption

The rate of solute movement relative to water can be determined by an adsorption coefficient expressing the partitioning of solute between the solid and solution phases. To calculate this partitioning, weighed soil samples are shaken for a predetermined time period with a series of increasing solution concentrations at the fixed soil:solution ratio (usually 1 : 2 or 1 : 3). The final solution concentration (at equilibrium) is determined, and the difference between the initial and final concentrations yields the amount adsorbed.

The results are expressed in the form of an adsorption isotherm with the solution concentration in g cm^{-3} and the solid-phase concentration in $\mu g\ g^{-1}$ (Figure 4.6). The slope of the resulting isotherm yields the adsorption distribution coefficient.

For a linear isotherm the following relationship applies:

$$S = bC \tag{4.17}$$

where S is the amount of solute adsorbed ($g\ g^{-1}$), C is the solution concentration ($g\ cm^{-3}$), and b is the adsorption distribution coefficient. If the isotherm is non-linear (curve B, Figure 4.6), then Equation (4.17) is modified to:

$$S = bC^N \tag{4.18}$$

where N is a constant.

Once the distribution coefficients are known, retardation factors R may be obtained from:

(a) linear $\quad R = 1 + (\rho b/\theta) \qquad (4.19)$

(b) non-linear $\quad R = 1 + (\rho b/\theta)NC^{N-1} \qquad (4.20)$

For frontal displacement and a linear adsorption isotherm, R is often taken as equal to the point at which the effluent concentration equals $0.5C_e$ (MacDonald *et al*., 1976).

Equation (4.12) can be modified to include adsorption by including an additional term on the RHS. The general transport equation thus becomes:

$$\frac{\rho}{\theta}\frac{\partial S}{\partial t} + \frac{\partial C}{\partial t} = D'\frac{\partial^2 C}{\partial x^2} - U\frac{\partial C}{\partial x} \tag{4.21}$$

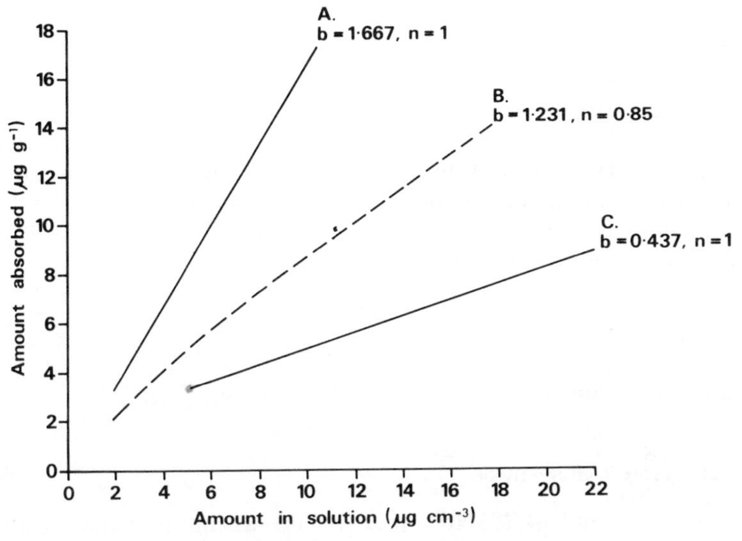

Figure 4.6. Adsorption isotherms

Substitution of Equations (4.19) or (4.20) into (4.21) yields:

$$\frac{\partial C}{\partial t} = \frac{1}{R}\left(D'\frac{\partial^2 C}{\partial x^2} - U\frac{\partial C}{\partial x}\right) \tag{4.22}$$

The effect of a small increase in R on the shape of a pulse-type BTC is shown in Figure 4.7. The curves were computed using a semi-infinite constant

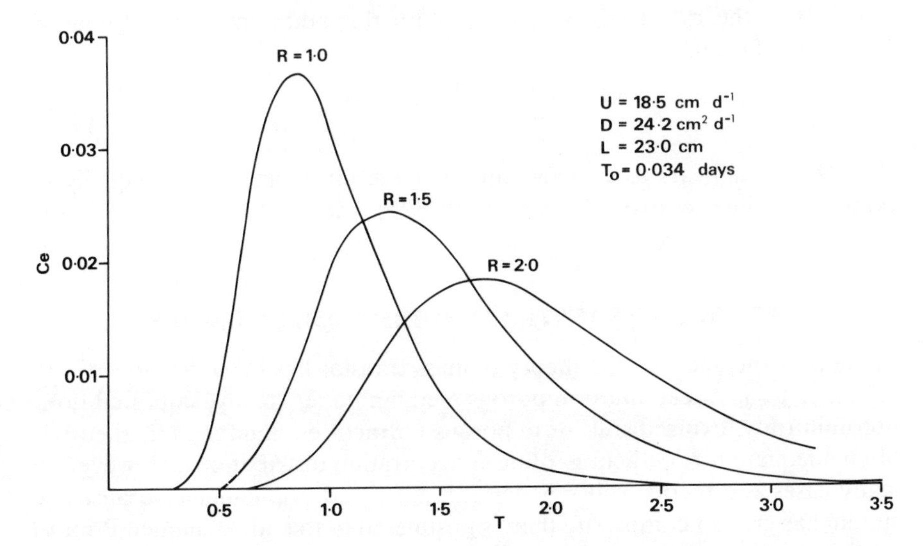

Figure 4.7. Breakthrough curves showing the effect of increasing solute adsorption

concentration-type boundary condition solution to Equation (4.22) given by Lapidus and Amundson (1952).

4.6.2 Kinetic adsorption

If the rate of adsorption is slow compared with the rate of flow, then an approach based on reaction kinetics may better describe the situation. One of the most widely used models is that of Lapidus and Amundson (1952), who proposed the following first-order rate expression:

$$\frac{\partial S}{\partial t} = \alpha(bc - S) \tag{4.23}$$

where α is a rate constant (t^{-1}) (usually the desorption rate).

4.6.3 Irreversible adsorption

Equations (4.21) and (4.22) are based on the assumption that adsorption (described as an equilibrium or kinetic process) is reversible. The chemical is thus seen as spending a certain amount of time in solution and a certain amount on the solid phase (in general, the fraction of time spent in solution is given by $1/(1 + b)$). In some cases, however, irreversible adsorption may occur. Alternately, the solute may be transformed or otherwise lost to the system. For either case, a term is required which will describe this loss. One possible sink term is (Misra and Mishra, 1977):

$$Q = MC \tag{4.24}$$

where M is the rate coefficient (t^{-1}). With this additional term, Equation (4.21) now becomes:

$$\frac{\rho}{\theta}\frac{\partial S}{\partial t} + \frac{\partial C}{\partial t} = D'\frac{\partial^2 C}{\partial x^2} - U\frac{\partial C}{\partial x} \pm Q \tag{4.25}$$

Note that Q appears as $\pm$, indicating that the term can be used equally to express a solute source (Rose *et al.*, 1982b) and applies to the bulk soil volume.

4.7 MASS TRANSFER IN MORE COMPLEX SOILS

In the previous sections the theory of mass transfer has been developed with reference to an 'ideal' uniform porous medium under steady, saturated flow. For uniformly textured soils or in porous unfractured aquifers, this theory is often adequate for predicting solute concentration distributions. However, in many cases the theory will not describe observed phenomena because the system has greater complexity than is postulated in the 'ideal' model. Prior to examining the problem at the field scale it is pertinent to consider three extensions of dispersion theory to account for more complex systems.

4.7.1 Layered soils

Some soils exhibit profiles with abrupt textural changes between, say, the A and B horizons. Thus, for example, a sandy A horizon might overlay a sandy-clay B horizon with a clay C horizon below this (or vice versa). This is shown schematically in Figure 4.8. In predicting the solute concentration distribution within such a system it is important to distinguish between effluent BTCs and solute distributions in the soil profile. The former case has been studied extensively by Selem *et al*. (1977) for steady flow conditions in both saturated and unsaturated soils. They concluded that the order of soil layering did not influence the effluent concentration distribution. For the case of linear adsorption, Equation (4.22) could be used with average values, to describe the BTC. For example, an average retardation factor $\bar{R}$ for n-layered soil can simply be obtained from:

$$\bar{R} = \frac{1}{L} \sum_{i=1}^{n} R_i L_i \tag{4.26}$$

Although an average retardation factor $\bar{R}$ cannot be employed in the case of non-linear or kinetic adsorption, these are also unaffected by the layer sequence, as is the sink term (Equation (4.24)).

If the solute distribution in the soil profile is desired then the use of average values is no longer valid and the problem must be treated as a multilayered

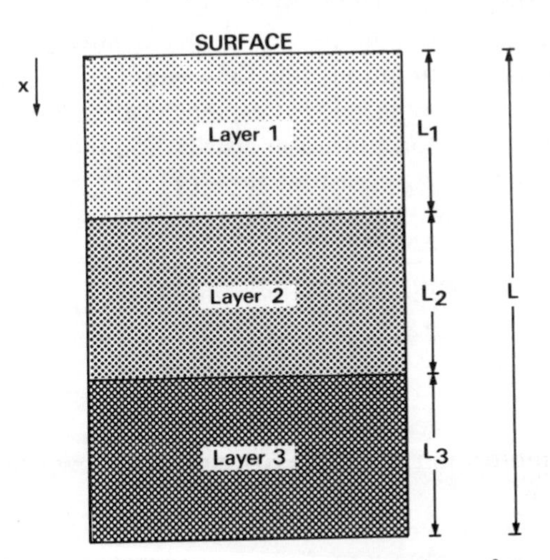

Figure 4.8. Schematic representation of a layered soil

case (Selem *et al*., 1977; Ghuman and Prihar, 1980). This involves solving Equation (4.25) for each soil layer with additional boundary conditions imposed at the layer interfaces in order to maintain solute concentration continuity at the boundaries.

4.7.2 Unsaturated soils

Under many field conditions the steady, saturated flow of water will be a rare occurrence in the surface soil layers. Most frequently, solute transfer will occur under transient, unsaturated flow. To simulate such conditions extensive use has been made of numerical techniques, particularly finite difference solutions of the appropriate differential equations (e.g. Bresler, 1973; Kirda *et al*., 1973). Although such solutions are not limited by restrictive boundary conditions and have application to a wide variety of situations, they require extensive input information and are rather complex. This limits their use for solving practical problems.

A particular difficulty is that exact solutions require extensive information concerning both hydraulic-conductivity/moisture-content and pressure-head/moisture content relationships, which are often difficult and time-consuming to determine. (Recent advances in calculating such properties from the soil moisture characteristic curve may simplify this restriction; e.g. Mualem, 1976; Mualem and Dagan, 1978.) This problem has led a number of workers to investigate the application of approximate analytical solutions which do not require such complex input information (Warrick *et al*., 1971; De Smedt and Wierenga, 1978; Rose *et al*., 1982a,b).

For a non-reactive solute, an analogy to the steady-state process is made. This is based on the assumption that the overall infiltration time is much longer than the time required to establish a steady infiltration rate. Calculation of the solute concentration distribution then follows in two parts. Firstly, using the analogy of 'piston' flow (§4.5.1), the solute penetration depth is estimated using one of the following methods (De Smedt and Wierenga, 1978).

For steady-state constant uniform infiltration:

$$\S = (q_1/\theta_1)t \tag{4.27}$$

where q_1 is the constant flux, and θ_1 is the moisture content of the upper soil profile.

For saturated infiltration when the *i(t)* curve is known:

$$\S = i(t)/\theta_s \tag{4.28}$$

where $i(t)$ is the cumulative infiltration, and θ_s is the moisture content at saturation.

During redistribution, determination of §(t) is more problematic. However, after several days the water content will approach the so-called 'field capacity' value θ_{fc} and can be calculated from:

$$\S = I/\theta_{fc} \tag{4.29}$$

where I is the total amount of water infiltrated (cm). The effect of dispersion around the estimated penetration depth is then calculated from a suitable approximate analytic solution to the dispersion-convection equation (Equation 4.12)). For frontal displacement and semi-infinite boundary condition, the theory of De Smedt and Wierenga (1978) leads to:

$$C_e = 0.5 \operatorname{erfc} \frac{x - \S}{2(D_L t + \varepsilon\S)^{\frac{1}{2}}} \tag{4.30}$$

During the redistribution (post-infiltration) phase the pore water velocity will decline towards zero and the influence of molecular diffusion on solute spreading will consequently increase. In the above case, therefore, dispersion is assumed to be proportional to the square root of the sum of two terms, with a relationship given by:

$$D' = D_L + \varepsilon U \tag{4.31}$$

where ε is the proportionality factor (cm).

The molecular diffusion coefficient is assumed constant and causes increased spreading with time, while the coefficient of mechanical dispersion is proportional to the pore water velocity and causes increased solute spreading with depth. For a pulse-type displacement, Rose *et al*. (1982b) have used a similar theoretical approach to give:

$$C_e = 0.5 \left\{ \operatorname{erfc}\left[\frac{x - \S}{2(D_L t + \varepsilon\S)^{\frac{1}{2}}} \right] - \operatorname{erfc}\left[\frac{x - \iota}{2(D_1 t + \varepsilon\iota)^{\frac{1}{2}}} \right] \right\} \tag{4.32}$$

where $\iota = (\S - \Delta F)$ and ΔF is the thickness of the solute pulse (cm).

With suitable modifications Equations (4.30) and (4.32) will also account for solute sources and sinks (Rose *et al*., 1982b) and could equally be extended to account for anion exclusion (De Smedt and Wierenga, 1978) or linear equilibrium adsorption. The latter two cases would be reflected in the solute penetration depth, thus:

$$R\S = I/\theta_{fc} \tag{4.33}$$

with $R > 1$ for adsorption and $R < 1$ for anion exclusion. In a recent field study under both irrigation and winter rainfall conditions, Cameron and Wild (1982) found Equation (4.32) to be more successful than the layer models of Burns (1974) or Addiscott (1977) in predicting chloride movement in a field soil.

One of the main limitations appears to be in adequately characterizing the dispersivity ($\varepsilon = (D' + D_L)/U$) of field soils under differing input and moisture regimes. This is considered subsequent to be following section.

4.7.3 Aggregated soils

The mass transfer theory developed above is based on the application of average values to account for the distribution of microscopic variables. By adopting this viewpoint the soil is conceptualized as a continuous porous medium resulting from the random packing of mineral soil particles. In natural soils these primary soils particles are often aggregated or clustered to form peds, which are described by Hodgson (1974) as 'natural, relatively permanent aggregates, separated from each other by voids or natural surfaces of weakness.' As a result, the soil may be composed of slowly and rapidly conducting pore sequences, with solute transfer dominated by displacement and hydrodynamic dispersion in the larger inter-aggregate pores and by diffusion in the smaller intra-aggregate pores. A common characteristic of such systems is that the concentration distribution may be strongly non-sigmoid or asymmetric (Key and Elrick, 1967; van Genuchten and Wierenga, 1977; Rao *et al*., 1980; Nkedi Kizza *et al*., 1982) either with depth or via the BTC.

The initial problem is that general solutions to Equation (4.12) predict nearly sigmoid or symmetrical concentration distributions (except when U/D is small). This has led workers to propose the concept of 'immobile' water (Coats and Smith, 1964) in order to account for asymmetry by application of a term accounting for sideward (diffusion-controlled) exchange. For a non-reactive solute under steady flow, the resulting equations are (Coats and Smith, 1964):

$$\theta_m \frac{\partial C_m}{\partial t} + \theta_{im} \frac{\partial C_{im}}{\partial t} = \theta_m D'_m \frac{\partial^2 C_m}{\partial x^2} - U_m \theta_m \frac{\partial C_m}{\partial x} \tag{4.34}$$

$$\theta_{im} \frac{\partial C_{im}}{\partial t} = \omega (C_m - C_{im}) \tag{4.35}$$

Where θ_m and θ_{im} are the mobile and immobile moisture contents, C_m and C_{im} are the concentrations in both regions, U_m is the average 'mobile' pore water velocity, and ω is a rate constant for sideward exchange (day^{-1}).

A consequence of the presence of 'immobile' water is that the average interstitial velocity has to increase for a given flux (van Genuchten and Wierenga, 1976; Gaudet *et al*., 1977), resulting in an early appearance of the solute front. van Genuchten and Wierenga (1976) noted that Equations (4.34) and (4.35) were mathematically similar to Equations (4.21) and

(4.23), and therefore combined them to present a solution for mass transfer in aggregated media that included adsorption. The reader should refer to their original article for further details.

It is important to note that asymmetric solute profiles are not solely a characteristic of aggregated media. Nielsen and Biggar (1961) presented experimental evidence to indicate that under unsaturated conditions the larger pores are eliminated and the proportion of water that does not readily move is increased. Under such circumstances the approximate analytical solutions presented in the previous section may be inappropriate. Asymmetry may also arise from channelling flow phenomena through 'macropores' (Germann and Beven, 1981) or vertically continuous planar voids (Anderson and Bouma, 1977). These influences are considered below.

4.8 MASS TRANSFER IN FIELD SOILS

In the field, the pattern of solute movement may be strongly influenced by two additional factors: the presence of well-connected large pores and spatial variability of parameters influencing the flow process.

These will be considered in turn, but are closely linked. The occurrence of large localized fluxes resulting from preferential flow is one particular process by which variations in the water flux distribution can arise.

4.8.1 Influence of macrostructure

Although difficulties associated with exact definition exist (Luxmore, 1981; Beven, 1981; Bouma, 1981; Skopp, 1981), it would appear that preferential flow may occur in pores as small as 0.02 cm diameter (Scotter, 1979). There is little doubt that, provided the pores are continuous and open to the source of solute supply (usually the soil surface), they strongly influence the pattern of solute movement under saturated flow conditions (Ritchie *et al*., 1972; Anderson and Bouma, 1977) or under intense rainfall (Omoti and Wild, 1977). However, few studies have been undertaken to assess their effect on solute movement at a scale compatable with hydrodynamic dispersion theory.

McMahon and Thomas (1974) presented BTCs for chloride and tritiated water ($^3H_20$) flow, through disturbed and undisturbed cores from three soils under steady, saturated flow. Although their observations were essentially qualitative, they concluded that disturbing the soil removed the influence of well-connected structural pores, thereby influencing the pattern of displacement. It is interesting to re-analyse their data quantitatively using a solution to Equations (4.34) and (4.35). In order to simplify the discussion, only their BTCs for tritiated water are considered.

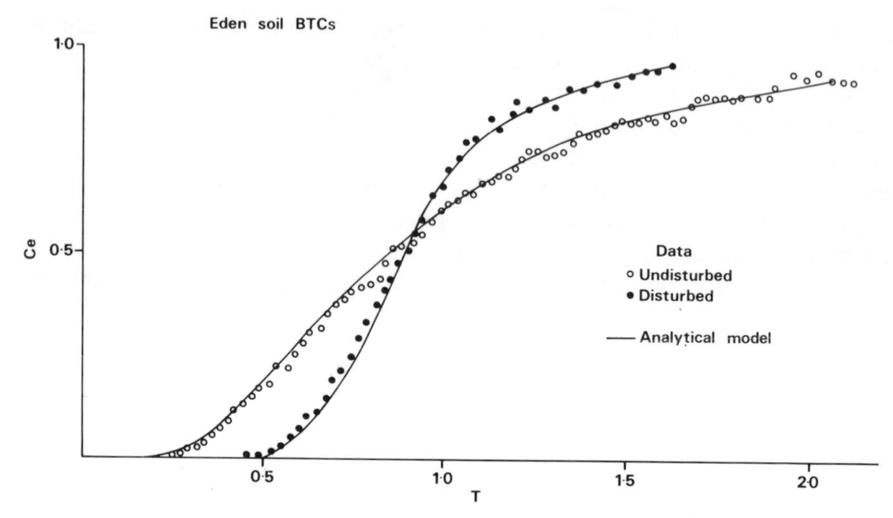

Figure 4.9. Eden soil breakthrough curves. *Reproduced by permission of the Soil Science Society of America*

Introduce the following dimensionless variables:

$$\phi = \theta_m/\theta = \theta_m/(\theta_m + \theta_{im}) \tag{4.36}$$

$$C_1 = (C_m - C_i)/(C_0 - C_i) \qquad C_2 = (C_{im} - C_i)/(C_0 - C_i) \tag{4.37}$$

$$P = U_m L/D' \qquad Z = x/L \tag{4.38}$$

$$\Omega = \omega L/q = \omega L/(\theta_m U_m) \tag{4.39}$$

Equations (4.34) and (4.35) become:

$$\phi(\partial C_1/\partial T) + (1 - \phi)(\partial C_2/\partial T) = P^{-1}(\partial C_1/\partial Z^2) - (\partial C_1/\partial Z) \tag{4.40}$$

$$(1 - \phi)(\partial C_2/\partial T) = \Omega(C_1 - C_2) \tag{4.41}$$

These can be solved subject to the boundary conditions and procedure of van Genuchten (1981), detailed in Smettem (1984).

The analytical model appears to describe the experimental data extremely well (Figures 4.9, 4.10, 4.11). The resulting dimensionless and transformed transport parameters are given in Table 4.2. The pore water velocity increases slightly in the undisturbed cores because the value of ϕ decreases. This suggests that channelling flow may emphasize the effects of a smaller number of transmission pores than more uniform flow in the disturbed cores. This is supported by the observation that the dispersion coefficient is much larger in the undisturbed cores and appears to deviate from a linear relationship with the average pore water velocity. A similar observation has been reported by Smettem (1984), who suggested that the effect was due to channelling flow

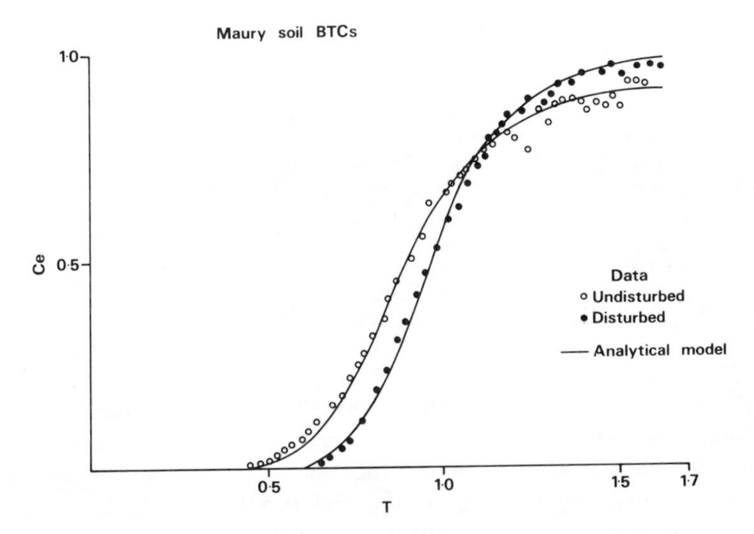

Figure 4.10. Maury soil breakthrough curves. *Reproduced by permission of the Soil Science Society of America*

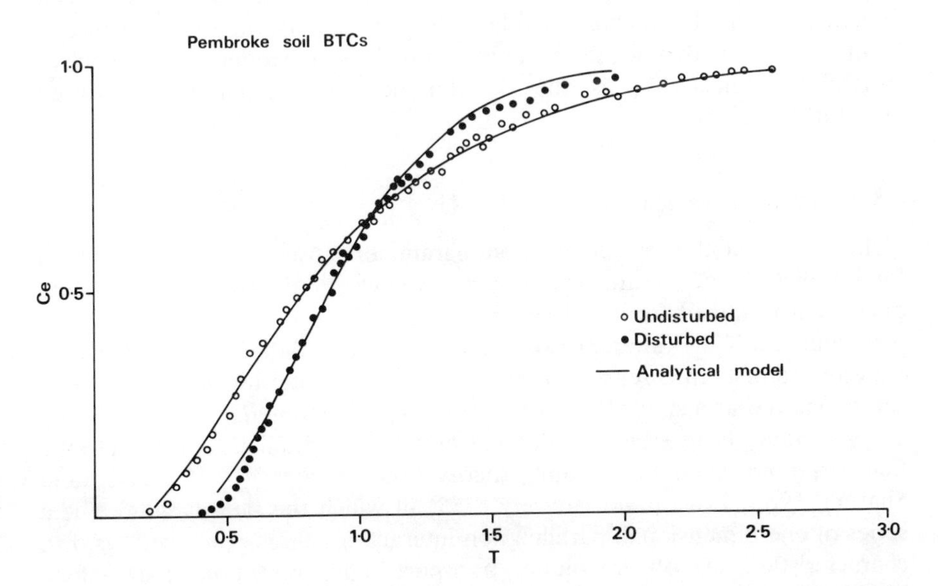

Figure.4.11. Pembroke soil breakthrough curves. *Reproduced by permission of the Soil Science Society of America*

Table 4.2. Dimensionless and transformed breakthrough curve parameters

	P	ϕ	Ω	V_m (cm d^{-1})	D' (cm^2 d^{-1})	ω (d^{-1})	D/V_m (cm)
Pembroke undisturbed	4.11	0.78	0.00	16.52	102.10	0.00	6.18
Pembroke disturbed	11.50	0.89	0.00	14.03	30.99	0.00	2.21
Eden undisturbed	5.21	0.82	0.05	13.27	64.69	0.006	4.87
Eden disturbed	31.60	0.89	0.08	11.87	9.54	0.013	0.80
Maury undisturbed	36.90	0.85	0.21	14.55	22.04	0.016	1.51
Maury disturbed	53.90	0.96	0.09	12.84	13.32	0.007	1.04

producing a skewed pore water velocity distribution in undisturbed soil-cores with visible macrostructure.

It should be noted that in Table 4.2 the Maury soil core was 56 cm and the Eden and Pembroke cores 25 cm. In the analytical models the hydrodynamic dispersion coefficient D' is assumed to be independent of sample length. In the case of structured soils this may be incorrect. Bouma (1982) has demonstrated a reduction in saturated hydraulic conductivity with increased sample length, due to decreasing connectivity of the larger pores. A similar argument could be advanced for the hydrodynamic dispersion coefficient, although at present there is a paucity of experimental information with which to test this hypothesis. Although the case study presented here refers to a limited amount of data, it is evident that dispersivity in the undisturbed Maury core is less than for the other two soils. This may reflect an influence of structural connectivity, as the dispersivities are similar for all three undisturbed soil cores.

4.8.2 Solute movement at the field scale

Field scale variability of soil and input parameters poses a particular problem for prediction of solute movement. Analytical discriptions of solute movement obtained from soil columns can, strictly, only be applied to the field situation if the inherent variation of generated data is zero at the field scale of interest. In a sense, therefore, models of deterministic phenomena can be viewed as a special case of probabilistic phenomena.

Some attempts have been made to extend the deterministic approach to the field scale. an example is scaling theory (Peck *et al.*, 1977; Luxmore and Sharma, 1980; Dagan and Bresler, 1979) in which the field is viewed as a series of one-dimensional, parallel, non-interacting regions each with its own characteristic set of hydraulic transport and retention parameters. Theoretically, a set of geometric scaling factors is obtained by making, say, infiltration or matric potential–water content measurements through the

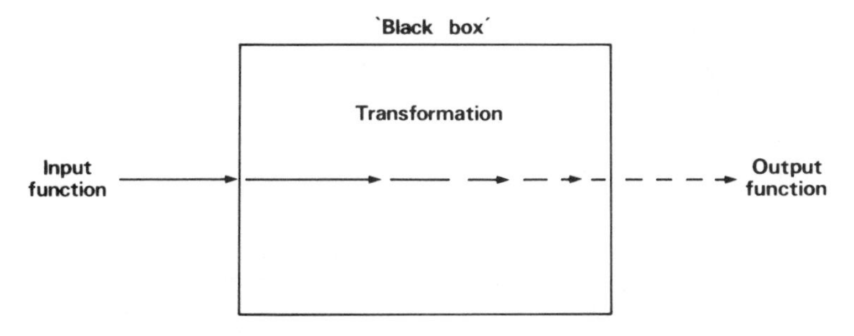

Figure 4.12. Schematic representation of a transfer function system

study area. These scaling factors are then used to calculate hydraulic and retention functions at different parts of the field from a standard set of such functions measured at one particular point. The assumption of similarity is, however, only approximately valid at best (Warrick *et al.*, 1977), and the approach also suffers from the presupposition of no lateral interaction between adjacent flow regions. This problem has led Jury (1982) to propose the application of a transfer function approach based on the assumption that, at the field scale, the soil transport properties are a 'black box' (unknown or unknowable). The envisaged system is depicted in Figure 4.12.

The output function is influenced by two distinct sources of random variation: variations in the input and variations in the transformation. For a non-reactive solute, the second will depend on the travel time density distribution and on variations in dispersion. Dagan and Bresler (1979b) have demonstrated that solute spread due to field scale heterogeneity is much larger than conventional pore-scale hydrodynamic dispersion. As an approximation, therefore, they have suggested that solute spread can be predicted by neglecting hydrodynamic dispersion and considering convection coupled with heterogeneity. This approximation is also implicit in the model of Jury, which requires a calibration of either the travel time density function, or a distribution function of net water input or displacement volume (I) required to transfer solute to a given depth.

Comparatively little information is available on the distribution of soil water and solute transport parameters in the field. Most have generally described log–normal distributions (Nielsen *et al.*, 1973; van de Pol *et al.*, 1977; Biggar and Nielsen, 1976). Jury, therefore, used the following log–normal travel time density function in his model calculations (based on data from van de Pol *et al.*):

$$f_L(I) = \exp\left[\frac{-(\ln I - \mu)^2/\sigma^2}{2\pi^{\frac{1}{2}}\sigma I}\right] \qquad (4.42)$$

Where μ is the mean and σ is the standard deviation. Several analytical solutions to Equation (4.42) are available for different input functions $C_{in}(I)$. Jury gives the following for a step function change in input, analogous to frontal displacement:

$$C_{in} = 0 \qquad I \leqslant 0 \tag{4.43}$$

$$C_{in} = C_0 \qquad I > 0 \tag{4.44}$$

$$C_{out}(Z,I) = C_0/2\left(1 + \text{erf}\left\{\frac{[\ln(IL/x) - \mu]}{(2)^{\frac{1}{2}}\sigma}\right\}\right) \tag{4.45}$$

It is interesting to not that when the travel time coefficient of variation is less than 1, the resulting cumulative density function (Equation (4.45)) or probability density function has an equivalent shape to deterministic BTC solutions of Equation (4.12). Therefore, as pointed out by Simmons (1982), the transport process could equally be viewed as resulting from a log–normal distribution of hydraulic conductivity.

Purely on the basis of 'fitting a curve', the log-normal distribution has often been used in engineering problems owing to its highly flexible shape (Bury, 1975). As a result, care must be taken to avoid process inferences simply because the distribution of certain field scale flow parameters can be described by the log–normal model.

Nevertheless, further development of mechanistic-stochastic models incorporating, for example, retarded or accelerated travel-time density functions may be a worthwhile line of enquiry.

4.9 CONCLUSIONS

Successful characterization of solute movement in soils can often be achieved by solving a suitable diffusive-type governing equation. The solution of the equation, however, entails an input of specified physical parameters, and it is in this area that he need for further research is most apparent. In particular, more information is needed on the distribution of flow parameters at the field scale and the manner in which these distributions alter temporally.

It is also now widely accepted that in many cases disturbance of a soil sample (e.g. repacking) will alter the observed transport parameters. An explanation for this is that the connectivity of the larger soil voids has been disrupted. Our understanding of the role played by larger soil voids in solute redistribution, the conditions under which they operate and the spatial scale of their operation, are all areas in which further research is needed. Such research may ultimately lead to a closer unification of laboratory and field scale modelling concepts and thus improve our ability to quantitatively predict the fate of both surface applied and soil derived solutes.

REFERENCES

Addiscott, T. M., (1977). A simple computer model for leaching in structured soils. *J. Soil Sci.*, **28**, 554–563.

Anderson, J. L., and Bouma, J. (1977). Water movement through pedal soils. 1: saturated flow. *Soil Sci. Soc. Am. J.*, **41**, 413–418.

Bear, J. (1969). Hydrodynamic dispersion. In *Flow Through Porous Media* (J. M. de Wiest, Ed.), Academic Press, New York.

Beven, K. (1981). Micro-, Meso-, Macroporosity and channeling flow phenomena in soils. *Soil Sci. Soc. Am. J.*, **45**, 1245.

Biggar, J. W., and Nielsen, D. R. (1962). Miscible displacement. II: Behaviour of tracers. *Soil Sci. Soc. Am. Proc.*, **26**, 125–128.

Biggar, J. W., and Nielsen, D. R. (1976). Spatial variability of the leaching characteristics of a field soil. *Water Resour. Res.*, **12**, 78–84.

Bouma, J. (1981). Comment on 'Micro-, Meso-, and Macroporosity of Soil. *Soil Sci. Soc. Am. J.*, **45**, 1244.

Bouma, J. (1982). Measuring the hydraulic conductivity of soil horizons with continuous macropores. *Soil Sci. Soc. Am. J.*, **46**, 438–441.

Bresler, E. (1973). Simultaneous transport of solutes and water under transient unsaturated flow conditions. *Water Resour. Res.*, **9**, 975–986.

Burchill, S., Hayes, M., and Greenland, D. J. (1981). Adsorption of organic molecules. In: *The Chemistry of Soil Processes*. (D. J. Greenland and M. Hayes, Eds.), Wiley, London.

Burns, I. G. (1974). A model for predicting the redistribution of salts applied to fallow soils after excess rainfall or evaporation. *J. Soil Sci.*, **25**, 165–178.

Bury, K. V. (1975). *Statistical Models in Applied Science*, Wiley, New York.

Cameron, K. C., and Wild, A. (1982). Prediction of solute leaching under field conditions: an appraisal of three methods. *J. Soil Sci.*, **33**, 649–659.

Coats, K. H., and Smith, B. D. (1964). Dead-end pore volume and dispersion in porous media. *Soc. Pet. Eng. J.*, **4**, 73–84.

Crank, J. (1975). *The Mathematics of Diffusion* (2nd edn.) Clarendon Press, Oxford.

Crank, J., McFarlane, J. C., Paterson, G. D. and Peddey, J. B. (1981). *Diffusion processes in Environmental Systems*, MacMillan.

Dagan, G., and Bresler, E. (1979). Solute dispersion in unsaturated heterogeneous soil at field scale. 1: Theory. *Soil Sci. Soc. Am. J.*, **43**, 461–467.

De Smedt, F., and Wierenga, P. J. (1978). Approximate analytical solution for solute flow during infiltration and redistribution. *Soil Sci. Soc. Am. J.*, **42**, 407–412.

Elprentice, A. M., and Day, P. R. (1977). Fitting solute breakthrough equations to data using two adjustable parameters. *Soil Sci. Soc. Am. J.*, **41**, 39–42.

Gardiner, W. R. (1965). Movement of nitrogen in soils. In: *Soil Nitrogen* (W. V. Bartholomew and F. E. Clark, Eds.), Madison, Wisconsin, Am. Soc. Agron.

Gaudet, S. P., Jegat, H., Vachaud, G., and Wierenga, P. J. (1977). Solute transfer with exchange between mobile and stagnant water through unsaturated sand. *Soil Sci. Soc. Am. J.*, **41**, 665–671.

Germann, P., and Beven, K. (1981). Water flow in soil macropores. I: An experimental approach. *J. Soil Sci.*, **32**, 1–13.

Ghuman, B. S., and Prihar, S. S. (1980). Chloride displacement by water in layered soil columns. *Aust. J. Soil Res.*, **18**, 207–214.

Hodgson, J. M. (1974). *Soil Survey field handbook*. Soil Survey, Harpenden, Herts, U.K.

Jury, J. (1982). Simulation of solute transport using a transfer function model. *Water Resour. Res.*, **18**, 363–368.

Kay, B. D. and Elrick, D. E. (1967). Adsorption and movement of lindane in soils. *Soil Sci.*, **104**, 314–322.
Kirda, C., Nielsen, D. R. and Biggar, J. W. (1973). Simultaneous transport of chloride and water during infiltration. *Soil Sci. Soc. Am. Proc.*, **37**, 339–345.
Kirkham, D., and Powers, W. L. (1972). *Advanced soil physics*, Wiley, New York.
Lapidus, L., and Amundson, N. R. (1952). *Mathematics of adsorption in beds. J. Phys. Chem.*, **56**, 984–995.
Luxmore, R. J. (1981). Comments on 'Micro-, meso-, and macroporosity of soil'. *Soil Sci. Soc. Amer. J.*, **45**, 671–673.
Luxmore, R. J., and Sharma, M. L. (1980). Runoff responses to soil heterogeneity: experimental and simulation comparisons for two contrasting watersheds. *Water Resour. Res.*, **16**, 675–684.
MacDonald, K. B., McKercher, R. B., and Mayer, J. R. (1976). Percloran displacement in soil. *Soil Sci.*, **121**, 94–102.
McMahon, M. A., and Thomas, G. W. (1974). Chloride and tritiated water flow in disturbed and undisturbed soil cores. *Soil Sci. Soc. Amer. Proc.*, **38**, 727–732.
Mistra, C., and Mishra, B. K. (1977). Miscible displacement of nitrate and chloride under field conditions. *Soil Sci. Soc. Amer. J.*, **41**, 496–499.
Mualem, Y. (1976). A new model for predicting the hydraulic conductivity of unsaturated porous media. *Water Resour. Res.*, **12**, 513–522.
Mualem, Y., and Dagan, G. (1978). Hydraulic conductivity of soils: unified approach to statistical models. *Soil Sci. Soc. Amer. J.*, **42**, 392–395.
Nielsen, D. R., and Biggar, J. W. (1961). Miscible displacement. I: Experimental information. *Soil Sci. Soc. Amer. Proc.*, **25**, 1–5.
Nielsen, D. R., Biggar, J. W., and Erh, K. T. (1973). Spatial variability of field-measured soil water properties. *Hilgardia.*, **42**, 215–260.
Nkedi-Kizza, P., Rao, P. S. C., Jessup, R. E., and Davidson, J. M. (1982). Ion exchange and diffusive mass transfer during miscible displacement through an aggregated oxisol. *Soil Sci. Soc. Amer. J.*, **46**, 471–476.
Nye, P. H., and Tinker, P. B. (1977). *Solute movement in the soil root system*, Blackwell, Oxford.
Omoti, U., and Wild, A. (1979). Use of fluorescent dyes to mark the pathways of solute movement through soils under leaching conditons. 2: Field experiments. *Soil Sci.*, **128**, 88–104.
Peck, A. J., Luxmore, R. J., and Stolzy, J. L. (1977). Effects of spatial variability on soil hydraulic properties in water budget modeling. *Water Resour. Res.*, **13**, 348–354.
Porter, L. K., Kemper, W. D., Jackson, R. D., and Stewart, B. A. (1960). Chloride diffusion in soils as influenced by moisture content. *Soil Sci. Soc. Amer. Proc.*, **24**, 460–463.
Rao, P. S. C., Green, R. E., Ahuja, L. R., and Davison, J. M. (1976). Evaluation of a capillary bundle model for describing solute dispersion in aggregated soils. *Soil Sci. Soc. Amer. J.*, **40**, 815–820.
Ritchie, S., Kissel, D., and Burnett, E. (1972). Water movement in undisturbed swelling clay soil. *Soil Sci. Soc. Amer. J.*, **36**, 874–879.
Rose, C. W., Chichester, F. W., Williams, J. R., and Ritchie, J. T. (1982a). A contribution to simplified models of field solute transport. *J. Env. Qual.*, **11**, 146–150.
Rose, C. W., Chichester, F. W., Williams, J. R., and Ritchie, J. T. (1982b). Application of an approximate analytic method of computing solute profiles with dispersion in soils. *J. Env. Qual.*, **11**, 151–155.

Rose, D. A. (1973). Some aspects of the hydrodynamic dispersion of solutes in porous materials. *J. Soil Sci.*, **24**, 284–295.
Rose, D. A., and Passioura, J. B. (1971). The analysis of experiments on hydrodynamic dispersion. *Soil Sci.*, **111**, 252–257.
Scotter, D. R. (1978). Preferential solute movement through larger soil voids. I: Some computations using simple theory. *Aust. J. Soil Res.*, **16**, 257–267.
Scotter, D. R., and Kanchanasut, P. (1981). Anion movement in a soil under pasture. *Aust. J. Soil Res.*, **16**, 67–77.
Selem, H. M., Davidson, J. M., and Rao, P. S. C. (1977). Transport of reactive solutes through multilayered soils. *Soil Sci. Soc. Amer. J.*, **41**, 3–9.
Shamir, U. Y., and Harleman, D. R. F. (1967). Dispersion in layered porous media. *Proc. Am. Soc. Civil Eng. Hydr. Div.*, **93**, 237–260.
Simmons, C. S. (1982). A stochastic-convective transport representation of dispersion in one-dimensional porous media systems. *Water Resour. Res.*, **18**, 1193–1214.
Skopp, J. (1981). Comment on 'Micro-, meso-, and macroporosity of soil'. *Soil Sci. Soc. Amer. J.*, **45**, 1246.
Smettem, K. R. J. (1984). Soil water residence time and solute uptake. 3: Mass transfer under simulated winter rainfall conditions in undisturbed soil cores. *J. Hydrol.*, **67**, 235–248.
Taylor, G. I. (1953). Dispersion of soluble matter in solvent flowing slowly in a tube. *Proc. Roy. Soc. London.*, **219A**, 186–203.
Thomas, G. W., and Swoboda, A. R. (1970). Anion exclusion effects on chloride movement in soil. *Soil Sci.*, **110**, 163–166.
van Genuchten, M. Th., (1981). Non-equilibrium transport parameters from miscible displacement experiments. *U. S. Salin. Lab-Dep. Environ. Sci. Univ. of California, Riverside, Calif.*, Res. Rep. No. 119.
van Genuchten, M. Th., and Wierenga, P. J. (1974). An evaluation of kinetic and equilibrium equations for the prediction of pesticide movement through porous media. *Soil Sci. Soc. Amer. J.*, **38**, 29–35.
van Genuchten, M. Th., and Wierenga, P. J. (1976). Mass transfer studies in sorbing porous media. I: Analytical solutions. *Soil Sci. Soc. Amer. J.*, **40**, 473–480.
Wild, A. (1981). Mass flow and diffusion. In: *The Chemistry of Soil Processes* (D. J. Greenland and M. H. B. Hayes, Eds.), Wiley, London.
Wild, A., and Cameron, K. C. (1981). Soil nitrogen leaching. In: *Soils and Agriculture* (P. B. Tinker, Ed.), Blackwells, Oxford.

Solute Processes
Edited by S. T. Trudgill

CHAPTER 5

Mineralogical control of the chemical evolution of groundwater

D. A. Spears

Department of Geology,
University of Sheffield

5.1 INTRODUCTION

The presence of seepages and springs is a demonstration of the movement of groundwater. This concept of groundwater flow is an extremely old one dating back at least to the time of the Babylonians, although their belief in the source of the water would now be considered misplaced. It was not until the seventeenth century that the cycle was demonstrated to be from the sea via the atmosphere and not underground as had hitherto been imagined. Present day absorption of groundwater from aquifiers requires careful aquifer management, involving detailed analysis of the flow, in order not to deplete the resource, which, although finite, is generally rechargeable. An analysis of this type usually deals with the bedrock as a chemically inert medium, but this is rarely the case. Although flow characteristics may change as a result of reaction between bedrock and groundwater, it is the influence of reactions on the quality of groundwater which is of greatest interest in this chapter.

In recent years there has been a growing realization that the widespread use of chemicals in modern agricultural practice and the tipping of industrial and domestic waste do pose serious problems for groundwater quality, and that we do have the capability of sterilizing aquifers. The possibility of anthropogenic input has been one of the factors stimulating research in the field of groundwater composition.

5.2 THE HYDROLOGICAL SYSTEM

Groundwater is only part of the subsurface water, just as the latter is only one part of the hydrological cycle. Figure 5.1 demonstrates that the groundwater is that part of the subsurface water located in the saturated zone—that is, below the water table. In the unsaturated zone the contained moisture is referred to as porewater. Downwards infiltration of porewater leads to aquifer recharge. The volumetric importance of this fraction will depend on the total precipitation and the loss by other means, including evapotranspiration. Groundwater, therefore, cannot be divorced from other parts of the hydrological system, and the groundwater composition is influenced by reactions taking place in the soil and in the porewaters in the unsaturated zone as a whole. The atmospheric solute input into the system also has to be taken into account, including elements transferred from the oceans. In the development of landforms the loss of elements from the soil and unsaturated zone into the groundwater is an essential factor in the solute budget. The individual element concentrations will depend on which minerals are undergoing decomposition, the composition of the solution itself, with the possibility of exceeding saturation levels, the time available for reaction and the volume of water passing through the system. These are aspects which we will consider in this chapter. In the development of soils interest tends to focus quite naturally on the solid products of weathering rather than on the elements lost in solution and certainly not on that part which disappears downwards.

The groundwater derived from atmospheric input is described as meteoric, whereas the groundwater which was originally trapped in sedimentary rocks at the time of formation is described as connate. This water may initially have had a low solute concentration, but usually it would be seawater. Irrespective

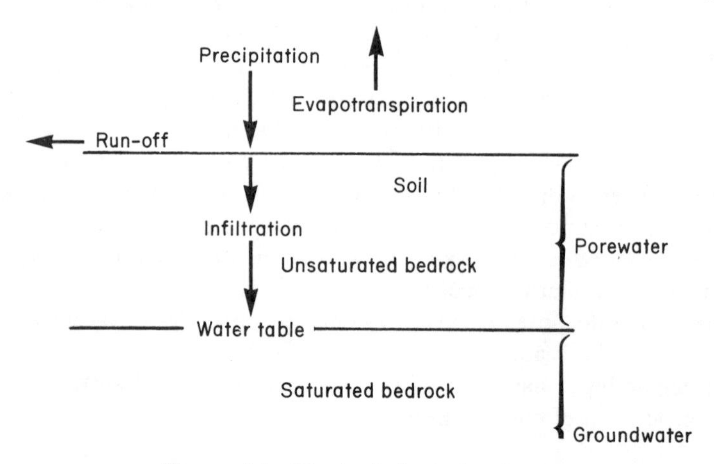

Figure 5.1. The hydrological system

of the initial composition the reactions taking place during the conversion of sediment to rock lead to high salinities in the residual fluids. These fluids are not static, but their migration is usually considered on a geological time scale. High-salinity groundwaters are encountered in deep boreholes for hydrocarbons and in underground mining operations. Their presence at depth may be a limiting factor on groundwater abstraction and delimits the zone of meteoric circulation.

It has been estimated (Garrels and Mackenzie, 1971) that the mass of the hydrosphere amounts to some $17{,}200 \times 10^{20}$g. The unit of 10^{20}g is often referred to as the geogram, which, as Garrels and Mackenzie point out, after a little use becomes familiar, if not comprehensible. Most of the hydrosphere is contained by the oceans (80%) as would be expected. Perhaps a little surprising is the 18.8% of the total mass contained in rocks and sediments. The rate of circulation through the rain–river–ocean–atmosphere is relatively fast and certainly fast compared with the groundwater. Nevertheless solute transfer from the groundwater directly into the oceans is important and has been estimated at 4.3×10^{14}g yr^{-1}. The stream and river contribution is much greater, 39.3×10^{14}g yr^{-1}, but some part of this is fed from groundwater itself. If we also include what Garrels and Mackenzie term the subsurface flux, which is probably mainly connate water, it becomes clear that the solute budget of groundwater is important not only on the small scale, as in a soil system, but also on a much larger, global scale.

5.3 MINERAL STABILITY—PRIMARY MINERALS

Most rocks consist of an assemblage of minerals, and to understand how rocks react in the presence of water we need to know how the minerals themselves react. In certain cases the rocks are monomineralic and the problem resolves itself into, for example, the dissolution of calcite (some limestones) or the dissolution of quartz (quartzites). The array of naturally occurring minerals is truly enormous, but many are of infrequent occurrence and are certainly not common rock-forming minerals. We are therefore able to simplify this discussion by considering only those minerals which are either important by virtue of abundance in common rock types or those minerals which, although a minor component in many rocks, greatly influence the composition of the pore fluids and the groundwaters because of rates of reaction and high equilibrium activities of the ions in solution.

Most of the minerals in crystalline igneous rocks are included in the framework outlined in Figure 5.2. The basic building block of these structures is an $[SiO_4]^{4-}$ tetrahedra which is either linked to other tetrahedra via a cation, usually Mg^{2+}–Fe^{2+}, or the oxygens of the tetrahedra are shared with adjacent tetrahedra. There is an increase in the polymerization of silica which takes place in a number of steps. The free tetrahedra is the arrangement in the

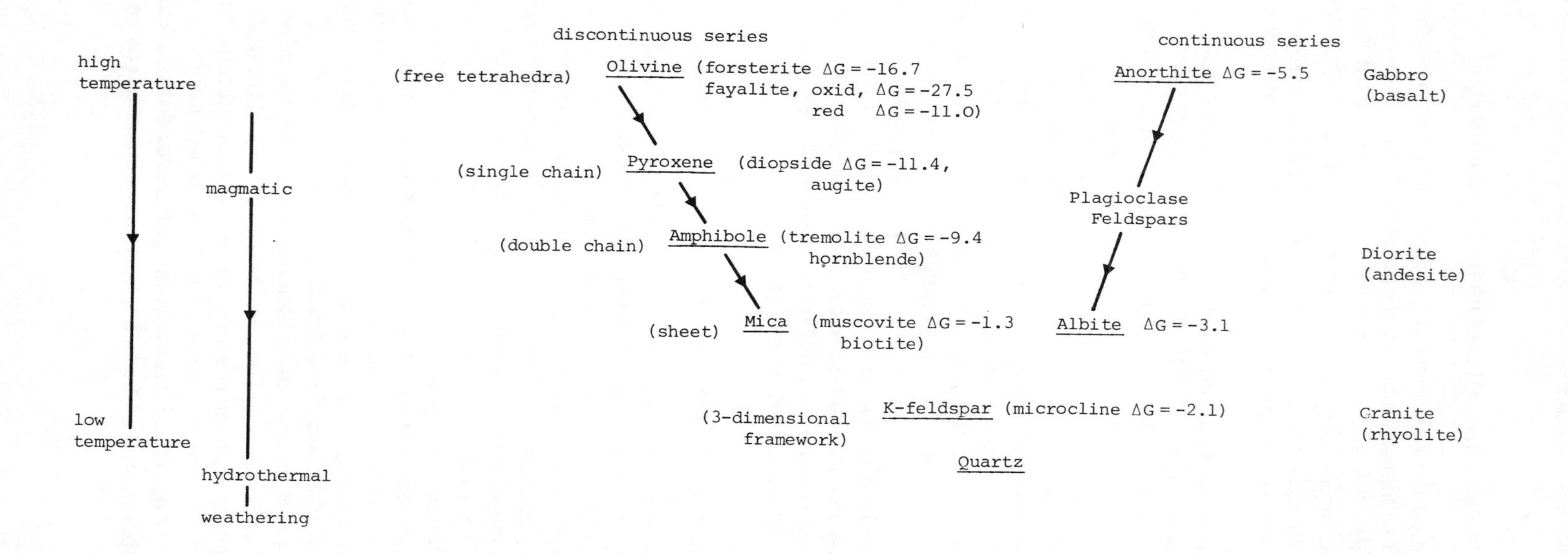

ΔG values are energy changes in weathering reactions, expressed as G^{o}_{r}, kJ g $atom^{-1}$, after Curtis (1976)

Figure 5.2. Framework of minerals in crystalline igneous rocks

olivine group of minerals. If one of the oxygens is shared between adjacent tetrahedra a single chain results which is then linked to other similar chains via cations. This is the structure of the pyroxene group of minerals. The Si–O bond is essentially covalent in character, whereas the cation–O bond is mainly ionic. Sharing two oxygens leads to a double chain and the amphibole group of minerals. In the silicates classified as phyllo-silicates, three of the oxygens are shared to give a sheet structure. The sheets are linked via cations between the sheets to produce a number of extremely important minerals. In igneous rocks muscovite and biotite are the common examples. Sharing of all the oxygens in the tetrahedra produces a three-dimensional structure best illustrated by quartz itself. These steps in the formation of structures are related to temperature. Based on observations down the microscope and supported by experimental evidence, Bowen's Reaction Series was established. This is the framework given in Figure 5.2, and the discussion has dealt with the discontinuous reaction series expressing the structural steps. As a magma cools and minerals crystallize, so there is a tendency for the early formed minerals to react with the melt and to convert to structures stable at lower temperatures. Major structural reorganizations are required and reactions are frozen by the progressive crystallization. The right-hand side of the diagram is the continuous reaction series involving the plagioclase feldspars. These are framework structures, but at high temperature the plagioclase has a composition of $CaAl_2Si_2O_8$. This is the mineral anorthite. Greater substitution of Al for Si is possible at higher temperatures because the thermal energy increases the vibrational energy of the atoms, producing an increase in the mean interatomic distance and thus in the ability to accommodate a larger cation. As the temperature falls the proportion of Al decreases and the charge balance is maintained by incorporation of Na rather than Ca. Albite, $NaAlSi_3O_8$, is the low-temperature plagioclase feldspar.

The basis of the mineral grouping shown on Figure 5.2 is one of reaction and conversion to a stable form. In weathering reactions, which may be considered as an extension of the reaction series to lower temperatures, it would be expected that the minerals higher in the reaction series would be unstable and should react. This was observed to be the case by Goldich (1938), and the order of persistence was found to correlate with position in the reaction series. Temperature is not the only factor, however. Structures stable over the same elevated temperature range are not necessarily going to have equal stability—or perhaps we should say persistence—in weathering environments. Quartz and K-feldspar (microcline) provide a good example, with the preferential loss of the latter due to its greater proportion of ionic bonds, which disrupt more readily in water than covalent bonds. It will be noted that in the reaction series (Figure 5.2) the degree of covalent bonding increases with falling temperature. Some of the minerals contain Fe^{2+}, particularly the olivine, pyroxene and amphibole groups. Oxidation of the

Fe^{2+} during weathering leads to charge inbalance and structural instability. The reaction series may therefore be projected to temperatures existing in weathering environments, with the proviso that other aspects of the chemical environment are not constant and modifications do result. Nevertheless the reaction series provides us with an overall, qualitative understanding of weathering behaviour of crystalline igneous rocks. Many of the minerals are common to metamorphic rocks and the discussion is not therefore limited to igneous rocks but covers in fact the 'hard rocks'. It is also not surprising to learn that the aluminium silicates which form during weathering, that is the clay minerals, have sheet structures. The order of persistence observed in soils is linked to the rates of reaction and hence the rates of supply of elements into solution. Fine-grained volcanic igneous rocks will behave in a comparable manner if they are crystalline, although the finer grain size will lead to greater surface area for reaction and consequently faster reaction rates. If, however, the volcanic rocks are non-crystalline, that is composed of glass, reaction rates will be very much faster as is demonstrated by the short period of time that may elapse before lavas and volcaniclastics are cultivated.

The order of persistence of minerals in low-temperature environments may also be established from the mineralogy of sandstones. Very rapid erosion is capable of producing a sand deposit with the mineral assemblage very similar to that of the source rocks. Although converted with time to rocks, these deposits remain in a relatively low-temperature environment and elimination of unstable minerals is observed. Pettijohn (1975) has noted that the order of persistence with age shows a close correspondence with the stability order as determined by Goldich (1938). In sandstones the maturity index is based on the quartz-K feldspar proportion. Although the ratio changes considerably from source rock to final sediment, the survival of K-feldspar and the persistence of other minerals in sandstones demonstrates that the rates of reaction for the primary minerals are slow, and hence the rate of transfer of elements into solution—and thus into the groundwater—will also be slow. Concentrations are only likely to be high from primary minerals if the residence period is measured in thousands of years or longer, and if equilibrium activities are not exceeded. The inclusion of sandstones in a discussion on mineral stability related to an igneous rock reaction series demonstrates its general applicability, which thus encompasses not only the 'hard rocks' but also the detrital sedimentary rocks.

A more rigorous approach to the mineral stability follows from an analysis of free-energy changes in the weathering reactions (Curtis, 1976). Thus the decomposition of forsterite may be represented as:

$$Mg_2SiO_4(s) + 4H^+(aq) = 2Mg^2(aq) + 2H_2O(l) + SiO_2(s)$$

where s = solid; aq = aqueous; and l = liquid.

The standard free-energy change of the reaction is the sum of the free energies of formation of the products in their standard states minus the free energies of formation of the reactants in their standard states:

$$\Delta G_r^\circ = G_f^\circ(2Mg^{2+}) + \Delta G_f^\circ(2H_2O) + \Delta G_f^\circ(SiO_2) - \Delta G_f^\circ(Mg_2SiO_4) - \Delta G_f^\circ(4H^+)$$

The free-energy change for this reaction is -184.0 kJ mol^{-1}. The negative value indicates that energy is liberated in the reaction and therefore the reaction does proceed from left to right as we have implied. The reaction involving fayalite provides an interesting contrast because fayalite crystalizes at a lower temperature than does forsterite. The reaction may be written as:

$$Fe_2SiO_4(s) + 4H^+(aq) = 2Fe^{2+}(aq) + 2H_2O(l) + SiO_2(s)$$

and the free-energy change for this reaction is -121.3 kJ mol^{-1}. Again the products are stable with respect to the reactions and forward spontaneous reaction is predicted. The conditions specified, however, do differ from most weathering environments in that conditions are not oxidizing. Rewriting the reaction for more normal conditions gives us:

$$Fe_2SiO_4(s) + \tfrac{1}{2}O_2(g) = Fe_2O_3(s) + SiO_2(s)$$

for which $\Delta G_r^\circ = 220.0$ kJ mol^{-1}. The free-energy change is much greater and the implication is that the Fe^{2+} silicate breaks down more rapidly than its Mg counterpart because of oxidation. If oxidation is excluded it would be expected that the rate of breakdown of the lower-temperature form, that is the Fe variety, would be slower. In making these comparisons between the weathering of magnesium and iron olivines it might be correctly suspected that, whereas the reactions with Fe^{2+} and Mg^{2+} as products are directly comparable, the oxidation reaction is different because the number of product and reactant atoms is different. The problem becomes critical if we are to compare the breakdown of more complex silicates (Figure 5.2); double up the structure and we double up the free-energy change. In order to compare the reactions the free-energy changes should be expressed on the basis of the number of product atoms, that is as gram atomic values (Curtis, 1976). The free-energy changes for other silicate reactions expressed in this way are given on Figure 5.2. There is a direct relationship between these values and the order of persistence of minerals in soils as determined by Goldich and others. There is not generally a quantitative relationship between free-energy changes and rates of reaction. The fact that for the silicates there does appear to be a qualitative relationship suggests that the rate determining step is comparable for all the silicates in weathering environments.

In the reactions dealt with above the free-energy changes are all negative, but differ in magnitude. It is this variation which is linked to the order of mineral persistence. If the values of the equations turned out to be positive, reaction would be in the opposite direction, that is of 'reverse weathering' and soils would be rather different. The third possibility is that the free-energy change is zero, in which case there is no reaction. Thus for a system at equilibrium $\Delta G_r = 0$.

The survival of feldspars and other unstable minerals in soils and in the derived sediments bears witness to the slowness of reaction rates in the low-temperature aqueous environment. The kinetics of silicate weathering and the mechanisms involved have been investigated in the laboratory. One of the early workers to adopt a kinetic approach to weathering was Wollast (1967), who determined the release of silica with time from ground K-feldspar in buffered aqueous solutions. The concentration of silica decreased with time, and this was attributed to the formation of a protective aluminium-rich surface layer on the feldspar grains. The calculated diffusion rates through this surface layer were found to be of the same order as for solid diffusion. Working with similar materials other workers have verified the experimental results and have made further suggestions as to the nature of the surface layer; possibilities are a clay layer or a residual feldspar layer with a thickness of several tens of nanometers. The resolution of modern analytical techniques is greater than this, but the residual layer has not been detected (Petrovic *et al*., 1976). Furthermore, in laboratory experiments using HF treated grains a constant rate of silica dissolution has been observed (Berner and Holdren, 1979; Berner, 1981). The reason for this is thought to be that fine grinding produces some submicron sized particles and strained regions on larger grains which dissolve faster. Treatment with HF eliminates these surfaces.

Dissolution of minerals involves the detachment of the ions or molecules from the mineral surface and transport away from the mineral grains. If the detachment from the surface is sufficiently rapid the rate of dissolution is governed by the rate of transport of ions away from the surface until the equilibrium concentration is attained. Transport may be by a combination of porewater movement and diffusion. In the case of a static porewater only diffusion is possible, and therefore the rate of dissolution is at a minimum (for a given departure from equilibrium). If the dissolution of feldspar was controlled by transport in aqueous solution, calculations demonstrate (Berner, 1981) that grains of 200 μm radius would survive only 14 months in soils. The calculation assumes that the concentration in the porewater is very much smaller than the equilibrium concentration, which is reasonable for dilute soil solutions. If the movement of the porewater is also taken into account the survival period is reduced. The survival of feldspars for periods many thousands of times longer is ample evidence that the dissolution of feldspar is not transport controlled; furthermore there could not be an order of mineral persistence as the dissolution would be a function of the fluid and not of the crystal structure. The alternative is that the rate of detachment of ions or molecules from grain surfaces is considerably slower than the rate of transport in solution. One feature of surface-reaction controlled dissolution is that there is selective dissolution in areas of greatest energy associated with crystal defects and that etch pits result. The essential point is that the crystal structure influences the dissolution surface. This contrasts with transport-controlled dissolution which results in smooth and rounded surfaces. From the foregoing discussion involving mineral persistence it would be expected that the silicates should dissolve by a surface-reaction control, and that the more ionic in character the structure the greater the solubility and probability of transport control being dominant. A note of caution must, however, be sounded; the dissolution kinetics of too few minerals have been studied, and in the laboratory it is difficult to reproduce natural conditions and in particular to evaluate the role that foreign ions have on the inhibition of surface reactions (Berner, 1978).

5.4 WEATHERING REACTIONS OF PRIMARY MINERALS

In considering the free-energy changes involved in the breakdown of the primary minerals the reactions were written with H^+ exchanging for a metal cation. Organic acids produced in soils are one source of hydrogen ions, but generally the most common source is dissolution of carbon dioxide in the following manner:

$$H_2O(l) + CO_2(g) \rightleftharpoons H_2CO_3(aq) \rightleftharpoons H^+(aq) + HCO_3^-(aq)$$

Silica was also represented as quartz, whereas in feldspathic igneous rocks it is well known that the groundwaters are concentrated in dissolved silica together with Na^+, Ca^{2+} and HCO_3^-. The dissolution of quartz may be written as:

$$SiO_2 + 2H_2O = Si(OH)_4$$

which is a neutral molecule (or sometimes written as H_4SiO_4). The Mg–olivine breakdown is therefore written as:

$$Mg_2SiO_4 + 4CO_2 + 4H_2O = 2Mg^{2+} + 4HCO_3^- + Si(OH)_4$$

If we take diopside as an example of a pyroxene, the weathering equation is:

$$CaMg(Si_2O_6) + 4CO_2 + 6H_2O = Ca^{2+} + Mg^{2+} + 4HCO_3^- + 2Si(OH)_4$$

Substitution of Fe^{2+} for Mg^{2+} in the pyroxene structure produces an insoluble Fe_2O_3 residue under normal weathering conditions. Tremolite, which is an amphibole, is similar to diopside:

$$Ca_2Mg_5Si_8O_{22}(OH)_2 + 14CO_2 + 22H_2O$$
$$= 2Ca^{2+} + 5Mg^{2+} + 14HCO_3^- + 8Si(OH)_4$$

There are again usually Mg–Fe substitutions in the structure, and under oxidizing weathering conditions there will be a residue of Fe_2O_3. If the groundwaters remain in contact with the rock for a very long time, leading to high concentrations in solution, saturation with respect to quartz would be predicted.

Reactions involving the aluminosilicates follow the same pattern as above except that aluminium has a very low solubility. The minimum solubility is less than 0.01 mg/l at a pH of near 6.0 (Hem, 1970), which is within the weathering range. The products will therefore include a solid form of aluminium, the nature of which will depend on the concentrations of other ions. It is very common to find that analyses of stream and meteoric groundwater fall in the kaolinite stability field. This was illustrated by Garrels and Christ (1965, Figure 10.6) with a plot of groundwater analyses from feldspar-rich sediments. The weathering reaction of a K-feldspar, as an example of an aluminosilicate, may therefore be sensibly written with kaolinite as a product:

$$2KAlSi_3O_8 + 2CO_2 + 11H_2O = Al_2Si_2O_5(OH)_4 + 2K^+ + 2HCO_3^- + 4Si(OH)_4$$

The plagioclase feldspar reactions are very similar; thus one end member, albite, gives:

$$2NaAlSi_3O_8 + 2CO_2 + 11H_2O = Al_2Si_2O_5(OH)_4 + 2Na^+ + 2HCO_3^- + 4Si(OH)_4$$

and the other end member, anorthite, is:

$$CaAl_2Si_2O_8 + 2CO_2 + 3H_2O = Al_2Si_2O_5(OH)_4 + Ca^{2+} + 2HCO_3^-$$

Note that in this reaction the Al–Si proportions in the anorthite are accommodated by the kaolinite and silica is not lost in solution.

5.5 WEATHERING REACTIONS AND MASS BALANCE

5.5.1 Granites in the Sierra Nevada

Having written typical, so we hope, weathering reactions for a range of silicate minerals, we are now in a position to examine groundwaters in many igneous and metamorphic rocks in order to: (a) test the equations; (b) provide evidence from the natural environment on relative, and if possible absolute, rates of reaction; and (c) establish a material balance. An extremely good example of this approach is described by Garrels and Mackenzie (1967). Groundwater emerging as ephemeral springs from granitic rocks in the Sierra Nevada of California and Nevada had the average composition shown in Table 5.1. Deducting from this the atmospheric input into the system (determined from snow analyses) gives the rock contribution to the solutes,

Table 5.1. Composition of springwater in Sierra Nevada and calculation of mineral contribution[a] (Garrels and Mackenzie, 1967)

	Spring water	Snow water	Δ Rock-derived	Minus plagioclase	Minus biotite	Minus K-feldspar
Na^+	13.4	2.4	11.0	—	—	—
Ca^{2+}	7.8	1.0	6.8	—	—	—
Mg^{2+}	2.9	0.7	2.2	2.2	—	—
K^+	2.8	0.8	2.0	2.0	1.3	
HCO_3^-	32.8	1.8	31.0	6.4	1.3	—
SO_4^{2-}	1.0	1.0	—	—	—	-
Cl^-	1.4	1.4	—	—	—	—
SiO_2	27.3	0.3	27.0	5.0	3.5	1.2
pH	6.2					

[a]Concentrations in mmol dm^{-3}. 1 mol dm^{-3} solution is formed by dissolving 1 mol of substance in enough solvent to produce 1 dm^3 of solution. 1 litre ≡ 1 dm^3 ≡ 1000 cm^3. Molality denotes the amount of substance divided by the mass of solvent, e.g. 1 mol kg^{-1}.

which is then expressed in terms of the rock minerals. These are quartz, plagioclase, feldspar, K-feldspar and accessory biotite or hornblende. The Na^+ and Ca^{2+} are re-expressed as plagioclase, the composition of which is found to agree with the actual plagioclase composition. The next step is the determination of the biotite loss from the Mg^{2+}, and then the K-feldspar loss from the K^+. A small amount of silica remains (Table 5.1) which is within the error of the original analyses and need not be attributed to quartz dissolution. The above analysis of the data (Table 5.1) demonstrates that plagioclase and biotite are the main minerals contributing to the solutes, with a smaller contribution from K-feldspar and almost nothing from quartz. The relative contribution of these minerals agrees with the persistence order described earlier. It is also noteworthy that HCO_3^- is the main anion in the springwaters corresponding to the predicted weathering reactions and the source of hydrogen ions. A possibility that has not been considered is hydration; thus for albite the reaction would be written as:

$$2NaAlSi_3O_8 + 11H_2O = Al_2Si_2O_5(OH)_4 + 2Na^+ + 2OH^- + 4Si(OH)_4$$

The resulting solution would be alkaline, whereas measured pH values in groundwater systems often have a pH less than 7, as in the Sierra Nevada springwater (Table 5.1).

5.5.2 Pelites in Maryland

The approach adopted by Garrels and Mackenzie (1967) has been successfully used by other authors. Thus Cleaves *et al*. (1970) investigated a small (95 acres) forested watershed in Maryland. Rainwater, stream and groundwater samples were collected over a 2-year period and the input–output budget determined. Seasonal variations were observed, as would be expected, and the average annual loss from the watershed by chemical weathering in this time period was estimated as 16.9 tons per square mile. The bedrock in the watershed is a pelitic schist consisting of quartz, plagioclase feldspars, muscovite, biotite and staurolite. Accessory minerals include zircon, garnet, tourmaline and kyanite which are resistant to chemical weathering as is staurolite. The same major minerals are present in the schist as in the granite considered earlier, except that K-feldspar is absent, and comparable weathering reactions may be written. Plagioclase and biotite are thought to contribute the bulk of the dissolved solids to the streamwater. The area investigated is thought to be watertight; that is there is negligible loss via groundwater. Evidence for this is provided by a balance in the input Cl (precipitation) and output Cl (stream). The role played by groundwater is nevertheless important because the topography ensures that there is groundwater flow into the stream. In the tower part of the watershed the valley is narrow and steep-sided; hence the lateral groundwater movement and the discharge into the stream of the solutes from the weathering reactions.

The weathering reactions are thought by Cleaves *et al.* (1970) to be: (1) plagioclase → kaolinite; (2) biotite → vermiculite → kaolinite; and (3) muscovite → illite and/or kaolinite. The extent of the reactions involving muscovite could not be quantified. In addition gibbsite was recorded in well-drained areas in the watershed, indicating that reactions were proceeding beyond the stage of kaolinite. Low concentrations of H_4SiO_4 in solution are required, which is compatible with good drainage, high throughput with short residence time, and consequently porewaters with low ionic strengths.

The average annual loss of 16.9 tons per square mile by chemical weathering is about five times the estimated loss by mechanical erosion measured over the same 2-year period. However, if this were the long-term ratio resistate quartz should have accumulated and no evidence of this was found. Cleaves *et al.* (1970) suggested the long-term ratio may be nearer to 1:1 and that mechanical (fluvial) erosion is greatly influenced by high-intensity low-frequency events. Nevertheless, the work on this watershed does demonstrate the importance of chemical weathering in terrain in which mechanical erosion would previously have been considered the dominant process.

5.6 MASS-BALANCE METHOD AND REACTION-PATH SIMULATION

In the examples considered above the element concentrations added to the groundwater are determined by differences between input and output concentrations. The element concentrations in solution are then related to the minerals present in the proportions demanded by the composition of reactant and product phases. The initial mineral composition might be determined either from the optical properties of minerals using a microscope, or more accurately by chemical means, either *in situ* or using separated grains. The selection of kaolinite as a product phase might seem to be inspired guesswork, but it certainly is not. The selection is based on field observations and thermodynamic stability. Failure to achieve a balance in the water chemistry could be due to an incorrect choice of product phase. Gibbsite is a possibility if the groundwater system is very dilute, as noted in the preceding section. The product phases must therefore be stable in the groundwater system. Thus once a mass balance has been achieved the thermodynamic reality of the reactant phases would be assessed.

The groundwater evolution in a granite could be expressed in the following terms:

initial solution composition + 'reactant phases' →
final solution composition + 'product phases'

which expresses the mass balance and is referred to as the mass balance method (Plummer *et al*., 1983). Although the examples considered are relatively simple, it will be clear that a number of possible reactions could be tested before a satisfactory answer was obtained. Additional calculation is then required to check on the stability of the products with respect to the solution. Such modelling and speciation calculations are ideally suited for computer application. The concepts, assumptions and possible limitations are comprehensively described by Nordstrom *et al*. (1979). At this stage it is worth mentioning a rather different approach which starts with an initial groundwater composition and follows the evolution of the water through a series of hypothetical reactions, checking the aqueous phase for saturation at each step and adjusting masses by precipitation or dissolution of appropriate minerals to retain equilibrium. This is reaction-path simulation (Plummer *et al*., 1983). Projected compositions may be obtained for systems for which there is very little information, or alternatively used to check the thermodynamic feasibility of mass balance deductions.

To illustrate these concepts, Plummer *et al*. (1983) use analytical data from three wells in the principal aquifer in central Florida. The changes in groundwater composition along the flow path represented by the three wells are attributed to a reaction model involving incongruent dissolution of dolomite driven irreversibly by gypsum dissolution and accompanied by calcite precipitation, oxidation of a small amount of organic matter, and sulphate reduction. The net reaction between Polk City and Wauchula, the wells furthest apart, is Polk City Water + $0.96CaMg(CO_3)_2$ +

$1.6CaSO_4.2H_2O + 0.17CH_2O + 0.53CO_2 + 0.03FeOOH \rightarrow 1.84CaCO_3 + 0.03FeS_2$ + Wauchula Water (mmol $kg^{-1}H_2O$).

The procedures briefly described above relate essentially to the elements in solution and are commonly expressed as either mols dm^{-3} or mols kg^{-1}. In favourable situations it may be possible to determine the actual mass of each element lost from the system. Given the rate of flow the masses may be determined. Alternatively if the latter is known the rate of flow can be calculated.

In this section we started by applying some of the weathering reactions appropriate for the common primary minerals and have seen, glimpsed perhaps, how the chemical modelling approach may be extended into the realm of solution equilibria and thermodynamics. This is a realm into which we should venture, possibly not deeply, but far enough to judge the feel of the water.

5.7 THE ROLE OF SEDIMENTARY MINERALS

The discussion so far has dealt mainly with the silicate minerals occurring in igneous and metamorphic rocks—the so-called 'hard rocks'. The minerals in sedimentary rocks, the 'soft rocks', will now be considered. The minerals fall into three categories, resistate, hydrolyzate and precipitate. The resistate minerals survive the weathering process and are transported as detrital grains. Quartz is the most important of these, together with lesser amounts of feldspar. The proportion of the latter is a function of the content in the source rocks and the severity of weathering. The more rapid the erosion, the higher the content of reactive primary minerals in the resulting sediment. In the sediment these minerals remain in a low-temperature environment in potential contact with pore fluids, and therefore after burial weathering reactions may continue. The net effect of this is that the mineralogy of sandstones becomes progressively simpler back through the geological column.

The resistate minerals are predominantly of sand and silt size. In the finer grained clastic sediments, that is in the mudrocks, the products of weathering reactions are to be found, and this is the hydrolyzate fraction. It therefore follows that when these minerals, mainly clay minerals, are re-exposed to the influence of another weathering cycle that structural changes are likely to be minor and certainly slow. A rapid response will be observed, however, in the case of surface and ion exchange reactions, and redox reactions will be in the same category. If organic matter is present in the sediment dissolved oxygen in the groundwater may be consumed, followed by denitrification, reduction of MnO_2 and Fe_2O_3 and possibly even sulphate reduction. Thus the hydrolyzate fraction has an important influence on the chemistry of groundwater because of redox and surface reactions. In the long term changes in the clay mineral structures will take place, with the transfer of elements to and from solution. The weathering of clay minerals is extremely important and particularly in soils. Thus, for example, micas are the most important natural source of potassium in most soils. The potassium is released from interlayered positions with attendant structural changes. The economic importance of such reactions in soils has resulted in considerable research and a large volume of literature. Probably the most comprehensive account is to be found in *Minerals in Soil Environments*, edited by J. B. Dixon and S. B. Weed and published by the Soil Science Society of America (1977). Thus the weathering reactions involving the resistate minerals are slow, but are nevertheless important in groundwater evolution if the time period is long ($>10^3$ years). The same is also true of the hydrolyzate fraction if changes in clay structures are involved. If the reactions are ion-exchange or redox controlled they are important on a time scale much shorter than 10^3 years.

A time period in excess of 10^3 years is short on a geological time scale but long in the history of aquifer recharge and extraction. It is the third group of minerals, namely the precipitates, which are particularly important in this context of aquifer management because of the much more rapid rates of reaction. The minerals in this group include the carbonates (mainly calcite and dolomite in ancient sediments), evaporites (gypsum, anhydrite and halite; other evaporite minerals are quantitatively less important), and sulphide minerals (mainly pyrite and particularly in marine black shales). It will be noted that the emphasis is on rates of reaction and not just solubility. It is important not only that the minerals should react—that is they are not in equilibrium in the aqueous environment—but also that the rate of reaction should be finite with respect to the residence time of pore fluids.

5.8 SEDIMENTARY MINERALS AND EQUILIBRIUM THERMODYNAMICS

The evolution of groundwater entails reaction between pore fluid and surrounding rock, and in this chapter we have attempted to break down the rocks into mineral groups. The concept of equilibrium is an underlying theme, and in the earlier discussion of silicate weathering reactions it was noted that a zero free-energy change defines an equilibrium condition. More familiar to the reader will probably be the Law of Mass Action in which the reaction in a system is related to the concentrations of reactants and products. Equilibrium is achieved when the rates of reaction to the left and to the right are equal. Thus for the following reaction at equilibrium:

$$\alpha A + \beta B = \gamma C + \delta D$$

that is αmoles of A and βmoles of B have come to equilibrium with γmoles of

C and δmoles of D; then:

$$\frac{a_C^{\gamma} a_D^{\delta}}{a_A^{\alpha} a_B^{\beta}} = K_{eq}$$

Where a is the activity raised to the appropriate power and K_{eq} is the equilibrium constant. The equilibrium constant is related to the standard free-energy change of reaction by the relationship:

$$\Delta G_r^{\circ} = -RT \ln K$$

which gives, at 25°C, $\Delta G_r^{\circ} = -5.708 \log K$ $(= -1.364 \log K$ if ΔG_r° is expressed in kcal).

The solubility of a sparingly soluble salt MX can be related to the equilibrium:

$$MX(s) = M^+(aq) + X^-(aq)$$

$$K_{eq} = \frac{a_{M^+} a_{X^-}}{a_{MX}}$$

The activity of the pure solid is unity, and the equilibrium constant $K_{eq} = a_{M^+} a_{X^-} = K_s$, where K_s is the solubility product. In actual solutions the state of saturation is often expressed by the ion-activity product (IAP, namely $a_{M^+} a_{X^-}$) and K_s. If IAP $= K_s$ the solution is saturated, if IAP $> K_s$ it is supersaturated, and undersaturated if IAP $< K_s$. As the equilibrium constant is related to the free-energy change of the reaction, then so too is the solubility product.

Gypsum is an important evaporite mineral in the context of groundwater composition. The reactions are similar to the general case, thus:

$$CaSO_4\,2H_2O(s) = Ca^{2+}(aq) + SO_4^{2-}(aq) + 2H_2O(l)$$

$$K_{eq} = \frac{a_{Ca^{2+}} a_{SO_4^{2-}} a^2_{H_2O}}{a_{CaSO_4\,2H_2O}}$$

The activity of the solid is again unity. The activity of water, which is a new term in the equation, approximates to unity provided that the solution is not concentrated. In seawater, for example, $a_{H_2O} = 0.98$. Therefore, $K_{eq} = a_{Ca^{2+}} a_{SO_4^{2-}} = K_{sp}$, which can be calculated from the ΔG_r° value.

It will also be noted that in the equations activities are used rather than concentrations. The activity is related to concentration molality m by the relationship $a = \gamma_m$. Only as the molality becomes very low do the values of the individual activity coefficients approach unity. Values are a function of the ions involved and the ionic strength of the solution. In the case of the gypsum saturated solution, for example, the $\gamma Ca^{2+} = 0.51$ and $\gamma SO_4^{2-} = 0.46$, and hence use of molalities instead of activities would have provided an unreliable estimate of the state of saturation.

The dissolution of calcite is a little more complicated than gypsum; thus:

$$CaCO_3(s) = Ca^{2+}(aq) + CO_3^{2-}(aq)$$

and $K_{calcite} = a_{Ca^{2+}}\, a_{CO_3^{2-}} = 10^{-8.3}$ at 25°C. The complication arises from the species present in carbonate system, namely P_{CO_2}, $a_{H_2CO_3}$, $a_{HCO_3^-}$, $a_{CO_3^{2-}}$, a_{H^+}, and a_{OH^-}. These are defined in four equations:

$$K_{CO_2} = \frac{a_{H_2CO_3}}{P_{CO_2}} = 10^{-1.47} \quad \text{at } 25°C$$

$$K_{H_2CO_3} = \frac{a_{H^+} a_{HCO_3^-}}{a_{H_2CO_3}} = 10^{-6.35} \quad \text{at } 25°C$$

$$K_{HCO_3^-} = \frac{a_{H^+}\, a_{CO_3^{2-}}}{a_{HCO_3^-}} = 10^{-10.3} \quad \text{at } 25°C$$

$$K_{H_2O} = \frac{a_{H^+}\, a_{OH^-}}{a_{H_2O}} = 10^{-14.0} \quad \text{at } 25°C$$

The application of these equations to a number of specific situations is elegantly described in a classic text by Garrels and Christ (1965). Thus one example describes, in a step-by-step calculation, how to determine if a natural water is in equilibrium with calcite from the chemical analyses. Another example represents the situation where rainwater in equilibrium with atmospheric CO_2 descends through the soil before coming in contact with a calcite-bearing material. The solubility of calcite in this example is low and is only about 10% higher than in porewater. Higher solubilities are encountered in practice, which demonstrates the importance of the P_{CO_2} in the soil atmosphere. Soil-derived CO_2 is an important component in limestone and other aquifers (Plummer, 1977). The calculations described by Garrels and Christ are highly recommended and might be considered a prerequisite before applying the available computer programs.

5.9 GROUNDWATER EVOLUTION IN TRIASSIC SANDSTONES, ENGLAND

The Triassic sandstones are the second most important aquifer system in the United Kingdom and are also important elsewhere in northern Europe. The economic importance of the aquifer has resulted in a number of hydrogeochemical surveys, two of which are described below.

5.9.1 Triassic sandstones in the Vale of York

In this area in the north of England the aquifer is extensively covered with Quaternary deposits of glacial, lacustrine and fluviatile origins. Much of the recharge of the aquifer is through these deposits, although recharge also takes place in the limited areas of sandstone outcrop. The superficial deposits were thought to contribute to the geochemistry of the groundwater in the aquifer because shallow wells in the superficial deposits yielded extremely hard water ($\geq$500 mg/l). The composition of these deposits and their role in the evolution of the groundwater was investigated by Spears and Reeves (1975a).

Samples were taken from shallow boreholes along a 2 km traverse some 20 km north-west of York. In addition to the analysis of porewaters extracted from the borehole samples, the whole-rock geochemical and mineralogical analyses demonstrate the progressive addition of elements to solution, or removal as the case may be, along the infiltration pathway. The mineralogical information together with the porewater analyses provides the data for the mass balance and reaction path simulation calculations referred to earlier. An alternative approach, however, is to combine the mineralogical information with the whole-rock geochemistry in order to determine quantitatively the loss of elements from the system over a long period of time. The whole-rock analyses should also agree with the porewater analyses; thus if a mineral, such as calcite, is dissolving, then the gain in Ca in the porewaters should be accompanied by a loss of Ca from the rock. If calcite is absent then the porewater Ca concentrations will be low provided that there is not another Ca source. The possibility of a source outside the system should not be overlooked. If calcite is present in the rock the porewater Ca concentrations should be high. Statistical analysis of porewater, mineralogical and whole-rock geochemistry should therefore reveal the mineralogical controls on the porewater composition. Such a statistical approach was adopted by Spears and Reeves (1975a).

The samples analysed from the Vale of York (Spears and Reeves, 1975a) ranged from glacial boulder clays to water laid sands and clays. The mineralogy is related to the grain size of the sediment, with quartz dominant in the coarser sediments and clay minerals in the finer sediments. The order of clay mineral abundance is kaolinite, illite and chlorite. K-feldspar and plagioclase were detected in all samples, and although systematic variations in abundance were noted these are inherited characteristics and not due to weathering reactions. Calcite and dolomite were present in some samples, but not in others. Carbonates are absent in the more permeable near-surface sediments and this is attributed to infiltration. Pyrite was also detected, but again not in the near-surface sediments. The mineralogy is typically sedimentary, which may appear surprising for glacial deposits, but it reflects the dominance of sedimentary rocks at outcrop in the areas over which the ice sheet passed. The mineralogy is, in fact, comparable with that in the underlying Triassic sandstones.

Statistical analysis of all the data revealed significant relationships between porewater Ca and Mg concentrations and carbonates, and also between porewater SO_4^{2-} and pyrite. The two important reactions taking place are the dissolution of carbonates and the oxidation of pyrite. Carbonates and pyrite are preserved at depth and within impermeable sediments. Although pyrite is destroyed the iron liberated is immobilized as Fe_2O_3, which also demonstrates the continued presence of dissolved O_2 in the infiltrating porewaters. Reactions involving other minerals, such as feldspars, are difficult to identify from the whole-rock data. There is a small, but significant increase in the whole-rock K_2O percentage in the direction of infiltration from 1.23% ($SD = 0.42, n = 18$) to 1.55% ($SD = 0.17, n = 8$), which is probably due to a combination of K-feldspar weathering and loss of interlayer K from illite. These reactions illustrate the important contribution of the precipitate minerals to the groundwater composition with a smaller contribution from the resistate (feldspars) and hydrolyzate (illite) fractions, thus bearing out the earlier discussion. In the Vale of York there has been insufficient time to remove entirely the precipitate minerals. Radiocarbon dating gives a figure of 12,400 years B.P. The additional time required is at least another 10^4 years, provided that infiltration conditions remain constant. Average compositions for the porewater from the superficial deposits and the groundwater in the sandstone aquifer are shown on Table 5.2. The porewater analyses demonstrate (a) the presence of reactions in the superficial deposits; and (b) the importance of these deposits on the groundwater based on the similarity between porewater and groundwater compositions.

Omitted from the discussion so far is the possibility of an external source for any of the elements. An indication of such an origin is provided by the statistical analysis of the porewater and whole-rock data, because some elements do not appear to have a mineralogical control. The lack of

Table 5.2. Porewater, groundwater and rainwater compositions in the Vale of York (Spears and Reeves, 1975a)

Element (mg/1)	Porewater		Groundwater		Rainwater Median	Porewater/ rainwater ratio
	Median	Mean	Median	Mean		
Na^+	15	19	13	38	2.3	6.5
K^+	4	5	2	4	0.5	8
Ca^{2+}	71	71	89	152	2.7	26
Mg^{2+}	18	27	22	37	1.0	18
SO_4^{2-}	170	193	76	256	12	14
Cl^-	33	58	22	42	5.3	6.2
HCO_3^-	—[a]	—[a]	326	348		

[a]The HCO_3^- determination on extracted porewaters is difficult because: (1) the volume is usually limited; and (2) the extraction procedure may modify HCO_3^- concentrations.

correlation could, however, mean either that the mineral does not exhibit a significant variation due to infiltration, or that more than one mineral is involved. Cl does not correlate with any of the minerals or with any of the mineral related elements present in the porewater. The only correlation involving Cl in the porewater was found to be with Na, suggesting a rainwater source. Average rainwater concentrations are included on Table 5.2. A concentration factor due to evapotranspiration of about 6 could account for the recorded porewater values of Na and Cl, whereas for the carbonate derived elements such as Ca and Mg the required concentration factors are unrealistically high, as would be predicted.

5.9.2 Triassic sandstones in the East Midlands

The Triassic sandstone aquifer in the East Midlands, approximately 100 km south of the area described above, is the subject of a recent investigation by Edmunds *et al.* (1982). The aquifer dips uniformly to the east and the presence of impermeable mudrocks above and below makes it a single hydraulic unit. It is thus possible to study the evolution of the groundwater over a distance of some 20 km down gradient and at much greater depths than in the Vale of York. Groundwaters were taken from 32 abstraction boreholes and the mineralogy determined on core material from two of the boreholes. The latter provide a mineralogical control for the computer speciation and equilibrium calculations using the WATEQF program (Plummer *et al.*, 1976) and thus the approach is similar to that adopted in the Florida aquifer described earlier. It does differ from the Vale of York investigation in that very limited whole-rock information is available because of the non-availability of borehole core material. The volume of water available for analysis is, however, much greater than for extracted porewaters and more chemical parameters can be determined.

Isotopic and inert gas data (Bath *et al.*, 1979; Andrews and Lee, 1979) for the aquifer demonstrates that there are three depth zones of groundwater composition. The shallowest zone is interpreted as predominantly modern, with ages not exceeding a few tens or hundreds of years. The next zone is thought to have ages in the range 1000–10,000 B.P. and the deepest zone, ages from 10,000 up to 35,000 years B.P. The zones are defined on ^{14}C, ^{4}He, δO^{18}_{SNOW}, Ar/Kr palaeotemperatures, thermonuclear tritium and chloride. The latter is above 20 mg l^{-1} in zone 1, 8 mg l^{-1} in zone 2 and in the 7–21 mg l^{-1} range in zone 3. The low values (zone 2) are not consistent with modern recharge allowing for concentration during evapotranspiration and assuming chloride is conservative, i.e. not lost by reaction in the aquifer. To account for the low Cl concentrations, Bath *et al.* (1979) envisaged recharge by meltwaters during periglacial conditions and later the dominance of precipitation from continental air masses rather than from the maritime air masses which dominate modern precipitation.

The depth zones provide a framework in which the chemical composition of the groundwaters are discussed (Edmunds *et al*., 1982). The groundwater evolution is dominated by carbonate and sulphate mineral solution and precipitation and by redox reactions. Cation exchange reactions are negligible and residual saline waters are not present. In the shallowest zone congruent dissolution of dolomite is important and the system remains oxidizing. The atmospheric input from rain is important for Na^+, Cl^-, F^-, B and probably HPO_4^{2-}. The anthropogenic input is only important in zone 1 because of the groundwater age spectrum and is obtained by comparing zone 1 with zones 2 and 3. The (modern) anthropogenic input is thought by Edmunds *et al*. (1982) to probably account for enhanced SO_4^{2-} and NO_3^- and possibly Cu, Zn and Pb concentrations. In zone 2 the groundwaters are predominantly reducing, and important reactions are the incongruent dissolution of dolomite and solution of gypsum. Reducing conditions are also encountered in zone 3, but sulphate reduction involving organic matter is quantitatively unimportant. Again the sulphate source is the dissolution of gypsum, which also adds Sr^{2+} and Mn^{2+} to the groundwaters. The addition of SO_4^{2-} to the groundwater limits the Ba^{2+} in solution, and groundwaters are mostly at equilibrium with respect to barite. Dissolution of carbonates are a probable source of Ba^{2+}. In the older groundwater, K^+ concentrations are rather constant and equilibrium with respect to illite may be the control. The uranium concentrations in the oxidizing groundwater are related to the stability of the uranyl carbonate, but as the Eh values fall so too do uranium concentrations in solution indicating U^{6+}–U^{4+} reduction and implied precipitation of urannite. In zones 2 and 3 the concentrations of Sr^{2+}, Mn^{2+}, Li^+, SO_4^{2-}, Ca^{2+} and As increase in the groundwaters as a result of reaction. The As source could be solution of interstitial ferric oxide associated with falling Eh values.

In this investigation of the Triassic sandstone aquifer the importance of the precipitate sedimentary minerals is again highlighted. Water samples are available for analysis from boreholes and the equilibrium calculations reveal the mineralogical control. In such an investigation it is usually not practicable to do complementary whole-rock analyses. There are a number of sampling problems in dealing with an aquifer, not the least of which is the lack of rock samples. This necessitates an approach to the more inaccessible, deeper parts of the aquifer different from that adopted where the interest lay in the recharge through the unsaturated zone and borehole core was readily available.

5.10 RATE OF LIMESTONE REMOVAL

Evidence of rapid attainment of equilibrium with respect to calcite in the Vale of York investigation is provided by the IAP values calculated from the porewater analyses, which are greater than or equal to K_s calcite for those samples containing carbonates. Further evidence comes from the whole-rock analyses, which show that there is a sudden jump in the carbonate percentages, rather than a gradual down-hole increase. This is due to the infiltrating porewaters reaching the level at which carbonate is present, reacting rapidly, reaching equilibrium and moving on with no further reaction. There is, therefore, what may be termed a weathering front, which with time extends further along (down) the infiltration pathway.

The data on the removal of carbonates due to weathering reactions in the above example permits a total mass balance calculation to be made. Given that the carbonate concentration in weathered and unweathered sediments was originally comparable, which appears to be the case from the other minerals, then the mass of Ca lost from the system is known. This Ca is removed in the porewaters at concentrations dictated by the carbonate equilibrium and which were measured in the extracted porewaters. Dividing the total amount of Ca lost from the system by the Ca concentration in the porewaters gives the total volume of water that has flowed through the system. If the present conditions are typical of the last 12,400 years (the age of the sediments), the rate of infiltration into the aquifer may be calculated. The values obtained from the geochemical data do correspond with those determined by other means (Table 5.3), although it should be noted that no method gives particularly precise recharge rates.

The average loss of Ca from the weathered sediments in the Vale of York was 3.6% Ca(9.0% $CaCO_3$) and the depth of weathering was recorded as 7.3 m. This corresponds to an equivalent surface lowering of 660 mm, or an annual dissolution rate of 0.05 mm. A number of other methods have been suggested

Table 5.3. Recharge rates of the Triassic sandstone aquifer in Vale of York (Spears and Reeves, 1975b)

Method	Recharge ($10^3 m\ d^{-1}$)	Range
Chemical data[a]		
Ca	41	26–65
Mg	24	15–41
K	90	12–265
Meteorological data	50	36–64
Groundwater level fluctuations	68	28–108
Groundwater gradients and aquifer properties	70	45–95
River baseflow analysis	70	0–140

[a]The Ca value results from the dissolution of calcite and dolomite, the Mg value dissolution of dolomite and the K value from minor weathering of K-feldspar and illite; hence the increased uncertainty expressed in the range.

to determine the rate of surface lowering. One approach is to determine the amount of Ca removed in solution by surface runoff. This was the method used by Sweeting (1965) in the North Pennines. The limestones are underlain by impermeable basement, hence losses from the area by groundwater flow are negligible. The annual rate of surface lowering calculated by Sweeting (1965) was 0.04 mm. In the same upland limestone area are to be found perched glacial erratics (Penny, 1974). The limestone immediately below an erratic is protected from erosion and now appears as a pillar. The height of the pillars is a measure of the overall surface lowering (450 mm) which, coupled with the age of the glacial event, gives an annual rate of 0.05 mm. Thus the same rate is obtained by a very different method of calculation. An alternative calculation is to date the age of the glacial event given the rate of limestone removal. This was in fact suggested almost a hundred years ago (Hughes, 1886), long before absolute age dating.

Another variation on rate determinations is based on the amount of insoluble residue remaining after dissolution of a limestone. The main difficulty with this approach is one of erosion of the insoluble residue. This difficulty did not arise in an example described by Piggott (1965), again in the north of England where the limestone was overlain by superficial deposits with an intervening thin (10 mm) uncontaminated layer interpreted as the protected insoluble residue. The superficial deposits were thought to post-date the last episode of frost heaving; hence the age is known. The residue is equivalent to a thickness of 5000 mm and the annual rate of surface lowering is therefore 0.04 mm. It is noteworthy that the rate values derived by very different means are very similar.

5.11 CONCLUSIONS

The increasing demands of modern society on the environment have necessitated a fuller understanding of the hydrological cycle. The water which enters the soil system is the important additive in promoting chemical reactions which continue down through the unsaturated zone and into the underlying aquifer. In order to predict and preserve water resources, more data on both quality and quantity of aquifer groundwater are required. This chapter has focused attention on the chemical evolution of groundwater and therefore relates to water quality. Although attention tends to focus on the solid products of weathering, the water which disappears downwards must not be forgotten. Indeed the water present in the ground is about 20% of the hydrosphere.

Reaction of primary silicate minerals in the aqueous weathering environment is an extension of Bowen's Reaction Series to low temperatures. A qualitative view of stabilities is provided by the relative positions of the minerals in the series. More quantitative information is obtained from the free-energy changes in the silicate reactions, including the effects of oxidation. In the silicate reactions the rate determining step is probably similar; hence the general agreement between magnitude of free-energy change and rate of reaction. The reaction rates are not controlled by transport of ions in solution from the mineral surface, which would lead to smooth and rounded surfaces. Rather, the weathered surfaces, for example of feldspars, are characterized by etch pits suggestive of surface-reaction-controlled dissolution. Although residual weathered layers and protective clay coatings were proposed to explain the reaction rates, the presence has not been confirmed.

The primary silicate minerals contribute to sedimentary rocks, but only the most persistent survive weathering and diagenetic processes. These minerals, mainly quartz and to a lesser extent low-temperature feldspars, are concentrated in sandstones. In mudrocks the fine-grained products of weathering accumulate; these are the clay minerals. Rates of reaction involving structural changes will be slow for these minerals, and hence significant concentrations will only be added to groundwaters over a long time period, such as 1000 years. Exchange and redox reactions will, however, be much quicker. In the depositional and diagenetic environments minerals precipitate from solution, mainly carbonates, sulphides and sulphates. These minerals are present in sandstones and mudrocks and are, of course, the major component in chemical sediments. The rainwater entering the unsaturated zone will initially be undersaturated with respect to the precipitate minerals. Reaction will be relatively rapid with the attainment of saturation. Dissolution of one mineral may lead to precipitation of another as in the aquifers in Florida (Plummer *et al*., 1983) and in eastern England (Edmunds *et al*., 1982). In the unsaturated zone oxidation of pyrite will be the rule unless waterlogging leads to anaerobic conditions, whereas at depth in the aquifer sulphate reduction is a possibility, particulary in 'old' groundwaters ($>10^3$ years).

The chemical evolution of groundwater in rocks, such as granite, composed of primary silicates, may be treated by writing weathering reactions in terms of reactant and product phases and solution compositions. This is the mass-balance approach. The choice of product phases may be governed in the first instance by observations in the natural environment. An incorrect choice will lead to either mass unbalance or unlikely reactant phases. The product phases must be stable in the final solution and this should be checked by calculation. The reaction path simulation method is an extension of this idea. The porewater composition is followed through a series of hypothetical reactions and at each stage equilibrium is maintained by adjusting masses. The calculations are complicated and require the use of a computer. This method may be used in a groundwater system for which there is very little analytical data, or alternatively to check the thermodynamic reality of mass-balance calculations. If borehole rock samples are available, which may mean shallow boreholes and hence the unsaturated zone, it is possible to extend the mass-balance method to include total mass change and hence volume of water passing through the system. This proved possible in the Vale of York (Spears and Reeves, 1975b) where precipitate minerals are being removed from dominantly clastic sediments. From the data may also be calculated the equivalent rate of limestone surface lowering.

REFERENCES

Andrews, J. N., and Lee, D. J. (1979). Inert gases in groundwaters from the Bunter Sandstone as indicators of age and palaeoclimatic trends. *J. Hydrol.*, **41**, 233–252.

Bath, A. H., Edmunds, W. M., and Andrews, J. N. (1979). Palaeoclimatic trends deduced from the hydrogeochemistry of Triassic sandstone aquifers, United Kingdom. In: *Isotope Hydrology, Proc. Symp. Vienna, 1978*, **11**, International Atomic Energy Agency, Vienna, pp. 545–568.

Berner, R. A. (1978). Rate control of mineral dissolution under earth surface conditions. *Amer. J. Sci.*, **278**, 1235–1252.

Berner, R. A. (1981). Kinetics of weathering and diagenesis. In: *Kinetics of Geochemical Processes* (A. C. Lasaga and R. J. Kirkpatrick, Eds.), Reviews in Mineralogy, vol. 8, Min. Soc. Am., pp. 111–134.

Berner, R. A. and Holdren, G. R. (1979). Mechanism of feldspar weathering. II: Observations of feldspars from soils. *Geochim. Cosmochim. Acta*, **43**, 1173–1186.

Cleeves, E. T., Godfrey, A. E., and Bricker, O. P. (1970). Geochemical balance of a small watershed and its geomorphic implications. *Geol. Soc. Am. Bull.*, **81**, 3015–3032.

Curtis, C. D. (1976). Stability of minerals in surface weathering reactions: a general thermochemical approach. *Earth Surface Processes*, **1**, 63–70.

Dixon, J. B., and Weed, S. B. (1977). *Minerals in Soil Environments*, Soil Science Soc. of America, Madison, Wisconsin.

Edmunds, W. M., Bath, A. H., and Miles, D. L. (1982). Hydrochemical evolution of the East Midlands Triassic sandstone aquifer, England. *Geochim. Cosmochim. Acta*, **46**, 2069–2081.

Garrels, R. M., and Christ, C. L. (1965). *Solutions, Minerals and Equilibria*, Harper and Row, New York.

Garrels, R. M., and Mackenzie, F. T. (1967). Origin of chemical compositions of some springs and lakes. In: *Equilibrium Concepts in Natural Water Systems* (W. Strumm, Ed.), Advan. Chem. Ser. **67**, 222–242.

Garrels, R. M., and Mackenzie, F. T. (1971). *Evolution of Sedimentary Rocks*, W. W. Norton & Co. Inc., New York.

Goldich, S. S. (1938). A study in rock weathering. *J. Geol.*, **46**, 17–58.

Hem, J. D. (1970). Study and interpretation of the chemical characteristics of natural water. *U. S. Geol. Surv. Water Supply Paper 1473*.

Hughes, T. McK (1886). On some perched blocks and associated phenomena. *Q.Jl geol. Soc. Lond.*, **42**, 527–39.

Nordstrom, D. K., Plummer, L1. N., Wigley, T. M. L., Wolery, T. J., Ball, J. W., Jenne, A. E., Bassett, R. L., Creras, D. A., Florence, T. M., Fritz, B., Hoffman, M., Holdren, Jr. G. R., Laton, G. M., Mattigod, S. V., McDuff, R. E., Morel, F., Reddy, M. M., Sposito, G., and Thrailkill, J. (1979). A comparison of computerised chemical models for equilibrium calculations in aqueous systems. In: *Chemical Modelling in Aqueous Systems* (R. F. Gould, Ed.), A.C.S. Symposium Series, Am. Chemical Society, 875–891.

Penny, L. F. (1974). Quaternary. In: *The Geology and Mineral Resources of Yorkshire* (D. H. Rayner and J. E. Hemmingway, Eds.), Yorkshire Geological Society, pp. 405.

Petrovic, R., Berner, R. A., and Goldhaber, M. B. (1976). Rate control in dissolution of alkali feldspar. 1: Study of residual feldspar grains by X-ray photoelectron spectroscopy. *Geochim. Cosmochim. Acta*, **40**, 537–548.

Pettijohn, F. J. (1975). *Sedimentary Rocks*, Harper and Row, New York.

Piggott, G. D. (1965). The structure of limestone surfaces in Derbyshire. *Geogrl. J.*, **131**, 4–41.

Plummer, L. N. (1977). Defining reactions and mass transfer in part of the Flordian Aquifer. *Water Resources Res.*, **13**, 801–812.

Plummer, L. N., Jones, B. F., and Truesdell, A. H. (1976). WATEQF—A FORTRAN IV version of WATEQ, a computer program for calculating chemical equilibria of natural waters. *U.S. Geol. Survey, Water Resources Invest., 76–13*, NTIS Tech. Rept. PB-261027, Springfield, VA 22161, 61 pp.

Plummer, L. N., Parkhurst, D. L., and Thorstenson, D. C. (1983). Development of reaction models for groundwater systems. *Geochim. Cosmochim. Acta*, **47**, 665–686.

Spears, D. A., and Reeves, M. J. (1975a). The influence of superficial deposits on groundwater quality in the Vale of York. *Q. Jl. Engng Geol.*, **8**, 255–269.

Spears, D. A., and Reeves, M. J. (1975b). The infiltration rate into an aquifer determined from the dissolution of carbonate. *Geol. Mag.*, **112**, 585–591.

Sweeting, M. M. (1965). Introduction to symposium on denudation in limestone regions. *Geogrl. J.*, **131**, 34–37.

Wollast, R. (1967). Kinetics of the alteration of K-feldspar in buffered solutions at low temperatures. *Geochim. Cosmochim. Acta*, **31**, 635–648.

Solute Processes
Edited by S. T. Trudgill

CHAPTER 6

Runoff processes and solutional denudation rates on humid temperature hillslopes

T. P. Burt

School of Geography,
University of Oxford

6.1 INTRODUCTION

Recent investigations of hillslope solute systems have tended to polarize around two main themes: the relationship of solute leaching processes to the runoff processes operating on the hillslope; and the study of long-term denudation rates. Clearly the timescales involved differ markedly. Many leaching studies concentrate solely on individual storm events or on annual nutrient cycles; in contrast, the estimation of solutional denudation rates must necessarily consider much longer timescales, particularly if the development of hillslope form is to be predicted. This dichotomous state of solute studies within hillslope geomorphology reflects the schismatic state of geomorphology as a whole (Lewin, 1980). It is a major purpose of this chapter to demonstrate that such a division need not exist as far as hillslope geomorphology is concerned. It has been argued elsewhere (Burt and Walling, 1984) that modelling studies in fluvial geomorphology may involve two stages:

1. development and verification of models based on observation of the current operation of the process response system;
2. the extension of the model to longer timescales to examine system evolution.

Successful (i.e. verifiable) process-based prediction of the evolution of hillslope forms is not currently possible; even so, current field studies should be directed towards the calibration of predefined models. Once verified, such models may then be extended to longer timescales, at which stage alternative field evidence may also be required for model evaluation. Such a methodology demands, however, that an experimental design must be specific to testing a given model; it is therefore unlikely that one field experiment can formally study both nutrient cycling and process–form relationships, except in preliminary studies where initial hypotheses about either (or both) systems may be formulated.

Attention to the strict requirements of scientific methodology (Anderson and Burt, 1981) will ensure that some elements of the 'schism' noted above must persist, although the development of model structures and field techniques common to both types of study will mean that much common ground will still exist. Clearly interest such as solute levels in streams, nutrient cycling and geochemistry must continue to represent legitimate subject matter for hillslope geomorphologists. Kirkby (1976) has noted the value of overlap between models of hillslope evolution and other environmental models such as those used in hillslope hydrology or nutrient cycling. In the initial development of theories, a broad approach to modelling is clearly advantageous. Only at a later stage must the *specific* model structure and experimental design used to study hillslope process and form necessarily differ from those used to investigate runoff–nutrient systems.

6.2 SUBSURFACE RUNOFF PROCESSES AND SOURCE AREAS

6.2.1 Conditions for lateral subsurface flow

Over the last twenty years, the dominant focus of research within hillslope hydrology has involved the study of lateral water flow *within* the soil layers. While the concept of subsurface stormflow is not new (e.g. Hursh and Brater, 1941), the dominance of runoff theories based on the occurrence of infiltration-excess overland flow (Horton, 1933, 1945) meant that research into subsurface runoff processes was neglected. However, it gradually became apparent that in areas of permeable soils, where there is either a gradual or abrupt decrease in hydraulic conductivity with depth, subsurface runoff within the regolith can account for much if not all of the storm runoff leaving a catchment (Whipkey, 1965; Weyman, 1973). The primary mechanism in the production of significant quantities of subsurface flow involves the development of saturated conditions (often termed the 'saturated wedge') within the soil profile (Weyman, 1973; Anderson and Burt, 1978). When the soil profile becomes completely saturated, saturation overland flow will also occur. The generation of both subsurface stormflow and saturation-excess overland flow is limited to certain source areas (as described below) and both may occur at rainfall intensities much below those required to produce infiltration-excess overland flow. Since there is a clear link between the occurrence of subsurface runoff and solute removal from hillslopes, it is of some relevance to detail the processes and source areas involved, since such an analysis may form the basis for an understanding of hillslope solutional denudation.

Classical infiltration models have assumed a semi-infinite soil where storage effects are unimportant (e.g. Horton, 1933; Philip, 1957); the generation of overland flow is predicted to occur only when rainfall intensities exceed the soil's infiltration capacity (i.e. 'infiltration-excess' or 'Hortonian' overland flow). Knapp (1978) and Kirkby (1978) have both considered infiltration processes in layered soils. The effect of a lower horizon of reduced hydraulic conductivity is to limit the storage capacity of the upper layer so that even though rainfall intensities may be well below infiltration capacity, surface runoff will still be generated once the upper layer store becomes filled. Consideration of such storage models is also beneficial with respect to subsurface runoff generation, since lateral flow within the soil also depends on the existence of a layer of reduced hydraulic conductivity at depth. Given a soil with a basal percolation rate K_2, for a rainfall intensity I, and assuming that intensity is less than the surface infiltration capacity, the soil moisture will increase at a rate:

$$I - K_2 \tag{6.1}$$

If the total storage in the upper layer is H_1, then the time taken to reach saturation is

$$t = \frac{H_1}{I - K_2} \tag{6.2}$$

The total infiltrated during this time will be

$$V = \frac{I H_1}{I - K_2} \tag{6.3}$$

From Equation (6.3) it follows that, as rainfall intensity increases, the total volume which infiltrates before the upper layer store becomes filled decreases, as demonstrated in Figure 6.1. For low rainfall intensities and high upper-layer storage, it is clear that considerable quantities may infiltrate before saturation occurs, even when basal impedance exists. Under such conditions, lateral flow is likely to occur since a saturated zone will build up above the impeding layer, provided that input I exceeds basal percolation K_2 from the upper profile (Kirkby and Whipkey, 1978). The vertical drainage model described in Equations (6.1)–(6.3) require modification, therefore, to include this lateral component. Figure 6.2 illustrates a simple storage model in which both lateral and vertical flow within and between soil layers is accommodated. Runoff simulations made using this model provided the

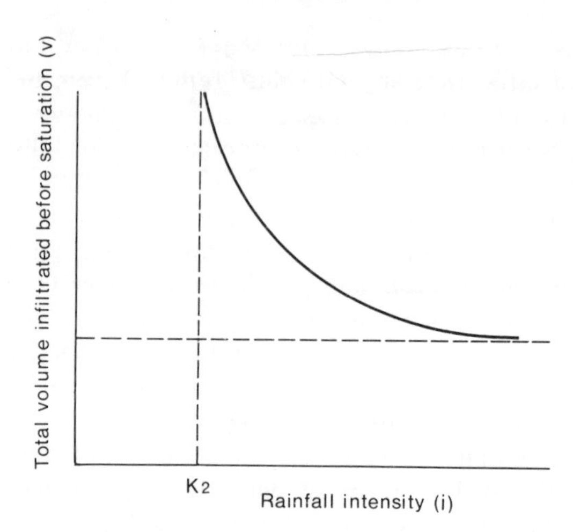

Figure 6.1. The effect of limited soil storage on the total volume of rainfall infiltrating before saturation for different rainfall intensities (Knapp, 1978)

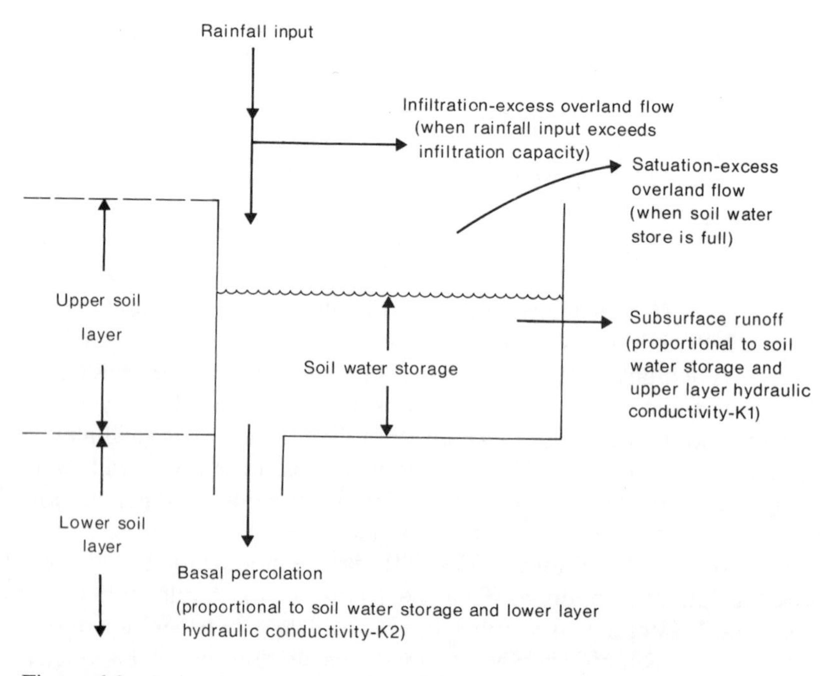

Figure 6.2. A simple two-layer soil model for generating lateral hillslope flow

results displayed in Figure 6.3. All simulations assumed a constant rainfall intensity and a fixed storage capacity for the upper layer with the constant rate of basal percolation from the upper layer. The figures plotted indicate the total discharge produced after twenty iterations. It is clear that production of subsurface runoff, and associated saturation excess overland flow, is strongly favoured by a low basal percolation rate K_2 in combination with high upper-layer hydraulic conductivity (i.e. a high K_1/K_2 ratio). Where basal percolation rates become significant, even with a high upper-layer hydraulic conductivity, production of subsurface runoff will be limited. If the basal percolation rate exceeds rainfall intensity, or if the upper layer conductivity is less than that of the lower layer, drainage through the upper layer will be unsaturated; and since no saturated layer builds up at the base of the upper layer, no lateral subsurface flow is generated. Figure 6.4 shows the volumes of subsurface runoff produced using the same layered soil model, but with a much larger upper store capacity. This avoids the 'plateau' effect evident in Figure 6.3(a) (which reflected complete saturation of an upper layer of limited storage, and thus a maximal volume of throughflow). In this latter case the upper layer is never completely saturated and subsurface runoff is maximized, being limited only by the magnitude of the rainfall input. Once again, the combination of higher upper layer and minimal lower layer hydraulic conductivity is seen to be not conducive to the production of saturated subsurface runoff.

It is now appropriate to introduce a greater element of realism into the conceptual two-layer soil model described above, using the analytical framework suggested by Zaslavsky (1970). Consider infiltration into a soil column consisting of two soil layers, permeable above less permeable; no lateral flow can occur and pressure is atmospheric at the base of the lower column. Figure 6.5 illustrates such a column, indicating the variation in soil water potential through the column for five different rainfall intensities:

Curve 1: $I < K_2$—a uniform distribution of soil water potential throughout most of the upper layer; I equals the unsaturated hydraulic conductivity of the soil at this potential. An increase in potential at the column base denotes the effect of the less permeable lower layer.

Curve 2: $I = K_2$—the lower layer is saturated with zero potential being reached at the base of the upper layer.

Curve 3: $K_1 > I > K_2$—saturation of parts of the upper layer begins.

Curve 4: $I_4 = K_2\ (D_1 + D_2)/D_2$—upper layer is now saturated to the surface where incipient ponding occurs. Note that the downward flux q exceeds K_2 since from Darcy's law $q = ki$, flux is dependent on both hydraulic conductivity *and* hydraulic gradient (the latter being maximized through the lower layer in this case).

Curve 5: $I > I_4$—surface ponding and saturation overland flow occur.

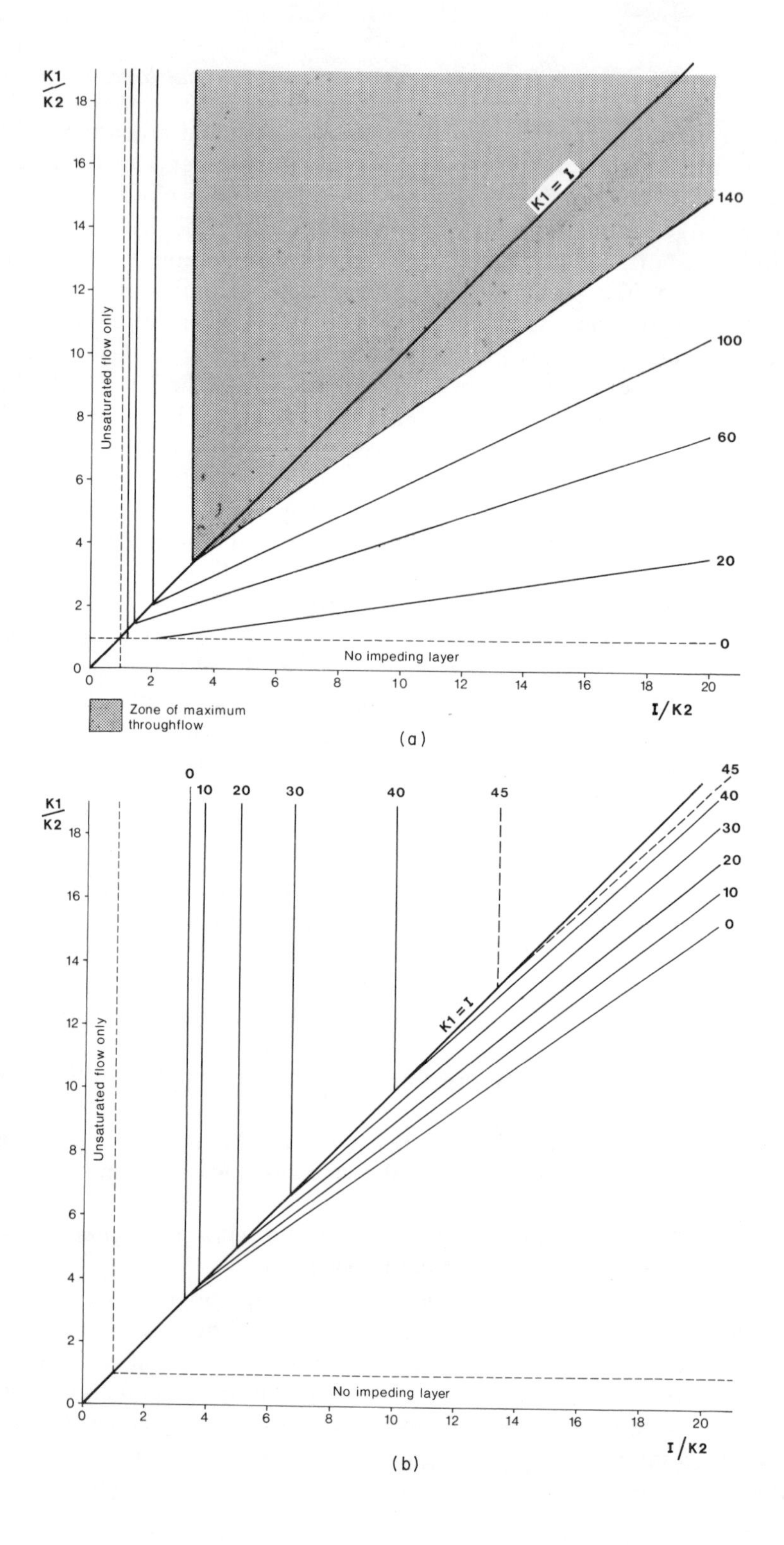
K1/K2
I/K2
Unsaturated flow only
No impeding layer
K1 = I
140
100
60
20
0
Zone of maximum throughflow
(a)
0
10
20
30
40
45
45
40
30
20
10
0
K1 = I
Unsaturated flow only
No impeding layer
(b)

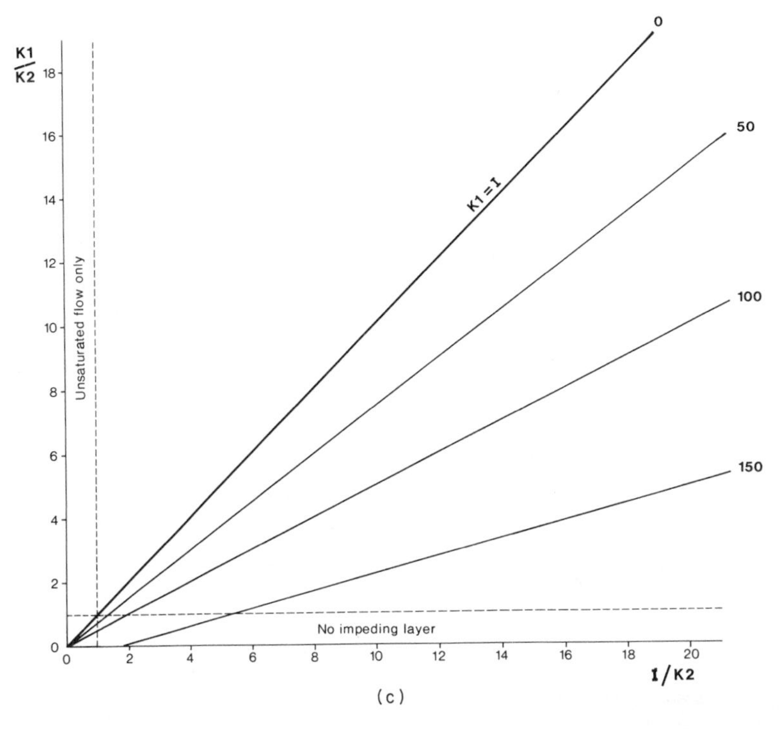

Figure 6.3. The effect of soil layering on runoff production as indicated by computer simulations using the model described on Figure 6.2: (a) total throughflow; (b) total saturation-excess overland flow; (c) total infiltration-excess overland flow. For all simulations, constant values of soil storage and rainfall intensity were used; model run times were also identical in each case

If we now relate conditions prevailing in the draining soil column to those existing on a hillslope of similar characteristics (Figure 6.6), the development of a horizontal flow component can be defined. Using the soil water potential distribution for curve 4, a downslope flow direction in the upper layer is predicted from the distribution of total hydraulic potential (ϕ—the summation of soil water and gravitational potentials). In the lower layer, the flow direction is marginally upslope, although since the lower layer has a much reduced hydraulic conductivity, the resultant flux is of little consequence. Line AB shows that at a given elevation within the upper layer, the soil water potential is higher at the upslope profile, so that a horizontal flow component must develop. In the lower layer, line CD demonstrates the reverse situation. Zaslavsky (1970) has assessed for anisotropic soils (i.e. those soils where conductivity is not uniform in all directions) the effect of hillslope gradient on resultant flow direction. For those cases where the

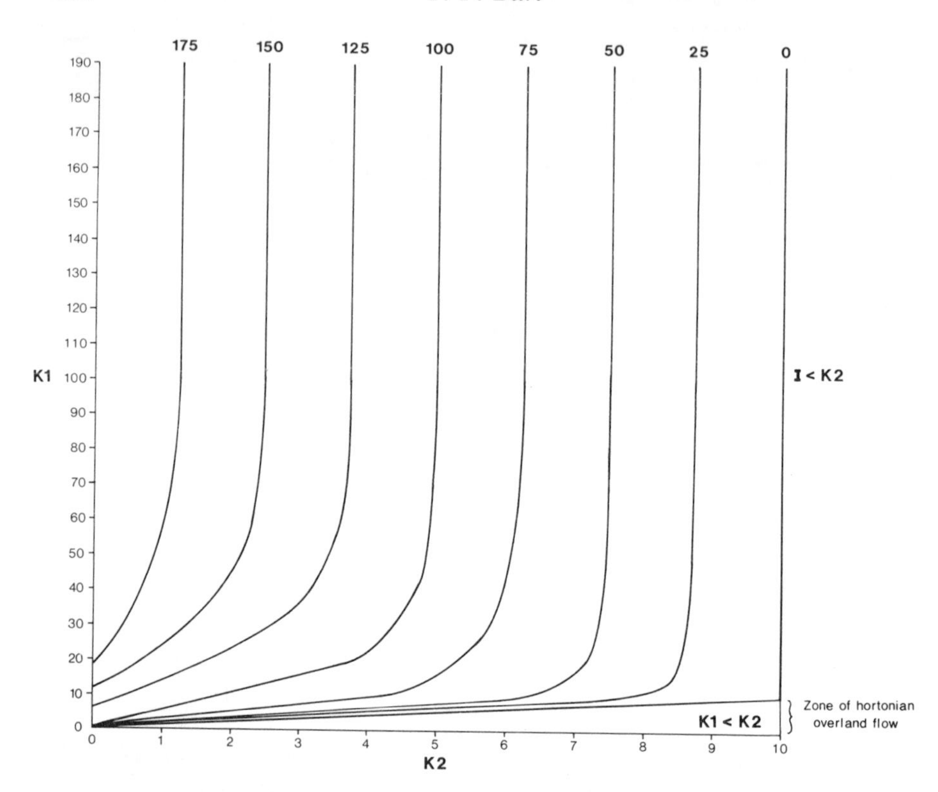

Figure 6.4. Volumes of subsurface runoff produced by the two-layer soil model for a situation with unlimited surface layer storage. K_1 and K_2 are the upper and lower layer hydraulic conductivity respectively

downslope hydraulic conductivity is greater than down-profile conductivity, he shows that streamlines will converge on a convexo-concave slope even for a low ratio of K_1/K_2 (Figure 6.7(a)). Where the ratio is increased, the lateral deflection of streamlines will be even more marked (Figure 6.7(b)); increased narrowness of streamlines denotes higher flux per unit area. Only on low-angle slopes with K_1/K_2 ratios close to unit does near-vertical drainage occur. As slope angles increase, convergence will occur closer to the soil surface (Figure 6.7(c)). The effect of soil anisotropy is therefore to engender lateral subsurface runoff; moreover, on all but the gentlest convex slopes, convergence of streamlines will occur, leading to ponding of water and the generation of surface runoff.

The discussion so far has assumed that only matrix flow has been occurring, and no reference has been made to soil water flow through macropores (i.e. pores above capillary size). The presence of macropores may alter the timing and solutional load of subsurface runoff (see Smettem, 1982; also this volume,

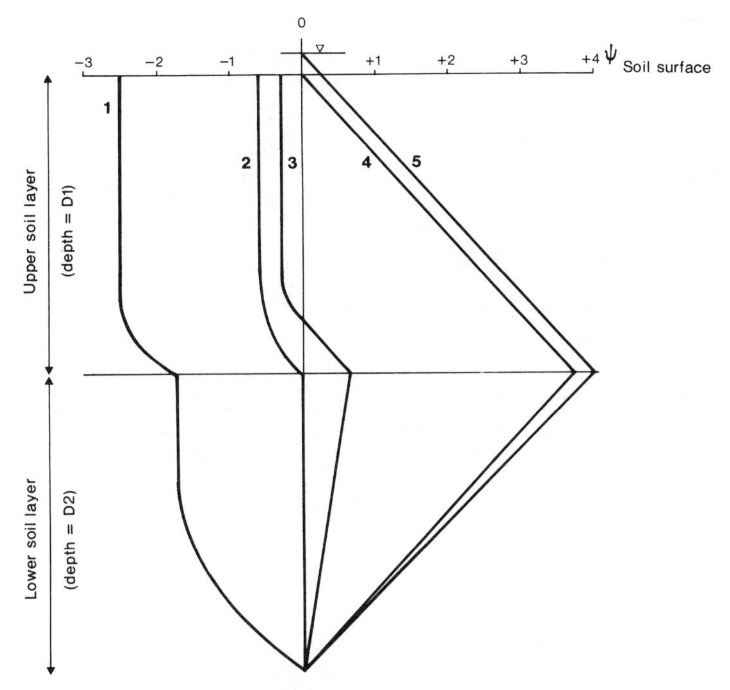

Figure 6.5. Distributions of soil water potential in a double layer soil column (permeable above less permeable) for different rainfall intensities (Zaslavsky, 1970)

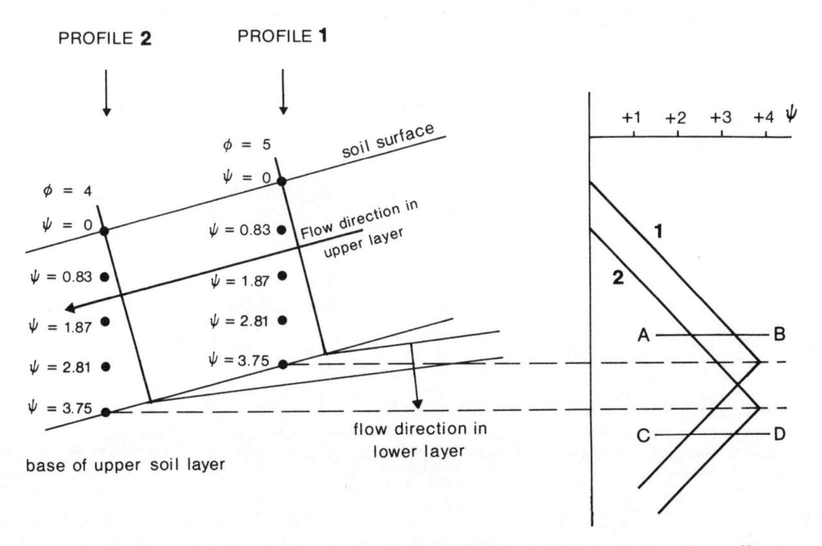

Figure 6.6. The distribution of soilwater potential on a hillslope using the soilwater potential distribution for a double-layer soil column as given as Figure 6.5 (case 4)

Chapter 4). However, their influence on hillslope flow may be adequately contained *within* the models presented above, since they effectively increase the bulk hydraulic conductivity of the soil layer, at least at a scale relevant to hillslope runoff (Beven and Germann, 1981). The size and connectivity (both lateral and vertical) of the macropores will themselves help to determine the variations in soil anisotropy which result from their presence in the soil.

6.2.2 Production of subsurface runoff on hillslopes

Where deep permeable soils overlie less permeable soil or bedrock, and where steep hillslopes border a narrow valley floor, subsurface runoff dominates the storm hydrography volumetrically, and in certain cases also provides the peak storm discharge. As slope angles become shallower, soils thinner, and the valley floodplain more extensive, saturation overland flow

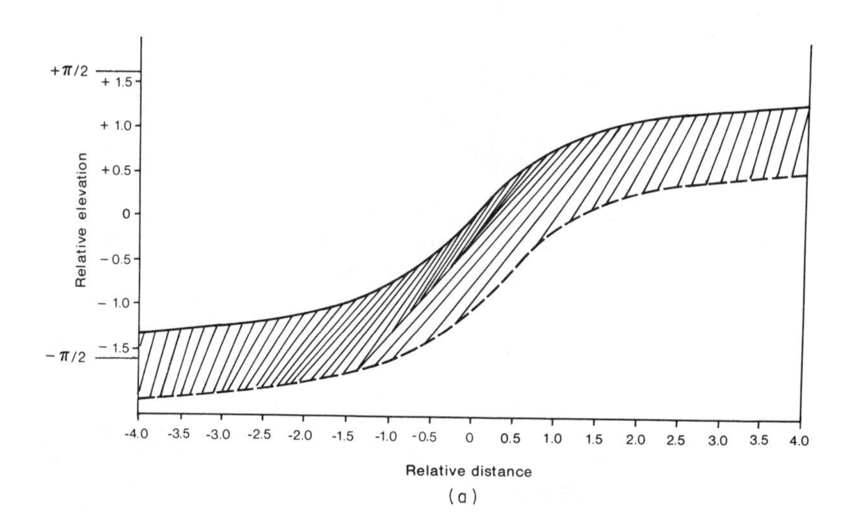

(a)

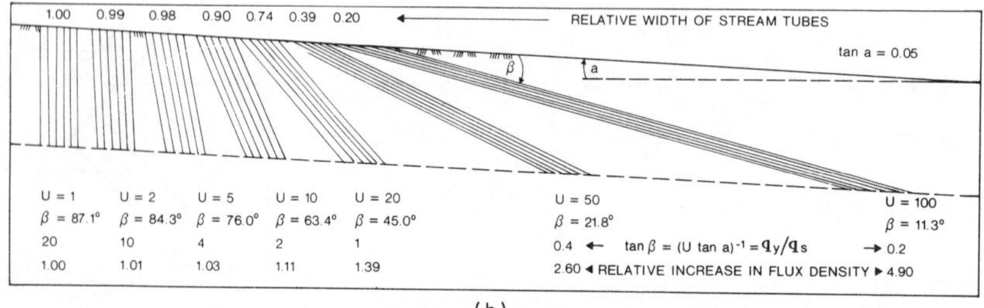

(b)

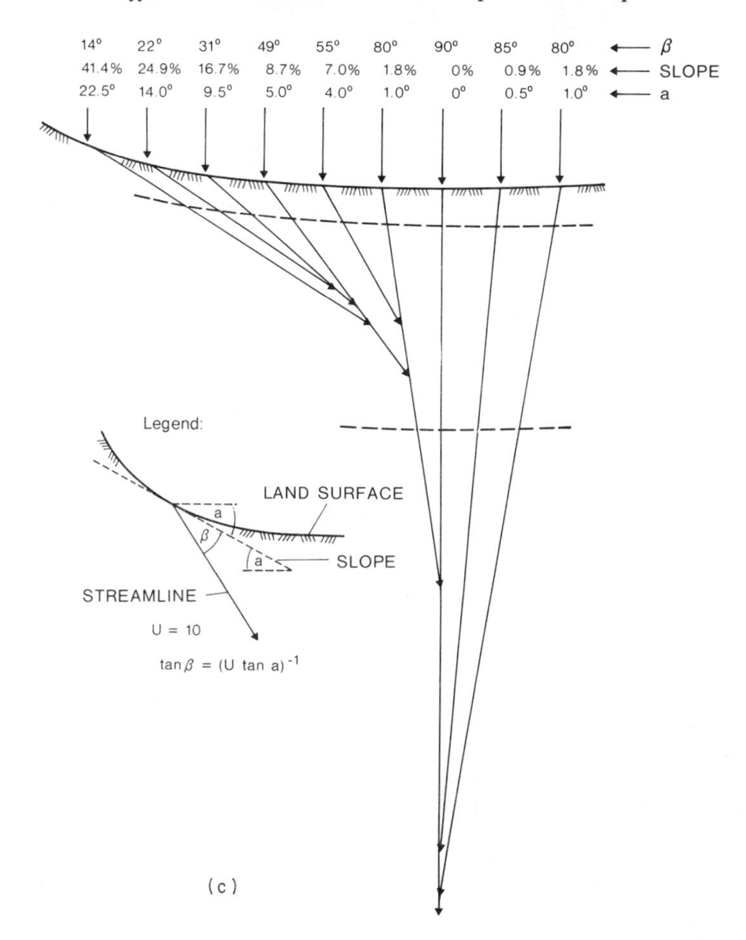

Figure 6.7. The effect of soil anisotropy on the flow direction of soil water: (a) steamlines on a convexo-concave hillslope; (b) the effect of the degree of anisotropy on flow the deflection laterally; (c) the effect of slope angle on flowline convergence (Zaslavsky, 1970)

and return flow (subsurface flow which returns to the land surface having flowed through the upper soil horizon) become dominant. Even where subsurface runoff volumetrically is most important, the peak storm discharge is often provided by a combination of channel precipitation and direct runoff from permanently saturated zones at the base of the steep hillslopes. The development of a second delayed peak in stream discharge denotes the major volumetric contribution of subsurface stormflow; in addition some subsurface runoff may also contribute to the first peak (Anderson and Burt, 1982). Figure 6.8 shows a range of storm hydrographs where soil anisotropy has

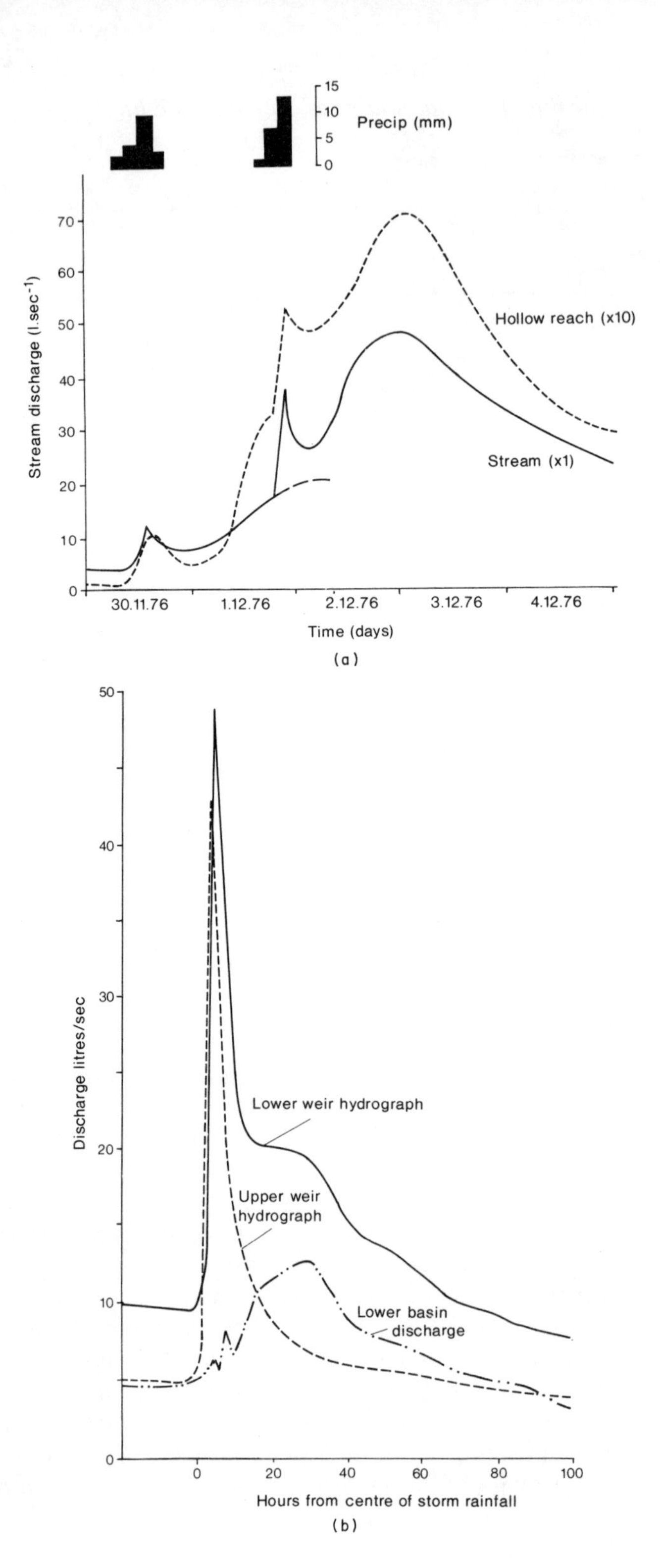
15
10
5
0
Precip (mm)
70
60
50
40
30
20
10
0
Stream discharge (l.sec-1)
Hollow reach (x10)
Stream (x1)
30.11.76
1.12.76
2.12.76
3.12.76
4.12.76
Time (days)
(a)
50
40
30
20
10
0
Discharge litres/sec
Lower weir hydrograph
Upper weir
hydrograph
Lower basin
discharge
0
20
40
60
80
100
Hours from centre of storm rainfall
(b)

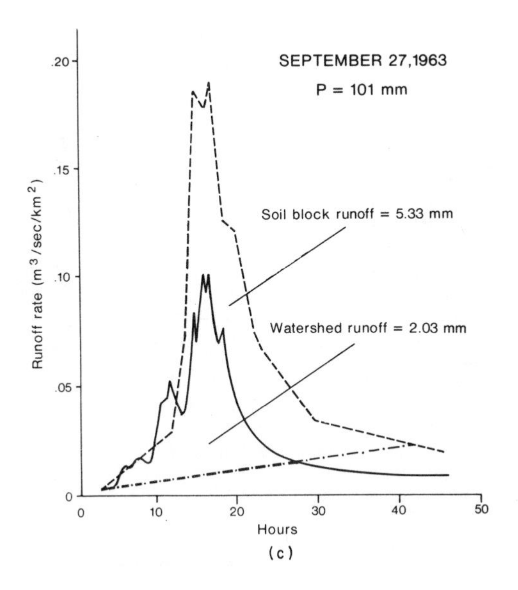

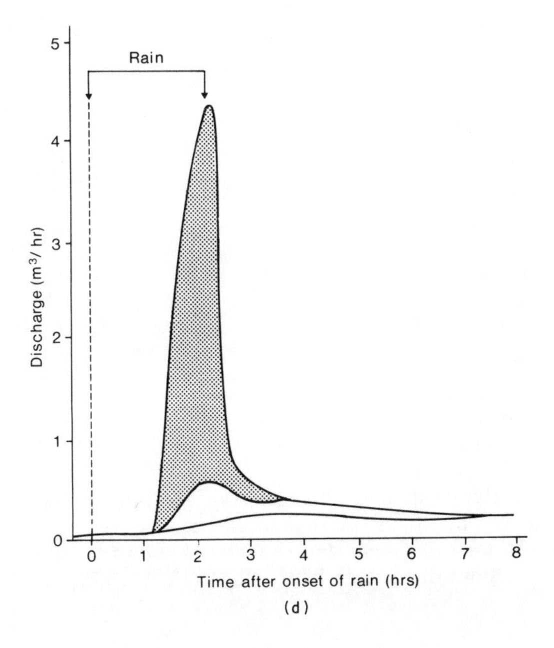

Figure 6.8. Storm hydrographs for four catchments where subsurface runoff is dominant: (a) Bicknoller Combe; (b) East Twin Brook; (c) Coweeta; (d) Sleepers River

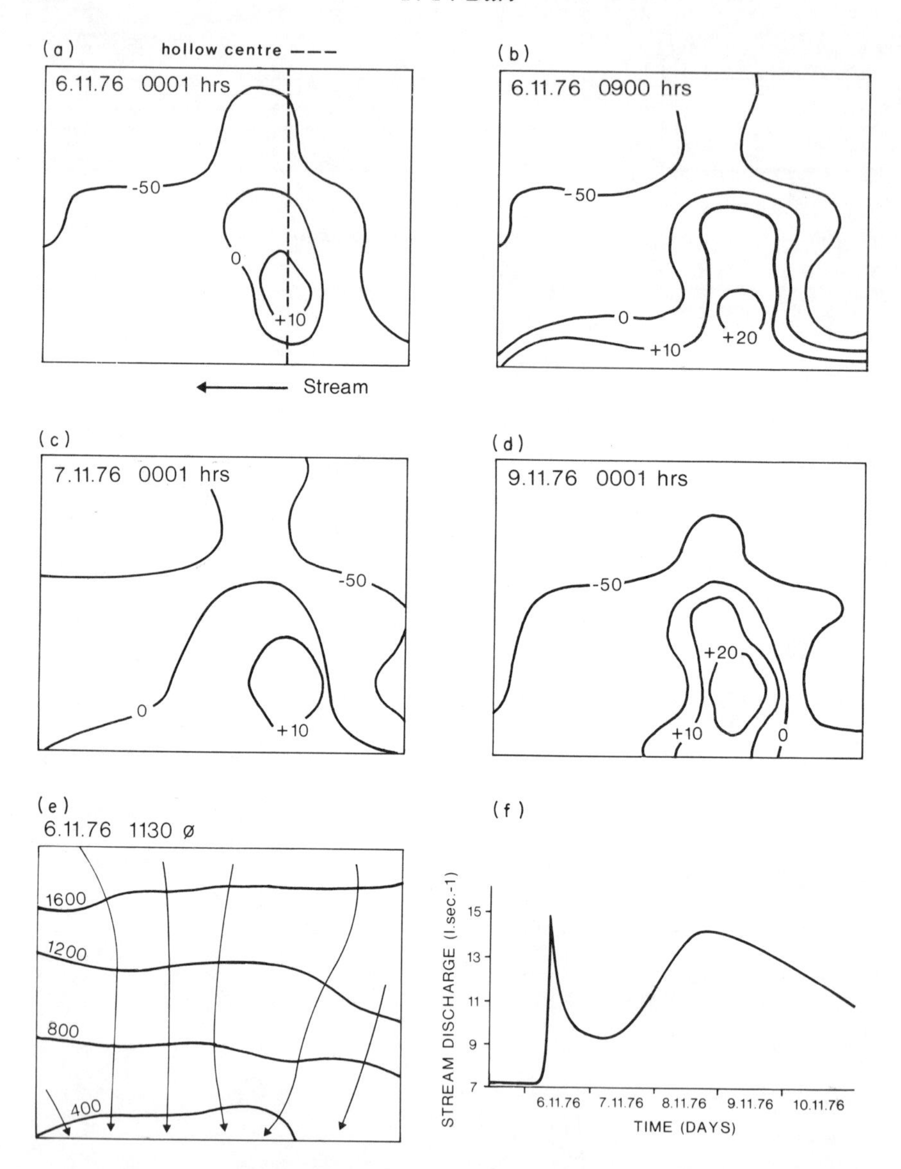

Figure 6.9. Soil water potential changes on an instrumented hillslope at Bicknoller Combe during the period of a double-peaked hydrograph: maps (a–d) show soilwater potential at 60 cm depth; map (e) shows the flow net; map (f) shows the stream hydrograph (Burt, 1978)

provided lateral subsurface runoff:

Bicknoller Combe: Two separate storm events, the second rainfall period coinciding with the delayed throughflow response of the first event. The initial 'storm' runoff peak is provided by Hortonian overland flow from a limited area of the catchment (footpaths); the second delayed peak is larger both in terms of volume and absolute peak discharge. Discharge for the 'hollow reach' indicates the important function of hillslope hollows as sources of subsurface runoff. Note that some 'rapid' throughflow occurs also at the time of the rainfall input. Soils are permeable, 2 m deep, over impermeable Devonian sandstone; valley-side slope angles average 25° (Burt, 1978).

East Twin Brook: Peak storm discharge is dominated by runoff from the peat-covered bowl-shaped headwaters. 'Lower basin discharge' is provided by subsurface runoff from gentle convex slopes (up to 23° close to the stream) covered by permeable Brown Earth soils. Hydrographs for individual hillslope sections correlate closely with the lower basin response (Weyman, 1973).

Coweeta: Runoff from a 15.2 hectare watershed is largely reflected by runoff from a 0.0074 hectare experimental plot, except that 'direct' runoff, probably channel precipitation, contributes to the watershed response (Hewlett and Nutter, 1970).

Sleepers River: Runoff from a 44 mm 2 hour artificial storm on a concave hillslope plot. The size of storm, in combination with shallow soils and gentler slopes, favours a runoff response dominated by return flow, with subsurface stormflow being limited by complete profile saturation (Dunne and Black, 1970).

The changes in soil moisture conditions on the hillslope associated with the generation of subsurface stormflow have been described in detail elsewhere (Weyman, 1973; Anderson and Burt, 1978; Burt *et al.*, 1983), and it is sufficient here merely to summarize the salient features.

Figure 6.9(a)–(d) show soil water potential changes during the period of a double-peaked storm hydrograph (Figure 6.9(f)) for the Bicknoller Combe catchment. The hillslope section mapped measured 70 m × 50 m. The map shows moisture conditions at 60 cm depth for a hillslope hollow and its adjacent spurs. The maps show that during the rainfall (6 November, 0900 hrs—Figure 6.9(b)), a widespread but shallow zone of saturation develops at the base of the slope. The generation of the delayed peak (9 November) is associated with convergence of soilwater into the lower part of the hollow from further upslope and from the adjacent spurs. This provides

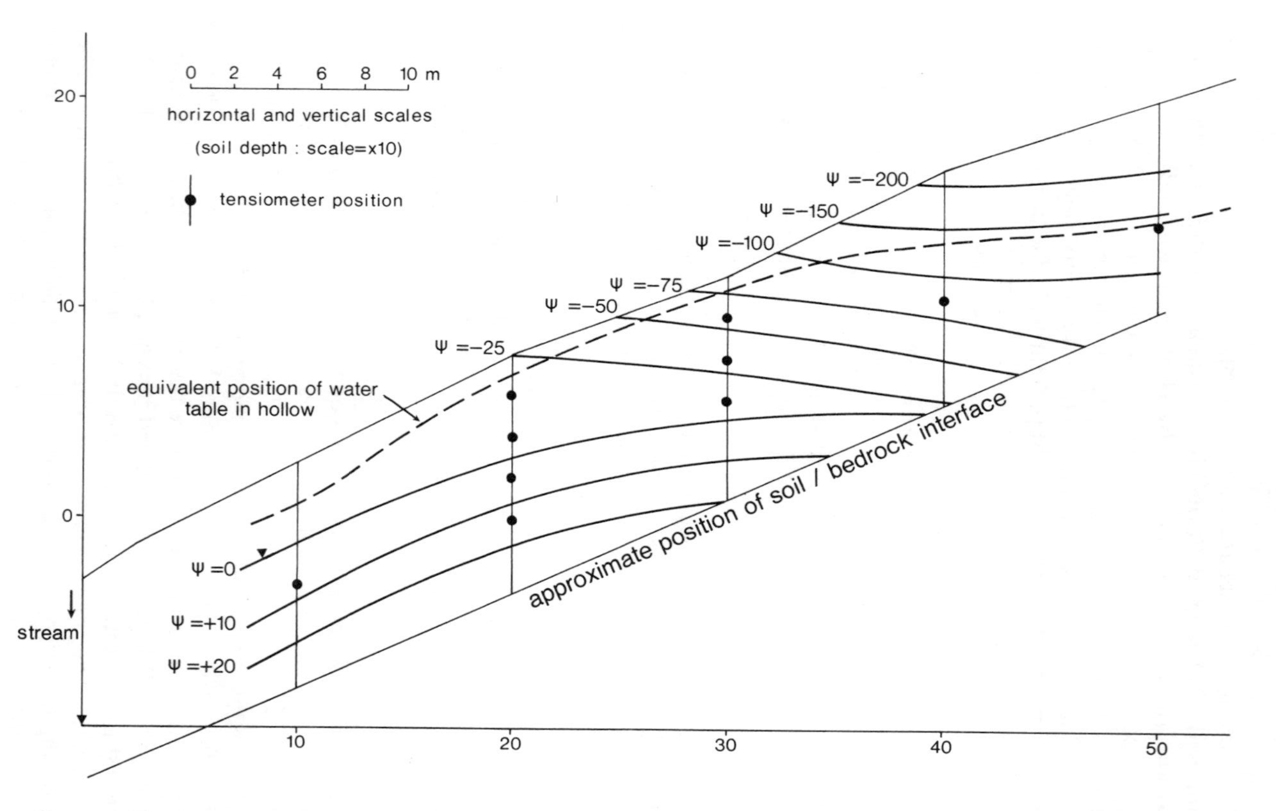

Figure 6.10. Contrast between hollow and spur soil water conditions at the time of maximum subsurface runoff production (Burt, 1978)

the deepest saturation during the study period, although the saturated wedge is now spatially less extensive than during the rainfall (Anderson and Burt, 1978b). Following the period of maximum throughflow, the wedge drains rapidly to reach its pre-storm dimensions some four days later. The flow-net (orthogonals drawn at right-angles to the line of equal total potential) indicates convergent flow into the lower hollow (Figure 6.9(e)); such convergence is maintained at all stages of flow, being controlled largely by topography at such high slope angles (Anderson and Burt, 1978a; Anderson, 1982). This contrast between hollow and spur drainage is emphasized in Figure 6.10. On the spur only a small saturated wedge exists at the slope base (where soil water potentials are positive); for the rest of the slope length, the soil is unsaturated. In contrast, a large saturated wedge is maintained in the hollow. Moreover, the conditions described in Figure 6.10 represent the most favourable period of saturation on the hillslope; at most other times the contrast between hollow and spur is even more marked. During times of rainfall, the unsaturated spur soil will be dominated by vertical infiltration processes (see also Weyman, 1973); by contrast, flow within the hollow, even during rainfall, will be dominated by lateral movement. Such a contrast may well have implications for differences in the pattern and rate of solutional erosion on the two hillslope profiles (see §6.3).

At Bicknoller, hillslope runoff is generated almost entirely from the valley-side slopes, the interfluve areas being too distant to contribute to the main stormflow response. In other locations, where valley-side slopes are shorter, interfluve areas may also contribute significant volumes of subsurface runoff, as in the Slapton Wood catchment (Devon, England), where major contributions of 'interfluve' runoff are routed through valley-side hollows to add to the saturated wedge in the hollow base (Burt *et al.*, 1983). Such effects help emphasize the hydrological contrast between the hollow and spur, as well as extending the potential source areas for solute teaching to those interfluve zones peripheral to such valley-side hollows.

The subsurface stormflow is controlled by a combination of antecedent soil moisture and storm rainfall input. Figure 6.11 indicates the effect of antecedent conditions on the magnitude of the delayed runoff response; for almost identical rainfall inputs (20 mm), the second storm provides a greatly enhanced runoff rate, the earlier storm having 'primed' the hillslope system by providing initial enhancement of the saturated wedge. Butcher (1985) has examined 13 years of data for the Slapton Wood catchment and has identified the major controls of the delayed subsurface runoff response for that basin. Double-peaked hydrographs occur only between the months of December and March; during the summer, soil moisture deficits are too great to allow lateral flow to be transmitted downslope. Throughflow peak discharge and runoff volume correlates strongly with storm rainfall as well as with indices of antecendent catchment wetness such as prestorm discharge and API. Butcher

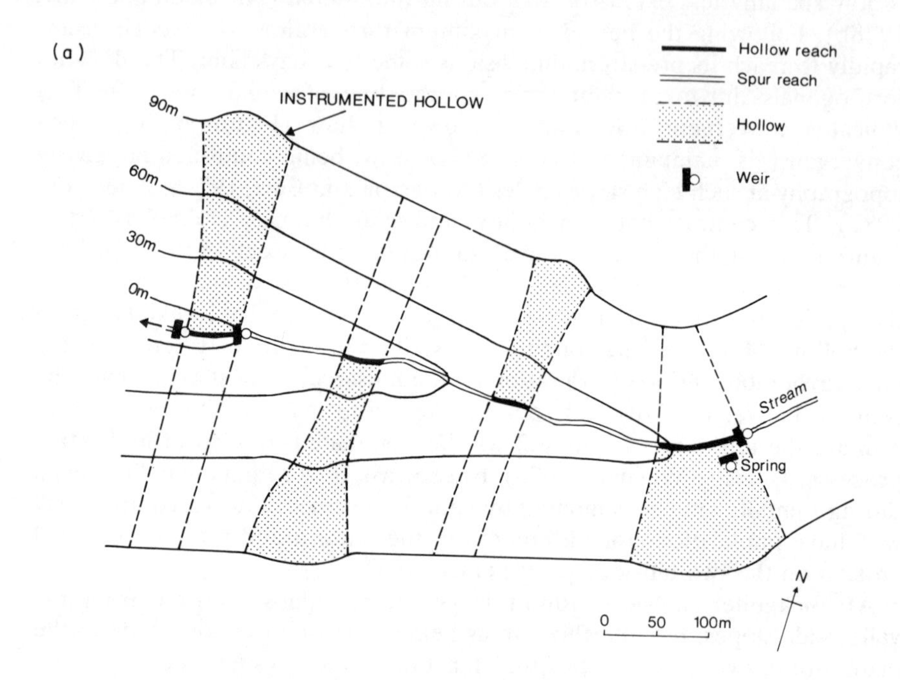
(a)
Hollow reach
Spur reach
Hollow
Weir
INSTRUMENTED HOLLOW
90m
60m
30m
0m
Stream
Spring
N
0 50 100m

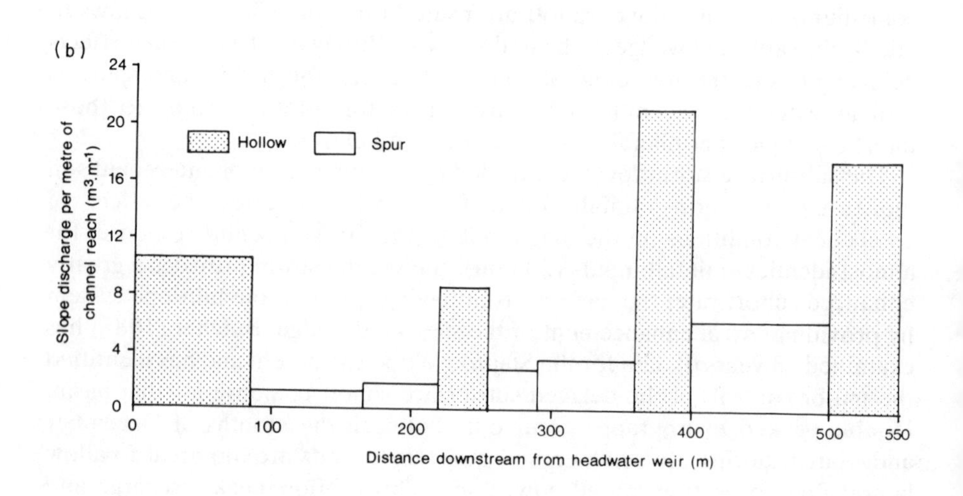
(b)
Hollow
Spur
Slope discharge per metre of channel reach (m3.m-1)
24
20
16
12
8
4
0
0 100 200 300 400 500 550
Distance downstream from headwater weir (m)

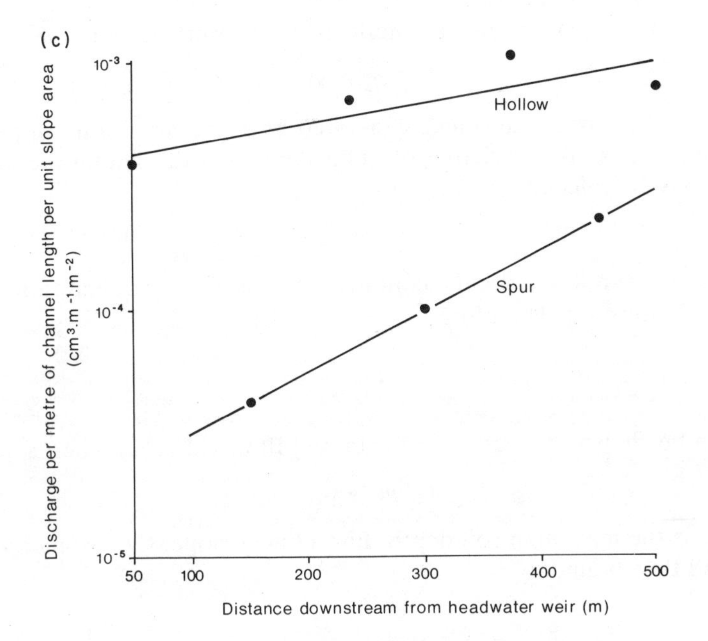

Figure 6.11. Hollow and spur reach discharge inputs for the Bicknoller Combe catchment: (a) map of stream reaches; (b) discharge per unit length of channel; (c) discharge per unit length of channel per unit area of hillslope (Anderson and Burt, 1978)

also suggests that two subsets of throughflow events may occur, the delayed peak being larger in those events where significant interfluve runoff contributes to the hydrograph.

6.2.3 Source areas for subsurface runoff generation

The areas of a catchment contributing stormflow, whether as subsurface runoff or as saturation overland flow, are much more limited than Horton assumed. Nevertheless, one may model subsurface runoff using Horton's equation for overland flow as a basis for prediction. Horton showed that overland flow discharge Q_x at distance x from the divide would be:

$$Q_x = x(i - f) \text{ per unit contour width} \tag{6.4}$$

where i is rainfall intensity and f is infiltration capacity. Where plan convergence or divergence occurs, then:

$$Q_x = a(i - f) \tag{6.5}$$

where a is the area draining downslope to the unit contour width in question.

The Horton equation may be modified for subsurface runoff q_x:

$$q_x = ja \tag{6.6}$$

where j is the subsurface runoff generated by a unit area of hillslope.

From Darcy's law, discharge per unit contour width within the saturated wedge may be calculated:

$$q_x = ksd \tag{6.7}$$

where k is saturated hydraulic conductivity, s is slope gradient, and d is the depth of saturation. Therefore

$$d = \frac{ja}{ks} \tag{6.8}$$

It can be shown that saturation overland flow will occur where:

$$ja > skz \tag{6.9}$$

where z is the maximum soil depth. For a linear slope, the distance x_s to the overland flow boundary is:

$$x_s = \frac{skz}{j} \tag{6.10}$$

From Equation (6.9), it follows that saturation overland flow will occur where:

$$\frac{a}{s} > \frac{kz}{j} \tag{6.11}$$

Kirkby (1978) has mapped the paramter a/s to denote source areas of saturation overland flow for the Sleepers River catchment. While there are some problems with the model—j is difficult to define, and some 'potential' contributing (interfluve) *areas* may not, in fact, be hydrologically active in producing throughflow; the model assumes equilibrium flow conditions—nevertheless the model does serve as a basis for the delineation of source areas in a catchment. Saturated depth is directly proportional to a/s, and to unit runoff, and inversely proportional to hydraulic conductivity. It is likely that surface saturation will therefore occur where drainage from large areas becomes concentrated, particularly if slope gradient also decreases; the lower part of hillslope hollows, where both plan and profile curvature tend to be concave, is a particularly favourable location for the development of a deep saturated wedge, and thus for the occurrence of surface runoff. Such contributing areas will be variable in extent, since during rainfall the zone of surface saturation will extend upslope. Dunne (1978) has described seasonal and within-storm variations of the extent of surface saturation in the Sleepers

River catchment, where as expected most source areas are confined to the hillslope hollows (see also Butcher, 1985).

Three zones may be identified where maximum subsurface flow will occur:

1. *Areas of streamline convergence*: Where the shape of the hillslope profile (Figure 6.8) or plan (Figure 6.9(e)) favours the convergence of streamlines, accumulation of soil water will occur, particularly when soil anisotropy is high. Equation (6.6) shows that concentration of subsurface drainage into hillslope hollows must maximize flow rates as well as engender surface saturation. O'Loughlin (1981) has analysed the effects of hillslope geometry on soil moisture distributions, using idealized plan and profile shapes. Anderson and Burt (1978) showed that hillslope hollows were the major source of stream discharge for the Bicknoller Combe catchment (Figure 6.10). Taking into account both the (hillslope) contributing area and the length of each stream reach, hollows are seen to produce the major portion of the subsurface runoff response, despite the greater area of the catchment being 'spur' zones.
2. *Zones at the slope base*: Immediately adjacent to the stream channel, subsurface runoff will reach its maximum volume for a given slope profile, since distance downslope must correspond to increasing discharge (Equation (6.6)), at least under equilibrium flow. Using a computer simulation model, Beven (1977) has shown that slope gradient and hydraulic conductivity control throughflow rates (Table 6.1) such that at the base of steeper slopes, or where soil hydraulic conductivity is greater, higher throughflow inputs to the stream will occur. Anderson and Burt (1978) showed that slope length can exert a significant control on slope discharge, particularly where convergent flow does *not* occur (Figure 6.11(b)), such that longer slopes provide a greater input of subsurface runoff into the stream (Equation (6.6)). Butcher (1985) has shown that increased slope length in the interfluve zone leads to greater hillslope runoff.
3. *Areas of reduced soil moisture storage*: Any reduction in total soil moisture capacity (i.e. a reduction in the soil's transmissivity) will be accompanied by an increase in soil moisture content, since from Equation (6.7) total storage is reduced. Whether this is caused by a reduction in profile thickness or in some other soil parameter, such as saturated moisture content, sites of reduced storage favour the development of increased throughflow production and surface saturation. Beven (1977) shows that saturated moisture content is inversely proportional to throughflow production, where other soil parameters are held constant (Table 6.1). Increased storage means that the saturated wedge is shallower and thus less throughflow is generated.

The existence of less permeable layers in the soil profile causes lateral

Table 6.1. Simulations of hillslope throughflow discharge for different topographic and antecedent moisture conditions (Beven, 1977)

(a) *The influence of slope configuration*

	Peak time (min)	Peak discharge (cm s^{-1})	Total discharge 0–2000 min (litres)
Straight slope (1 in 5)	900	21.96	14.84
Divergent slope	880	21.38	14.35
Convergent slope	1560	45.91	53.03
Straight slope (1 in 10)	920	26.16	29.82
Convex slope	880	13.28	7.64
Concave slope	810	29.38	34.94

(b) *Runs with different soil parameters*

A_K	K_S	Sat *MC*		Peak time	Peak *Q*	Total *Q*
0.0008	0.02	0.04		900	21.96	14.84
0.008	0.02	0.4		1100	3.62	3.77
0.00008	0.02	0.4		7.10	30.72	36.8
0.0008	0.02	0.5		1260	3.87	4.75
0.0008	0.02	0.3		680	36.4	44.35
0.0008	0.2	0.4		670	35.1	34.11
0.008	0.002	0.4	(TQ)	1410	3.39	3.29
			(OF)	600	52.86	10.84

soil water movement to occur. This subsurface runoff is spatially variable, being dependent on hillslope geometry and soil conditions. Where soil storage is exceeded, surface saturation will occur, causing saturation-excess overland flow. Were it not for the development of lateral soil water flow, all solutional processes within the soil would be controlled by (vertical) infiltration. Lateral subsurface runoff, being itself spatially variable, is therefore directly responsible for providing the major source of spatial variability in solutional denudation rates on hillslopes of uniform parent material.

6.3 MODELLING SOLUTIONAL DENUDATION

6.3.1 Approaches to modelling

Much research, both theoretical and empirical, has been conducted into subsurface runoff processes operating on humid temperate hillslopes. In contrast, the study of solute removal from hillslopes has received much less attention. Carson and Kirkby (1972, p. 238) note that importance of solute removal from humid and temperate hillslopes in that the rate of solutional erosion may exceed that of all mechanical processes combined. However, they also note that the geomorphological literature contains few measurements which are *specifically* related to solutional erosion and slope development, but remains limited to detailed discussions of soil formation which are not immediately applicable. In parallel to this lack of field observations, few geomorphologists have attempted to model the development of hillslopes by solute removal, the process–response model of Carson and Kirkby (1972, p. 258) being a notable exception. However, even in that case, the model has remained largely untested, because of the unavailability of field data. Two related tasks therefore face the hillslope geomorphologist: to formulate and actualize models of the solutional development of hillslopes, and at the same time to undertake field experiments designed specifically to test those models.

Possible approaches to hillslope modelling are illustrated in Figure 6.12. Empirical models, which arise from the results of field work, will be discussed in §6.4. Such models may be the direct result of observations on soil solutes, or arise indirectly from the study of soil catenas, or from the measurement of solute concentrations of streamwater or throughflow output from the base of the slope. Such empirical evidence will help us to formulate our concept of how hillslope solutional models should be structured, and will perhaps provide some indication of the rate of process operation. However, such 'case studies' remain site-specific illustrations of the system and do not produce unequivocal conclusions about the operation and controls of the processes (Church, 1984). Efficient experimental progress must be preceded by the development of a model. The models are likely to be of two types—conceptual or deterministic—both of which are essentially mass-balance models constrained by an equation of storage or continuity respectively. The greater complexity of deterministic models (both theoretical and operational) has restricted their application. Coupled transport of soil water and solutes (vertically) within the soil profile has been successfully produced using process-based models (e.g. Bresler, 1973; Smettem, 1982), but such results have not been applied to hillslope systems. Similarly, the processes and products of chemical weathering have been described for a number of rock types (e.g. Waylen, 1979; Curtis, 1976; Verstraten, 1980). Such weathering models, while based on the concept of thermodynamic equilibrium in geochemical systems, represent a given geochemical process by a succession of partial equilibrium states, each reversible with respect to the next, but all irreversible in relation to the initial state of the system. Verstraten (1980) has compared theoretical predictions with chemical and mineralogical data of the soil and weathering materials in the Haarts catchment, Luxembourg; the predictions were generally in close accordance to the field situation, although the fact that such weathering systems are open,

Model Type	Dimension 1 (soil profile)	2 (slope profile)	3 (slope topography)
Empirical	Qualitative and statistical models based on field evidence of solutional processes (often inferred from stream records or from soil catenas) – see Section 6.4		
Conceptual	'storage' models of solute leaching in soil profiles (eg. Burns 1974–1980; Addiscott 1977)	Carson and Kirkby (1972) model – generalised storage model based on unit runoff and 'effective' solubility of oxides.	Soil plasma redistribution model of Huggett (1975)
Deterministic	Modelling simultaneous transport of solute and water using continuity equations (eg. Bresler, 1973). Soil weathering models (eg. Verstraten, 1980)	Mass balance modelling of runoff and solute transport, using continuity equations, possibly as developments of finite element (eg. Beven, 1977) or finite difference (eg. Bernier, 1981) approaches to hillslope hydrology.	

Figure 6.12. Approaches to modelling the solutional development of hillslopes

with particular respect to oxygen and carbon dioxide (gases), does cause some inaccuracies. However, Verstraten's conclusions extend only to soil genesis, and do not consider variable conditions over a hillslope. Such an approach remains typical of most other attempts to use soil weathering models in fluvial geomorphology (e.g. Kirkby, 1976). As with the study of hillslope hydrology, the use of simplified conceptual models for the study of solutional erosion on hillslopes seems at present to be a most productive use of research, although no doubt theoretical and computational advances will eventually favour the deterministic approach.

6.3.2 Conceptual models

Carson and Kirkby (1972) proposed a model of solutional denudation on hillslopes in which it is assumed that all the rainfall which infiltrates into the soil comes rapidly to equilibrium with the soil solutes. The soil water is now subject to evaporation such that, if the concentration of any substance in the soil water rises above its saturated concentration, then the excess must be precipitated within the soil. Given that lateral flow occurs to give the downslope increase in soil moisture, the redeposition may involve a net downslope transfer of solutes. The quantity of each solute removed at any point on the slope is taken to be equal to the rainfall input multiplied by the proportions of the oxide, p, present in the soil or to the product of the net runoff contribution of each unit area of hillslope times 1.0, whichever is less (Figure 6.13). Carson and Kirkby show that for limestones, the proportion p, of calcium carbonate present is almost 1.0, so that chemical removal decreases downslope (since evaporative losses increase downslope, thus reducing unit runoff), and slope decline will occur (Figure 6.14(a)). For igneous rocks in humid temperate areas, several components of the rock

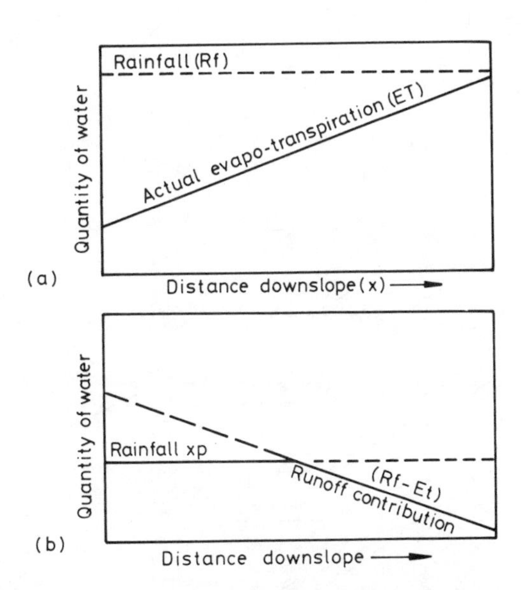

Figure 6.13. Conceptual model of hillslope solutional denudation proposed by Carson and Kirkby (1972): (a) distribution of rainfall and evaporation; (b) runoff contribution, (see text for explanation). *Reproduced by permission of Cambridge University Press*

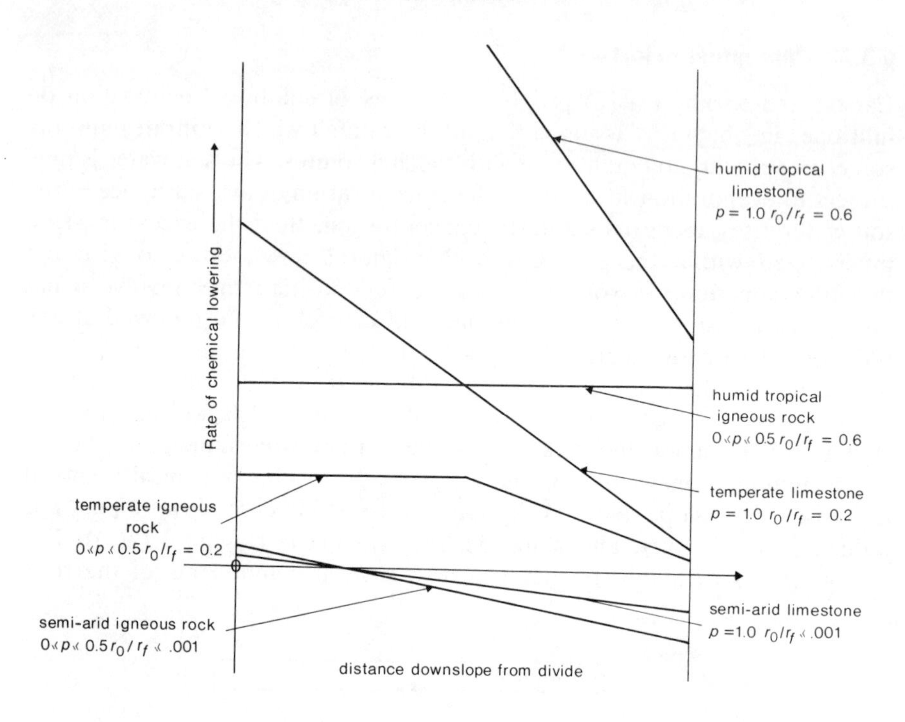

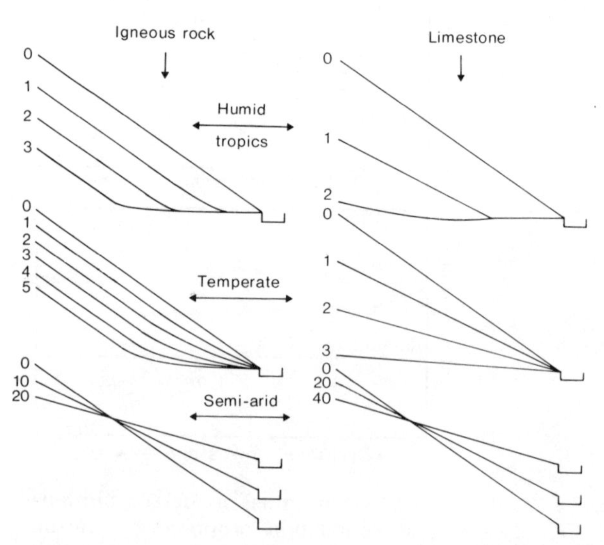

Figure 6.14. Patterns of solutional denudation, as predicted by the Carson and Kirkby (1972) model, for different rock types and climates: (a) downslope variations in solution rates; (b) slope profile evolution. *Reproduced by permission of Cambridge University Press*

satisfy the case where rainfall $\times p <$ runoff for the whole slope, in which case parallel restrict occurs, while other components satisfy the case where rainfall $\times p >$ runoff for the whole slope, giving slope decline. The result is that solutional erosion is constant in the upper part of the slope, but decreases progressively in the lower part of the slope, providing a dog-legged slope profile which will be converted over time into a smooth concavity (Figure 6.14(b)).

The Carson and Kirkby model requires that the initial solution process is short compared with the total residence time of water in the soil, so that erosion is proportional to unit runoff, rather than to total runoff passing a given point on the slope. In some cases, however, the dissolution process may be more gradual, with the implication that water moving downslope maintains some ability to acquire solutes. Figure 6.15 illustrates idealized slope development sequences for an initially straight slope. In the extreme case, water remains equally aggressive at all times (Figure 6.15(a), curve 1). It is much more likely that the ability to dissolve will decrease downslope, i.e. as residence time increases (Figure 6.15(a), curve 2). When total runoff at a point is multiplied by the mean 'aggressiveness' of the soil water, the product indicates the spatial distribution of solution erosion (Figure 6.15(b)). Case 1 indicates increasing erosion downslope which would provide slope steepening (Figure 6.15(c), curve 1). Case 2 indicates a maximum erosion rate in midslope, which would lead to the formation of a hillslope hollow (Figure 6.15(c), curve 2). Cases 3 and 4 correspond to the examples provided by Carson and Kirkby for parallel retreat throughout the entire slope (Case 3) or for decreased erosion in the lower slope (Case 4). Revisions of the models are presented in Chapter 10.

The two models described are both extremely general and represent end-members of a series governed by the rate of solution. A general approach is valid, given that slope development occurs slowly; thus the decision to avoid unnecessary detail of runoff processes by effectively concentrating on annual totals of water and solute transport may be a fair one (Kirkby, 1975). In this context, the assumption of unit runoff is also much more meaningful than for individual runoff events. Some models of solute leaching in soil profiles have also adopted a conceptual approach (e.g. Addiscott *et al*., 1978; Burns, 1974–80), but incorporate greater realism with respect to both runoff and the acquisition of solutes by the runoff. Of necessity, such models take a basic time unit much shorter than one year; but it is possible nevertheless to use such structures to model solutional erosional on hillslopes and thereby increase the realism of the model with respect to the acquisition of solutes by the throughflow.

Burns (1974) describes a soil profile leaching model in which water infiltrating into the surface layer (I) is added temporarily to the soil water already present (H), whose solute load is X grams. Rapid equilibrium within

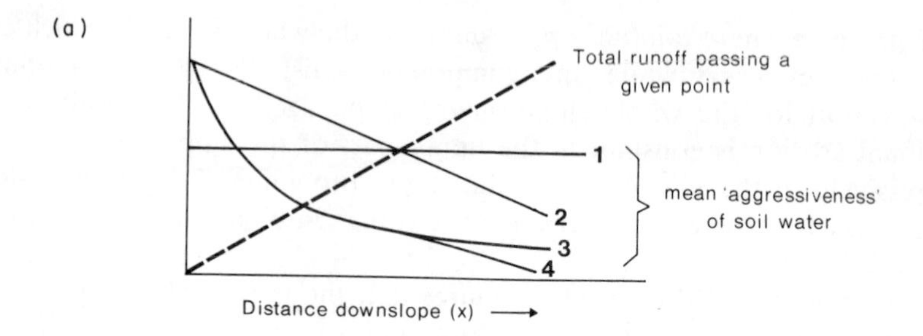

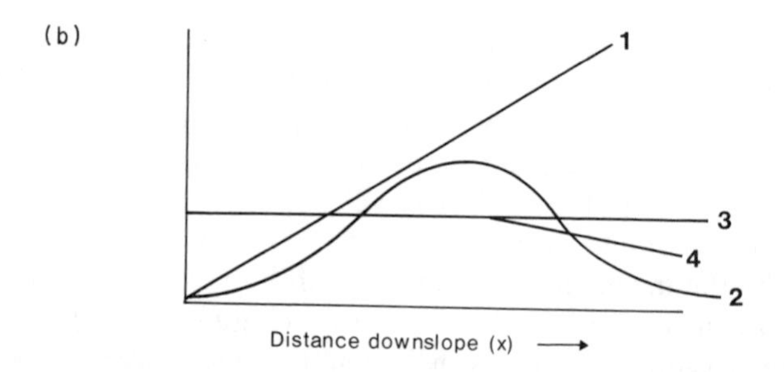

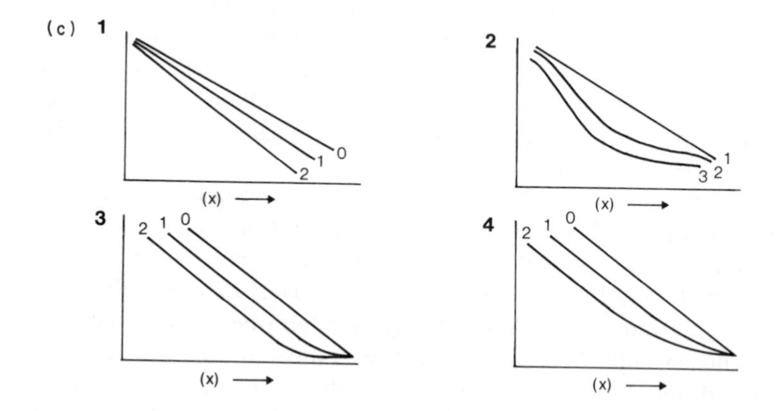

Figure 6.15. Idealized slope development for an initially straight hillslope profile under different conditions of soilwater 'aggressiveness' down the hillslope: (a) generalized distribution of soil water 'agressiveness'; (b) the product of total runoff and mean 'aggressiveness'; (c) patterns of hillslope denudation for the four cases described

the soil water is assumed to occur, giving a new concentration of $X/(I + H)$. As the soil is assumed to be at field capacity before infiltration, the temporary oversaturation is removed by transferring input I to the next layer down, thereby removing $IX/(I + X)$ grams of solute. If the proportion of runoff leaving each horizon $I/(I + X)$ is p, it follows that:

$$\text{Loss from horizon 1} = pX$$

$$\text{Loss from horizon 2} = (pX + X)p = p^2X + pX$$

$$\text{Loss from horizon 2 itself} = p^2X$$

$$\text{Loss from horizon } n = p^nX + p^{n-1}X \ldots {}^{+}pX$$

$$\text{Loss from horizon } n \text{ itself} = p^nX$$

Since $p < 1$, $p^2X < pX$. Therefore the rate of leaching decreased down-profile.

The Burns' profile model may be 'rotated' to provide a hillslope runoff model consisting of adjacent stores linked in a downslope direction. The subsurface runoff lost from each store is proportional to gradient, as in the model described in Figure 6.2. Runoff and solute output from the adjacent upslope store, plus rainfall, are input to a given store, with outputs of runoff and solute to the next store downslope. Given the importance of lateral flow on hillslopes, relative to vertical flow within the soil profile, the use of a single store at each point on the hillside is justified.

Such a model may be used to study solute leaching in individual storm runoff events, but may also be applied to long-term hillslope evolution. Figure 6.16 shows throughflow discharge and nitrate concentration for a straight slope section, following an initial input of 10 units of rainfall for the first 5 time periods, with a nitrate input of 1000 units in time period one (equivalent to leaching of surface-applied fertilizer, therefore). Although no delayed throughflow discharge peak is produced for this simulation, a realistic delayed peak in nitrate concentration does occur. Huggett (1975) has shown a similar effect for a three-dimensional hillslope model, where a solute peak was shown to move as a wave downslope and eventually out into the stream. Clearly, different combinations of hillslope shape, antecedent conditions and rainfall input would produce a range of dynamic leaching responses, which, while of great relevance to nutrient budgeting, are of too short a timescale to be relevant to the development of hillslope form.

If the model is operated under equilibrium flow conditions (i.e. input = output; no change in soilwater storage at the end of each iteration), then the progressive loss of solutes from the hillslopes over time may be studied. For four slope profiles, equilibrium flow was established for an input of one rainfall unit per time period. The model was then run with an equal input of solute to each store in each time period. In this time form the model simulates

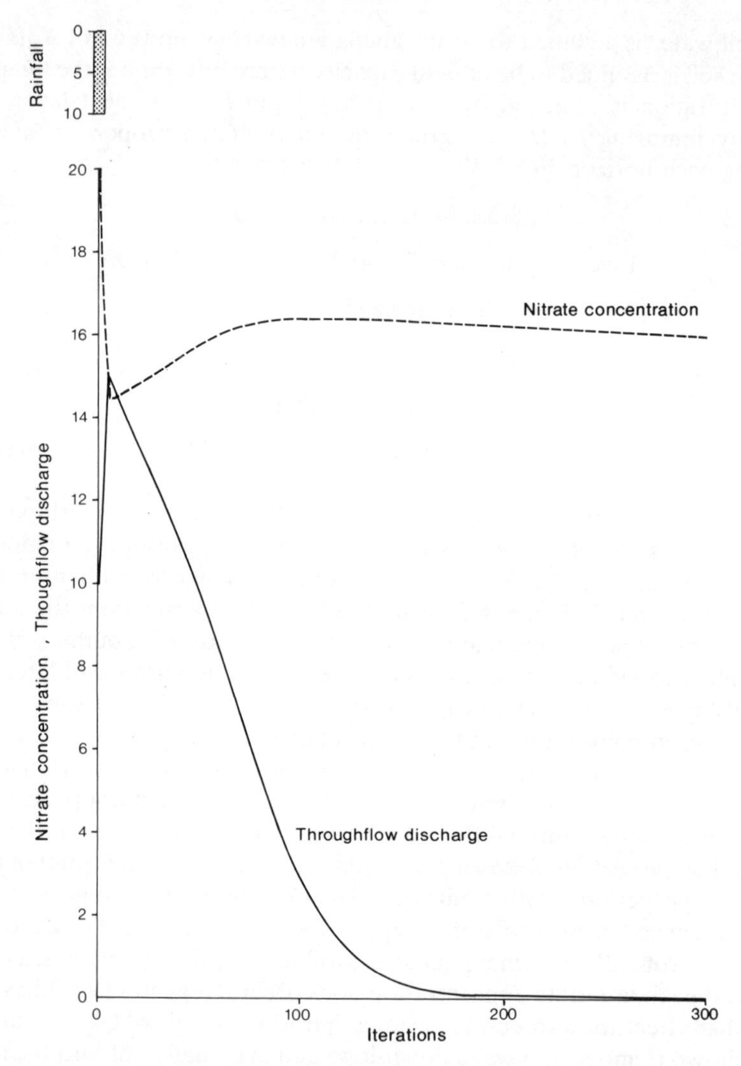

Figure 6.16. Throughflow discharge and nitrate concentration over time for a straight-slope section as predicted by the coupled hillslope runoff–solute leaching model

one *annual* cycle of weathering and runoff each iteration; an equal amount of solute is made available for leaching at all points on the slope; the assumption of equilibrium flow is acceptable when modelling annual runoff since, on average, soil water storage should remain constant at this timescale. Results are given in Figure 6.17: the graph plots the total amount of solute removed from each store after 300 iterations, and is thus indicative of the relative

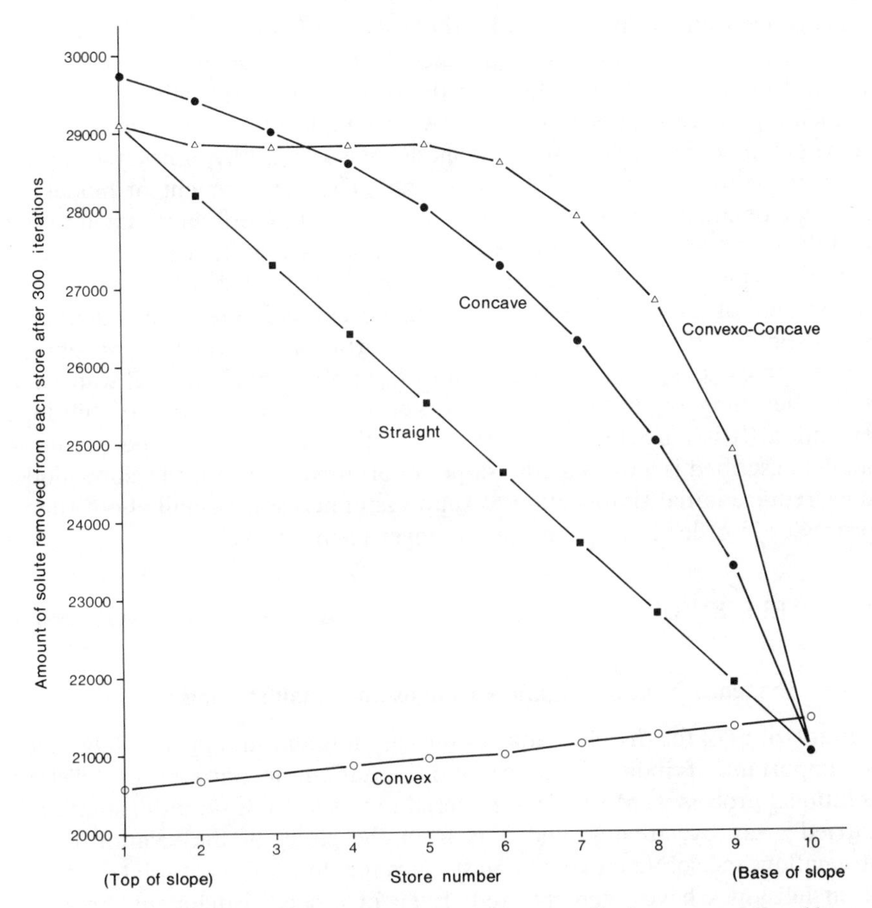

Figure 6.17. Relative erosion rates for four slope profile forms as predicted by the coupled hillslope runoff–solute leaching model

amounts of erosion occurring down the slope. On the initially *straight* slope, the erosion rate declines downslope, implying that slope decline would occur. The *concave* slope shows a marked decrease in erosion at the slope base, implying once again that slope decline would occur, effectively reducing the gradient of the upper slope. The *convex* slope shows a slight tendency for erosion to increase downslope, implying that the convexity would become accentuated. On the *convexo-concave* slope erosion rates are almost constant until below the inflexion point (between stores 5 and 6) where the rate falls rapidly; once again slope decline will occur, with the altitude of the upper slope and the gradient of the midslope being reduced.

All three conceptual models discussed here remain very simplistic in their structure, and are uncalibrated. No doubt, each of the models could be dimensioned and refined to make them more realistic, and to incorporate features already present in the other models. Even as they stand, two features of the models remain important: firstly, production of a realistic simulation model, regardless of its simplicity, allows the implications of the initial theory to be explored. Secondly, use of the model may suggest hypotheses which are susceptible to testing in field experiments. The development of models of hillslope denudation *must* be accompanied by field experiments aimed to test and develop such models; at present, as will be seen in the next section, such planned experimentation is generally lacking. Finally, it should be noted that the version of the Burns' soil profile model presented here, while admittedly simplistic in structure, does offer one particularly attractive possibility, namely to establish a link between short-term observations of soil water and solute leaching on hillslopes with longer-term considerations of hillslope evolution. It is particularly important that the *basis* of the slope evolution model described is a dynamic hillslope runoff model, even if the slope model does require initial simplication to suppress unnecessary detail about runoff processes in order to concentrate on longer-term effects.

6.4 FIELD EVIDENCE OF THE PATTERN OF SOLUTION PROCESSES

6.4.1 Evidence from throughflow troughs and small streams

Observations of the dissolved load of throughflow and streamwater have been an important, if indirect, source of information on patterns of hillslope solutional processes. Most of the material removed from slopes in solution is carried away by streams, and it is from the study of the relationship of streamflow and solute movement in streams that most ideas of solute removal from hillslopes have been inferred. Even in a small catchment, however, observations made at the basin outlet represent a 'lumped' average of the catchment response; where possible, results should relate to distinct subcatchment units (Foster, 1979), while direct observations of throughflow are preferred, since the response of an individual hillslope section can then be isolated from the overall catchment response.

The study of solutes in stream and throughflow water has three major areas of relevance for hillslope geomorphology.

Identification of runoff components

Variations in the solute concentration of runoff during flood flows (often termed the 'chemograph') may be used to estimate the relative contributions of components of flow with different residence times. Particular attention has been paid to those solute species which consistently exhibit a decrease in concentration during a storm event. Walling and Foster (1975) showed that something more than a simple dilution is involved, since the occurrence of the point of maximum dilution need not be coincident with the stream discharge peak; on the river Creedy the discharge peak preceded the dilution trough by as much as 14.5 hours. Shorter lags tended to be associated with the highest streamflow rises. The lag time was also inversely related to antecedent catchment wetness, implying that different runoff processes, as well as variability in solute supply, may control the timing of the lag effect. In many cases, the dilution effect results from the mixing of dilute surface runoff with more concentrated subsurface runoff. Surface flow itself may have a variable solute load. Walling and Foster argue that the incorporation of new solute sources, as contributing areas expand, can maintain the solute concentration of surface runoff, so that the dilution trough will occur late in the runoff event. The dilution effect may also be produced by variations in the source and pathways of subsurface runoff. Figure 6.18 shows, for the Slapton Wood catchment, that throughflow concentrations are not reduced during the period of storm runoff, the dilution of streamwater being entirely related to inputs of dilute overland flow in this case. The increase in throughflow discharge, without dilution, shows that a 'shunting' effect has occurred: 'new' (dilute) soilwater is added to the top of the saturated wedge, with long residence time 'old' (concentrated) soilwater leaving the base of the slope. Where macropores occur, or where rapid transmission of subsurface runoff through very permeable surface horizons is possible, a dilution effect may be produced for the hillslope chemograph. Pilgrim, Huff and Steele (1979) show that flow through cracks, along roots, and in small holes resulting from animal and insect activity, provided a rapid increase in discharge with a coincident dilution effect (Figure 6.19). The gradual increase in the concentration of subsurface runoff during the recession period was related to the increased contact time of 'new' water in the soil, gradually acquiring a solute load; little or no 'old' soil water was thought to have contributed to the runoff response Trudgill *et al*. (1983b) have also shown that rapid flow through macropores can produce a dilution effect in the soil water outflow; this effect is not apparent if macropore flow does not occur, or is not transmitted throughout the entire soil system. Figure 6.20 illustrates the types of hillslope runoff which may occur, together with an indication of the likely solute concentration of that runoff. It is clear, given the variety of runoff processes occurring on a hillslope, combined with the problems of soil water residence time as discussed by Pilgrim *et al*. (1979), that the identification of different flow components on the basis of chemograph analysis is likely to yield ambiguous results, particularly where *only* streamflow chemographs are available.

The chemical mixing model has been frequently used to calculate the contributions of different flow components to the total storm hydrograph (Pinder and Jones, 1969; Sklash *et al*., 1976; Pilgrim *et al*., 1979; Anderson

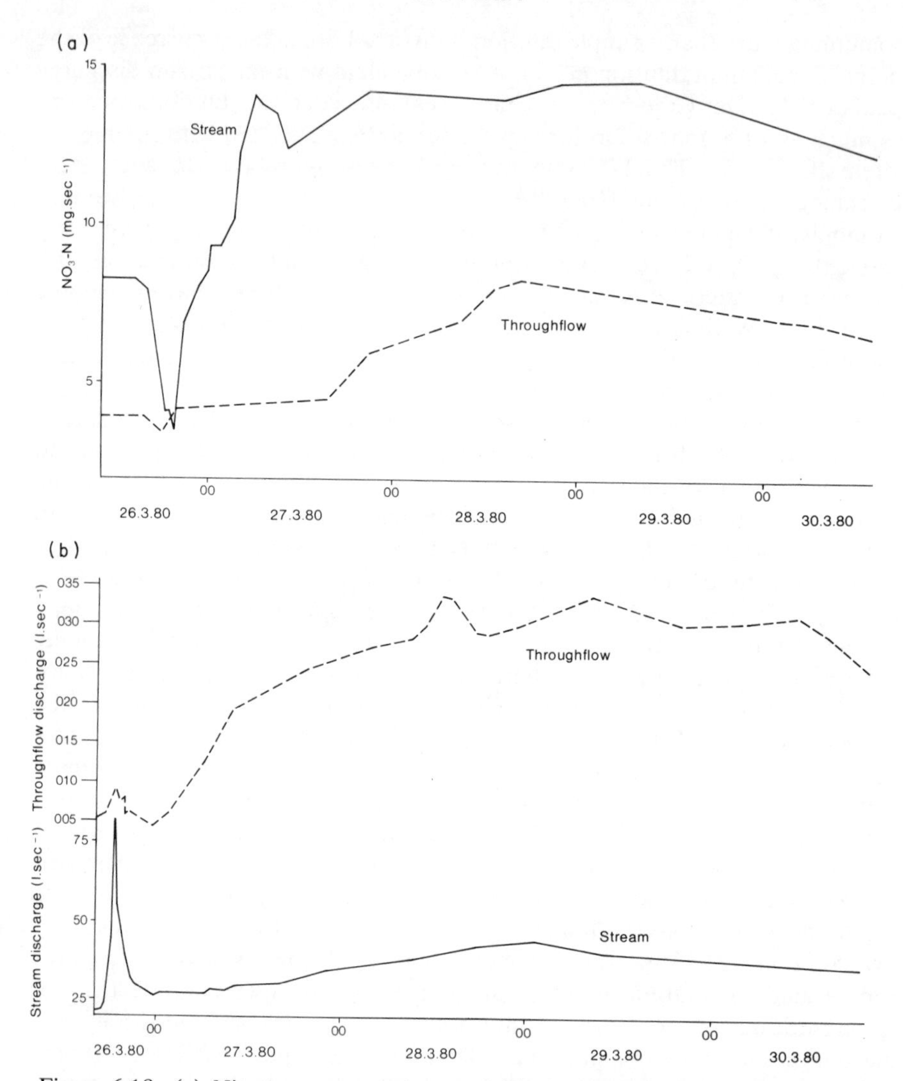

Figure 6.18. (a) Nitrate concentration, and (b) discharge rates throughflow and streamwater at the Slapton Wood catchment during a double-peak storm hydrograph (Burt *et al*., 1983). *Reproduced by permission of the Field Studies Council*

and Burt, 1982). Where two runoff components are involved, the mass balance model takes the form:

$$C_1Q_1 + C_2Q_2 = C_TQ_T \tag{6.12}$$

where C is the concentration of dissolved solids, Q the discharge rate, and 1, 2 and T are the two runoff components being mixed and the total runoff respectively. Where more than two runoff components are involved, the

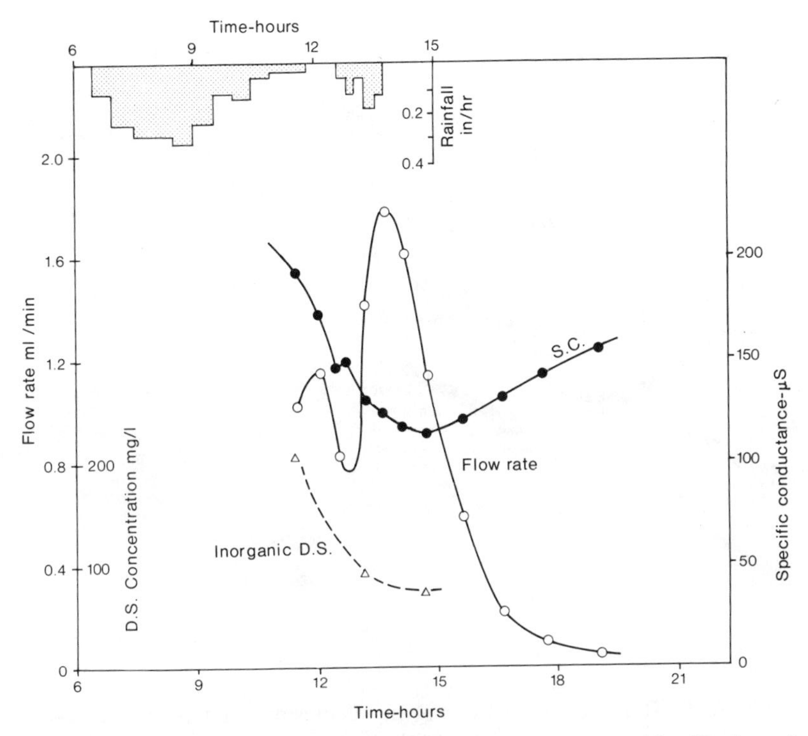

Figure 6.19. Evidence of rapid subsurface runoff as indicated by dilution of soilwater outflow from the hillslope (Pilgrim *et al*., 1979)

mass balance equation may be suitably expanded. Solution of the model to find Q_1 and Q_2 demands that C_1, C_2, C_T and Q_T are known or can be estimated through the storm event; clearly overall model accuracy will be improved as the frequency of input observations is increased.

The potential value of the mixing model lies in its ability to provide inferences about runoff processes in the absence of direct hydrological process observations. There are a number of problems with the use of the model, however: the model is 'lumped' and therefore ignores intracatchment variability; this may pose a particular problem if values of C_1 or C_2 are not spatially representative. Pilgrim *et al*. (1979) note that the concentration of 'new' soil water will rapidly increase; the effect is to provide overestimates of the amounts of 'old' soil water contributing to the storm runoff response; this effect must be allowed for either by directly sampling new soil water or by developing an algorithm to predict the increase in concentration based on laboratory contact-time experiments, in the manner described by Pilgrim *et al*. Finally, Anderson and Burt (1982) have shown for the Bicknoller Combe catchment that model predictions, while *generally* correct, may not correspond in detail with supporting hydrological data. They conclude that

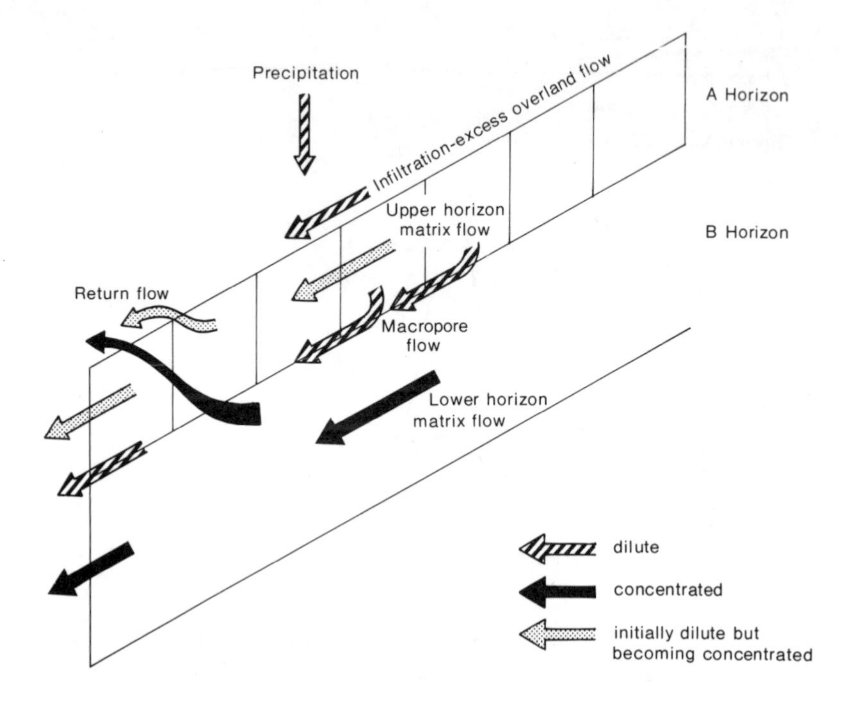

Figure 6.20. Relationship between type of hillslope runoff and its solute concentration

initial use of the chemical mixing model may provide some indication as to the direction later field research may develop; however, in the absence of field data (either detailed water sampling throughout the catchment, or direct hydrological measurements), *detailed* interpretation of mixing model results must be questioned since it is unlikely that a simple, lumped model can accurately predict a complex, spatially variable runoff response, even in a small catchment.

Estimation of denudation rates

The primary aim of many investigations has been to estimate rates of chemical denudation. However, in many cases such estimates relate only to relatively large drainage basins; even where a small catchment has been studied, the results still only provide an average value, giving no idea of the spatial variation between different hillslope sections. It is clear that this 'lumped' approach also ignores the effect of geochemical, biological and land-use variations within the catchment area. Few data are available which consider intracatchment variability at the small scale: Foster (1979a) considered the effects of land-use on the solute response of three source areas in a small catchment. He showed that contrasts in the solute response of streams draining unmodified areas of mixed deciduous woodland compared with surface or subsurface streams draining subcatchments under cultivation could be related to a complex interaction between the biological, geochemical and hydrological characteristics of the three source areas studied. Such results exemplify the need for a carefully stratified experimental design in field programmes designed specifically to distinguish between different hillslope sections in order to take account of variations in land-use, lithology, and so on. A number of studies have examined the short-term solute response of individual hillslope sections (e.g. Burt, 1978; Crabtree, 1981; Trudgill *et al.*, 1980), but usually with the purpose of comparing hillslope and catchment response. Such studies invariably include a sample of only one runoff plot, and so provide no information about the solute yield of different hillslope sections.

Where the estimation of chemical denudation rates is attempted, in order to distinguish between individual hillslope sections, two considerations demand particular attention: the quality of the solute data and the estimation of the 'non-denudational' component. Walling and Webb (1978) have reviewed procedures for estimating the portion of the solute load which originates from non-denudation sources. Where a detailed geochemical inventory is attempted, solute contributions from atmospheric precipitation (Gorham, 1961; Cryer, 1976), fertilizer application (Troake *et al.*, 1976), and biochemical processes must be evaluated. In the absence of such measurements, it has been customary to follow the procedure of Reesman and Godfrey (1972), to subtract the contributions of sodium and chloride (devised largely from precipitation), and nitrate (derived from natural and man-made organic wastes) from the gross solute yield estimates. Walling (1978) and Foster (1980) have reviewed the reliability and accuracy of calculations of total stream solute loads. In addition to errors inherent in laboratory determinations, Walling shows that errors arise in the estimation of dissolved loads because of the use of rating curves, and that results may not be comparable where different methods have been used. Table 6.2 compares the measured annual total dissolved load for the River Creedy with rating curve estimates. Similar errors were obtained where regular sampling programmes were operated, the load being calculated as the product of the sample concentration and the total water discharge in the intervening period. Predictably, larger errors resulted for individual ionic species (Mg: ±15%; NO_3: ±20%). Foster (1980) has shown that the form of the rating curve adopted can affect the accuracy of load calculations, and that even with rating models explaining over 80% of the variance in the data, errors in excess of 50% for monthly estimates still occur. Foster (1980) and Webb and Walling (1982) both show that the total length of observations is also particularly

Table 6.2. Comparison of measured annual total dissolved loads for the River Creedy with rating curve estimates (Walling, 1978)

Method of evaluating load	1973–4	1974–5	1975–6
Continuous record	22,794 t	19,549 t	6743 t
Instantaneous rating and hourly flow series	+1.3%	+3.7%	−2.4%
Instantaneous rating and daily mean flow series	+2.0%	−4.2%	−2.0%

important. Since high annual variability may occur in response to variations in the hydrological regime. While the errors associated with solute load calculations are generally low, it is clear that reliable conclusions about the rate of chemical denudation depend on a consistent and rigorous experimental design, entailing, where possible, continuous water quality monitoring in combination with an extensive geochemical analysis of samples obtained.

Inferences about the evolution of hillslope form

To infer the development of hillslope form from observations made at the slope base is clearly a difficult task. Few researchers have attempted to do more than consider the solute removal processes in such cases. At Bicknoller Combe, throughflow observations may provide at least an initial hypothesis about differences in the solutional erosion of hillslope hollows and spurs (Burt, 1979). The development of a delayed discharge peak is associated with an *increase* in the solute concentration of throughflow and streamwater (Figure 6.21). The delayed runoff therefore represents a major period of solute removal; this has been noted on other catchments too (e.g. Burt *et al*., 1983; Walling and Foster, 1978). This flushing effect may be one possible way in which hillslope hollows are eroded more quickly than adjacent spurs. The rise in solute concentration, which occurs concurrently with the delayed discharge pulse, represents an increase in the solute load over and above that gained during infiltration. Most of this increase will be derived from the hillslope hollows since this is where the saturated wedge deepens at the time of the delayed discharge peak (Anderson and Burt, 1977, 1978a). The continued presence of saturated wedges in the hollows (in contrast to the spurs where saturation is rare) may also contribute to hollow formation, since there could be continuous removal of solutes in the throughflow, although this could be offset by precipitation due to evaporation or to other changes in the chemical state of the soil water (Carson and Kirkby, 1972). Clearly the

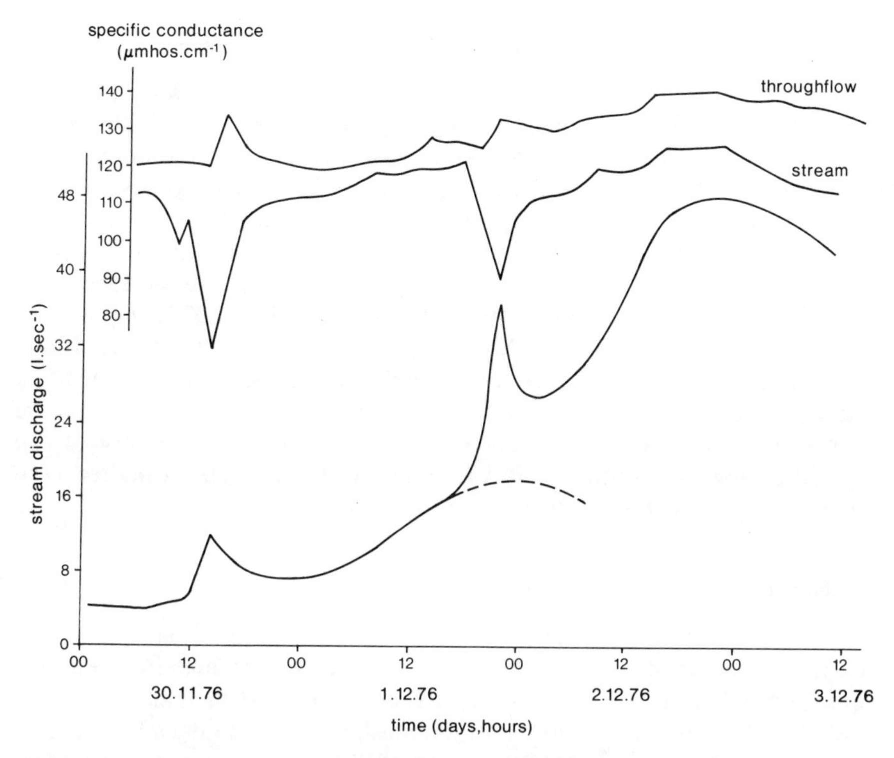

Figure 6.21. Increase in solute concentration of throughflow and stream water at the time of the delayed peak in stream discharge (Burt, 1979). Data from Bicknoller Combe

evidence is not conclusive, and a thorough analysis of soil water changes on the slope itself, in combination with the investigation of soil profiles, is required to verify such ideas. Nevertheless it would be surprising if there was not a link between the contrasting hydrological conditions and the erosional development of hollow and spur slope sections. The straight-slope spur section implies an equal, or regularly decreasing erosion rate, while the concave hollow suggests a focus of erosional activity in the upper and mid-part of the slope, to produce the basal concavity. It may be that spur development is governed by process activity as described by the model in curve 3 of Figure 6.15, while hollow development is augmented by rapid transmission of aggressive water into the midslope section, to increase erosion rates at that point (curve 2 of Figure 6.15). Certainly the hollow represents a relatively mobile environment for solute transport throughout its length (Burt *et al*., 1984). It is clear that such inferences require direct observations from the hillslope itself if they are to remain more than untested hypotheses.

6.4.2 Direct evidence of the patterns of solutional denudation

In contrast to the study of hillslope runoff processes, the study of solute removal from hillslopes is much less clearly defined. In general, studies of soil moisture movement on the hillslope have not been accompanied by observations of soil water solutes. Thus it remains difficult to describe the pattern of solute removal for a given hillslope, let alone begin to develop a physically based model of solutional denudation. Only a few experiments have indicated the longer-term pattern of solutional erosion over a hillslope using techniques such as micro-weight loss rock tablets (Trudgill, 1975) emplaced in the soil. However, much evidence of *potential* value to the hillslope geomorphologist is not in fact immediately applicable to the problem of defining hillslope evolution. It is clear that important links must be established with geochemists, pedologists and ecologists if evidence of soil nutrient cycling, soil profile analyses, and other similar subject matter, is to become relevant to the *hillslope* geomorphologist.

Weathering materials and products

In order to describe fully the chemical weathering process it is necessary to study *both* the solid phases of the rock and weathering materials, and the liquid phase (precipitation-, soil- and stream water). If such evidence is to be related to the form and evolution of the hillslope, the data must be collected in a controlled manner so as to sample the full range of conditions prevalent on the slope. Unfortunately, few field experiments have satisfied all these criteria; most available information provides only partial coverage, therefore, but may include the study of soil water solutes within the nutrient cycle, the nature and rate of weathering processes operating, and the (solid) weathering sequence which results. Solute budgeting studies usually take the entire catchment as their accounting unit, within which the nutrient balance is provided as a 'lumped' calculation. Studies of water–rock interaction and soil genesis have been largely aspatial also, relating to single or typical soil profiles. By contrast, spatial variations of geochemical processes and products across individual hillslope units have received little attention.

The study of nutrient cycles in natural ecosystems usually includes measurement of soil water solutes, as well as other components of the hydrological cycle. Even if sampling is not related to hillslope form, such information will define the changing status of individual ions as the water moves through the system, and may be interpreted in relation to other geochemical evidence, particularly soil/rock weathering sequences. A number of studies have examined the chemistry of water moving through the ecosystem, for each hydrological state—precipitation, throughfall, stemflow, soil water, stream water (e.g. Likens *et al.*, 1967; Borman *et al.*, 1969;

Crabtree, 1981). The various methods used to collect such water samples have been fully reviewed in Cryer and Trudgill (1981). Atmospheric contributions of solutes may be divided into wet and dry fallout; in some studies it may be desirable to distinguish between these, in which case more sophisticated collection gauges will be required (e.g. Benham and Mellanby, 1978). Crabtree and Trudgill (1982) have described the use of ion-exchange resins in rain gauges; this allows solutes to be collected efficiently, particularly for remote gauges which are visited infrequently. Stemflow is usually measured by affixing a collar around the tree trunk; the adhesives used to seal the collar should be inert. Soil water samples are normally collected using suction samplers, sometimes termed 'lysimeters'. The sampler consists of a tube with a porous cup at its base; the sampler is evacuated to cause water movement into the tube. The use, efficiency and zone of influence within the soil of such samplers has been discussed by Parizek and Lane (1970), Hansen and Harris (1975), Chow (1977), Talsma *et al*. (1979), and Crabtree (1981). Soil water may also be collected at the base of lysimeters (e.g. Crabtree, 1981) and using throughflow troughs (e.g. Burt, 1979; Burt *et al*., 1983a). Throughflow sampling, since continuous flow is often involved, may involve the use of automatic water samplers, and the sampling considerations are therefore much the same as for stream water sampling (Walling, 1978).

Dye tracing techniques provide direct evidence of soil water flow paths. Such information may be used to augment observations of soil water and throughflow solutes. Adsorption presents a major limiting factor on dye use in soils; because of adsorption, time to peak concentration and arrival times may show a lag behind other tracers. Dyes may be used in the field to identify mobile soil water at an outflow point, using short trace distances, or to show soil water flow pathways. Tracer dyes have been used by Mosley (1979) and by Trudgill *et al*. (1983a,b) to demonstrate rapid travel times for soil water movement in woodland soils. Trudgill *et al.* (1980) used dyes to assist in the interpretation of Ca/Mg ratios in hillslope drainage waters on dolomitic limestone. Coles and Trudgill (1985) used dye tracing to aid interpretation of penetration depth of nitrate fertilizers in agricultural soils. It is clear that the use of dyes in soils can provide qualitative evidence of soil water movement, although other tracers, such as tritium, are required if reliable qualitative results are required (for example, in the calculation of solute pentration through a soil profile).

Foster (1979b) has detailed monthly mean concentrations for several solutes for a small catchment in Devon, England (Table 6.3). The relative enrichment of solutes passing through the system is expressed by the ratio of the mean concentration of the solute at a given sample location to the concentration of that solute in bulk precipitation. The results indicate enrichment in throughfall samples, particularly for potassium (although the

Table 6.3. Mean monthly solute concentration of precipitation, throughfall, soil water and stream water (Foster, 1979b)

(a) *Mean and standard deviations of monthly solute data (mg l^{-1}) (April 1975–September 1977)*

Solute	Bulk precipitation $\bar{X}$	S	Throughfall $\bar{X}$	S	Soil water (120 cm) $\bar{X}$	S	Soil water (60 cm) $\bar{X}$	S	Stream water $\bar{X}$	S
K	1.9	1.5	13.7	9.0	2.0	0.2	2.6	0.9	2.1	1.0
Ca	2.4	1.2	5.9	3.2	69.2	10.1	73.0	47.7	39.8	50.5
Na	3.7	2.5	8.6	5.1	32.6	2.0	25.6	7.2	20.5	7.0
Mg	0.9	0.4	2.2	1.5	18.5	1.0	18.1	9.0	11.2	12.7
Cl	6.8	4.2	15.4	9.0	25.7	2.3	24.7	6.9	28.8	9.4
NO_3–N	0.9	0.5	1.2	1.5	–	–	–	–	3.0	2.4

(b) *Ratio of mean solute concentration to concentration in bulk precipitation*

Solute	Throughfall	Soil water (120 cm)	Soil water (60 cm)	Stream water
K	7.20	1.05	1.37	1.11
Ca	2.45	28.83	30.42	16.58
Na	2.32	8.81	6.92	5.54
Mg	2.44	20.55	20.11	12.44
Cl	2.26	3.78	3.63	4.24
NO_3–N	1.33	–	–	3.33

enrichment of potassium is not translated to the stream, since it is incorporated in the nutrient cycle following absorbance in the soil). Calcium and magnesium concentrations increase greatly in soil water, probably as a result of the high levels of both ions on exchange sites within the soil. As Foster points out, the interpretation of such data relies on a combination of hydrological, pedological and biogeochemical factors. Such data may indicate the locus of solutional erosion within the catchment system, but still may not indicate spatial variations within the hillslope system. This is indicated in Foster's results by the example of nitrate concentrations: this was detectable in soil water solutions for only short periods (maximum observed concentration = 0.5 mg per litre; no mean value given) and yet was always detectable in the streamwater (mean = 3.0 mg per litre). This suggests that spatial variations in soil water solutes were not detected by Foster's sampling regime, both within and between hillslope sites.

The work of Verstraten (1980) in the Haarts catchment, Luxembourg, represents one attempt to link field studies of water chemistry and rock/soil weathering sequences to the predictions of theoretical weathering models. Nevertheless this study, like most others of its type, focuses on the soil profile and total catchment scales, and avoids within-slope variations. Verstraten studied weathering processes and resultant soil formation in low-grade metamorphic rocks. As well as the solid phases, the liquid phase was also studied, since small undetectable changes in the composition of rock and soil may provide relatively large changes in the composition of the liquid phase (e.g. between precipitation and soil water). The predictions of the weathering models proposed by Verstraten compared favourably with the clay mineralogy of the regolith (which was established using XRD). However, the composition of soil water and spring water suggested that the reaction rates of the models used were incorrect, and should be changed if the correct succession of formation of weathering products, including the simultaneous formation of kaolinite, vermiculite and/or smectite, was to be predicted. Kirkby (1976; and Chapter 10, this volume) has incorporated the concept of thermodynamic equilibrium in geochemical reactions into a soil development model, which may then be used as a component of a slope development model. It is clear that such a *slope* model requires field verification using an experimental design which includes the range of measurements and analyses undertaken by Verstraten (1980). However, as Dunne (1981) points out, such an approach may well require the combined talents of several geomorphologists if all aspects of the scientific investigation (modelling, fieldwork, laboratory analysis) are to be successfully achieved.

Crabtree (1981) provides an example of a field experiment where measurement of spatial variations of solutional denudation over a hillslope has been the primary aim of the investigation. A wooded lowland hillslope, underlain by Magnesian limestone, at Whitwell Wood to the east of Sheffield, England, was studied in detail to assess the pattern of water movement and solute uptake in the soil over the slope. Soil water conditions were established on the hillslope using manual and automatic tensiometers, with samples of precipitation, throughfall, stemflow, soil water and stream water being analysed to establish the pattern of solute uptake on the hillslope. The short-term solute data were set in a longer-term context by the use of micro-weight loss tablets (see also next section). Figure 6.22(a) shows the form of the hillslope section studied; the dominant direction of water flow in the soil was vertical, except at the slope base where solute-rich groundwater moved laterally towards the stream. Figure 6.22(b) shows the change of soil chemistry upslope. As Crabtree points out, solute uptake and the capacity of the soil to neutralize percolating acidic water will be related to the chemistry and level of readily available solutes in the soil; because of the soil changes shown in Figure 6.22(b), solute uptake in the soil decreases upslope, with a consequent increase in solutional denudation at the soil/bedrock interface. Only in the valley bottom do the alkaline conditions prevent acidic water in the soil matrix from reaching the soil–bedrock interface. Observations of soil water solutes confirmed that soil water concentrations increased towards the slope base.

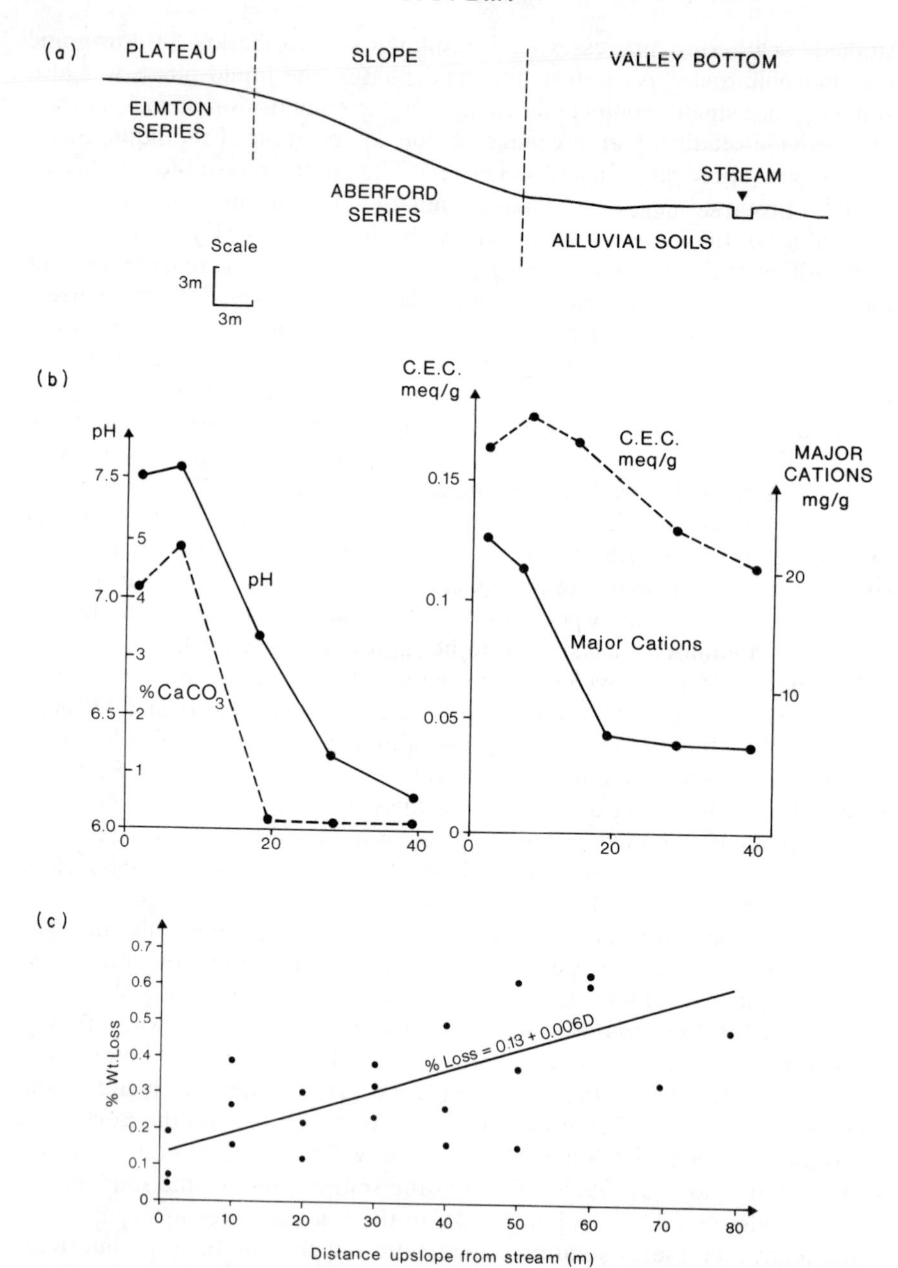

Figure 6.22. Patterns of solutional denudation on a Magnesian limestone hillslope: (a) the hillslope section studied; (b) changes of soil chemistry up the slope; (c) relative rates of solutional denudation provided by weight loss of rock tablets buried in the soil (Crabtree, 1981). *Reproduced by permission of the author*

Micro-weight loss tablets were used to establish relative rates of solutional denudation up the slope (Figure 6.22(c)); while there is a fair degree of variability in the results obtained, particularly since each point is itself a mean value, the results still suggest an upslope increase in solutional erosion which confirms the observations of soil chemistry and soil water made over the hillslope. Crabtree suggests, since only vertical percolation is involved (except on the alluvial soils of the floodplain), that the solutional model of Carson and Kirkby (Figure 6.13) is invalidated, showing rather that in this case at least, soil chemistry not slope hydrology controls the solution denudation. This does, however, beg the question as to what controls the distribution of soil chemistry. Carson and Kirkby's model does not necessarily involve lateral flow (although this is the most likely cause of downslope increases in soil moisture, and therefore evaporation); if evaporation does increase downslope on the Whitwell Wood slope, then, given the limestone bedrock, solutional denudation should decrease downslope, with the result that, over time, solutes would tend to remain more available in the soil towards the slope base. However, Crabtree (1981, p. 522) suggests that the parent materials of the Elmton soil series on the plateau (acid clay glacial drift) and the Aberford soil series on the slope (fluvioglacial and solifluction material) are different. Thus, the availability of solutes on the hillslope may be controlled by initial conditions as well as by the distribution of hydrological processes. The value of Crabtree's experiments lies, not so much in the results obtained, but in the experimental design itself. The integrated nature of the study, incorporating observations of soil chemistry, water chemistry and longer-term solutional experiments, combined with a sampling regime deliberately aimed at elucidating process variations over the hillslope, provides an example of the experimental framework needed if solutional denudation on hillslopes is to be investigated thoroughly in the field. Once again, the need for such experiments to be conducted in concert with the use of theoretical models must be stressed.

Use of micro-weight loss techniques

As noted above, emplacement of rock tablets in the soil provides a means of comparing relative rates of solutional erosion (Trudgill, 1975; Crabtree, 1981). Since freshly cut bedrock is used to manufacture the tablets, there are problems of preferential solution of fresh-cut rockfaces, so that the weight-loss technique should not be used to measure absolute solutional denudation rates. The tablets are cut from a core of bedrock and are buried in the soil, in pits preferably at the soil–bedrock interface. After a period of time (at least one year) the tablets can be excavated and reweighed. By using a large number of tablets in each soil pit, an average percentage weight loss per tablet may be calculated. From this a spatial pattern of relative solutional denudation can be derived by comparing the mean values for each pit.

Micro-weight loss tablets were used in an experiment at Bicknoller Combe, to investigate the spatial pattern of relative solutional denudation for the hillslope hollow and adjacent spurs previously described (Burt, 1978; Crabtree and Burt, 1982). Tablets prepared from Old Red Sandstone weighed about 12.5 g, with a diameter of 31 mm and a thickness of 7 mm. Weight loss of 0.2–0.6% was observed over a period of fifteen months (May 1979 to August 1980); the detectable limit of weight loss obtainable using a microbalance is approximately 0.001%. Twenty pits were dug, with ten tablets emplaced in each. Owing to the stony nature of the soil, pits could not be dug down to the soil–bedrock interface (below 1m); tablets were buried at a depth of 0.4 m. Five profiles were selected for tablet emplacement—one up the centre of the hollow, with two profiles at either side, spaced at 10 m intervals. Four pits were dug on each profile at distances of 15, 43, 75 and 115 metres from the stream.

The measured weight loss of the tablets was considered to be entirely due to solution processes as described by Waylen (1979). Neutralization of infiltrating rainfall takes place by dissolution of soil and bedrock material in acid soils. There was an overall increase in relative solutional denudation upslope, which may be related to the general increase in soil acidity upslope (Table 6.4a). As throughflow moves downslope, increasing solute uptake reduces the potential for causing solutional erosion since only long residence-time solute-rich water reaches the slope base. Table 6.4 shows the distribution of erosion by slope profile. Weight loss is greatest in the hollow: at the base of the slope, the hollow has the least weight loss, reflecting the presence of solute-rich throughflow in the saturated wedge; upslope the hollow shows maximum weight loss so that the hollow-spur contrast increases further upslope. The hydrological contrast between hollow and spurs also increases upslope (Figure 6.9). The higher erosion rate in the upper hollow may relate to the convergence of large quantities of short-residence-time acid throughflow into the hollow from the upper spurs, and especially from the interfluve areas. Such results imply that the additional solute load noted in the delayed throughflow pulse (Figure 6.9) may be derived, not from the hollow base, but from further upslope. The results suggest that the topographic contrast between spur and hollow will continue to develop within the general pattern of hillslope decline caused by rates of solutional denudation increasing upslope. The hydrological contrast between the spurs and hollow is seen to be the most important control on erosion rates: the increased volumes of throughflow, of shorter residence-time than soil water infiltrating on the spurs, produces this increased solution.

Rock tablet experiments provide a relatively long-term view of hillslope evolution, but give little indication of short-term processes acting on the

Table 6.4. Micro-weight loss tablet results for the Bicknoller hollow site (Crabtree and Burt, 1982)

(a) *Upslope changes in mean tablet micro-weight loss from 5 pits across the slope combined*

	Distance upslope from stream (m)	Mean tablet micro-weight loss (%)
4	115	0.42 ± 0.17
3	75	0.38 ± 0.13
2	43	0.35 ± 0.08
1	15	0.30 ± 0.08

(b) *Mean micro-weight loss for each upslope profile*

	Profile	Mean micro-weight loss, 4 pits combined (%)
A	Downstream spur	0.34 ± 0.07
B	Downstream spur flank	0.35 ± 0.07
C	Hollow	0.42 ± 0.09
D	Upstream spur flank	0.33 ± 0.07
E	Upstream spur	0.31 ± 0.15

hillslope. Such results do at least yield the (relative) pattern of solutional denudation over the hillslope. It seems most desirable that micro-weight loss experiments should be combined with an evaluation of the short-term hydrochemical processes operating on the hillslope, in order to allow calibration of deterministic models of hillslope solutional denudation.

Observations of soil properties

Given the temporal and spatial variability of hillslope runoff processes, their erosional effect might perhaps be more easily evaluated, not by measurement of the processes themselves, but by an analysis of their responses. The general relationship between soils and position on the hillslope has long been recognized (see, for example, the review in Gerrard (1981)). The notion of a soil catena was adopted by Milne (1935) to group soil profiles, very different in their fundamental structure, which occurred in a predictable sequence along the hillslope profile. The soil catena is a complex response to a combination of spatially variable erosional processes—surface wash, subsurface solution, soil creep and rapid mass movement. If the catena is to provide evidence of *individual* process–response systems, the contribution of each process to soil profile formation must be established at each point on the hillslope. In most available studies such information is lacking. Nevertheless, observations of soil changes within catenary sequences can provide extremely useful

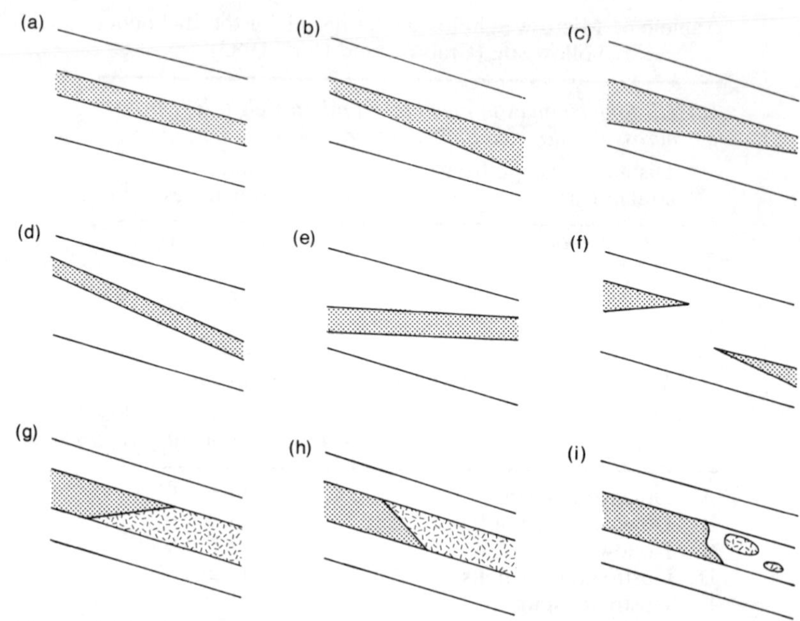

Figure 6.23. Possible horizon changes in a catenary sequence (Gerrard, 1981, after Young, 1976). *Reproduced by permission of Cambridge University Press*

information on soil formation processes, particularly on the relationship between soilwater conditions and iron mobility.

Young (1976) and Gerrard (1981) have described the possible horizon changes which may occur in a catenary sequence (Figure 6.23). A given horizon may remain much the same downslope (a), or it may thicken (b) or thin (c). At the same time the depth of the horizon may change (d, e). The horizon may be lost completely (f), perhaps resuming further downslope, or it may be replaced by a different horizon, starting at the base or the top (g, h). Even if the horizon remains intact, some of its properties may alter (i). It is clear that such changes require a process-based explanation; in providing such process information, it seems likely that much could be learned about the overall pattern of solutional denudation on the hillslope at the same time. As Gerrard's book suggests, there is much to be gained from an integrated approach to the geomorphology and pedology of hillsopes.

In some cases, other pedological and/or hydrological information may be available to the interpretation of changes in soil horizons downslope. For the instrumented hillslope at Bicknoller, Fielder (1981) described the soil catena from the interfluve to the lower hollow and related the varying distributions of extractable iron in the soil profiles to changing drainage conditions down the hollow. Such evidence provided support for the thesis that enhanced erosion of the hollow compared with the spurs was related in some way to the continued presence of acid soil water in the hollow (Burt *et al*., 1984). Eleven soil profiles were examined between the summit and the base of the slope (Figure 6.24). Pyrophosphate extractable iron (Fe(p)) and residual dithionate extractable iron (Fe(res)) were determined (Avery and Bascomb, 1974). Pyrophosphate extraction gives a good estimate of 'active' soil iron, while residual 'aged' iron oxides are extracted with sodium dithionate. The iron ratio Fe(p)/(Fe(p) + Fe(res)) may be used to classify soil horizons: in podzolic soils the iron ratio increases down the soil profile to exceed 0.5 in the illuvial B_s horizon; in periodically wet stagnogleyic soils the iron ratio becomes 'C' shaped; in brown earths the iron ratio is low and constant with depth, implying no redistribution of iron. Figure 6.24 shows the distribution of extractable iron, and the iron ratio, for four horizons. Profile 3 was a stagnogleyic brown podzolic soil with some rotting in the B horizon; the iron ratio showed a typical 'C' shaped distribution. Profile 5 was the wettest profile encountered being gleyed particularly at depth. Profile 9 was drier than profile 3; less frequent saturation may accord with the spur-edge location of this profile. Profile 11 illustrates a brown earth soil, typical of the well-drained spur soils; its Fe ratio plot implies that little redistribution of iron had taken place, in contrast to profile 9, for example. The distribution of iron in the soil profiles, together with the horizons identified, helped confirm the hydrological and solutional importance of the hollow. The presence of relatively poorly drained soils in the hollow confirm the continued presence of saturated conditions, a factor which may help explain the higher rates of solutional denudation observed in the hollow (Table 6.4). Redistribution of active iron in podzolic soils and removal from gleyic soils also confirms the frequent occurrence of soil saturation. Profiles near the base of the slope have the greatest extractable iron values: this suggests that iron removed in solution from gleyed horizons is accumulating further downslope. While distributions of extractable iron (and the soil horizons to which they relate) cannot be directly equated with rates of solutional denudation, such data are important since they integrate the long-term effects of both hydrological and hydrochemical processes. In these respects, the study of extractable iron distributions appears to be a most useful component to be included in any integrated approach to the study of solutional denudation on hillslopes.

The association of soil properties with topographic position, particularly slope gradient and distance from divide, has been investigated using statistical techniques. Furley (1971) found high correlations between some soil properties and slope angle for convex and straight segments, but much weaker relationships for concave elements. However, such associations could also be dependent on slope position. Slope gradient is highly correlated with distance on many slopes, particularly if the slopes are separated into concave and convex elements. Furley found that of 48 slopes, slope position was the

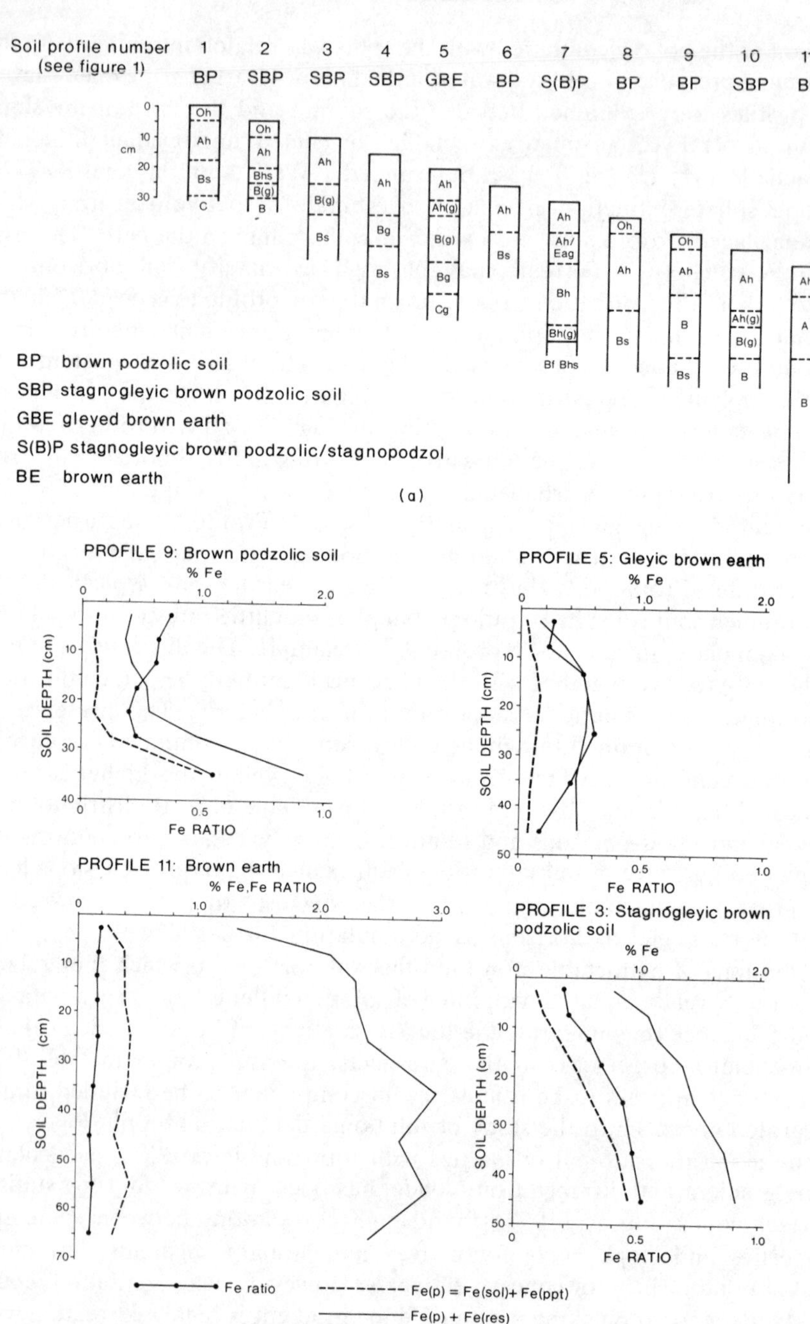

Figure 6.24. (a) Eleven soil profiles for the hillslope section at Bicknoller Combe. (b) The distribution of iron in the soil at profiles 3, 5, 9 and 11. (Burt *et al.*, 1984)

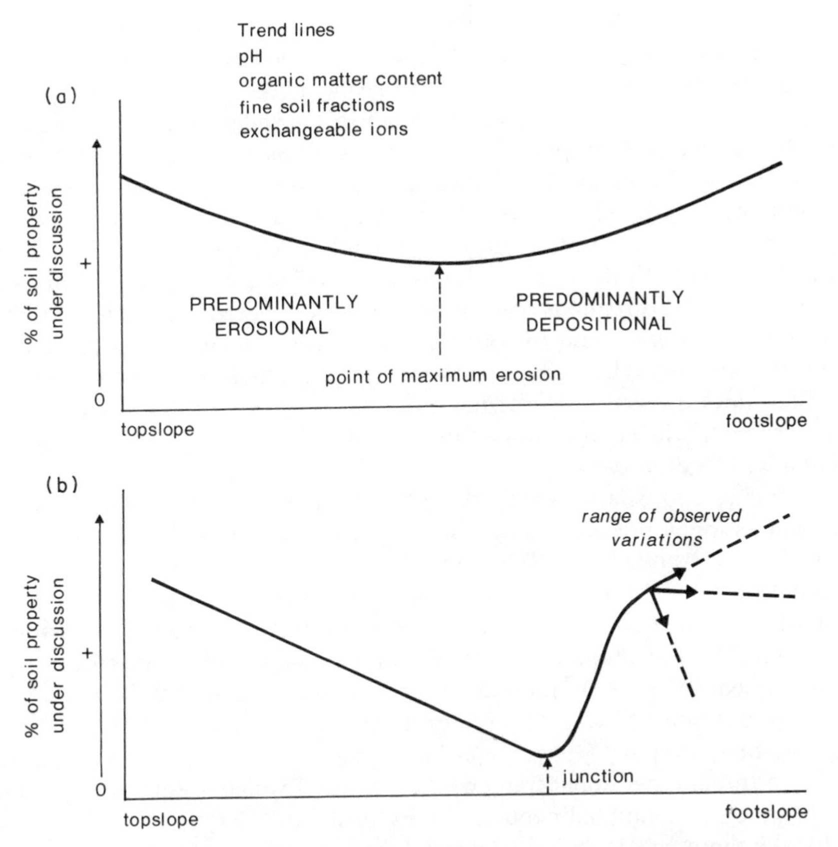

Figure 6.25. (a) Theoretical distribution of soil properties and slope form. (b) Distribution of certain soil properties on chalk slopes (Furley, 1971). *Reproduced by permission of the Institute of British Geographers*

dominant control on 34, while on only 7 slopes was slope angle most important. However, when slopes were subdivided into upper and lower section, position and angle assumed equal dominance on the upper section while position remained most important on the lower hillslope. Furley's model to explain the distribution of soil properties on slopes is shown on Figure 6.25. The upper slopes are envisaged as predominantly erosional with solutes being leached downslope. The lower slopes are essentially depositional, although as Gerrard (1981) points out, most basal slopes are not purely accumulation zones and some transport must occur. Nevertheless, the complexity of processes operating on the lower slopes probably accounts for variability of responses observed on those slopes. Some soil properties, such as organic matter, remain more closely related to gradient than to slope distance; such effects are superimposed on the general pattern controlled by distance downslope. Information on the spatial distribution of soil properties perhaps provides clearer evidence than soil catenas of the operation of erosional processes on a hillslope. While processes may only be *inferred* from such evidence, it is clear that such data provide a necessary component of any integrated study of solutional denudation on hillslopes. The changing distribution of soil properties and soil horizons on hillslopes has been considered by Zaslavsky (1970) and related to the theory of anisotropic soils as outlined in §6.2. In a layered soil with a less permeable lower layer, soil water will not infiltrate vertically but tend to flow downhill; this tendency increases with both the degree of anisotropy (Figure 6.7(b)) and with slope angle (Figure 6.7(c)) and is only absent on slopes of zero angle. Figure 6.7(a) shows that flowlines tend to converge on concave elements and diverge on convex elements. Thus for soils where infiltration is responsible for soil profile development, it seems reasonable to suppose that profile enhancement will be encouraged on concave slopes, while proceeding at a diminished rate on convex slopes.

Lotspeich and Smith (1953) describe the Palouse catena where the B_2 horizon is only present on the lower concave slopes. Similar evidence provided by Aandall (1948) relates slope position and the total nitrogen content of the soil, including a consideration of the role exerted by plan as well as by profile curvature. More recently, Huggett (1975) has demonstrated that lateral translocation of soil plasma can be a significant contributor to soil development. Moreover, the nature and degree of concentration can vary with local topography, with greater lateral movement of soil plasma in hollows than on spurs. Huggett concludes that pedogenesis should be studied within a three-dimensional framework, a concept which is equally applicable to the study of solutional denudation on hillslopes.

Ideally therefore, it should be possible in *one* experiment to establish flow directions for soil water movement on a hillslope and relate this to solute uptake and transport, rock/soil weathering sequences, changing soil properties, and the pattern and rate of solutional denudation on the hillslope. Such an integrated experimental design is at present lacking; several studies noted above partially cover the subject matter but, not surprisingly, none covers them all. Dunne (1981) and Burt and Walling (1984) have stressed the need for an integrated approach to fluvial geomorphology, particularly to combine the talents of modeller and field-worker. In the study of solutional denudation, talents from several disciplines may be required—geomorphology, hydrology, geochemistry, pedology—if a successful integrated programme is to be achieved.

6.5 CONCLUSIONS

There is much to be gained by such an integrated approach to the study of solutional denudation on humid temperate hillslopes, particularly with regard to the use of models and techniques already available in other branches of environmental science. As the investigations develop, then it will become necessary to formalize the experimental design, in the manner described by Church (1984). At this stage, however, the study of solutional denudation on humid temperate hillslopes seems only to have reached the stage of 'exploratory' experiments in which both theory and experimental methods are still evolving. It will perhaps be some time before it is possible to formalize 'confirmatory' experiments. Thus, at present, a functional approach to the study of hillslope solute systems is advocated in which broad overlap with many other disciplines is positively encouraged. This should not, however, be taken as an excuse for avoiding the strict demands of scientific method: of crucial importance is the theory being tested and developed, which must remain paramount, regardless of whatever overlap of ideas and techniques occur with studies of different purpose. Even at an 'exploratory' stage, the experimental design must be specific to the hypothesis being tested. Such an approach also demonstrates that the study of current process–response systems and landform evolution are *both* legitimate parts of fluvial geomorphology; indeed it is difficult to envisage an approach to modelling the solutional denudation of hillslopes which is not firmly process-based.

REFERENCES

Aandall, A. R. (1948). The characterisation of slope positions and their influence on the total nitrogen content of a few virgin soils of Western lowd. *Proceedings of the Soil Science Society of America*, **13**, 449–454.

Addiscott, T. M. (1977). A simple computer model for leaching in structured soils. *Journal of Soil Science*, **28**, 554–563.

Anderson, M. G. (1982). Modelling hillslope soil water status during drainage. *Transactions of the Institute of British Geographers*, **7**(3), 337–353.

Anderson, M. G., and Burt, T. P. (1978). The role of topography in controlling throughflow generation. *Earth Surface Processes*, **3**(4), 331–344.

Anderson, M. G., and Burt, T. P. (1982). The contribution of throughflow to storm runoff: an evaluation of a chemical mixing model. *Earth Surface Processes and Landforms*, **7**, 565–574.

Avery, B. W., and Bascomb, C. L. (1974). *Soil Survey Laboratory Methods*, Soil Survey Technical Monograph, No. 6.

Benham, D. G., and Mellanby, K. (1978). A device to exclude dust from rainwater samples. *Weather*, **33**, 151–154.

Bernier, P. Y. (1982). *VSAS2: A revised source area simulator for small forested basins*. Unpublished PhD thesis, University of Georgia.

Beven, K. J. (1977). Hillslope hydrographs by the finite element method. *Earth Surface Processes*, **2**, 13–28.

Beven, K., and Germann, P. (1981). Water flow in soil macropores, II: A combined flow model. *Journal of Soil Science*, **32**, 15–29.

Bormann, F. K., Likens, G. E., Siccama, T. G., Pierce, R. S., and Eaton, J. S. (1969). The export of nutrients and recovery of stable conditions following deforestation at Hubbard Brook. *Ecological Monographs*, **44**, 255–277.

Bresler, E. (1973). Simultaneous transport of solutes and water under transient unsaturated flow conditions. *Water Resources Research*, **9**(4), 975–986.

Burns, I. G. (1974). A model for predicting the redistribution of salts applied to fallow soils after excess rainfall or evaporation. *Journal of Soil Science*, **25**, 165–178.

Burt, T. P. (1978). *Runoff process in a small upland catchment with special reference to the role of hillslope hollows*. Unpublished PhD thesis, University of Bristol.

Burt, T. P. (1979). The relationship between throughflow generation and the solute concentration of soil and stream water. *Earth Surface Processes*, **4**, 257–266.

Burt, T. P., and Butcher, D. P. (1984). A basic hillslope runoff model for use in undergraduate teaching. Huddersfield Polytechnic, Department of Georgraphy, Occasional Paper No. 11.

Burt, T. P., Butcher, D. P., Coles, N., and Thomas, A. D. (1983). The natural history of Slapton Ley Nature Reserve. XV: Hydrological processes in the Slapton Wood catchment. *Field Studies*, **5**, 731–752.

Burt, T. P., Crabtree, R. W., and Fielder, N. (1984). Patterns of hillslope solutional demudation in relation to the spatial distribution of soil moisture and soil chemistry over a hillslope hollow and spur. In: *Catchment Experiments in Fluvial Geomorphology* (T. P. Burt and D. E. Walling, Eds.), GeoBooks, Norwich, 431–446.

Burt, T. P., and Walling, D. E. (1984). Catchment experiments in fluvial geomorphology: a review of objectives and methodology. In: *Catchment Experiments in Fluvial Geomorphology* (T. P. Burt and D. E. Walling, Eds.), GeoBooks, Norwich, 3–18.

Butcher, D. P. (1985). *Field Verification of Topographic Indices for Use in Hillslope Runoff Model.* PhD thesis, Huddersfield Polytechnic.

Carson, M. A., and Kirkby, M. J. (1972). *Hillslope Form and Process*, Cambridge University Press, 475pp.

Chow, T. L. (1977). A porous cup soil water sampler with volume control. *Soil Science*, **124**, 173–176.

Church, M. A. (1984). On experimental method in geomorphology. In: *Catchment Experiments in Fluvial Geomorphology* (T. P. Burt and D. E. Walling, Eds.), GeoBooks. 563–580.

Colos, N., and Trudgill, S. (1985). The movement of nitrate fertiliser from the soil surface to drainage waters by preferential flow in weakly structured soils, Slapton, S. Devon. *Agriculture, Ecosystems and Environments*, **13**, 241–259.

Crabtree, R. W. (1981). *Hillslope Solute Sources and Solutional Denudation on Magnesian Limestone*. Unpublished PhD thesis, University of Sheffield.

Crabtree, R. W., and Burt, T. P. (1982). Spatial variation in solutional denudation and soil moisture over a hillslope hollow. *Earth Surface Processes and Landforms*, **8**, 151–160.

Crabtree, R. W., and Trudgill, S. T. (1982). The use of ion-exchange resin in monitoring the calcium, magnesium, sodium and potassium contents of rainwater. *Journal of Hydrology*, **53**, 361–365.

Cryer, R. (1976). The significance and variation of atmospheric nutrients input in a small catchment system. *Journal of Hydrology*, **29**, 121–137.

Cryer, R., and Trudgill, S. T. (1981). Solutes. In: *Geomorphological Techniques* (A. S. Goudie, Ed.), 181–195.

Curtis, C. D. (1976). Stability of minerals in surface weathering reactions. *Earth Surface Processes*, **1**, 63–70.

Dunne, T. (1978). Field studies of hillslope flow processes. In: *Hillslope Hydrology* (M. J. Kirkby, Ed.), pp. 227–294.
Dunne, T. (1981). Concluding comments to the Christchurch symposium on 'Erosion and Sediment Transport in Pacific Rim Steeplands'. *Journal of Hydrology, N.Z.*, **20**, 111–114.
Dunne, T., and Black, R. D. (1970). An experimental investigation of runoff production in permeable soils. *Water Resources Research*, **6**, 478–490.
Fielder, N. (1981). *Distribution of Iron and Aluminium in some Acidic Soils*. Unpublished PhD thesis, University of Bristol.
Foster, I. D. L. (1979a). Intra-catchment variability in solute response: an East Devon example. *Earth Surface Processes*, **4**(4), 381–394.
Foster, I. D. L. (1979b). Chemistry of bulk precipitation, throughfall, soil water and stream water in a small catchment in Devon, England. *Catena*, **6**, 145–155.
Foster, I. C. L. (1980). Chemical yields in runoff, and denudation in a small arable catchment, East Devon, England. *Journal of Hydrology*, **47**, 349–368.
Furley, P. A. (1968). Soil formation and slope development. 2: The relationship between soil formation and gradient angle in the Oxford area. *Zeitschrift fur Geomorphologie*, **NF12**, 25–42.
Furley, P. A. (1971). Relationships between slope form and soil properties developed over chalk parent materials. In: *Slopes, Form and Process* (D. Brunsden, Ed.), Institute of British Geographers, Special Publication, No. 3, 141–164.
Gerrard, A. J. (1981). *Soils and Landforms*. George Allen and Unwin, 219pp.
Gorham, E. (1961). Factors influencing supply of major ions to inland waters, with special reference to the atmosphere. *Bulletin of the Geological Society of America*, **72**, 795–840.
Hansen, E. A., and Harris, A. R. (1975). Validity of soil water samples collected with porous ceramic cups. *Proceedings of the Soil Science Society of America*, **39**, 528–536.
Hewlett, J. D., and Nutter, W. (1970). The varying source area of streamflow from upland basins. *Proceedings of the Symposium on Interdisciplinary Aspects of Watershed Management*. Montana State University, Boseman, America Society of Civil Engineers, pp. 65–83.
Horton, R. E. (1933). The role of infiltration in the hydrological cycle. *Transactions of the American Geophysical Union*, **14**, 446–460.
Horton, R. E. (1945). Erosional development of streams and their drainage basins: hydrophysical approach to quantitative morphology. *Bulletin of the Geological Society of America*, **56**, 275–370.
Huggett (1975). Soil landscape systems: a model of soil genesis. *Geoderma*, **13**, 1–22.
Kirkby, M. J. (1976). Soil development models as a component of slope models. Department of Georgraphy, University of Leeds: Working Paper No. 145, 27pp.
Kirkby, M. J. (1978). Implications for sediment transport. In: *Hillslope Hydrology* (M. J. Kirkby, Ed.), pp. 325–363.
Kirkby, M. J. (1984). In: *Solute Processes* (S. T. Trudgill, Ed.). Wiley, pp. 439–495.
Kirkby, M. J., and Chorley, R. J. (1967). Throughflow, overland flow and erosion. *Bulletin of the International Association of Scientific Hydrology*, **12**, 5–21.
Knapp, B. J. (1978). Infiltration and storage of soil water. In: *Hillslope Hydrology* (M. J. Kirkby, Ed.), pp. 43–72.
Lewin, J. (1980). Available and appropriate time scales in geomorphology. In: *Time Scales in Geomorphology* (R. A. Lullingford, D. A. Davidson and J. Lewin, Eds.), Wiley, London, pp. 3–10.
Likens, G. E., Bormann, F. K., Johnson, N. M., and Pierce, R. S. (1967). The calcium, magnesium, potassium and sodium ludgets for a small forested ecosystem. *Ecology*, **48**, 712–785.

Lotspeich, F. B., and Smith, H. W. (1953). Soils of the Palouse Loess.1: The Palouse catena. *Soil Science*, **76**, 467–480.

Milne, G. (1935). Some suggested units of classification and mapping particularly for East African soils. *Soil Research*, **4**, No. 3.

Mosley, M. P. (1979). Streamflow generation in a forested watershed, New Zealand. *Water Resources Research*, **15**, 795–806.

O'Loughlin, E. M. (1981). Saturated regions in catchments and their relations to soil and topographic properties. *Journal of Hydrology*, **53**, 229–246.

Parizek, R. R., and Lane, B. E. (1970). Soil water sampling using pan and deep pressure vacuum lysimeters. *Journal of Hydrology*, **11**, 1–21.

Philip, J. R. (1957). The theory of infiltration. *Soil Science*, **83**, 345–357.

Pilgrim, D. H., Huff, D. D., and Steele, T. D. (1979). Use of specific conductance and contact time relations for separating storm flow components in storm runoff. *Water Resources Research*, **15**(2), 329–339.

Pinder, G. F., and Jones, J. F. (1969). Determination of the ground-water component of peak discharge from the chemistry of total runoff. *Water Resources Research*, **5**, 438–445.

Reesman, A. L., and Godfrey, A. E. (1972). Chemical erosion and denudation in Middle Tennessee. *Water Resources Research Series, Tennessee Department of Conservation*, No. 4.

Sklash, M. G., Farvolden, R. N., and Fritz, P. (1976). A conceptual model of watershed response to rainfall, developed through the use of oxygen-18 as a natural tracer. *Canadian Journal of Earth Science*, **13**, 271–283.

Smettem, K. R. J. (1982). Unpublished PhD thesis, University of Sheffield.

Smettem, K. R. J. (1984). In: *Solute Processes* (S. T. Trudgill, Ed.), pp. 140–165.

Talsma, T., Hallam, P. M., and Mansell, R. S. (1979). Evaluation of porous cup soil water extractors: physical factors. *Australian Journal of Soil Research*, **17**, 113–125.

Troake, R. P., Troake, L. E., and Walling, D. E. (1976). Nitrate loads of South Devon streams. *Technical Bulletin, MAFF, London*, **32**, 340–351.

Trudgill, S. T. (1975). Measurement of erosional weight-loss of rock tablets. *BGRG Technical Bulletin*, **17**, 13–20.

Trudgill, S. T., Pickles, A. M., Smettem, K. R. J., and Crabtree, R. W. (1983a). Soil water residence time and solute uptake. 1: Dye tracing and rainfall events. *Journal of Hydrology*, **60**, 257–279.

Trudgill, S. T., Pickles, A. M., and Smettem, K. R. J. (1983b). Soil water residence time and solute uptake. 2: Dye tracing and preferential flow predictions. *Journal of Hydrology*, **62**, 279–285.

Trudgill, S. T., Smart, P. L., and Laidlaw, I. M. L. (1980). Soil water residence time and solute uptake on a dolomite bedrock—preliminary results. *Earth Surface Processes*, **5**, 91–100.

Verstraten, J. M. (1980). *Water-rock interactions in (very) low-grade metamorphic shales.* Publication of Physical Geography and Soil Science Laboratory, University of Amsterdam, No. 29, 243pp.

Walling, D. E. (1978). Reliability considerations in the evaluation and analysis of river loads. *Zeitschrift fur Geomorphologie, Supplementeland 29*, 29–42.

Walling, D. E., and Foster, I. D. L. (1975). Variations in the natural chemical concentration of river water during flood flows and the lag effect: some further comments. *Journal of Hydrology*, **26**, 237–244.

Walling, D. E., and Foster, I. D. L. (1978). The 1976 drought and nitrate levels in the River Exe basin. *Journal of the Institution of Water Engineers and Scientists*, **32**(4), 341–352.

Walling, D. E., and Webb, B. W. (1978). Mapping solute loads in an area of Devon, England. *Earth Surface Processes*, **3**, 85–99.
Waylen, M. J. (1979). Chemical weathering in a drainage basin underlain by Old Red Sandstone. *Earth Surface Processes*, **4**, 167–178.
Webb, B. W., and Walling, D. E. (1982). The magnitude and frequency characteristics of fluvial transport in a Devon drainage basin and some geomorphological implications. *Catena*, **9**, 9–23.
Weyman, D. R. (1973). Measurements of the downslope flow of water in a soil. *Journal of Hydrology*, **20**, 267–288.
Whipkey, R. Z. (1965). Subsurface stormflow from forested watersheds. *Bulletin of the International Association of Scientific Hydrology*, **10**, 74–85.
Whipkey, R. Z., and Kirkby, M. J. (1978). Flow within the soil. In: *Hillslope Hydrology* (M. J. Kirkby, Ed.), pp. 121–144.
Young, A. (1976). *Tropical soils and soil survey*. Cambridge, Cambridge University Press.
Zaslavsky, D. (1970). *Some aspects of watershed hydrology*. USDA Research Paper, ARS 41–157, 96pp.

Solute Processes
Edited by S. T. Trudgill

CHAPTER 7

Solutes in river systems

D. E. Walling and B. W. Webb

Department of Geography,
University of Exeter

7.1 THE CONTEXT

Interest in the transport of material in solution by rivers can be traced back for well over 100 years. Analyses of the chemical composition of water from the Rhine undertaken in 1837 are reported by Livingstone (1963a), and the same author refers to similar data collected from the Seine, Loire, Rhone and Garonne rivers in France in 1848 and from the Vistula in Poland in 1863. Popp (1875) describes measurements made on the River Nile in Egypt in 1870. In most cases this early work focused on chemical analyses of individual water samples, rather than attempting to document temporal variations in solute levels or the loads transported during specific periods of time, since records of water discharge were generally lacking. Justification for these pioneering studies can be found in an essentially academic interest in geochemistry and also in a more practical concern for the quality of water for both drinking and irrigation.

From these early beginnings, interest in solute transport by rivers expanded rapidly. Attention was soon turned to the estimation of the loads involved, and Reade (1876) and Penck (1894) introduced a geomorphological perspective by using such data to estimate rates of chemical denudation. Solute transport data have since been used for a variety of purposes including studies of biogeochemical cycling in small drainage basins (e.g. Likens *et al*., 1977; Waylen, 1979); evaluations of material transport to the oceans (e.g. Clarke, 1924; Alekin and Brazhnikova, 1968; Meybeck, 1976, 1979); assessments of the environmental impact of non-point pollution and other human activity (e.g. Steele and Gilroy, 1971; Hotes and Pearson, 1977); and the use of hydrochemical properties of runoff to elucidate runoff processes and sources (e.g. Skakalskiy, 1966; Newbury *et al.*, 1969). Expansion of water sampling activity into many areas of the globe has also provided the basis for global compilations of information on riverwater composition (e.g. Clarke, 1924) and particular reference must be made to the reports of Livingstone (1963a) and Meybeck (1979) in improving our understanding of the global variations and patterns involved. Studies of solute transport by rivers have now been undertaken at scales ranging from small instrumented watersheds, through the larger basin and region to that of the globe. Detailed information on spatial and temporal variations of solute concentrations and loads encountered in rivers are currently available from a substantial number of investigations (e.g. Webb, 1983a) and there has been much associated interest in strategies for modelling these variations.

Against this background, the objective of this chapter is to review our current knowledge and understanding of solute transport in river systems. Attention will be restricted to the essentially natural conditions found in unpolluted rivers, and the perspective will be primarily that of the hydrologist and geomorphologist interested in spatial and temporal variations in river behaviour and their geomorphological implications, rather than that of the aquatic chemist concerned with chemical equilibria (e.g. Stumm and Morgan, 1970) or the hydrodynamicist concerned with dispersion and diffusion processes and travel times (e.g. Fischer, 1973; Keefer and McQuivey, 1974; Brady and Johnson, 1981).

In studying solute transport by rivers, it is essential to recognize that the chemical character of water flowing in a river channel will reflect the spatial and temporal integration of the complex sequence of pathways interposed between precipitation input to the drainage basin and output as channel flow and the associated evolution of the water chemistry (cf. Walling, 1980a; Eriksson, 1981). Thus, whereas a number of early workers attempted to account for the solute content of riverwater in terms of dissolution of the channel bed material, it is now clear that attention must be turned to processes operating throughout the whole of the basin. Processes related specifically to the channel environment, such as sediment/solute interactions (e.g. Green *et al.*, 1978; Cullen and Rosich, 1979; Casey and Farr, 1982), biotic uptake (e.g. Edwards, 1974; Casey and Ladle, 1976) and chemical precipitation (e.g. Blanc and Conrad, 1968), may exert a significant effect, but they must be viewed as subordinate to the basin-wide processes governing the evolution of runoff chemistry.

Figure 7.1 provides a simplified representation of the major hydrologic processes operating in a drainage basin and of the associated mechanisms influencing the quality of water moving through the system. Many of the processes discussed in Chapters 2–6 are therefore of major importance in governing the solute content and chemical composition of streamflow. The amounts of water moving through the system, the associated velocities, and

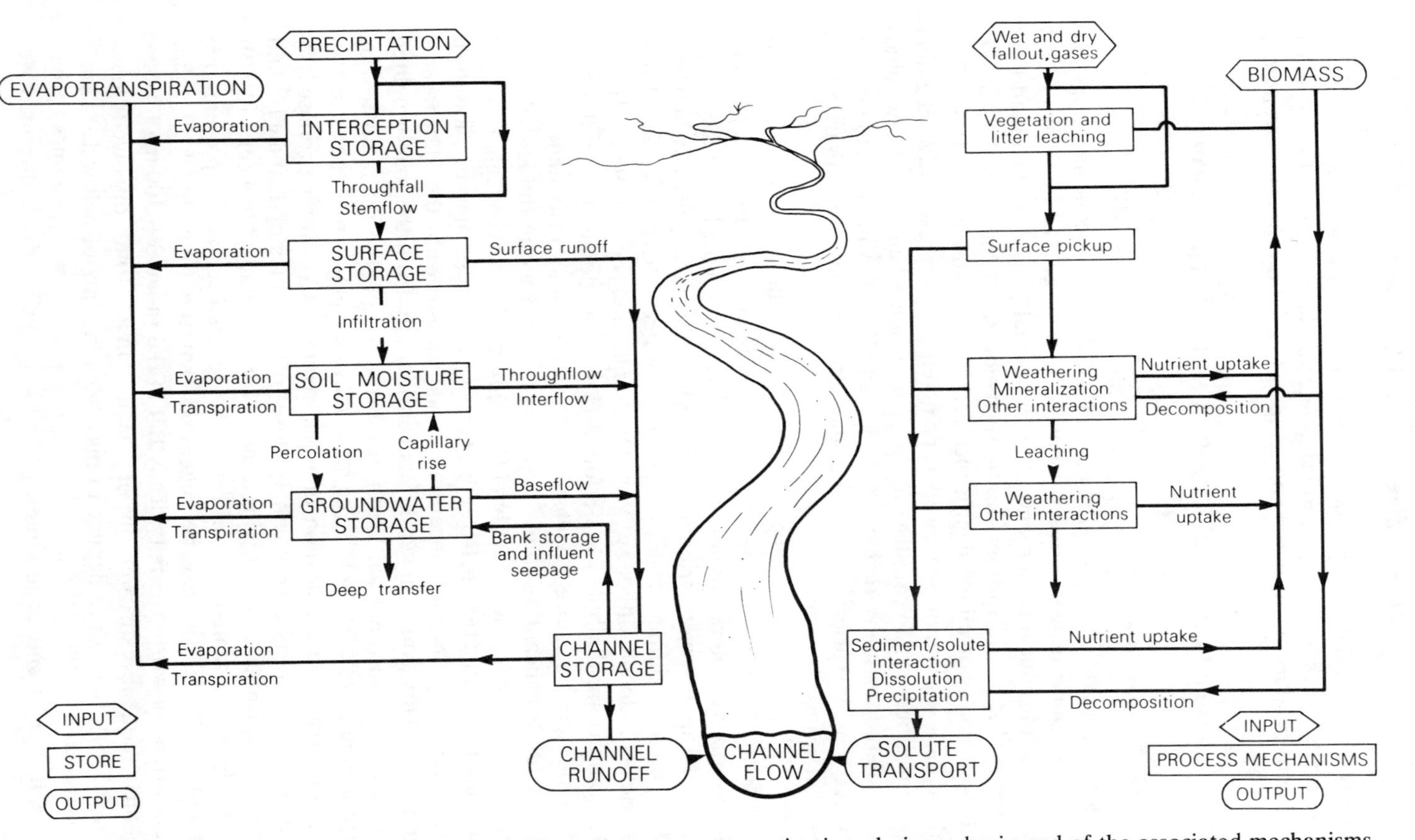

Figure 7.1. Simplified representation of the rainfall/runoff processes operating in a drainage basin and of the associated mechanisms governing solute levels in streamflow

residence times, the magnitude of the stores, as well as catchment characteristics such as lithology, soil character, relief and vegetation type, will combine to produce both spatial and temporal variations in solute transport.

7.2 SPATIAL VARIATIONS IN SOLUTE TRANSPORT

7.2.1 Global generalizations

In order to consider the general magnitude of the solute concentrations and loads to be found in world rivers, values of discharge-weighted mean total dissolved solids concentration and of mean annual total dissolved load have been assembled by the authors for 496 rivers located throughout the world in which pollution is of minor significance (cf. Walling and Webb, 1983). The resultant data set cannot be viewed as being entirely representative of global conditions, since certain countries are over-represented and only a few values were available for many regions of Africa and South America. However, it embraces a wide variety of climatic and physiographic conditions and of catchment scales.

Figure 7.2 portrays frequency distributions of the 496 values of mean annual total dissolved load and discharge-weighted mean concentration. Concentrations range between minimum values of 5–8 mg l^{-1} for the Negro, Xingu and Tapajos rivers, tributaries of the River Amazon, cited by Meybeck (1976), and a maximum of 4630 mg l^{-1} for the Pecos River near Red Bluff, New Mexico, documented by Van Denburgh and Feth (1965). Higher concentrations are doubtless to be found in streams dominated by drainage from saline deposits, since Livingstone (1963a) cites an instantaneous total dissolved solids concentration of 22,900 mg l^{-1} recorded in the La Sal Vieja River in Texas, USA, and Colombani (1983) refers to an even higher value of 60,000 mg l^{-1} in the Leben River in Tunisia, but these may be viewed as special cases. The mean concentration associated with the database is 255.6 mg l^{-1}. This simple mean differs considerably from the load-weighted mean concentration (i.e. total load transported by the 496 rivers divided by their total runoff) of 120 mg l^{-1}. The latter may be equated with the estimates of the total solute concentration of world average riverwater proposed by various workers and is encouragingly close to the values of 120 mg l^{-1} and 99.6 mg l^{-1} advanced by Livingstone (1963a) and Meybeck (1979) respectively. If this sample of world rivers is accepted as being representative of global conditions, it may be suggested that the total dissolved solids concentrations of world rivers typically fall in the range 30–300 mg l^{-1}.

In the frequency distribution of mean annual total dissolved loads (Figure 7.2(b)), the values depicted range from minima of below 1.0 t km^{-2} yr^{-1} associated with Dutch Creek (0.34 t km^{-2} yr^{-1}), James River (0.66 t km^{-2} yr^{-1}) and Sage Creek (0.84 t km^{-2} yr^{-1}), three intermediate-

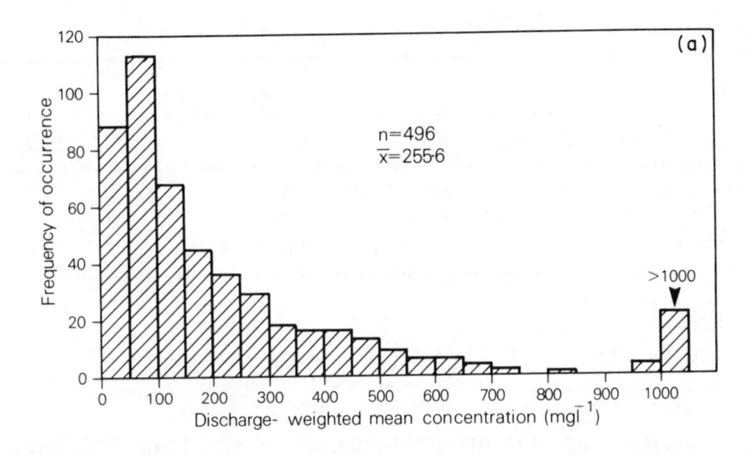

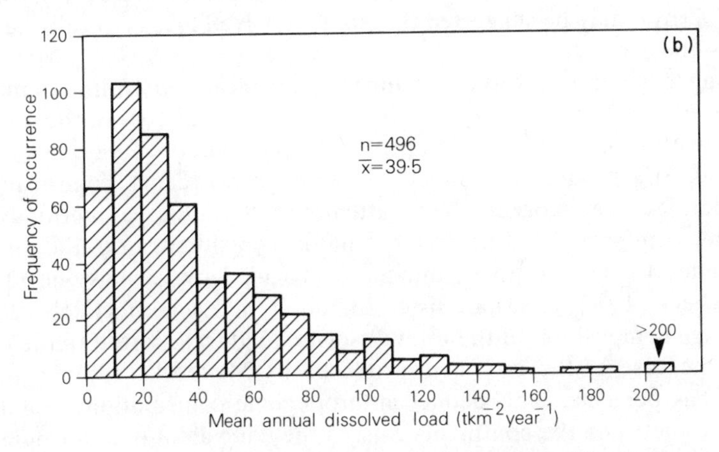

Figure 7.2. Frequency distribution of discharge-weighted mean total dissolved solids concentration and of mean annual total dissolved load for a sample of 496 world rivers

sized drainage basins in Southern Alberta, Canada, investigated by McPherson (1975), to a maximum of 500 t km^{-2} yr^{-1} associated with the 535 km^2 basin of the River Dranse in France, a tributary of Lake Geneva which drains an area of calcareous rocks, documented by Meybeck (1976). Again, higher values doubtless exist in streams draining saline deposits, and the definition of minimum values could be further complicated by the inclusion of extremely low loads associated with basins drained by ephemeral streams with low runoff totals and where solute transport occurs on only a few days of the year. The mean solute load associated with the global database is 39.5 t km^{-2} yr^{-1}, and this conforms closely to the estimate of mean dissolved

Table 7.1. Average composition of world riverwater

Author	Concentration ($mg\ l^{-1}$)								
	Ca^{2+}	Mg^{2+}	Na^{+}	K^{+}	Cl^{-}	SO_4^{2-}	HCO_3^{-}	SiO_2	Total
Livingstone (1963a)	15.0	4.1	6.3	2.3	7.8	11.2	58.4	13.1	120[a]
Meybeck (1979)	13.4	3.35	5.15	1.3	5.75	8.25	52.0	10.4	99.6[b]
Meybeck (1983)	13.5	3.6	7.4	1.35	9.6	8.7	52.0[c]	10.4	106.6[b]

[a]Total content of all solutes.
[b]Total of listed constituents.
[c]HCO_3^- not listed in Meybeck (1983); Meybeck (1979) value used.

load transport from the land surface of the globe of 37.2 t $km^{-2}\ yr^{-1}$ produced by Meybeck (1979). On the basis of the sample of global rivers represented in Figure 7.2(b) it may be suggested that total dissolved loads typically lie in the range 5–80 t $km^{-2}\ yr^{-1}$.

Turning to generalizations concerning the chemical *composition* of material transported in solution, it may be noted that more than 80% of the dissolved load of rivers is generally made up of just four components (HCO_3^-, SO_4^{2-}, Ca^{2+} and SiO_2) with a number of lesser constituents comprising the remainder. Several workers have attempted to define a world average riverwater composition in terms of major constituents on the basis of load-weighted concentrations, and those provided by Livingstone (1963a) and Meybeck (1979, 1983) are listed in Table 7.1. Each effectively refers to the average composition of the total dissolved load transported to the oceans and, in the case of the estimates produced by Meybeck (1979, 1983), an attempt has been made to deduct anthropogenic contributions. Significant differences between the continents exist in the equivalent data for individual continents produced by Meybeck (1979), which are listed in Table 7.2. The extent of contrasts between the continents varies for the different constituents, and further information concerning the typical range of concentrations associated with individual constituents at the global scale is presented in Table 7.3. This is based on the work of Meybeck (1981) and takes account of data assembled from major rivers representing 63% of the total water discharge to the oceans. The minimum and maximum values cited correspond to the 1% and 99% frequency values. Whereas there is only about one order of magnitude difference between the minimum and maximum values listed for SiO_2, K^+ and HCO_3^-, values of Cl^- span a sixty-fold range.

These and other contrasts in the behaviour of individual constituents will clearly be related to variations in the relative importance of particular sources contributing to the occurrence of an individual constituent in riverwater. These sources include rock weathering, atmospheric fallout of material of

Table 7.2. Average composition of riverwater from individual continents (Meybeck, 1979)

	Concentration (mg l^{-1})								
Continent	Ca^{2+}	Mg^{2+}	Na^{+}	K^{+}	Cl^{-}	SO_4^{2-}	HCO_3^{-}	SiO_2	Total[a]
Africa	5.25	2.15	3.8	1.4	3.35	3.15	26.7	12.0	57.8
North America	20.1	4.9	6.45	1.5	7.0	14.9	71.4	7.2	133.45
South America	6.3	1.4	3.3	1.0	4.1	3.5	24.4	10.3	54.3
Asia	16.6	4.3	6.6	1.55	7.6	9.7	66.2	11.0	123.55
Europe	24.2	5.2	3.15	1.05	4.65	15.1	80.1	6.8	140.25
Oceania	15.0	3.8	7.0	1.05	5.9	6.5	65.1	16.3	120.65

[a]Total of listed constituents.

both oceanic and terrestrial origin, atmospheric gases, and decomposition and mineralization of organic material. Thus the relatively low degree of variation associated with SiO_2 and K^+ could be ascribed to the dominance of silicate mineral weathering as a source, whereas the supply of Cl^- involves a major contribution from precipitation. Tentative estimates of the provenance of the major constituents of the total solute load transported to the oceans are listed in Table 7.4. The estimates refer to the natural component of this load and take no account of anthropogenic inputs either from effluent or from atmospheric pollution. With the exception of SiO_2, several sources provide significant contributions to the individual constituents, but there are important variations in their relative importance.

Table 7.3. Variability of the concentration of major dissolved constituents of riverwater at the global scale (Meybeck, 1981)

	World range (mg l^{-1})		
Constituent	minimum[a]	maximum[b]	maximum/minimum
SiO_2	2.1	21	10
Ca^{2+}	1.8	54	30
Mg^{2+}	0.8	15	18.8
K^{+}	0.5	4	8
Na^{+}	1.15	37	32
Cl^{-}	0.7	42	60
SO_4^{2-}	1.5	65	43
HCO_3^{-}	12	165	13.8

[a]1% point of frequency curve.
[b]99% point of frequency curve.

Table 7.4. Tentative estimates of the provenance of the major constituents of total solute transport to the oceans[a]

Source	Constituents (%)								
	Ca^{2+}	Mg^{2+}	Na^+	K^+	HCO_3^-	SO_4^{2-}	Cl^-	SiO_2	Total
Precipitation[b] (Oceanic Salts)	2.5	15	53	14	–	19	72	–	12
Chemical Weathering	97.5	85	47	86	43	81	28	100	60
Atmospheric CO_2	–	–	–	–	57	–	–	–	28

[a]Based largely on Meybeck (1979, 1983).
[b]Inputs of terrestial dust are included with chemical weathering.

To develop further the information on the average composition of global riverwater contained in Table 7.1., it is possible to identify dominant water types on the basis of the relative importance of the major constituents expressed as meq l^{-1}. Calcium (Ca^{2+}) and bicarbonate (HCO_3^-) ions represent the dominant cation and anion in most waters and, from his survey of major rivers representing 63% of the water discharge to the oceans, Meybeck (1981) indicated that more than 97% of global runoff was of this type. Sodium (Na^+) was the major cation in only 1.7% of the water and the alternative anions, Cl^- and SO_4^{2-}, were dominant in only 1.1% of the water. By considering the relative importance of the secondary anions and cations, it is possible to define sub-groups for the major water types, and these are listed in Table 7.5. According to this table, nearly 50% of riverwater may be characterized as having the composition $Ca^{2+} > Mg^{2+} > Na^+ > K^+$ and $HCO_3^- > SO_4^{2-} > Cl^-$. Gibbs (1970) suggested that it was possible to use such information on the dominant cation and anion to distinguish three major water types in terms of chemical evolution. Calcium and hydrogen carbonate (bicarbonate) were the dominant ions in the rock-dominated group where rock weathering was the major mechanism controlling solute content and where total ion concentrations typically ranged between 50 and 1000 mg l^{-1}. Sodium and chloride dominated the other two groups, which were further distinguished by their overall solute content. The rain-dominated group, where atmospheric precipitation provided the major controlling mechanism, was characterized by low total ion concentrations (<50 mg l^{-1}), whereas the evaporation–crystallization type representing waters in which calcium carbonate had precipitated was associated with high levels of mineralization (>1000 mg l^{-1}). Although this classification is attractive in its simplicity, it is clearly of limited value since Table 7.5 indicates that the latter two groups together only account for 0.1% of global runoff.

Compared with inorganic constituents, relatively little information is

Table 7.5. Major water types associated with world riverwaters (Meybeck, 1981)

Cations	Anions	Percentage of global sample
$Ca^{2+} > Na^{+} > Mg^{2+} > K^{+}$	$HCO_3^- \gg Cl^- > SO_4^{2-}$	33.1
	$HCO_3^- > SO_4^{2-} > Cl^-$	2.5
	$SO_4^{2-} > HCO_3^- > Cl^-$	1.0
$Ca^{2+} > Mg^{2+} > Na^{+} > K^{+}$	$HCO_3^- > SO_4^{2-} > Cl^-$	46.7
	$HCO_3^- > Cl^- > SO_4^{2-}$	15.0
$Na^{+} > Ca^{2+} > Mg^{2+} > K^{+}$	$HCO_3^- > Cl^- > SO_4^{2-}$	1.4
	$SO_4^{2-} > Cl^- > HCO_3^-$	0.1
	$Cl^- > HCO_3^- > SO_4^{2-}$	0.1

available on the transport of dissolved organic material by rivers. A few studies have included this aspect of solute transport within their scope, and Arnett (1978) reported mean dissolved organic solids concentrations of between 8.4 and 18.9 mg l^{-1} in 16 rivers in Yorkshire, UK. These concentrations constituted between 2.7% and 10.4% of the total dissolved load. Brinson (1976) cites similar concentrations for tropical streams in Guatemala, although it would seem that higher levels may sometimes be evident, for he reports a dissolved organic solids concentration of 36.4 mg l^{-1} for the Oscuro River.

Measurements of dissolved organic carbon (DOC) concentrations in riverwater may also be used to provide an indication of dissolved organic matter content. Values of the former may be converted to estimates of the latter by applying a conversion factor based on the carbon content of the constituent organic compounds, and a value of 2.0 is frequently used for this purpose. In his review of information on dissolved organic carbon transport by rivers, Meybeck (1982) states that concentrations commonly range between 1 and 20 mg l^{-1} with a median value of 5 mg l^{-1}, and that the loads involved vary between 0.2 and 14 t km^{-2}. These values should be doubled to provide estimates of total dissolved organic matter transport.

7.2.2 Global patterns

In looking for general spatial patterns of solute transport at the global scale, a number of workers have emphasized the importance of climate and drainage basin lithology as major controls. In the former case, it is clear that moisture availability, and to a lesser extent temperature, will exert a significant influence on rates of weathering and associated solute release through their influence on the kinetics of weathering reactions. Mineral availability and stability govern the nature of the weathering reactions and their products, and

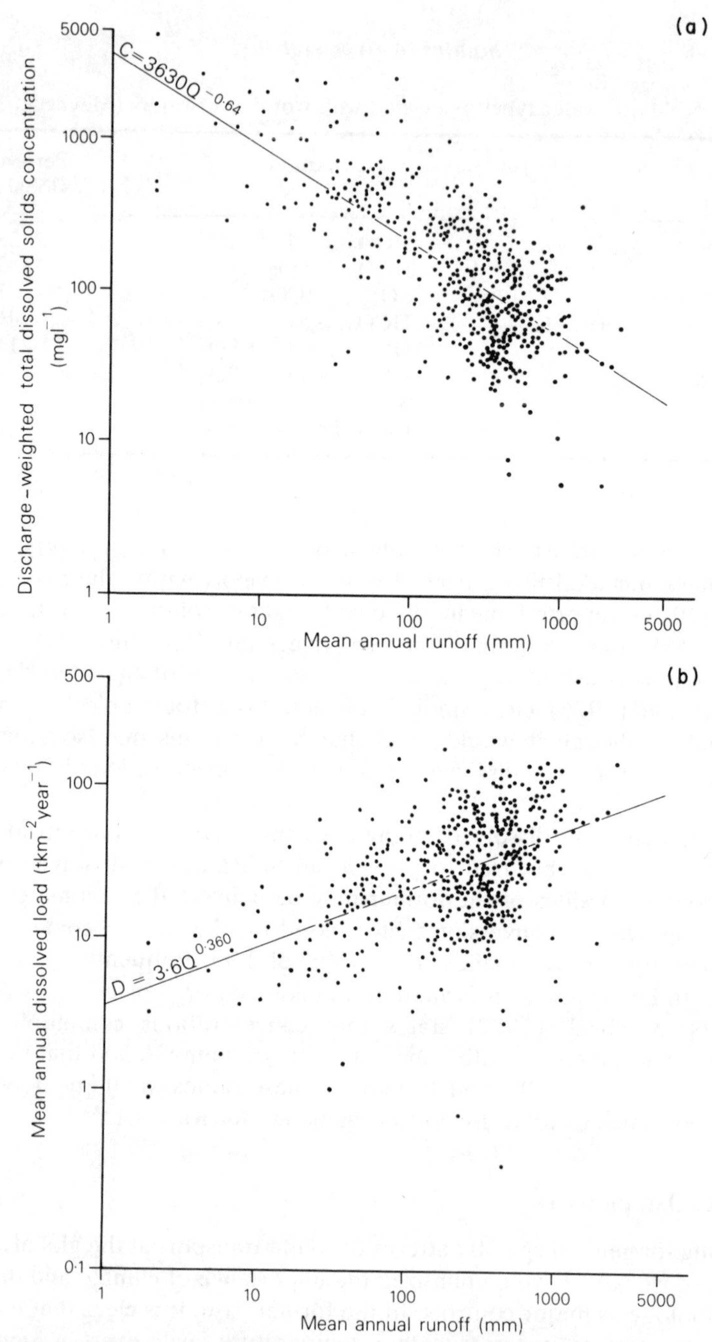

Figure 7.3. Relationship between mean annual runoff and discharge-weighted mean total dissolved solids concentration (a) and mean annual total dissolved load (b) for a sample of 496 world rivers

will be closely related to lithology. Since, as has already been indicated, approximately 60% of the total solue load of rivers is contributed by rock weathering, these controls should be reflected in global patterns of solute transport.

To demonstrate the influence of climate, as reflected in water availability, Figure 7.3 provides a plot of discharge-weighted mean total dissolved solids concentration and total solue load *versus* mean annual runoff for the 496 rivers included in the global database represented in Figure 7.2. Clear trends are apparent on the logarithmic plots, and straight lines have been fitted to these trends using least-squares regression. In the case of the plot of concentration versus mean annual runoff, the inverse relationship may be accounted for in terms of a general dilution effect, as runoff volumes increase. The positive relationship between load and annual runoff nevertheless indicates that increasing moisture availability provides an increase in the total quantity of dissolved material released or available for transport. Similar inverse relationships between total solute concentration and runoff have been presented by Langbein and Dawdy (1964), Durum *et al*. (1960) and Holland (1978) using data collected from the USA. Livingstone (1963b) and Meybeck (1976, 1977) have also demonstrated positive relationships between annual solute load and annual runoff for world rivers, and Langbein and Dawdy (1964) present a similar result for rivers in the USA. However, although Langbein and Dawdy (1964) suggest that the relationship between total solute load and annual runoff evidences a gradual flattening at annual runoff values in excess of 75 mm, there is little evidence of such flattening in Figure 7.3(b). Furthermore, the suggestion by Holland (1978) that dissolved load may be virtually independent of runoff at annual runoff values in excess of 100 mm is again refuted by Figure 7.3(b).

Figure 7.4, which is based on the work of Holland (1978), provides examples of the relationships between the concentrations of individual constituents of the total solute load and annual runoff. In this instance the plots are based only on data from the USA rather than the entire globe, but the wide range of climatic conditions represented ensures a general applicability. Negative trends exist for all four relationships, but the precise nature of these trends varies considerably between the individual constituents and Holland (1978) suggests that these contrasts may be related to the different sources of the constituents. The negative exponent of approximately -1 demonstrated by Cl^- in Figure 7.4 is similar to that exhibited by SO_4^{2-} and Na^+ for the same data set, and implies that loads remain relatively constant as runoff increases. Holland (1978) argues that this behaviour is characteristic of ions derived from atmospheric sources or highly soluble rocks. K^+ evidences a negative exponent of approximately -0.5 reflecting sources such as the weathering of silicate minerals in which total solute release increases with increasing water availability. The plot for Ca^{2+} which closely resembles that

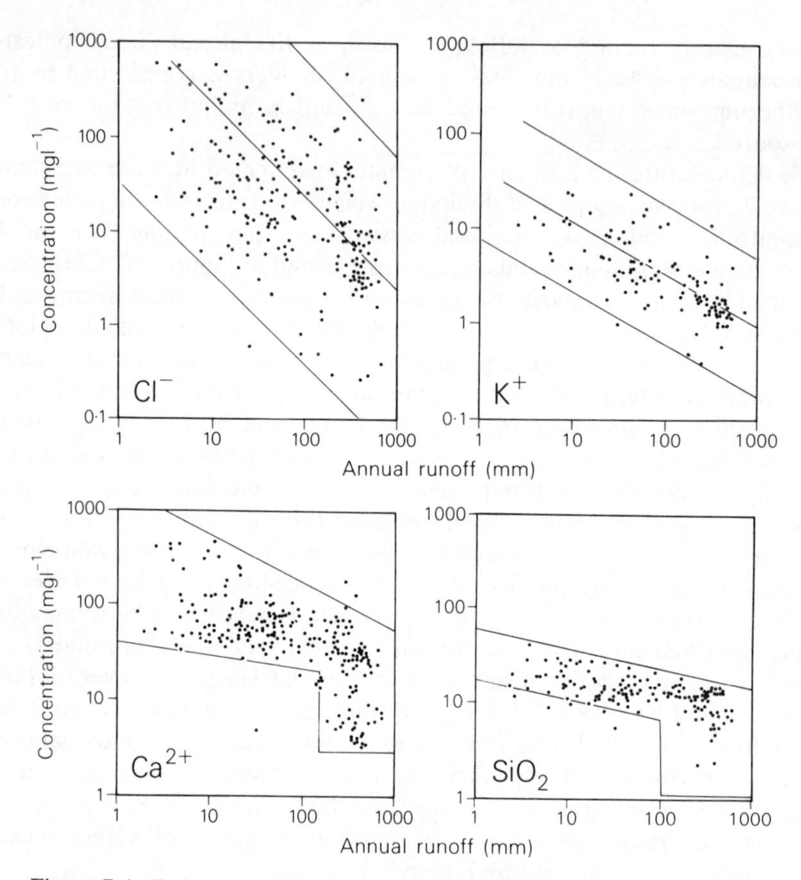

Figure 7.4. Relationship between mean concentrations of Na^+, K^+, Cl^- and SiO_2 and mean annual runoff for river waters of the USA (Holland, 1978)

for HCO_3^-, is rather different from those for Cl^- and K^+ in that there is evidence of near constant concentrations at an annual runoff of less than 100 mm. This feature may in turn be related to the saturation of riverwater with respect to calcite and the precipitation of $CaCO_3$ under arid conditions. Silica likewise evidences little variation in the concentrations associated with a wide range of annual runoff. This may point to weathering reactions which produce an equilibrium concentration which is independent of water availability.

Although general trends may be discerned for the plots illustrated in Figures 7.3 and 7.4, there is a very considerable degree of scatter in the relationships. This may be attributed to other climatic factors such as temperature, seasonality, and the rainfall/runoff ratio and to physiographic controls including rock type. Increased data availability, particularly from

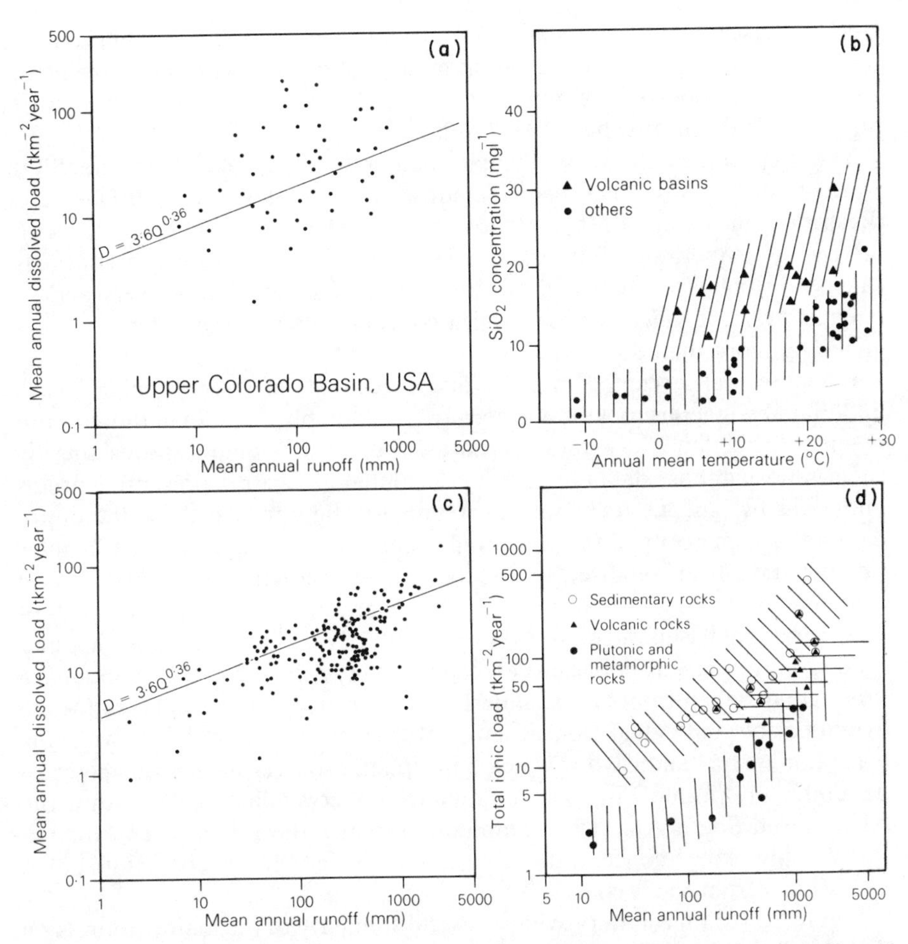

Figure 7.5. Relationships between dissolved load and mean annual runoff and temperature. (a) demonstrates the degree of scatter in the dissolved load/annual runoff relationship encountered in the Upper Colorado basin, USA; (based on data from Iorns *et al*. (1965)); (b) depicts the relationship between mean annual temperature and the SiO_2 content of river water proposed by Meybeck (1981); (c) superimposes those data points from the global data set representing drainage basins underlain by igneous rocks on the relationship developed for all rivers (cf. Figure 7.3); (d) presents the relationship between total ionic load and annual runoff for world rivers developed for rivers draining different rock types by Meybeck (1981)

small and intermediate-sized drainage basins, is required to elucidate the nature and relative importance of these various controls. An indication of the degree of scatter that may exist under relatively uniform climatic conditions is provided by Figure 7.5(a), which plots the relationship between total dissolved load and annual runoff for data representing 50 measuring stations in the 283,600 km^2 basin of the Upper Colorado River in the south-west

USA. The scatter of these values around the regression line established for the global data set is almost as great as that associated with the latter data, and a large proportion of this scatter may be related to the influence of physiographic controls, particularly lithology.

The importance of mean annual temperature in controlling the SiO_2 content of riverwater has been demonstrated by Meybeck (1981) and is illustrated in Figure 7.5(b). Concentrations show a clear increase with increasing temperature, but lithology is also significant since major contrasts can be distinguished between the behaviour of major world rivers and of streams draining volcanic areas. Silica concentrations in the latter areas are generally more than doubled.

It is now well accepted that the dissolved loads of river draining basins underlain by igneous rocks are generally considerably lower than those found in areas of sedimentary rocks. Figure 7.5(c) clearly demonstrates this by superimposing those data points from the global database representing basins underlain by igneous rocks, on the regression line produced for the entire data set. The majority of the points fall well below the regression line. Similar findings have been reported by Meybeck (1981) in an analysis of the total ionic load of major world rivers (Fig. 7.5(d)), and he indicates that the loads of rivers draining basins developed on sedimentary rocks are commonly about 5 times greater than from basins developed on crystalline rocks and about 2.5 times greater than from basins underlain by volcanic rocks. Similar ratios are exhibited by individual constituents of the total solute load, with the exception of Na^+ and Cl^- for which atmospheric sources have been shown to be dominant (Table 7.6). The dominance of crystalline rocks within the African and South American continents provides the principal explanation for the low ionic concentrations noted in Table 7.2 as characteristic of riverwaters from these continents.

Figure 7.5(d) therefore provides a useful summary of the major controls on the global pattern of inorganic solute transport exercised by annual runoff and underlying lithology. On the basis of this evidence it can be suggested that the effects of annual runoff are dominant in producing a range of variation in total solute loads of nearly two orders of magnitude whereas a range of about one order of magnitude is associated with contrasts in lithology. It is difficult to provide equivalent information for dissolved organic matter transport because of the lack of measurements of this aspect of fluvial transport. However, Meybeck (1982) suggests that dissolved organic carbon concentrations exhibit some general climatic control with taiga rivers evidencing relatively higher values (median 10 mg l^{-1}) than those in the humid tropics (median 6 mg l^{-1}) and in the temperate and semi-arid zones (median 3 mg l^{-1}), while lower concentrations (*c*. 2 mg l^{-1}) can be found in tundra rivers. Maximal natural levels of DOC, of approximately 25 mg l^{-1}, are found in rivers draining swamps or poorly drained soils. Because of the

Table 7.6. Influence of rock type on the average composition of world riverwaters (Meybeck, 1981)

	Average concentration (mg l^{-1})		
Constituent	Plutonic and highly metamorphic rocks	Volcanic rocks	Sedimentary rocks
SiO_2	1.5*x*	3.5*x*	*x*
Ca^{2+}	4	8	30
Mg^{2+}	1.0	3	8
K^+	1.0	1.5	1.0
Na^+	Oceanic influence dominant		
Cl^-	Oceanic influence dominant		
SO_4^{2-}	2	6	25
HCO_3^-	15	45	100
Total ions	30	70	175

x = Average SiO_2 content of water from rivers draining sedimentary rocks at a given temperature.

limited variation in DOC concentrations between different climatic zones, global variations in DOC loads (t km^{-2}) are closely related to those of annual runoff.

7.2.3 National, regional and local patterns

As the scale of attention decreases from one of broad global trends to a more detailed examination of spatial contrasts of solute transport within an individual country, region or a local area, the patterns become more complex and other physiographic factors assume increasing importance. Individual ions may also be expected to exhibit contrasting patterns and control by different factors in response to variations in their major sources. Figure 7.6 illustrates an attempt by the authors to produce generalized maps of the solute loads of small, essentially unpolluted, streams in Great Britain (cf. Walling and Webb, 1981). Maps of total dissolved solids, NO_3–N and Cl^- loads (t km^{-2}) are reproduced here. Total dissolved solids loads range between 10 and in excess of 200 t km^{-2} y^{-1}, but the national pattern is not closely related to that of annual runoff which has been superimposed on this map. The relatively low loads in the south-east of the country may be related to the low annual runoff totals in this area, but the areas of maximum runoff in the west of the country are associated with low rather than high load values. In this case, lithology exerts an overriding influence, since the large areas of minimum dissolved load in the high-runoff areas of the west of the country are underlain by resistant solute-deficient rocks. Furthermore, several of the

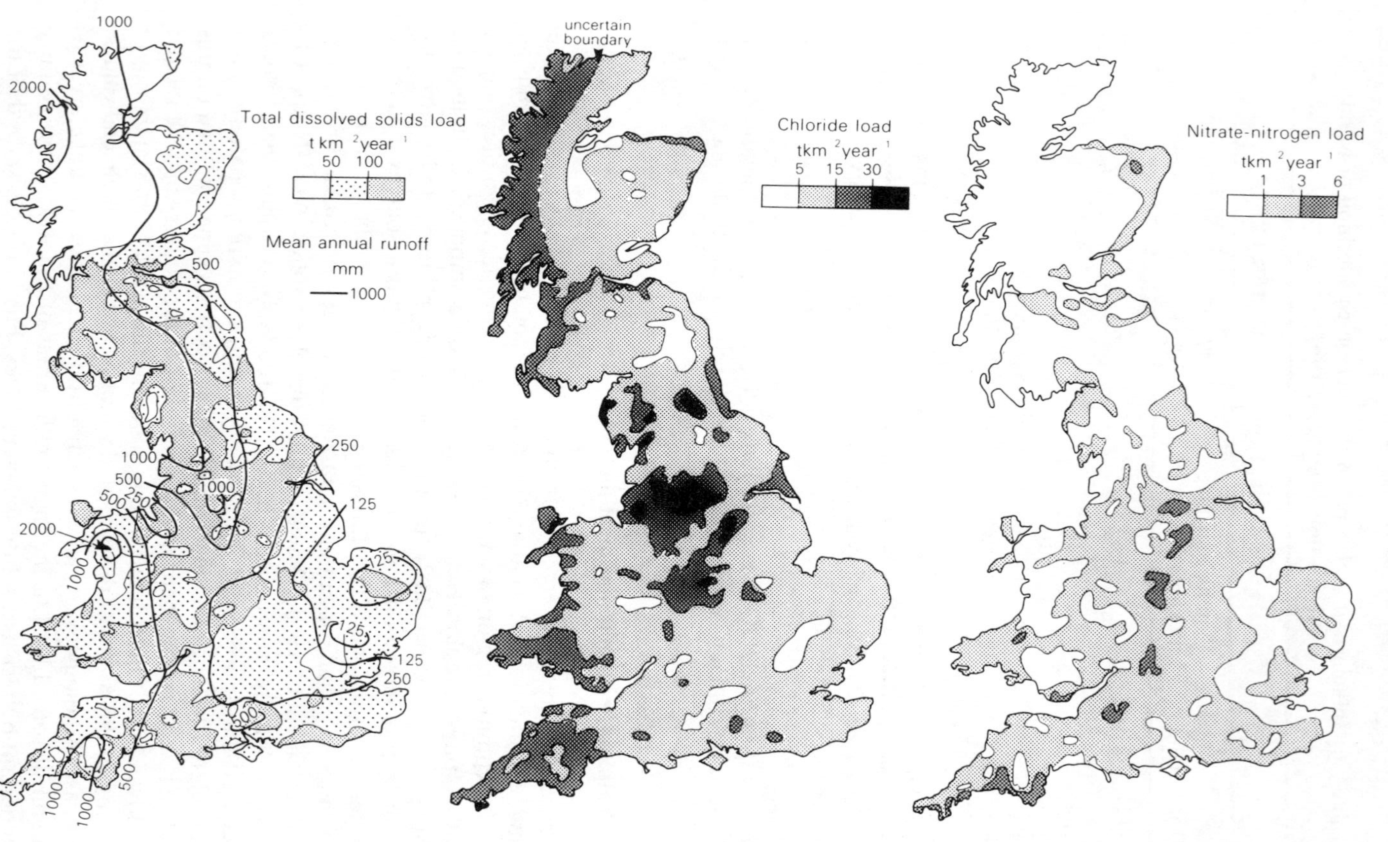

Figure 7.6. Generalized maps of total dissolved solids, NO_3–N, and Cl loads for small unpolluted streams in Great Britain. Isopleths of mean annual runoff based on Ward (1981) have been superimposed on the map of total dissolved solids load

mapped boundaries between load classes are clearly realted to the pattern of geological outcrops.

Comparison of the maps of NO_3–N and Cl^- loads with those of total dissolved solids reveals distinct contrasts between the patterns exhibited by the individual constituents. To a large extent, Cl^- loads demonstrate a positive relationship with annual runoff, since maximum loads occur in the western areas and minimum loads in the east. Because rock weathering is of minimal importance as a source of Cl^-, lithological controls are unimportant and the predominance of higher loads in the west of the country may be related to the high annual rainfall in these areas and to the importance of the prevailing westerly winds from the Atlantic Ocean in causing high concentrations of oceanic aerosols and therefore of Cl^- in the precipitation (cf. Stevenson, 1968; Walling and Webb, 1981). Cl^- concentrations in precipitation over western coastal areas frequently average 10–15 mg l^{-1} or even higher, whereas concentrations in eastern areas are typically in the range 5–10 mg l^{-1}.

In the case of nitrate-nitrogen the national pattern of loads reflects significant control by land-use. Nitrate-nitrogen concentrations in streamwater demonstrate a marked west–east gradient, with concentrations of less than 1.0 mg l^{-1} in western areas of Great Britain and levels above 10 mg l^{-1} in eastern areas, which may reflect the increased intensity of agricultural activity, particularly arable farming, in the east. This pattern of concentrations coupled with that of annual runoff provides maximum loads in the central areas of the country. Loads are lower in western areas because concentrations are low, and although high concentrations occur in eastern areas, these are offset by relatively low runoff totals to again produce lower loads.

Other studies have demonstrated the importance of a wide variety of factors in controlling regional and local patterns of solute transport. Meybeck (1983) has pointed to the importance of altitude and distance from the Atlantic coast in influencing the Cl^- concentrations encountered in unpolluted French streams, and this control may be related to the importance of oceanic salts as a source of Cl^-. Vitousek (1977) was able to demonstrate a similar control by catchment altitude in the White Mountains of New Hampshire. In this case rock type was relatively uniform throughout the area, and the concentrations of several ions showed a significant decrease with increasing altitude, which could be related to decreased evapotranspiration and increasing precipitation at higher altitudes. Another study in this same area by Leak and Martin (1975) provides a useful example of the effects of nutrient cycling by vegetation in regulating stream solute concentrations. In this study NO_3^- concentrations were measured in a variety of small forested watersheds, and the variations encountered appeared to be closely related to the degree of disturbance of the watershed in recent years and the age of the forest land. Lowest NO_3^- concentrations were associated with streams draining medium-aged stands and higher concentrations were found in over-mature forest stands, young stands and in areas subject to cutting and disturbance, where nitrogen uptake was less. Nitrogen and phosphorus concentrations and loads have also frequently been related to variables representing land-use conditions, since these constituents are significantly influenced by non-point pollution from agricultural activity (e.g. Omernik, 1976; Haith, 1976; Klepper, 1978; Hill, 1978). Kirschner (1975) was also able to demonstrate a very high correlation between drainage density and phosphorus export from forested streams in Ontario. In evaluating these and other results, however, the considerable degree of multicollinearity which may exist between physiographic variables must be clearly recognized. Furthermore, the potential validity of any proposed relationship should be tested against available knowledge concerning the dominant source of the constituent involved.

A final example of local-scale variation in solute transport and the associated controls may be drawn from the work of the authors in the 601 km^2 basin of the Middle and Upper Exe in Devon, UK (cf. Webb and Walling, 1983; Webb, 1976; Webb, 1983b). This catchment is free from major industrial or domestic pollution and contains a considerable variety of geological, topographical and land-use characteristics (Figure 7.7.). A network of more than 260 sampling sites was established on small streams within the basin in order to reflect water quality at source and to isolate the effects of different environmental factors. These sites were sampled during a stable baseflow period in May/June 1974, and an attempt has been made to evaluate the individual effects of geology, land-use and topography on the spatial variability encountered in specific conductance levels and cation concentrations.

Figure 7.8(a) presents frequency distributions of specific conductance levels associated with the major rock types. Clear contrasts are apparent between most of the distributions, and these may be related to the varying susceptibility of the major rock types to chemical weathering. Contrasts in rock type are associated not only with variation in the magnitude of total solute concentrations, as indexed by specific conductance, but also with changes in the balance between the concentrations of individual cations expressed as equivalents (Figure 7.8(b)). Strong contrasts in solute levels were also apparent in the study basin between areas of agricultural land and tracts of moorland which are much less disturbed by agricultural practices. However, because of the close interdependence between land-use and geology within the basin (cf. Figure 7.7), the effects of land-use can be more meaningfully distinguished by considering the variation of solute levels *within* an area underlain by a single rock type. In examining the data obtained from 75 small headwater tributaries (<3 km^2) underlain by Devonian strata,

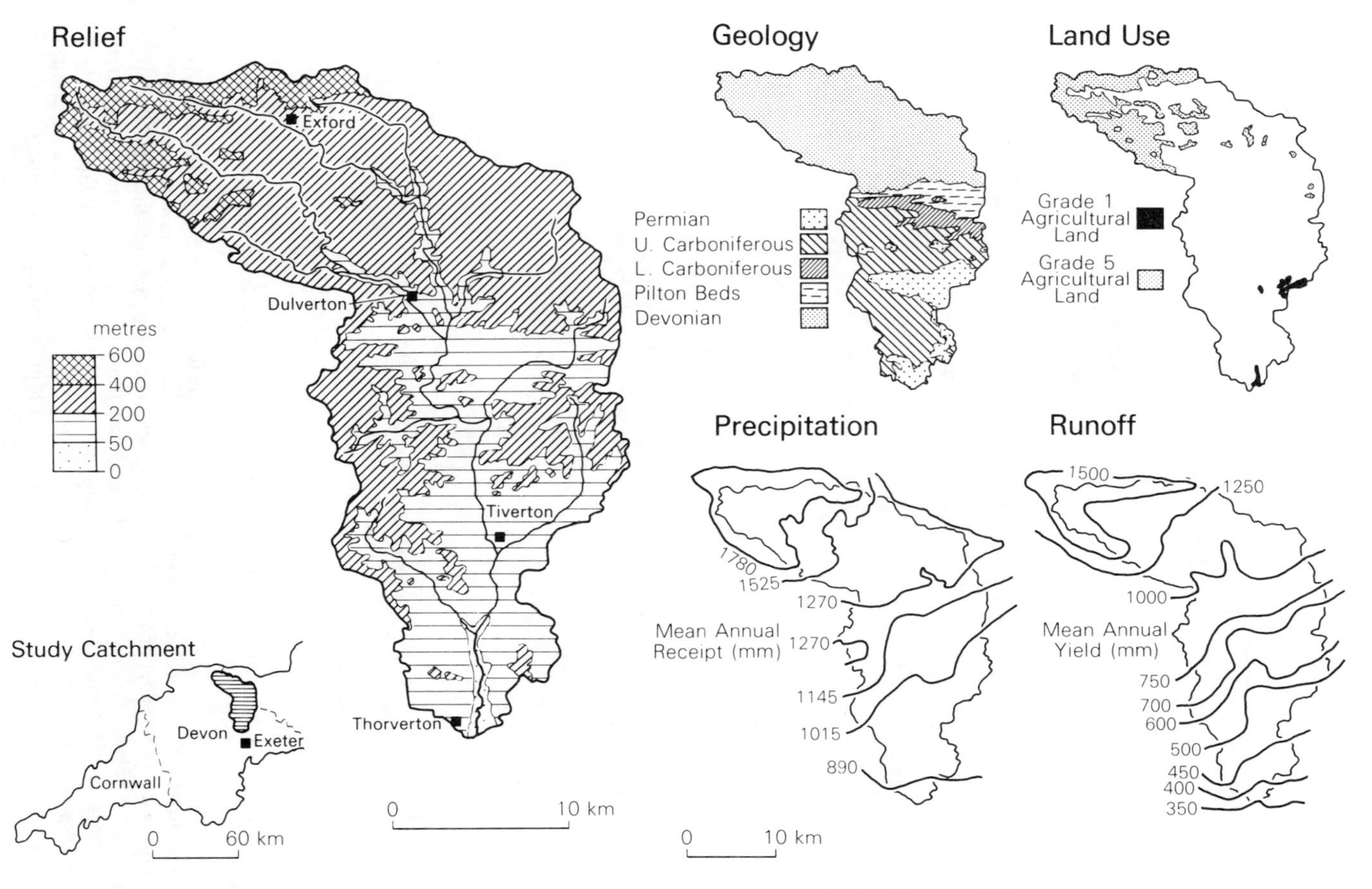

Figure 7.7. Characteristics of the Upper Exe Basin, Devon, UK

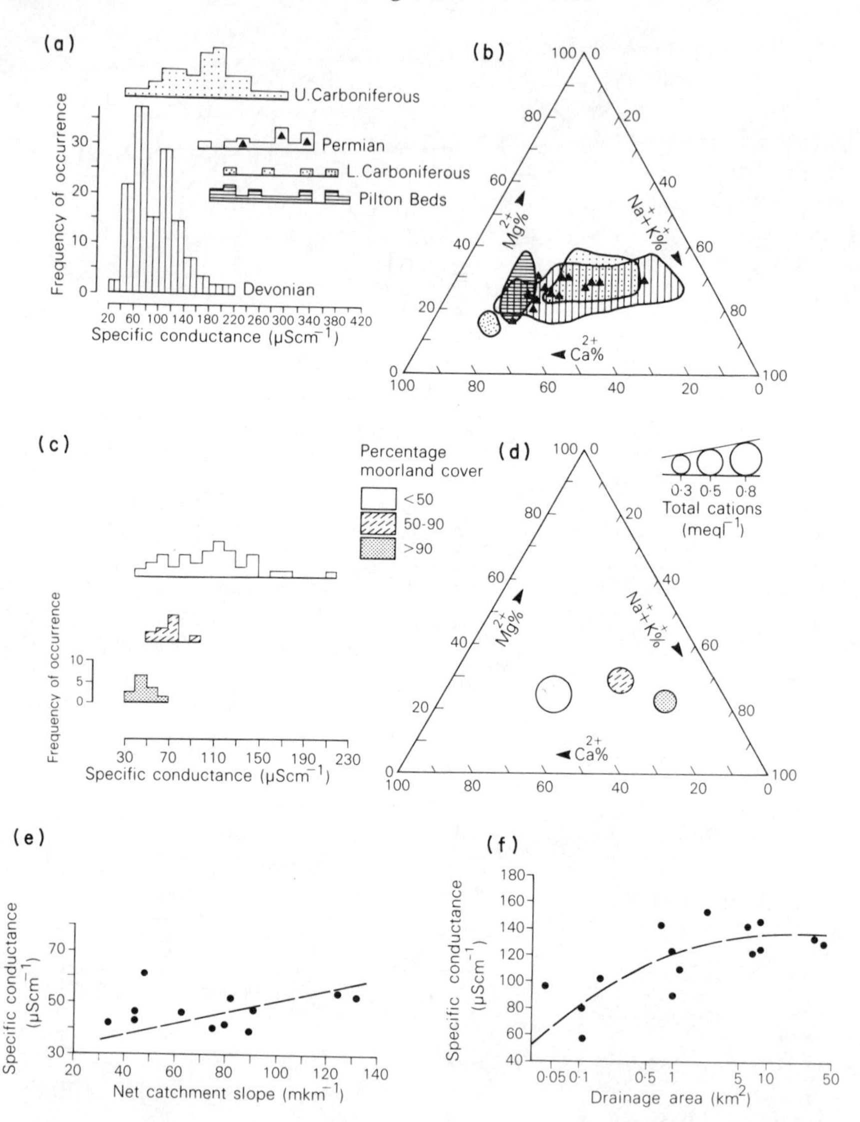

Figure 7.8. Controls on spatial variation of baseflow solute levels within streams of the Upper Exe Basin, Devon, UK. The influence of rock type on specific conductance levels and cation composition is depicted in (a) and (b); (c) and (d) illustrate the influence of land-use on specific conductance levels and cation composition of streams draining Devonian rock types; (e) presents the relationship between specific conductance levels and catchment slope for small streams underlain by Devonian rock types and dominated by moorland land-use; (f) indicates the influence of catchment scale on specific conductance values associated with streams within the subcatchment of the Iron Mill Stream, a west bank tributary of the River Exe

Figure 7.8(c) reveals that total solute levels progressively decline as the percentage of moorland cover increases from less than 50% to greater than 90%. The same trend is evidenced by individual cation concentrations, and Figure 7.8(d) indicates that increasing moorland cover is further associated with a change from calcic to sodic water types.

Topographic characteristics of a drainage basin, including slope, size and elevation, have been shown to influence stream solute levels (e.g. McCann and Cogley, 1971; Foggin and Forcier, 1977), and the effects of catchment slope and scale have also been investigated in the study area. Again problems of interdependence of controlling factors must be recognized, and Figure 7.8(e) attempts to isolate the effects of slope by considering the relationship between specific conductance levels and catchment net slope, calculated as the quotient of relative relief and maximum basin length, for small catchments underlain by a single rock type (Devonian) and possessing uniform land use (>90% moorland cover). A weak positive trend is apparent. This trend is opposite to that theoretically expected on the basis of more rapid runoff and reduced residence times in areas of steeper topography (e.g. Walling and Webb, 1975) and may indicate that steeper streams are supplied from groundwater of deeper origin and in turn higher solute content (cf. Ternan and Williams, 1979). However, the weak nature of this relationship suggests that this topographic factor is not of primary importance in controlling baseflow solute levels. Figure 7.8(f) also attempts to isolate the influence of catchment scale on baseflow solute levels by presenting the relationship between specific conductance and drainage area for samples collected within the subcatchment of the Iron Mill Stream. This basin is almost exclusively underlain by Upper Carboniferous strata and is dominated by agricultural land-use. Figure 7.8(f) indicates that baseflow conductance values rise with increasing catchment area until a threshold size of approximately 2 km^2 is attained. Beyond this threshold solute levels are essentially independent of basin scale. Attainment of near constant solute concentration on single rock types has been interpreted by Miller (1961) to result from the operation of weathering processes at uniform intensity over broad areas, and implies that underground flow paths of a minimum length are necessary in the Upper Carboniferous strata of this subcatchment before an equilibrium between weathering reactions and draining waters is obtained.

7.3 TEMPORAL BEHAVIOUR OF SOLUTE LEVELS

Stream solute levels vary through time as well as in space because the catchment processes which generate streamflow and its dissolved content (Figure 7.1) are dynamic rather than static in nature. The manner by which water moves through the system of stores and transfer routes in a drainage basin is conditioned by the sequence and magnitude of hydrometeorological events, and also by fundamental biological cycles of growth and decay. In turn, stream solute levels will vary according to the hydrological pathways in operation and in response to changes in the type and rate of solute mobilization processes at different stages of the catchment water cycle. Temporal behaviour of stream solute levels may therefore both reflect *and* illuminate solute processes operating in the contributing catchment, but it is also important to recognize that the nature of temporal solute behaviour will itself vary *through space* in response to changing environmental conditions (e.g. Biesecker and Leifeste, 1975).

The magnitude and controls of temporal variation have attracted serious investigation since pioneering studies undertaken more than 40 years ago (Butcher *et al*., 1939; Lenz and Sawyer, 1944). Many factors potentially influence temporal behaviour of solute levels and these include meteorological, geochemical, biological and anthropomorphic controls. However, most attention has been directed to the effects of hydrological conditions and more specifically to the nature of the relationship between water chemistry and streamflow discharge. It has often been argued that the volume of water available for dilution is one of the most direct and important influences on the concentration of dissolved solids (Hem, 1970) and this control provides a convenient starting point for a discussion of temporal solute behaviour.

7.3.1 Rating relationships

The response of solute levels to changes in stream and river discharge has been studied in several ways, but investigation of rating relationships between solute concentration and flow represents a traditional approach. Such relationships have been presented using various transformations of the dependent (solute concentration) and independent (discharge) variables (Douglas, 1964; Turvey, 1975) and have been fitted employing a number of statistical methods (e.g. Loughran and Malone, 1976). Most commonly, however, studies have used 'log–log' plotting in conjunction with least-squares regression techniques in order to develop a simple function of the form:

$$C = aQ^{\mathrm{b}}$$

where C is solute concentration, Q is discharge, a is the regression constant and b is the regression exponent. It has been found in many studies (Gregory and Walling, 1973) that the exponent b of the rating relationship has a negative value, which indicates a decrease in solute concentration with increasing discharge at a sampling station. The widespread occurrence of this dilution effect is clearly apparent from a survey of the relationships between total dissolved solids concentration and discharge reported from a global sample of 370 rivers (Walling and Webb, 1983). The frequency distribution of the associated exponent values (Figure 7.9(a)) reveals a dilution effect for more than 97% of the rivers investigated. The majority of the exponents fall into the range 0 to−0.4, and the overall mean of the distribution is −0.17. Correlation coefficients associated with rating relationships of this type often exceed −0.8 (Walling and Webb, 1981), and it is clear from the examples presented in Figure 7.9(b) that variation in solute levels can be very closely associated with discharge fluctuations. Clearly defined inverse rating relationships between total solute content and flow may be explained by reference to the sources and routes by which water is supplied to the river channel under different flow conditions in the following *general terms* (Walling, 1980a). At low flows, runoff is derived from the lower soil profile and the groundwater reservoir and has a relatively high concentration of dissolved solids because of long residence times within soil and rock, which promote solute release, and in response to evapotranspirational water losses, which concentrate dissolved material in baseflow. In contrast, during higher discharges much of the runoff is translated rapidly to the channel, has little opportunity for solute pickup and therefore has a much lower solute concentration.

It is also evident from Figure 7.9(b), and from Figure 7.9(c) which plots the range in slope of specific conductance-discharge rating lines for 12 sites in the Middle and Upper Exe Basin, Devon, UK (Webb, 1980) in the standardized form advocated by Edwards (1973), that exponent values vary both between regions and within a single drainage area. This variation can often be systematically related to the magnitude of low flow solute levels, since for most rivers, solute concentrations at high discharges will tend towards the limiting low values found in storm rainfall and the slope of the rating relationship will be governed by the concentration characteristic of baseflow. The occurrence of higher solute levels in baseflow of the Dolores River compared with the River Barle in Somerset (Figure 7.9(b)) therefore provides a greater potential for dilution during storm events and, in turn, produces a steeper rating line. No significant variation of total solute concentration (indexed by specific conductance) with discharge is apparent for the Dane's Brook (Figure 7.9(d)) because low solute levels in baseflow from this upland tributary in the Exe Basin offer little scope for dilution by rainfall and storm runoff.

Other factors, including catchment permeability and the nature of groundwater circulation, may also influence the exponent of rating relationships. Talsma and Hallam (1982), for example, have shown from a study of eight sites in the lower Cotter Valley near Canberra, Australia, that basins with the most permeable rock types have rating lines with the lowest slope because stream discharge is dominated in these circumstances by a groundwater component of near constant chemical composition. In contrast,

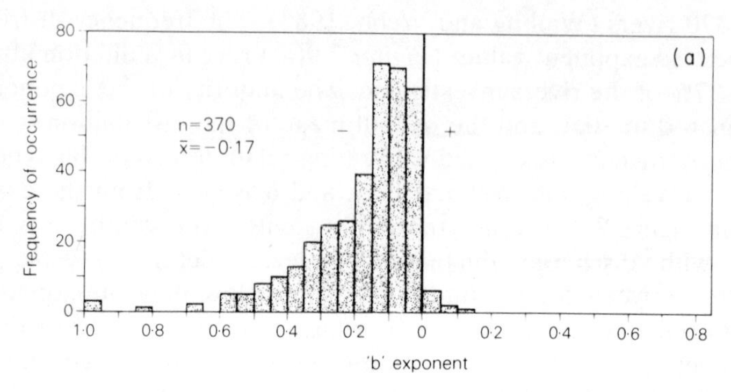
(a)
Frequency of occurrence
n=370
x̄=−0·17
'b' exponent

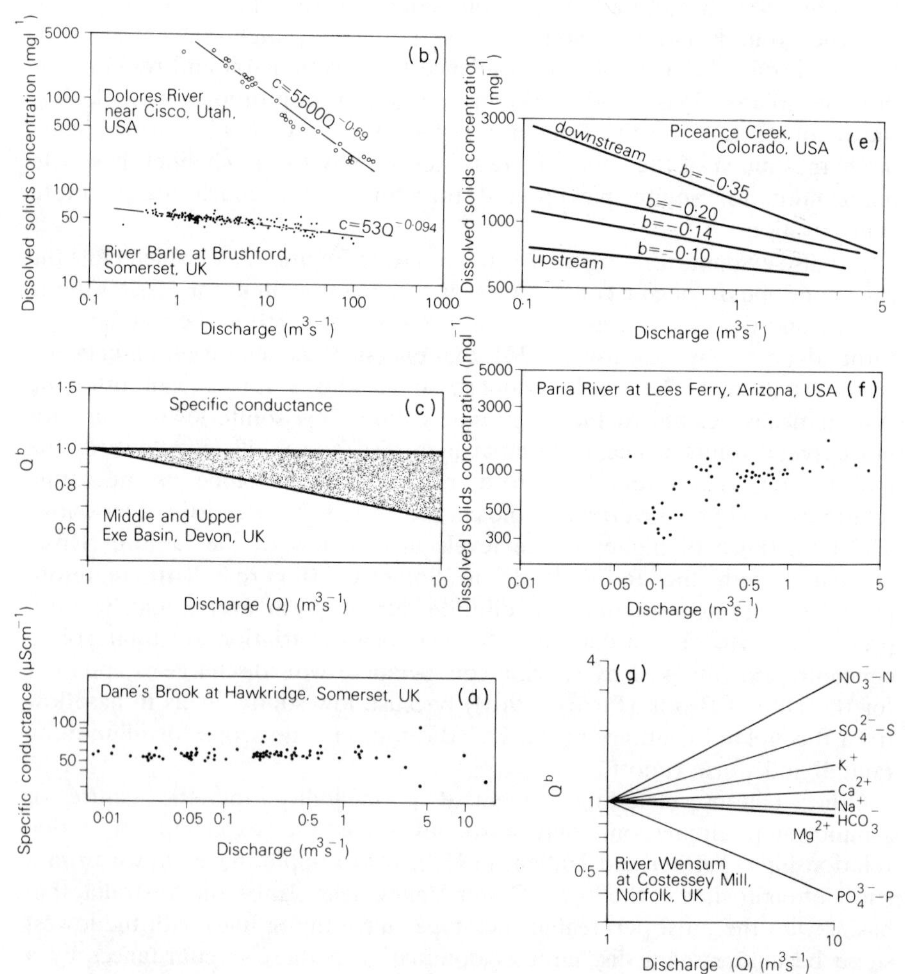
(b)
Dissolved solids concentration (mgl⁻¹)
Dolores River near Cisco, Utah, USA
c=5500Q^−0·69
c=53Q^−0·094
River Barle at Brushford, Somerset, UK
Discharge (m³s⁻¹)
(c)
Specific conductance
Q^b
Middle and Upper Exe Basin, Devon, UK
Discharge (Q) (m³s⁻¹)
(d)
Dane's Brook at Hawkridge, Somerset, UK
Specific conductance (µScm⁻¹)
Discharge (m³s⁻¹)
(e)
Piceance Creek, Colorado, USA
downstream
upstream
b=−0·35
b=−0·20
b=−0·14
b=−0·10
Dissolved solids concentration (mgl⁻¹)
Discharge (m³s⁻¹)
(f)
Paria River at Lees Ferry, Arizona, USA
Dissolved solids concentration (mgl⁻¹)
Discharge (m³s⁻¹)
(g)
NO3⁻−N
SO4²⁻−S
K⁺
Ca²⁺
Na⁺
HCO3⁻
Mg²⁺
PO4³⁻−P
River Wensum at Costessey Mill, Norfolk, UK
Q^b
Discharge (Q) (m³s⁻¹)

Andrews (1983) reports a situation from Piceance Creek, Colorado, USA, where the increasing influence of solute-rich groundwater from deep sources elevates baseflow solute levels at sites located progressively down the mainstream and, in consequence, causes a downstream steepening of the rating relationship with discharge (Figure 7.9(e)). It has been argued that the presence of readily soluble materials in the drainage basin can lead to a well-buffered system characterized by rating relationships of low slope (Carling, 1983), and it is also suggested by Edwards (1973) that ill-defined dilution patterns in the lowland and agricultural catchments of East Anglia reflect a relatively small range in discharge, the presence of highly soluble materials in soil profiles, a lack of overland flow and rapid solute pickup in runoff routed through drains and ditches.

The global data set (Figure 7.9(d)) exhibits positive exponents for only a few rivers, and the presence of a 'concentration' rather than a 'dilution' effect seems to arise only under special circumstances. Rivers characterized by very low solute concentrations and dominated by atmospheric solute sources are occasionally associated with positive exponent values, since in this situation maximum dissolved solids concentrations may occur during storm events when surface runoff mobilizes soluble material accumulated from atmospheric fallout or evaporation in a preceding dry period. Particular geological conditions, where baseflow runoff of relatively low solute concentration is derived from one rock unit and surface runoff with higher solute levels is generated from another, may also produce a concentration effect irrespective of the general level of dissolved solids within the river system. This situation is considered to account for the trend of increasing solute concentrations at higher flows recorded by Iorns *et al.* (1965) in the Paria River, Arizona, USA (Figure 7.9(f)). A combination of solute pickup by surface runoff and contrasting concentrations in pipeflow and peat water has been cited to explain the positive exponents of rating relationships established for the relatively dilute streamwaters of the Maesnant Catchment in mid-Wales (Cryer, 1980).

Figure 7.9. Solute concentration–discharge rating relationships. (a) depicts the frequency distribution of the exponent values exhibited by a global sample of 370 rivers; (b) presents rating relationships of contrasting slope, data for the Dolores River were obtained from Iorns *et al*. (1965); (c) plots the range in slope of specific conductance-discharge rating lines for the Middle and Upper Exe Basin in standardized form; (d) illustrates the rating plot for a stream in the Exe Basin which lacks a significant rating relationship; (e) indicates the changing nature of the rating relationship recorded along the mainststream of Piceance Creek by Andrews (1983); (f) reproduces the positive trend between solute concentration and discharge reported for the Paria River by Iorns *et al*. (1965); and (g) shows the range in slope of rating lines for individual solute parameters in standardized form for the River Wensum based on Edwards (1973)

Rating relationships not only display considerable contrasts between different streams but also may vary markedly in character for individual solute parameters at a single station as the standardized plot (Figure 7.9(g)) of exponent values associated with rating lines for individual ions in the River Wensum, UK (Edwards, 1973) clearly demonstrates. Much attention in this regard has been devoted to the response of the major inorganic constituents found in streamwaters (e.g. Troake and Walling, 1973; Buckney, 1977; Hughes and Edwards, 1977), but rating relationships have also been established for organic components (Arnett, 1978) and dissolved metal pollutants (Grimshaw *et al*., 1976).

Relationships between the concentration of individual dissolved constituents and discharge may be well-defined, less well-defined or non-existent (Feller and Kimmins, 1979), and although some ions, such as Mg^{2+}, are often strongly diluted in higher flows (Steele, 1968; Webb and Walling, 1983), other solutes, such as Na^+, frequently exhibit much less fluctuation through time and appear to be influenced by buffering mechanisms in the soil (Johnson *et al*., 1969). Furthermore, a well marked dilution effect is frequently absent for ions, including Na^+, Cl^- and SO_4^{2-}, which are dominated by atmospheric sources (Cryer, 1976; Foster, 1979a; Meybeck, 1983; Chapter 2 in this volume). Individual chemicals may also show little response to discharge fluctuations if they are derived from several sources which react differently to increasing flow. Edwards (1973) hypothesises that this situation applies in several rivers of Norfolk, UK, where Ca^{2+} ions are supplied from the solution of calcium carbonates and sulphates, which is respectively decreased and increased at higher runoff. In contrast, a clear concentration effect with increasing flow is often reported for ions, such as K^+ and NO_3^-, which are actively involved in the nutrient cycles of forested and moorland ecosystems. This behaviour is related to the storage of these ions in biological materials and to their mobilization in surface runoff through the action of vegetation leaching (Bond, 1979; Waylen, 1979).

The form of the rating relationship for individual solute parameters thus reflects the sources of runoff and the location and solubility of minerals in the vegetation, soils and rock (Edwards, 1973) of a drainage basin and may vary in nature with changing catchment characteristics (e.g. Keller, 1970). Furthermore, the considerable variety of individual solute responses which is characteristic of many rivers may also contribute to the ill-defined relationships between discharge and total dissolved solids concentration, or its surrogate, specific conductance, recorded at some sites (Kemp, 1971; Walling, 1978a).

In many plots relating total or individual solute concentrations and discharge it is possible to discriminate between samples taken at different times of the year (e.g. Figures 7.10(a) and (b)), and this seasonal differentiation may add considerably to the scatter about a rating relationship. It has in fact been suggested for some catchments and some solute parameters that it is more appropriate to define separate lines for different seasons of the year (Webb, 1980), wet and dry months (Al-Jabbari *et al*., 1983) or more prolonged wet and dry periods (Oborne *et al*., 1980). Seasonal contrasts in solute response have been isolated statistically (Foster, 1978a), and for some ions, such as K^+, a dilution effect has been reported for one season but a concentration effect for another (Loughran and Malone, 1976). However, the nature of seasonal contrasts and the explanations offered to account for them also vary between drainage basins and between solute species. For example, less pronounced dilution of Ca^{2+} concentrations during storms of summer and autumn months in a small agricultural catchment of East Devon, UK (Figure 7.10(a)) has been attributed to the accumulation of soluble material in the dry periods of summer (Foster, 1978a), whereas stronger dilution of total solute levels, indexed by specific conductance (Figure 7.10(b)), during the summer season in the River Quarme, a tributary of the Exe Basin, has been related to increased biological uptake, lower solute concentrations in precipitation and flashier discharge responses at this time of the year (Webb and Walling, 1983). It is apparent, therefore, that the local situation regarding the status of the hydrological pathways and the condition of the solute stores depicted in Figure 7.1 control the magnitude and nature of seasonal contrasts in solute responses to flow.

Scatter in rating relationships may also be attributed to contrasting hydrological conditions which occur within storm events, and in several studies separate rating lines have been established for samples taken during rising and falling or stable flows (Loughran and Malone, 1976). Total and individual solute concentrations are commonly higher during the rising than the falling limb of a flood wave, and this reflects flushing out of soluble material that has accumulated from weathering or farming activity in the period before a storm. A concentration effect during rising stage conditions but a dilution response in falling and steady flows has been demonstrated by Oxley (1974) for two small Welsh catchments (Figure 7.10(c)). In this case the difference in behaviour is attributed to the influence of pipeflow which has a relatively high dissolved content and supplies the rising limbs of hydrographs. It should be clear that the different components supplying streamflow have contrasting chemical properties because of their varied origins within a drainage basin (Hart *et al*., 1964; Schwartz and Milne-Home, 1982; Froehlich, 1983) and different contact times with soil and rock (Pilgrim *et al*, 1978, 1979), and in some studies (e.g. Walling, 1974) the influence of particular flow components on the scatter of solute rating plots has been determined (Figure 7.10(d)).

There is also increasing evidence that the simple power function is an oversimplified model of solute response to changing flow conditions. It was, for example, argued in a relatively early study of temporal behaviour

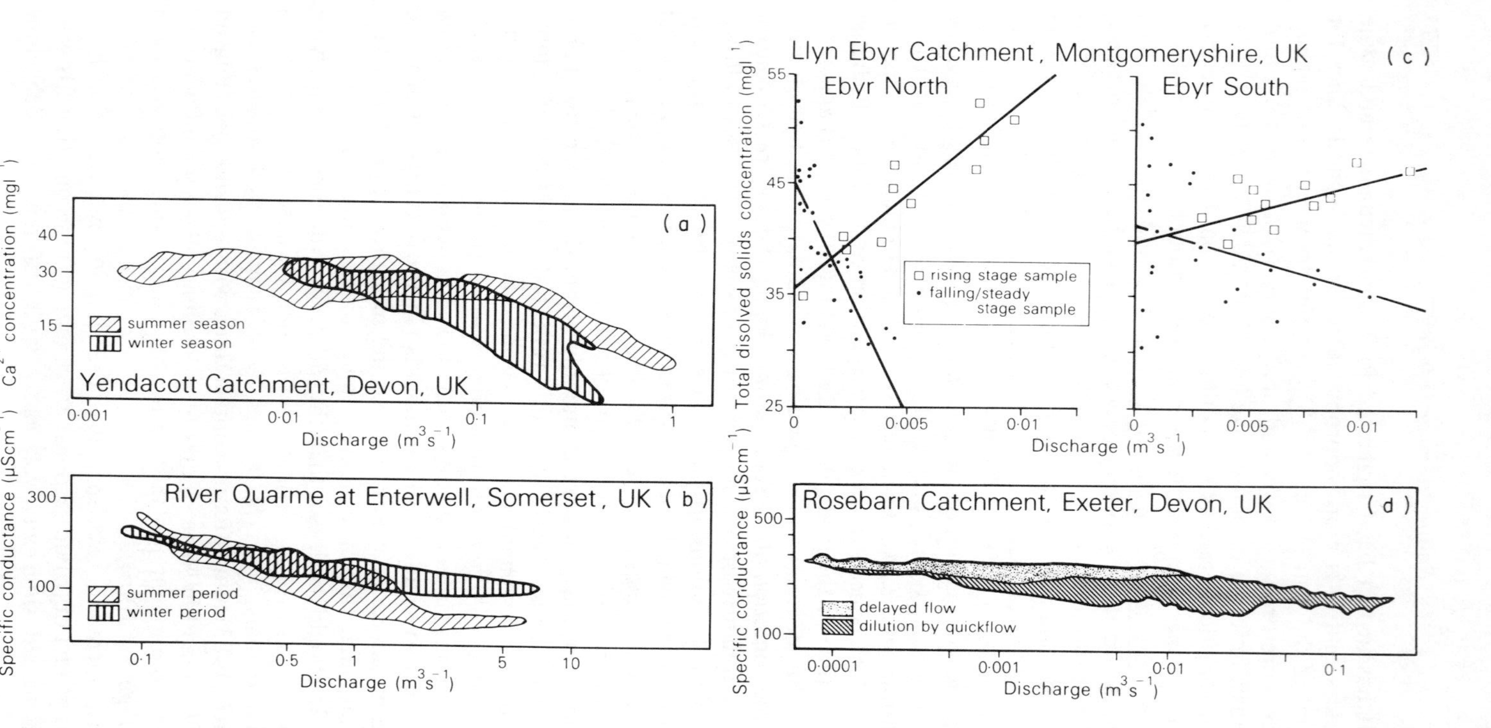

Figure 7.10. Scatter in solute concentration–discharge rating plots. (a) and (b) illustrate plots for catchments in the Exe Basin where separate relationships have been established for different seasons, data for the Yendacott Catchment were obtained from Foster (1978a); (c) presents separate rating relationships for rising and falling/steady stage conditions in the Llyn Ebyr Catchment based on Oxley (1974); and (d) differentiates samples on the rating plot for the Rosebarn Catchment according to components of flow present in the channel

(Ledbetter and Gloyna, 1964) that it may be more appropriate to continuously adjust the value of the exponent *b* in the rating relationship according to the level of discharge than to assume that it remains constant over the whole range of flows. Several subsequent investigations have shown that segmented (Steele, 1968; Troake and Walling, 1975; O'Connor, 1976) and polynomial (Pinder and Jones, 1969; Foster, 1980; Webb, 1980) functions rather than a straight-line trend characterize the relationship between solute concentration and discharge plotted on 'log–log' coordinates (Figure 7.11). A threshold at 1.4–2.8 m^3s^{-1} (50–100 cfs) in the rating relationships for Ca^{2+} and HCO_3^- concentrations in Pescadero Creek, California, USA (Figure 7.11(a)) has been attributed to the prevalence of different chemical processes under conditions of baseflow and storm runoff (Steele, 1968), whereas flattening of a rating line at lowest (Figure 7.11(b)) and highest (Figure 7.11(c)) discharges might be expected in response to the dominance in the channel respectively of baseflow in chemical equilibrium with soil and rock, which cannot be further increased in solute concentration, and stormflow with solute levels close to those in precipitation which cannot be further diluted (Gregory and Walling, 1973). In contrast, solute levels in some moorland streams such as the East Twin catchment, Mendip, UK (Finlayson, 1977) only react to the extremes of discharge, and concentrations are buffered and change little over a large part of the streamflow range (Figure 7.11(d)). The shape of solute concentration–discharge relationships has been demonstrated by O'Connor (1976) to depend on the absolute and relative magnitudes of dissolved content associated with surface flow and groundwater and on the ratio of groundwater to total flow. Furthermore, a number of mixing models (Johnson, 1969; Hem, 1970; Hall, 1970, 1971; Carling, 1983) also predict curvilinear rating lines. These models, which are described more fully later in this chapter, are however less successful in simulating scatter around rating relationships or in reproducing the more complex form of rating plots which occur in some situations (Figure 7.12).

These more complicated rating relationships include compound and 'U'-shaped curves (Figure 7.12(a)), which may arise when runoff is generated from contrasting terrain types in a catchment (Walling, 1971) or where sources of soluble material are distributed unevenly within a soil profile (Troake and Walling, 1973), and annual hysteretic loops, such as that described by O'Connor (1976) for the Snake River, Washington, USA (Figure 7.12(b)), which are prompted by seasonal change in hydrological and other conditions that will be discussed in more detail at a later stage in this chapter. A trapezoidal concentration–discharge relationship (Figure 7.12(c)) has been established for the meltwater stream draining from the snout of Gornergletscher, Switzerland (Collins, 1979), and reflects the special circumstances governing runoff and solute production in alpine areas (Zeman and Slaymaker, 1975). The trend of dissolved content with changing

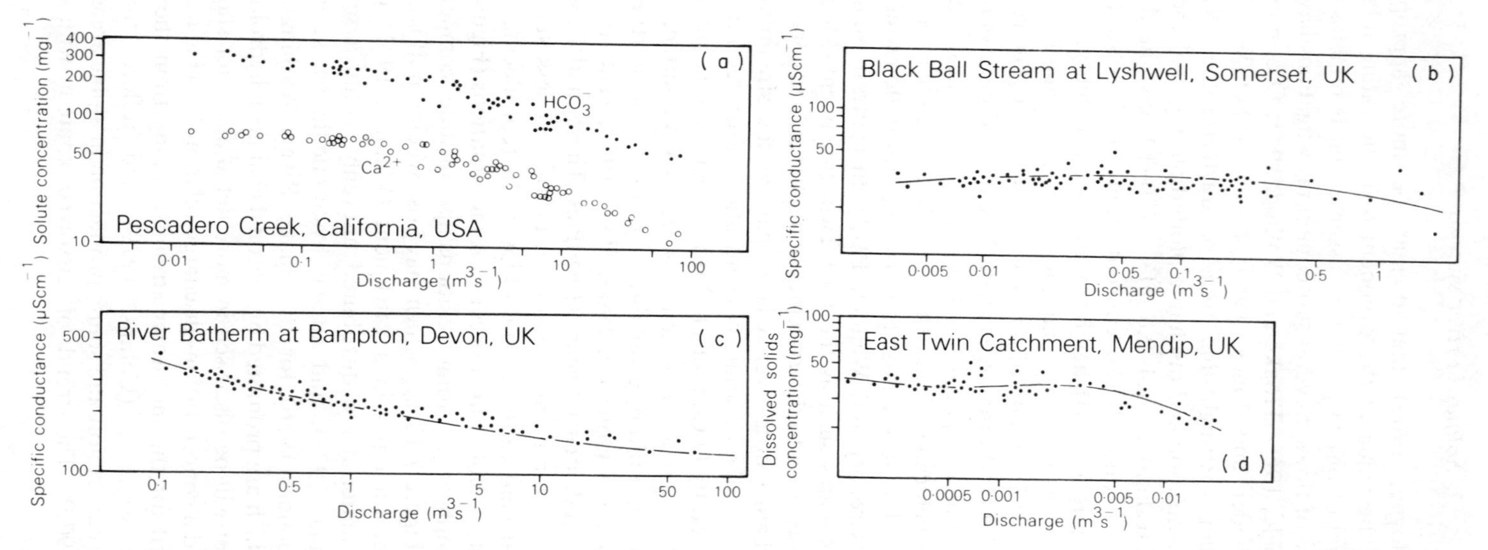

Figure 7.11. Segmented and polynomial solute concentration–discharge rating relationships. (a) illustrates rating plots recorded in Pescadero Creek by Steele (1968) which have a characteristic break in the rating line; (b) and (c) present rating relationships for rives in the Exe Basin which exhibit flattening at lowest and highest discharges respectively; (d) shows the cubic polynomial relationship employed by Finlayson (1977) to describe variation in solute concentration with discharge in the East Twin catchment

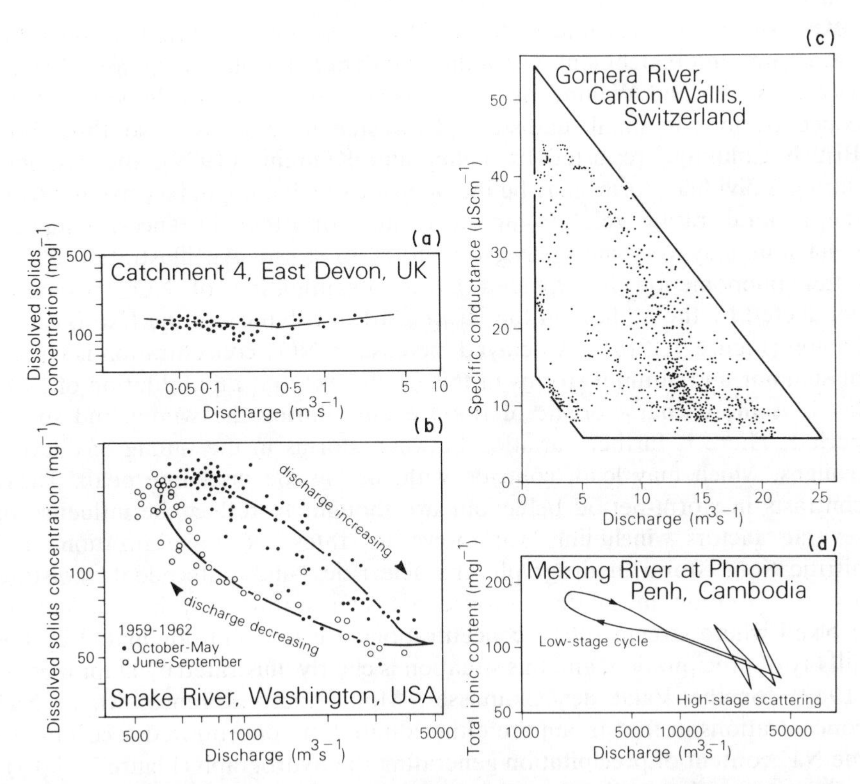

Figure 7.12. Complex solute concentration–discharge rating relationships. (a) depicts a compound rating relationship for a stream in East Devon; (b) presents a looped relationship between concentration and discharge for the Snake River based on O'Connor (1976); (c) illustrates the trapezoidal rating relationship for the Gornera Rivera reported by Collins (1979); and (d) shows the rating plot for the Mekong River which is based on Carbonnel and Meybeck (1975) and is substantially complicated by the impact of particular flood events

discharge in some rivers, such as the Mekong at Phnom Penh (Carbonnel and Meybeck, 1975), may be appreciably complicated (Figure 7.12(d)) by the impact of particular flood events, and results of this kind not only give truth to the assertion that mineral quality and streamflow are uniquely related at individual stations (Gunnerson, 1967) but also have stimulated more detailed studies of temporal solute behaviour during storm events.

7.3.2 Storm-period response

The adoption of intensive storm-period sampling programmes in a growing number of studies (e.g. Miller and Drever, 1977; Reid *et al.*, 1981a; Dupraz *et al.*, 1982) has not only highlighted the complexity of solute behaviour during storms at particular stations but has also demonstrated considerable intra- and intercatchment variability in detailed solute response (Foster, 1979b; Webb and Walling, 1983). In some cases, such as the study of K^+ concentrations in small, undisturbed forested watersheds of south-western British Columbia reported by Feller and Kimmins (1979), the response during individual storms may be the opposite of what might be expected from the general rating relationship at a site. In other instances, temporal behaviour may very markedly from storm to storm. An illustration of the latter phenomenon is provided by an examination of NO_3^- responses conducted by the authors in the River Dart, a tributary of the Exe Basin in Devon (Figure 7.13(a)). A delayed increase in NO_3^- concentration is typical of summer and autumn storms in this catchment, but rapid dilution of NO_3^- levels is more characteristic of flood events during the winter and spring period. There is further variation between storms in the timing of dilution troughs, which may lead, coincide with, or lag the discharge peak. These contrasts in storm-period behaviour are thought to reflect the influence of several factors, including variations in rates of mineralization and nitrification, storm hydrograph characteristics and antecedent moisture levels. Variation in solute response between storms may be particularly marked where stream solute concentrations are low and dominated by the quality of precipitation, and this situation is clearly illustrated by Dupraz *et al.* (1982) for the Valat des Cloutasses, Mont Lozère, France, where Na^+ concentrations in the stream are either diluted or concentrated according to the Na^+ content of precipitation generating the hydrograph (Figure 7.13(b)).

Storm-period sampling has also highlighted the contrasting behaviour of different solute parameters during flood events (e.g. Cleaves *et al.*, 1970; Douglas, 1972; Foster, 1978b). Responses of individual ions will differ in their detailed character because particular chemical species have varying origins, are stored at different locations within the vegetation, soil and rock of a drainage basin, and are accessed to different extents by runoff from various sources. These controls are evident in the moorland Glendye catchment of north-east Scotland (Reid *et al.*, 1981a) where dissolved species derived from chemical weathering, including Ca^{2+}, Mg^{2+}, Na^+, SiO_2 and HCO_3^-, have highest concentrations in baseflow draining from the lower mineral soil horizons but are strongly diluted during storm events by waters which originate in the upper organic and organo-mineral layers of the soil (Figure 7.13(c)). Dilution is, however, less pronounced for the cations because these are displaced from organic colloidal material by H^+ ions in percolating rainfall. In contrast, runoff from the surface horizons of the soils exhibits enriched concentrations of total organic carbon and of complexed Al, Fe and Mn, so that these parameters show an increase during flood events (Figure 7.13(c)). Concentrations of Fe and Mn peak relatively early during the storm because these elements are mobilized in the gleyed soils by reduction processes and are flushed into the stream with displaced soilwater at an early stage in a flood event. Although K^+ concentrations also increased during stream rises in the autumn months (Figure 7.13(c)), this ion was unaffected by storm events at other times of the year, and Reid *et al*. (1981a) argue that this behaviour is related to the leaching of potassium from decomposing plant litter.

Given the varied and often complex responses of individual ions, a clearer picture of changes in water chemistry during a storm event may often be obtained by considering the evolution of water composition which simultaneously takes into account the variation in concentrations of the main constitutents. Trilinear plotting, based on major cation and anion components expressed in equivalent weight units, provides a convenient means of investigating changes in water composition and has been employed by the authors to demonstrate the evolution of water chemistry which occurred over a 60 hours period in the River Clyst, Devon, UK, during a storm in December 1980 (Figure 7.13(d)). In this event, marked changes in water composition may be related to flushing of de-icing salt from road and other surfaces during the early stages of the stream rise, to strong dilution of HCO_3^- ions at peak flows, and to increasing nitrate concentrations in throughflow reaching the channel on the falling limb of the hydrograph.

A particularly important facet of storm-period solute behaviour is the occurrence of hysteresis whereby storm events are associated with well-defined variations in solute levels, but concentrations are markedly different at the same level of discharge on the rising and falling limbs of the hydrograph. This phenomenon has often been investigated in terms of a looped trend in the relationship between solute concentration and discharge for specific storm events (e.g. Hendrickson and Krieger, 1964; Collins, 1979) and the common occurrence of hysteresis severely limits the explanatory and predictive powers of simple rating lines when they are applied to storm periods. Basic clockwise and anticlockwise hysteretic loops were recognized in relatively early studies of streams in Kentucky and Georgia, USA (Hendrickson and Krieger, 1960; Toler, 1965), and a wide range of causes, often involving more than one factor for any one stream, have been proposed to account for this phenomenon. Miller and Drever (1977), for example, suggest that solution of material in the soil zone, dilution of baseflow, selective weathering of ferromagnesian minerals and leaching of biological materials all contribute to a marked hysteresis effect which occurred during a storm event in the North Fork of the Shoshone River, Wyoming, USA.

A full understanding of storm-period hysteresis, however, may only be obtained if attention is directed to the detailed nature of the chemograph and its relationship to the hydrograph, since hysteretic behaviour may be generated by the relative *timing* or the relative *form* of solute and discharge responses. Thus, a discrepancy in the precise timing of the chemograph and

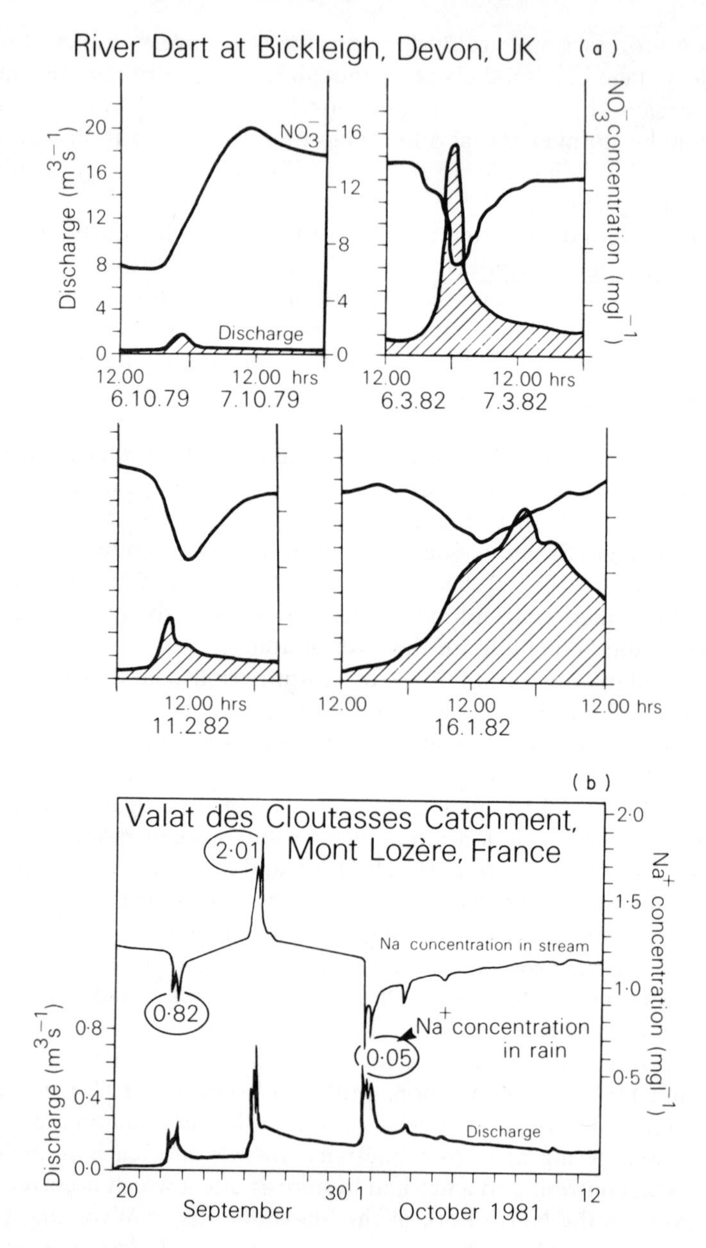

Figure 7.13. Storm-period solute response. (a) illustrates varying NO_3^- behaviour during storm events in the River Dart; (b) depicts variation in Na^+ response according to precipitation chemistry for the Valat des Cloutasses catchment based on Dupraz *et*

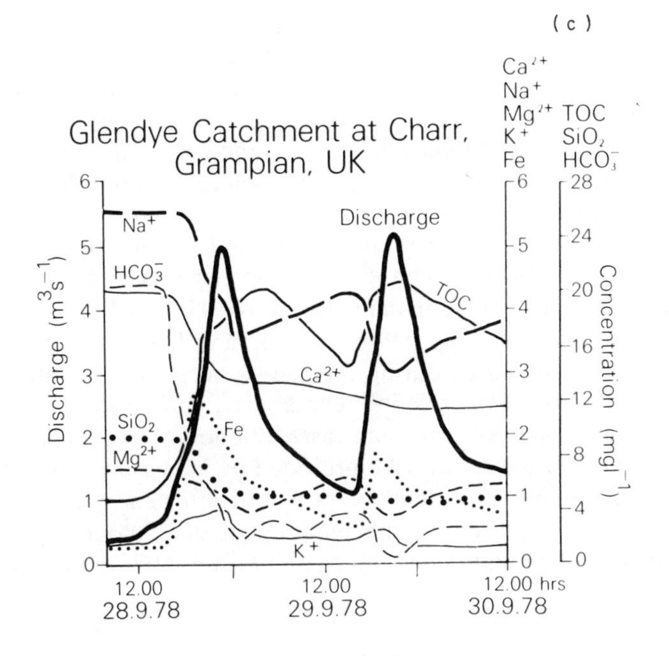

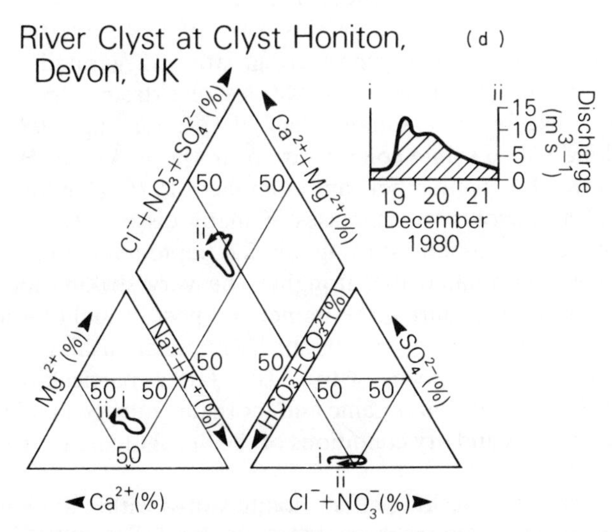

al. (1982); (c) presents storm-period responses of individual solute parameters recorded by Reid *et al*. (1981a) in the Glendye catchment; (d) shows the evolution of water composition during a storm event in the River Clyst hydrograph (Figure 7.14(a)(i)) will cause an hysteretic effect. However, a similar phenomenon can also be produced when the timing of solute and discharge behaviour is broadly in phase, but the form of the chemograph is not symmetrical with the hydrograph (Figure 7.14(a)(ii)). In practice, chemographs are far from simple phenomena, and hysteresis will often reflect both the detailed form and the timing of the solute response during a storm event (Figure 7.14(a)(iii)).

Several factors may influence the form and timing of storm-period chemographs and in turn, generate hysteretic behaviour. In particular, a 'flushing effect' has been noted in many rivers (e.g. Hendrickson and Krieger, 1960; Edwards, 1973; Walling, 1974; Walling and Foster, 1975) by which soluble material accumulated during the pre-storm period is mobilized and transported to the stream where it influences solute concentrations during the early stages of a storm event. In some cases, dilution in solute concentration associated with a storm even may be preceded by increasing concentrations. Solute accumulation before a flood event will be affected by evaporation of soil moisture, capillary rise and the build-up of dry fallout deposits, leaf residues and dead plant material (Walling and Foster, 1978). Flushing effects are particularly pronounced in autumn storms following solute accumulation over the summer period, and Klein (1981) has demonstrated that solute concentrations in overland and subsurface flows are systematically related to the length of the dry period preceding a storm.

Long dry spells will promote this form of hysteretic behaviour (Geary, 1981), and flushing phenomena were greatly exaggerated at the end of the 1976 drought in Britain when conditions of extreme dryness were rapidly changed by exceptional rainfall (Slack, 1977; Walling, 1980b). Records of total and individual solute concentrations in the Jackmoor Brook catchment, Devon, UK, for September and October 1976 (Figure 7.14(b)) reveal substantial and prolonged increases in many dissolved constituents when a period of heavy rainfall, starting on 22 September and continuing until mid-October, terminated the drought. The very striking increases in some ions, especially NO_3^-, during the immediate post-drought period are thought to have been generated not only by simple physical flushing processes but also by changes in soil biochemistry. The latter particularly involved the interrelationship between organic and inorganic nitrogen and were caused by high temperatures and dry conditions of the drought and the subsequent rapid wetting of the soil.

In contrast to the flushing effect, solute stores within a drainage basin may become exhausted during a sequence of closely spaced flood peaks, as is evident from the increasingly muted responses and progressively lower background concentrations of K^+ ions (Figure 7.14(c)) recorded through several storm events in the River Dart catchment during a seven-day period in

1975 (Walling, 1978a). In some circumstances, depletion of solute supplies may be so pronounced that a concentration response in the first storm of a series is replaced by a dilution effect in later events. Progressive exhaustion of solute supplies may also cause a systematic shift in the position of the hysteretic loop as reported by Cornish (1982) from an investigation of the German's Creek catchment in New South Wales, Australia (Figure 7.14(d)). Whether exhaustion or flushing phenomena occur will largely be influenced by the condition of a catchment at the onset of a storm, and a strong relationship has, for example, been demonstrated between the character of the chemograph and antecedent soil moisture status in certain Devon catchments (Walling and Foster, 1975).

The form and the timing of solute responses may also be appreciably affected by contrasts in the chemical concentration of surface and subsurface flow components and by the timing of these contributions from the hillslope system (Burt, 1979; Anderson and Burt, 1982; Chapter 6 in this volume). Water flowing by different routes through the soil at varying rates in separate phases of a storm event will have differential access to soluble material distributed unevenly within the soil profile, and Spraggs (1976), for example, refers to contributions from fissure, micropore and clay flow systems in order to explain multiple flushes of Ca^{2+} concentrations recorded during storms in the West Walk catchment of Hampshire, UK. It has also been argued by Johnson and East (1982) that, given specified antecedent, rainfall and recession conditions, hysteretic loops will assume one of four idealized forms (Figure 7.14(e)) and will proceed in a clockwise or anticlockwise direction *depending* on the relative contributions and solute concentrations of groundwater, interflow and surface runoff.

Further complexity is introduced to the understanding of storm-period solute response when the behaviour of river systems rather than small catchments is considered, and hysteretic effects are commonly found to be much more irregular in nature for larger and geologically less uniform drainage basins (Hendrickson and Krieger, 1960). Aggregation and transmission of solute responses from contrasting tributary areas will strongly influence the form and timing of chemographs at downstream sites in a river network. Studies undertaken by the authors in the Exe Basin (Walling and Webb, 1980; Webb and Walling, 1982a) have shown that completely different responses may be generated at the mainstream site of Thorverton depending on the spatial origins of the storm runoff (Figure 7.15(a)). Furthermore, routing of flows through the channel network in this basin may also produce a kinematic differential between floodwave and floodwater velocities (Glover and Johnson, 1974) which causes the chemograph trough to lag progressively behind the flood peak in a downstream direction (Figure 7.15(b)). In contrast, the effect of floodplain storage over a 13 km

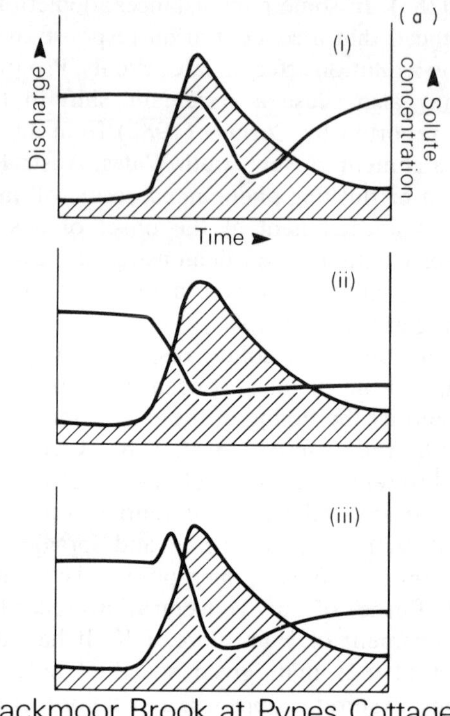
(a)
(i)
Discharge
Solute concentration
Time
(ii)
(iii)

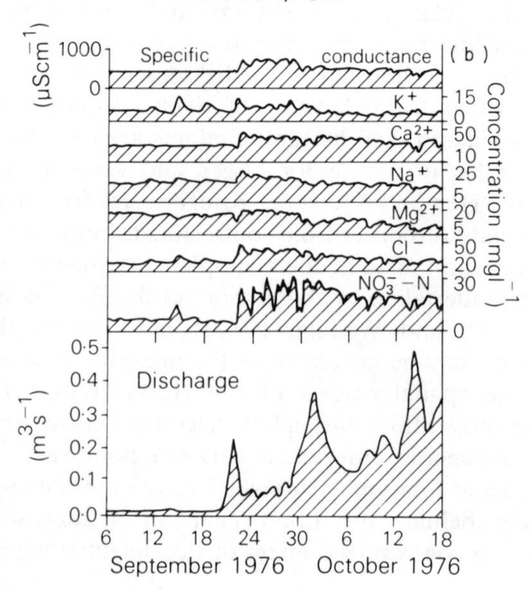
Jackmoor Brook at Pynes Cottage, Devon, UK
(b)
Specific conductance
(μScm⁻¹)
1000
0
K⁺
Ca²⁺
Na⁺
Mg²⁺
Cl⁻
NO₃⁻ −N
Concentration (mgl⁻¹)
15
0
50
10
25
5
20
5
50
20
30
0
Discharge
(m³s⁻¹)
0·5
0·4
0·3
0·2
0·1
0·0
6
12
18
24
30
6
12
18
September 1976
October 1976

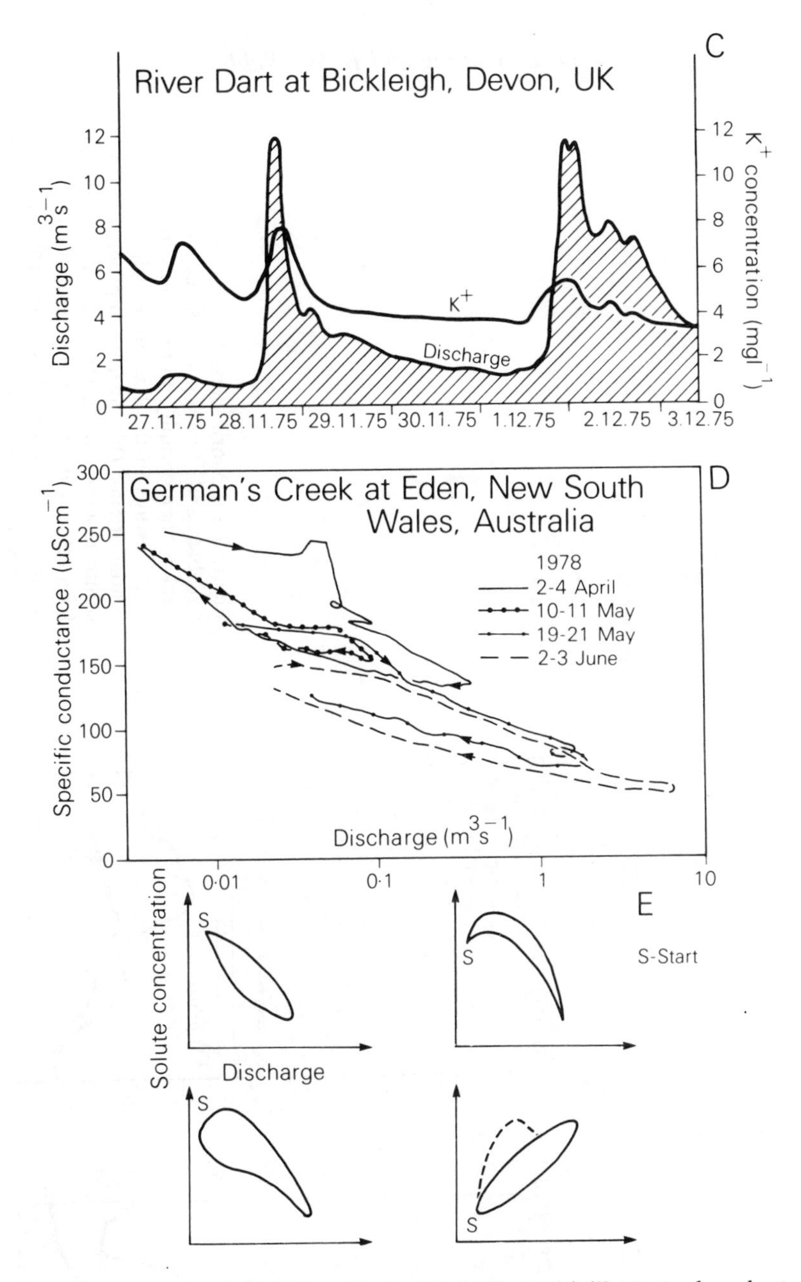

Figure 7.14. Hysteretic storm-period solute behaviour. (a) illustrates how hysteretic behaviour may be generated by the relative timing and relative form of solute discharge responses; (b) depicts solute behaviour in the Jackmoor Brook during the post-drought period of 1976; (c) depicts exhaustion of K^+ concentrations during a sequence of storm events in the River Dart; (d) shows the displacement of the hysteretic relationship between specific conductance and discharge recorded by Cornish (1982) in German's Creek; (e) presents the idealized forms of the hysteretic loop proposed by Johnson and East (1982)

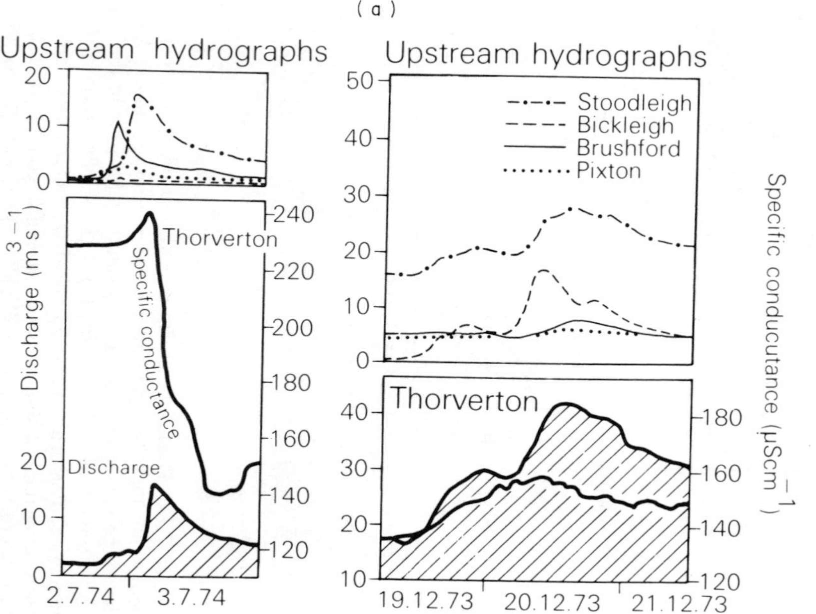

Figure 7.15. Storm-period solute responses in sizeable and heterogeneous catchments. (a) depicts chemograph and related hydrographs for storm events in the Exe Basin with contrasting spatial origins; (b) illustrates a progressive lag of solute response for a storm event in the River Exe; (c) reveals the development of a lead effect in the solute response during a storm event in the River Culm which was influenced by floodplain storage

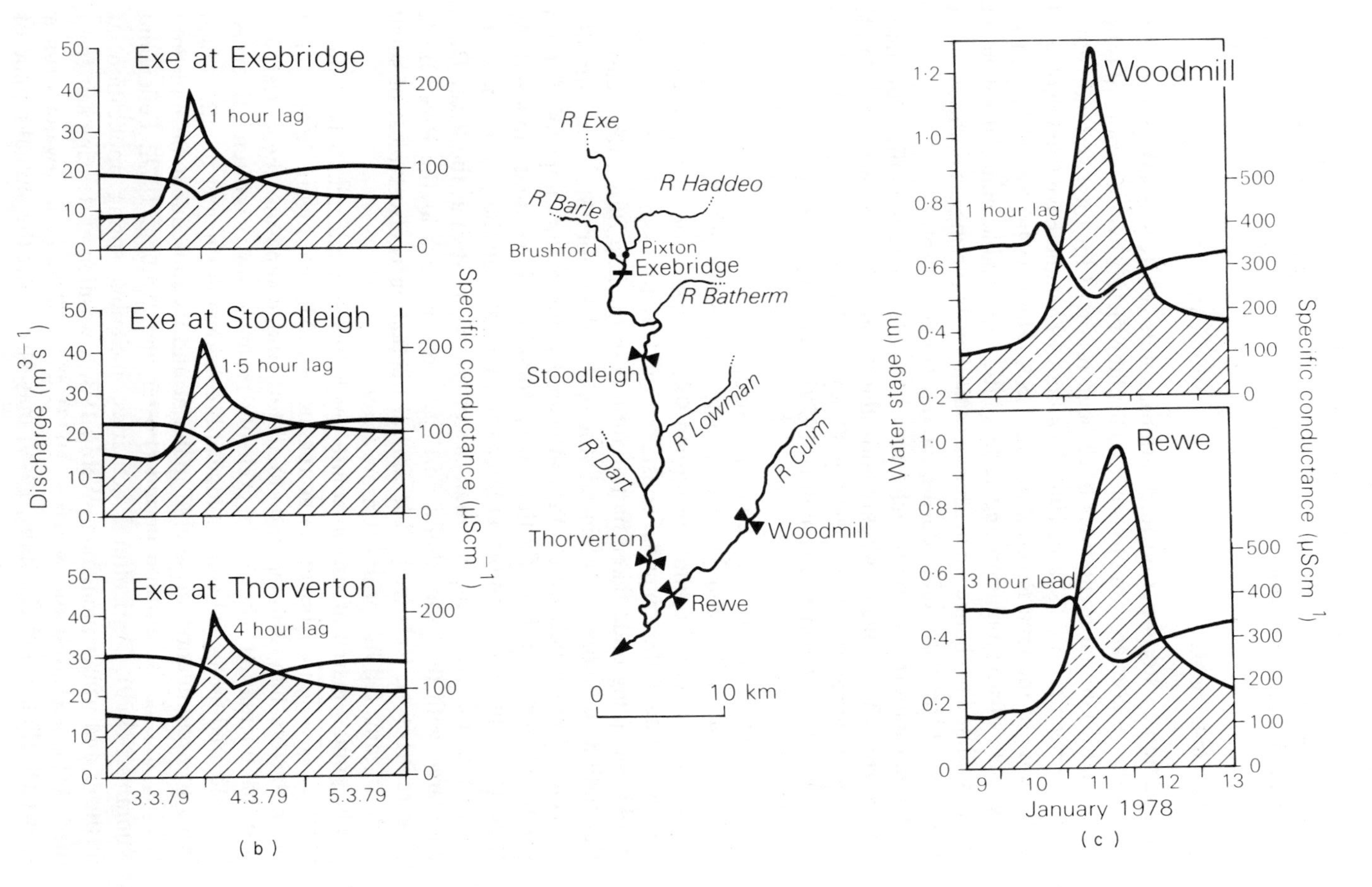
Exe at Exebridge
1 hour lag
Exe at Stoodleigh
1·5 hour lag
Exe at Thorverton
4 hour lag
Discharge (m3s−1)
Specific conductance (μScm−1)
3.3.79
4.3.79
5.3.79
(b)
R Exe
R Barle
R Haddeo
Brushford
Pixton
Exebridge
R Batherm
Stoodleigh
R Lowman
R Dart
R Culm
Thorverton
Woodmill
Rewe
0
10 km
Woodmill
1 hour lag
Rewe
3 hour lead
Water stage (m)
Specific conductance (μScm−1)
January 1978
(c)

reach of the River Culm in Devon, by delaying the hydrograph but not the solute response, may generate a 'lead' rather than a 'lag' effect in the downstream direction (Figure 7.15(c)).

7.3.3 Time trends

In addition to storm-period fluctuations, solute levels may also respond to variations in discharge and to other controlling factors over a variety of other time scales. Diurnal oscillations have been reported for total and individual solute concentrations (Sharp, 1969; Walling, 1975) and have been explained in the case of the Gila River, Arizona, USA (Hem, 1948) by reference to a daily cycle of accumulation and redissolution of soluble material which, in turn, is related to changes in discharge caused by increasing evaporation and a lowering of the water table during the day (Figure 7.16(a)).

Annual patterns of solute behaviour have also been identified for many rivers from the results of weekly and other sampling frequencies, by simple plotting of concentration values (e.g. Casey and Newton, 1972; Sutcliffe and Carrick, 1973a; Feller and Kimmins, 1979; Houston and Brooker, 1981) and by the application of more sophisticated statistical techniques including factor and principal components analysis (Imeson, 1973; Reid *et al*., 1981a; Williams *et al*., 1983) and time series analysis (Thornes and Clark, 1976). The nature of the annual regime for stream solute levels varies between stations and according to the ion under consideration. It is, for example, evident from weekly sampling undertaken by the authors in the Exe Basin, that seasonal fluctuation is more pronounced for Ca^{2+} than for Mg^{2+} or for K^+ concentrations, which are strongly affected by the random impacts of storm events (Figure 7.16(b)). Furthermore, total solute content, indexed by specific conductance, exhibits greater variation through the year in the more highly mineralized southern tributaries of this basin, such as the River Dart, compared with the more dilute northern streams, such as the River Barle (Figure 7.16(c)). Relatively little annual fluctuation is also typical of small chalk streams fed by springs (Casey, 1969).

The annual march of stream solute levels in some catchments is a simple inverse reflection of the discharge regime (e.g. Mg^{2+} in Figure 7.16(b)), but other factors may also operate to distort and complicate the pattern of seasonal fluctuation. The annual cycle of stream chemistry is often influenced by acclerated biological and geological breakdown in summer months which leads to accumulation of soluble material and its subsequent flushing into streams during the autumn period (Foster and Walling, 1978; Feller and Kimmins, 1979). Agricultural activities, including spring application of fertilizer and winter ploughing (Webb, 1980), washing of de-icing salt from road surfaces (Casey and Newton, 1973) and release of meltwater from a snowpack (Feller and Kimmins, 1979) may also affect the annual regime of

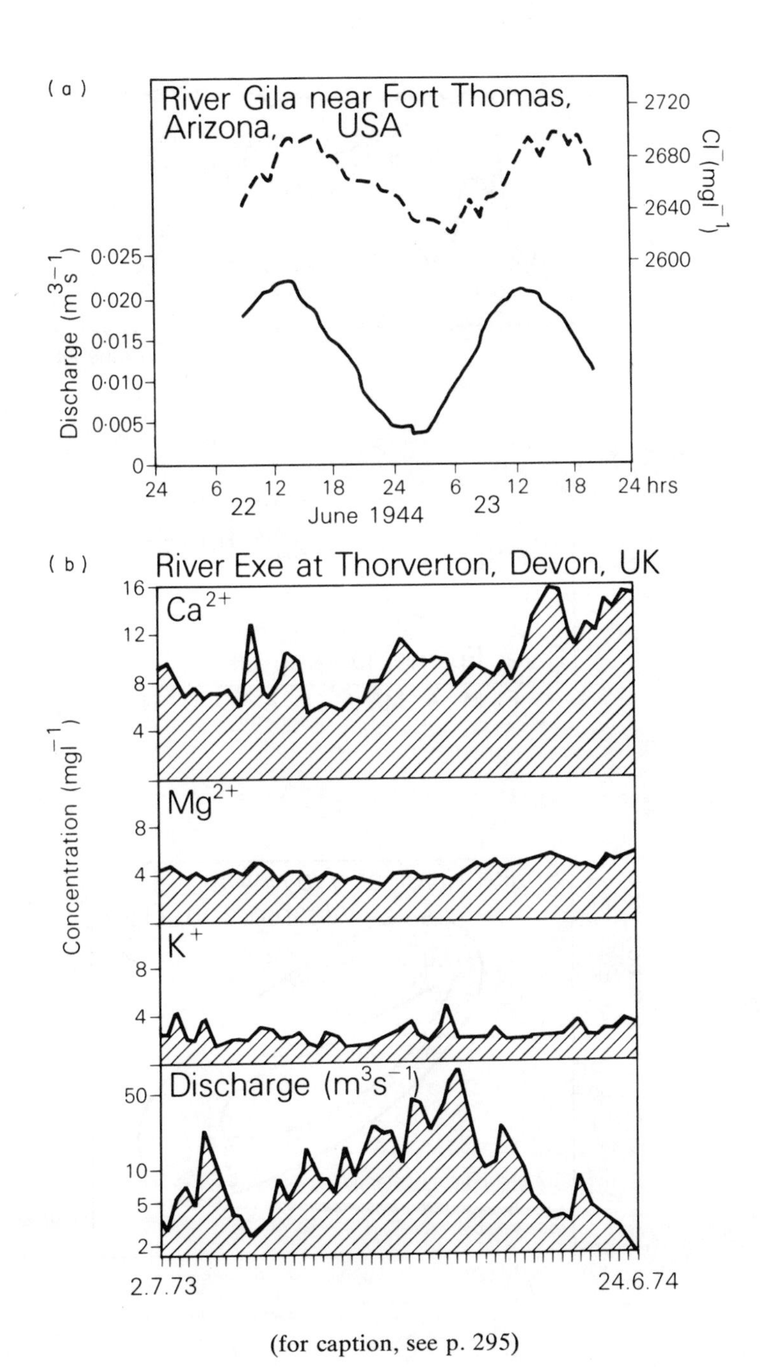

(for caption, see p. 295)

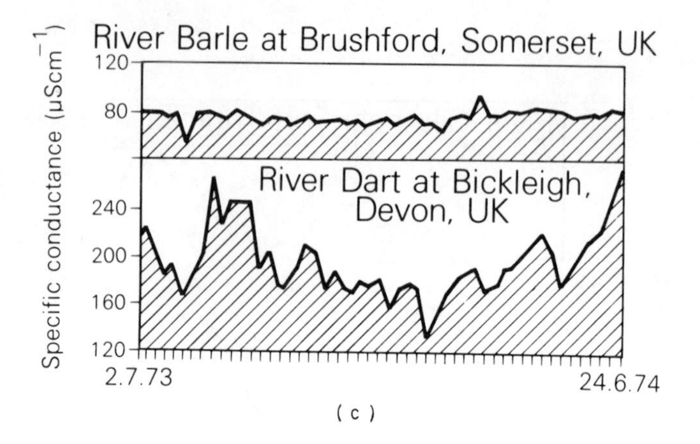
River Barle at Brushford, Somerset, UK
River Dart at Bickleigh, Devon, UK
Specific conductance (µScm^{-1})
120
80
240
200
160
120
2.7.73
24.6.74

(c)

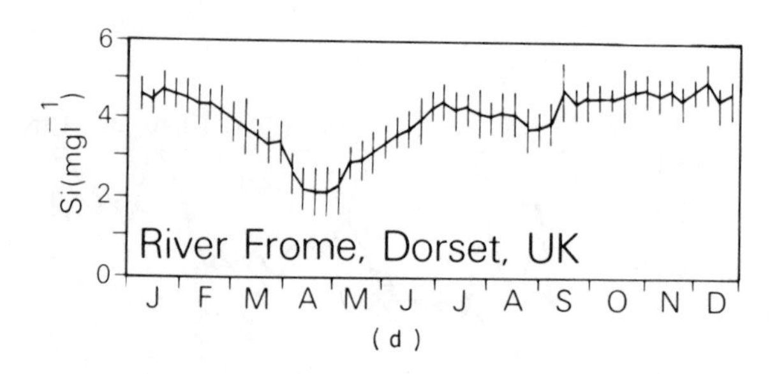
Si(mgl^{-1})
6
4
2
0
River Frome, Dorset, UK
J F M A M J J A S O N D

(d)

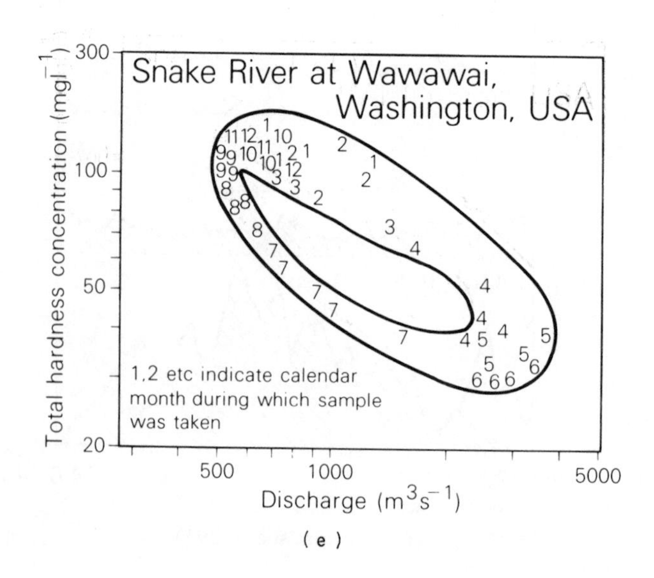
Snake River at Wawawai, Washington, USA
Total hardness concentration (mgl^{-1})
300
100
50
20
500
1000
5000
Discharge (m^{3}s^{-1})
1,2 etc indicate calendar month during which sample was taken

(e)

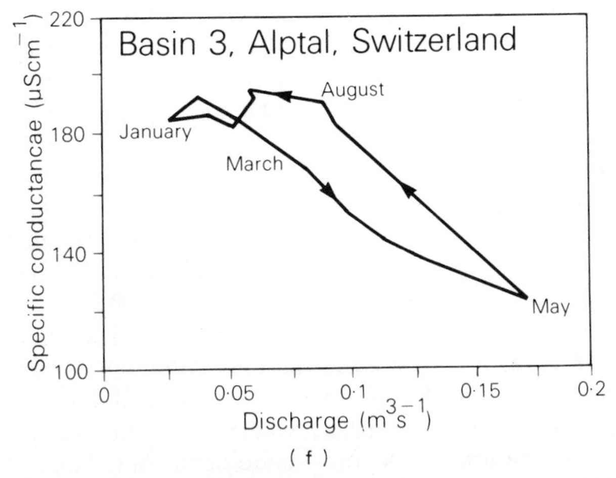

Figure 7.16. Diurnal and seasonal changes in solute levels. (a) illustrates the daily cycle in Cl^- concentration and discharge recorded for the River Gila by Hem (1948); (b) and (c) reveal contrasts in the annual regime of solute levels between individual parameters and sampling stations in the Exe Basin; (d) presents seasonal variation in silicon concentration in the River Frome derived from the means of weekly estimates for the period 1965–77, data are from Casey *et al*. (1981) and 95% confidence intervals are shown as vertical bars; (e) reproduces the 'elliptical doughnut' diagram employed by Gunnerson (1967) to portray the annual hysteresis in total hardness concentration recorded for the Snake River in the period October 1961 to September 1962; (f) shows a 'Q-c-t' diagram used by Davis and Keller (1983) to plot specific conductance as a function of flow in an Alptal basin

certain solute parameters. In some drainage basins, especially those of upland and moorland areas, seasonal variations in stream solute concentrations may be strongly related to annual variations in the chemistry of incoming precipitation (e.g. Sutcliffe and Carrick, 1973b; Cryer, 1980; Williams *et al*., 1983), whereas in lowland and more densely populated areas the annual regime of stream chemistry may reflect the varying dilution of sewage inputs (Davis and Slack, 1964).

The annual cycle of biological activity may also exert a strong influence on seasonal solute fluctuations in streams through the effects of vegetation die-back and leaf fall (Slack and Feltz, 1968) and the impact of seasonal uptake by plants and animals (Edwards, 1973; Casey and Ladle, 1976). It has been suggested that uptake by diatoms is the most important factor influencing the silicon content of streams (Edwards, 1974), and Casey *et al*.

(1981) have estimated a maximum Si uptake rate of 1–3 g m^{-2} day^{-1} for chalk streams in Dorset and have shown from a study extending over 13 years (Figure 7.18(d)) that diatom growth produces a substantial depletion of silicon in spring (March–May) and a secondary reduction in late summer (August–September).

Construction of 'elliptical doughnut' (Figure 7.16(e)) and 'Q-c-t' (Figure 7.16(f)) diagrams has been employed to highlight the general progression of solute concentrations and discharge through the year. An annual hysteresis in the concentration–discharge relationship may characterize total (Figure 7.12(b)) and individual solute content and may be clockwise or anticlockwise in direction (Figure 7.16). Clear annual cycles have been found for several tributaries in the Columbia River Basin, USA (Gunnerson, 1967) and for a range of small and large catchments in Switzerland (Davis and Keller, 1983), where cyclic variation is considered to be the dominant feature of temporal solute behaviour. The latter study also suggests that the nature of annual hysteresis differs between the nutrient and geochemical components of solute load and is likely to reflect in detail the origins of stream runoff.

Solute levels may also exhibit significant changes in the longer term, but evaluation of these time trends is restricted by the limited availability of records of sufficient quality and length. In particular, few studies have investigated long-term trends in the water chemistry of undistributed ecosystems, although work in the Hubbard Brook Experimental Forest, New Hampshire, USA (Likens *et al.*, 1977) provides a notable exception. Concentrations of most ions studied in the undisturbed forested watersheds of this experiment showed no significant variation from year to year. However, a trend of increasing annual weighted nitrate concentration was evident in the period 1964–74 (Figure 7.17(a)) and is attributed to the occurrence of widespread frosts in the forest soils, which are considered to have increased nitrification and mobilization of nitrate to drainage waters over the ten-year period.

More work has been undertaken to document the long-term changes in solute levels of streams draining areas disturbed by agricultural activity, and in Great Britain evidence has been assembled to suggest that nitrate levels and the concentration of other ions have been increasing in certain streams over the last 25 to 30, and perhaps 50, years (e.g. Tomlinson, 1970; Casey, 1975). White (1983) has presented data to show the upward trend of nitrate concentration in the River Thames and River Lee during the period 1928–80 (Figure 7.17(b)), and Casey and Clarke (1979) quantify the annual increase in NO_3–N concentrations in the River Frome, Dorset, at 0.11 mg l^{-1} over the 11 years from 1965 to 1975. A doubling of sulphate concentrations between 1931 and 1963 has been noted for the Rivers Chelmer and Blackwater in Eastern England (Davis and Slack, 1964), whereas Edwards and Thornes

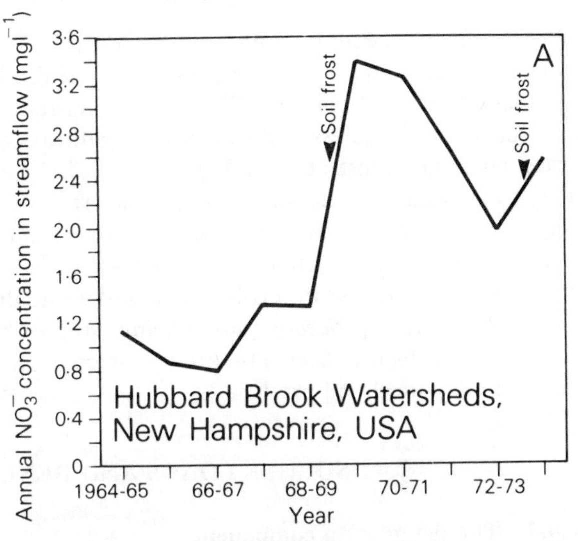

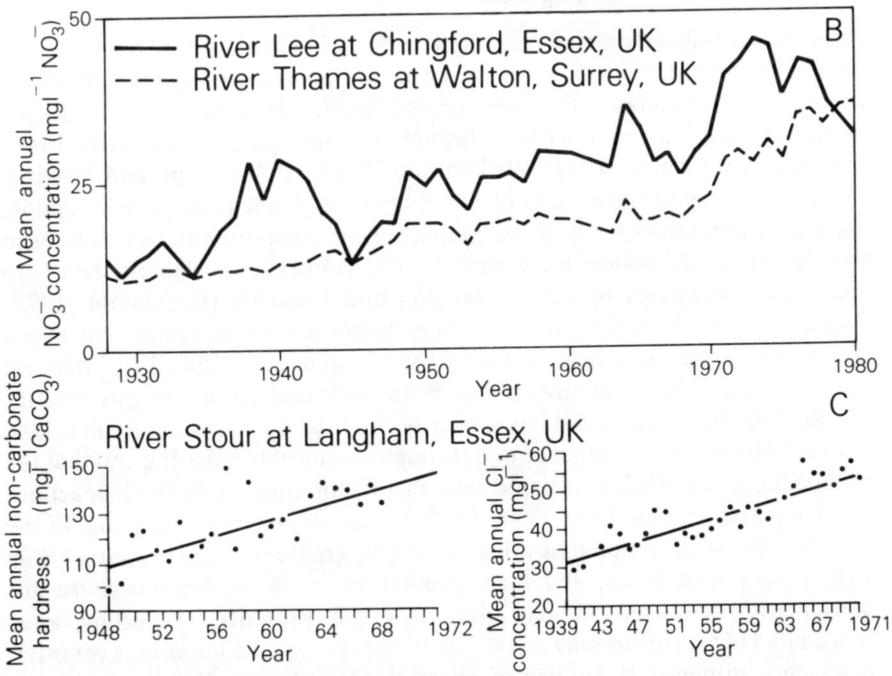

Figure 7.17. Longer-term trends in solute levels. (a) depicts the trend in weighted annual concentration of NO_3^- in streamwater for undisturbed watersheds at Hubbard Brook recorded by Likens *et al*. (1977); (b) shows changes in the mean annual NO_3^- concentration of two British rivers over a 52-year period reported by White (1983); (c) illustrates the long-term trends in mean annual non-carbonate hardness and Cl^- concentration for the River Stour based on Edwards (1975)

(1973) have demonstrated a significant increase in conductivity, nitrate-nitrogen and ammoniacal nitrogen for the River Stour in Essex over the period 1951 to 1970. The clear long-term trends in mean annual non-carbonate hardness and chloride concentration (Edwards, 1975) for the latter river are illustrated in Figure 7.17(c). Less clear are the causes of increasing solute concentrations in the rivers of southern and eastern England, since it is likely that these long-term trends in water chemistry are a complex response to changes in cultivation practices, improvements in drainage and increases in fertilizer use and waste discharges. Further work is required on the processes influencing solute behaviour in agricultural ecosystems (Edwards, 1975) before long-term trends in solute concentrations can be understood and predicted.

7.4 SOLUTE LOADS AND DENUDATION

7.4.1 The denudation component

Data concerning solute concentrations and loads of rivers may be employed to provide information on rates of chemical denudation (e.g. Miller, 1961; Gibbs, 1967; Douglas, 1972; Tricker and Scott, 1980), but this data source cannot be used in an uncritical fashion for this purpose (Meade, 1969; Goudie, 1970; Janda, 1971; Walling and Webb, 1978; Webb and Walling, 1980). Due regard must be paid to problems of inadequate field sampling, inaccurate laboratory analysis and inappropriate computational procedures in the derivation of solute loads and to the conversion of these data into meaningful estimates of surface erosion and lowering (Beckinsale, 1972; Waylen, 1979). However, it is also very important to recognize that by no means all of the dissolved load of a river system is produced by chemical denudation of rock and soil. It has been indicated earlier in this chapter (Table 7.4) that, on a global basis, only 60% of the total dissolved solids load delivered to the oceans is generated through chemical weathering and that the remainder is supplied as oceanic salts in precipitation or through reactions involving atmospheric CO_2. Janda (1971) suggests that the magnitude of the non-denudational component will generally be greater for igneous rather than sedimentary rock types, and that interactions involving the biosphere are particularly significant in accounting for the presence of certain ions, especially HCO_3^-, in streamwaters. The biotic factor is an important yet often neglected influence in studies of chemical denudation. If the role of the biosphere in generating non-denudational components is ignored, chemical denudation rates may be overestimated; but there are also certain environments where the rate of chemical denudation may be underestimated if the influence of vegetation is overlooked. The latter situation applies particularly to tropical environments with vigorously growing vegetation which may act as long-term sink for nutrients. A study of chemical budgets, undertaken by Bruijnzeel (1983) in a small drainage basin of central Java which supports fast-growing plantation forest, indicates that the true rate of chemical weathering may be underestimated by 75%, 33%, 17%, 70% and 12% for Ca^{2+}, Mg^{2+}, Na^+, K^+ and SiO_2 respectively if stream solute loads are not corrected for the effects of long-term vegetation uptake.

Several drainage basin studies have investigated the fundamental weathering reactions involved in a release of solutes to streamwaters (e.g. Cleaves *et al*., 1970; Verstraten, 1979; Waylen, 1979; Reid *et al*., 1981b; Afifi and Bricker, 1983), and the fate of particular ions in the nutrient cycles of ecosystems (e.g. Likens *et al*., 1977; Foster *et al*., 1983; Chapter 3 in this volume). Studies of this type have greatly assisted the apportioning of solute loads into denudational and non-denudational components, but unfortunately this approach is only feasible for very small catchments with relatively simple geology and vegetation in a steady-state condition. It would be impractical for a geomorphologist interested in the pattern of chemical denudation over sizeable areas to set up a series of experimental watersheds which are instrumented to establish weathering budgets, and in these circumstances, recourse must be made to extrapolation of solute concentration and load data collected from somewhat larger and more heterogeneous catchment areas. These data, however, must be carefully manipulated to account as far as possible for the denudational and non-denudational origins of dissolved constituents, if a meaningful estimate of chemical denudation is to be achieved.

A map of chemical denudation rates in mainland Britain, which has been produced by the authors, is presented in Figure 7.18 as an example of the way in which solute concentration and load data may be manipulated and extrapolated. The map is based on information which has been collated from the Regional Water Authorities of England, Scotland and Wales and relates to nearly 1600 sites located on unpolluted streams of small and moderate size. Values for total dissolved solids load and the loadings of certain ions at individual sites have been assembled by combining information on solute concentrations collected over the period 1977–79 with an estimate of mean annual discharge (Walling and Webb, 1981). Calculation of chemical denudation is based on a national grid of cells, each 100 km^2 in area, and the computational procedure first subtracts the non-denudational component from the solute load and then converts the remainder into terms of volumetric removal, or its equivalent surface lowering, by reference to rock density. Because the data collected on individual ion constituents by the statutory authorities are generally limited, the calculation of the non-denudational solute load has been relatively unsophisticated in this exercise. However, allowance has been made for Cl^- and Na^+ contributed in precipitation, the NO_3^- loading has been equated with biotic sources and non-point source

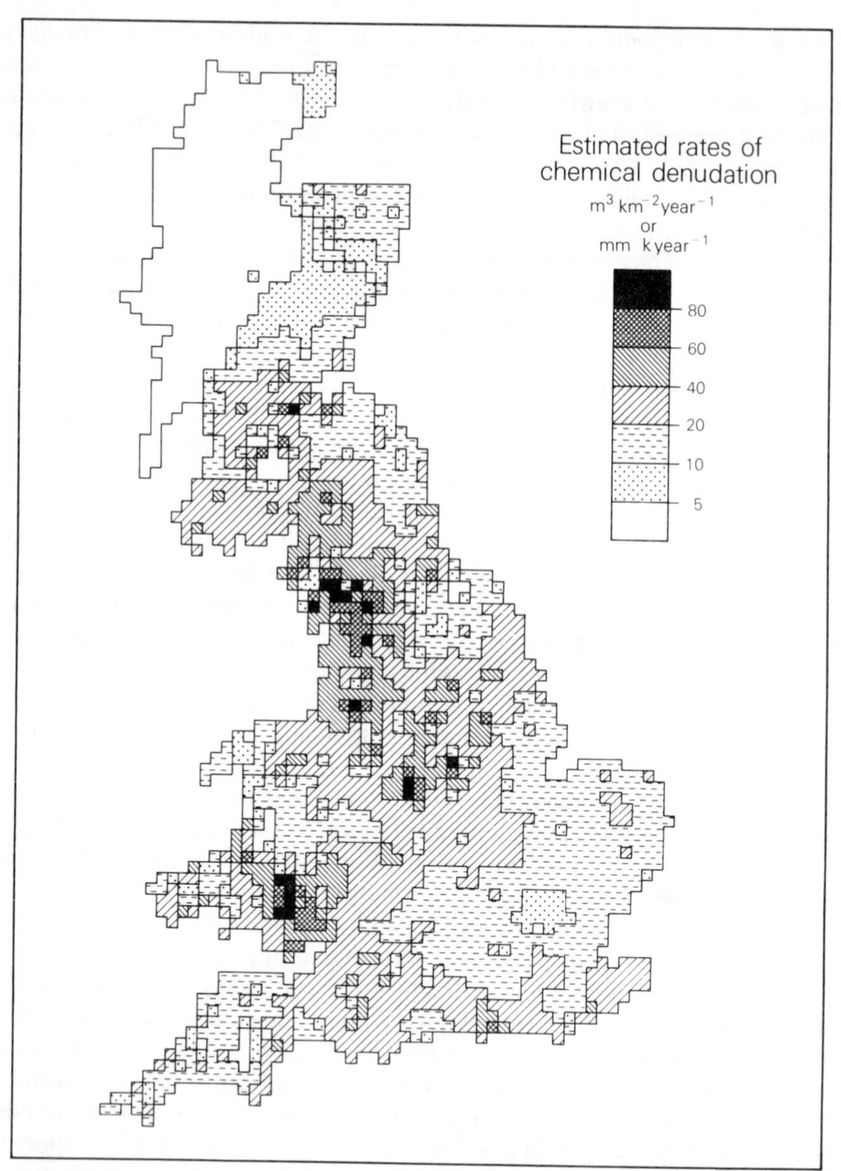

Figure 7.18. Estimated rates of chemical denudation in mainland Britain

pollution from agriculture, and information on alkalinity levels has been manipulated to account for the HCO_3^-, CO_3^{2-} and OH^- ions derived from interaction with the biosphere and atmosphere.

Figure 7.18 must therefore be viewed as a preliminary and tentative estimate of current chemical denudation rates in Great Britain, but the map also highlights two important features. First, the magnitude of chemical denudation is relatively low, and more than 50% of the country experiences an annual rate less than 20 m^3km^{-2}. A second feature of the national pattern of chemical denudation is its broad similarity with the countrywide variation in total dissolved solids load (Figure 7.6), and it may therefore be implied that variations in the non-denudational component of solute load do not have an overriding influence on the pattern of chemical denudation in Great Britain. In detail, the rate of denudation by chemical processes will reflect a complex interaction between solute supply, the volume of runoff and the magnitude of non-denudational contributions, so that simple coincidences between the countrywide patterns of denudation and major geological or topographic regions should not be expected and in fact do not arise.

7.4.2 Dissolved and particulate loads compared

Estimates of the total transport of dissolved and particulate material from the land surface of the globe to the oceans have clearly demonstrated the dominance of particulate transport. Thus if the estimate of the total annual suspended sediment load of 13.5×10^9t produced by Milliman and Meade (1983) is compared with an estimate for total dissolved load of 3.871×10^9t based on the work of Meybeck (1979), a ratio of 3.5:1 is evident. More recent work by Meybeck (1983) suggests that his earlier figure may underestimate the true value by about 3% but this would not produce a significant change in this ratio. The equivalent estimates for the individual continents are listed in Table 7.7. These indicate that this ratio reaches a maximum value of 10.45 for Oceania, which mainly reflects the high suspended loads transported by rivers draining such countries as New Zealand, New Guinea and Taiwan, and that dissolved loads only exceed suspended sediment loads in Europe. A

Table 7.7. A comparison of the dissolved and suspended sediment loads transported to the oceans from individual continents (Meybeck, 1979; Milliman and Meade, 1983)

Continent	Total transport to the oceans ($t \times 10^6 yr^{-1}$)		Ratio of suspended/ dissolved
	Dissolved	Suspended sediment	
Africa	201	530	2.64
North America	758	1462	1.93
South America	603	1788	2.97
Asia	1592	6433[a]	4.04
Europe	425	230	0.54
Oceania	293	3062	10.45

[a]Including value for Eurasian Arctic.

Table 7.8. Relative magnitude of the dissolved and suspended sediment loads transported by some major rivers[a]

River	Area ($km^2 \times 10^6$)	Load ($t \times 10^6 yr^{-1}$) Dissolved	Suspended	Ratio of sediment/dissolved
Amazon	6.15	290	900	3.10
Zaire	3.82	47	43	0.91
Mississippi	3.27	131	210	1.60
Porana	2.83	56	92	1.64
Yenesei	2.58	73	13	0.18
Lena	2.50	85	12	0.14
Ob	2.50	50	16	0.32
Amur	1.85	20	52	2.6
Mackenzie	1.81	70	100	1.47
Ganges/Brahmaputra	1.48	151	1670	11.0
Niger	1.21	10	40	4.0
Zambezi	1.20	15.4	48[b]	3.12
Orange	1.02	12	17	1.42
Danube	0.81	60	67	1.12
Orinoco	0.99	50	210	4.20
Indus	0.97	68	100	1.47

[a]Based largely on Meybeck (1976) and Milliman and Meade (1983).
[b]Includes estimate of sediment deposited in upstream reservoirs.

comparison of the suspended sediment and total solute loads of several major world rivers is provided by Table 7.8. This again demonstrates considerable variation in the relative magnitude of the two load components, with the Ganges/Brahmaputra exhibiting the maximum suspended sediment/dissolved load ratio of 11.0 and the Lena (USSR) a minimum ratio of 0.14.

Considering data derived from individual rivers, some controversy exists concerning the trend of the general relationship between the magnitude of suspended sediment and dissolved loads. Judson and Ritter (1964) used data representing the major drainage regions of the USA to suggest that an inverse relationship existed between mean annual dissolved and particulate load. Other workers such as Alekin and Brazhnikova (1962), Strakhov (1967), Gregory and Walling (1973) and Meybeck (1976, 1977) have, however, pointed to a positive relationship. Arguments to support these conflicting views have focused on the contrasting response of suspended sediment and dissolved loads to spatial variations in runoff in the case of Judson and Ritter (1964), and on the contention that increased mechanical erosion will in general by coupled with increased chemical erosion advanced by other workers.

Figure 7.19(a) illustrates the relationship between mean annual dissolved load and suspended sediment load for 302 rivers for which data on both load

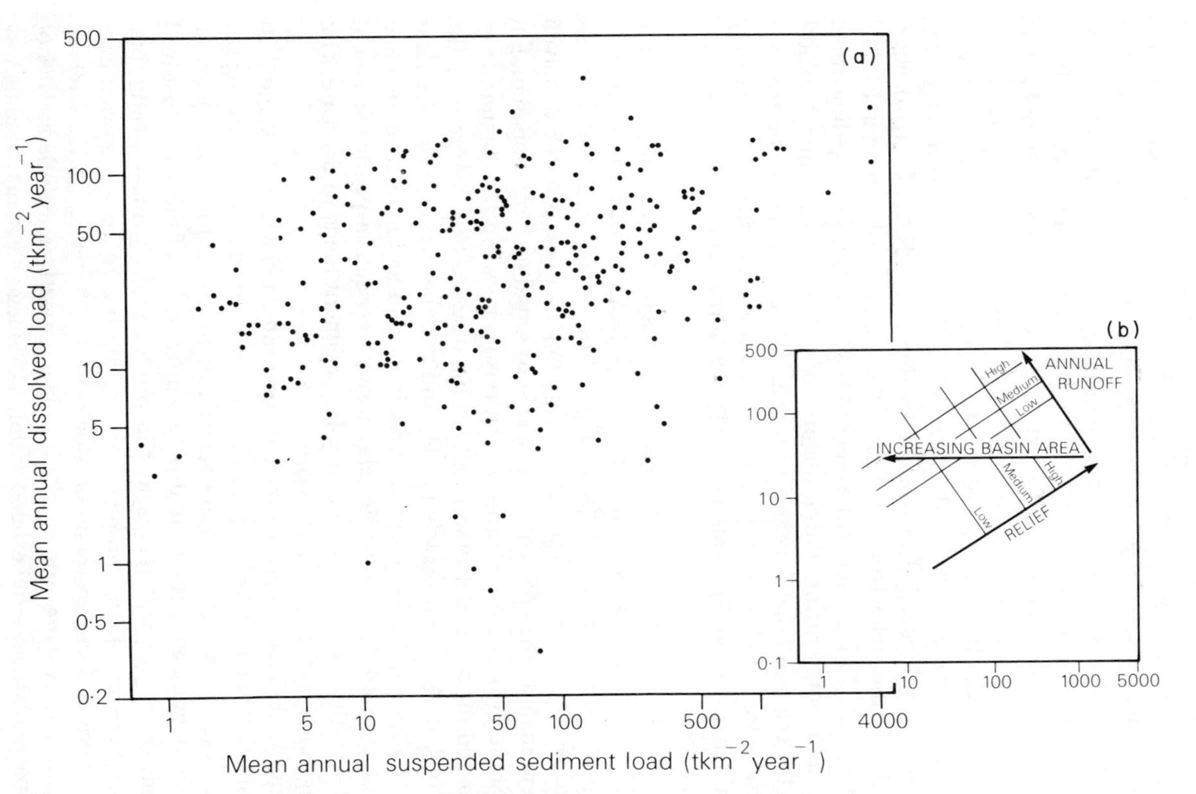

Figure 7.19. (a) Relationship between mean annual dissolved load and suspended sediment load for a sample of 302 world rivers. (b) Attempts to explain some of the scatter apparent in this relationship in terms of the effects of variations in annual runoff and catchment relief and of the influence of catchment area

components have been collected by the authors. No clear trend is apparent, although some evidence of a positive relationship exists. The absence of a clear relationship in Figure 7.19(a) can be readily accounted for in terms of the different sets of factors controlling the magnitude of the two load components and by contrasting response to individual controls. In the latter context, Walling and Webb (1983) suggest that several trends can be distinguished within the scatter apparent in this plot (Figure 7.19(b)). If the influence of basin relief is isolated, it may be suggested that both load components will increase as relief increases. Removing this effect, it is possible to consider the influence of annual runoff, and in this case it may be proposed that, whereas dissolved loads generally increase with increasing annual runoff (cf. Figure 7.3), suspended sediment loads frequently decline. By considering the yields from large drainage regions, rather than individual drainage basins, Judson and Ritter (1964) were effectively holding relief constant and emphasizing the influence of variations in annual runoff between the regions. Finally it can be suggested that increasing basin area will frequently cause a decline in suspended sediment load, owing to a decreasing sediment delivery ratio (cf. Walling, 1983), while the dissolved load remains effectively constant.

Figure 7.20 presents the relationship between the magnitude of the suspended sediment/dissolved load ratio and mean annual runoff for the rivers represented in Figure 7.19(a). Values of the ratio range from in excess of 100 to less than 0.05, and the particulate component exceeds the dissolved component in more than 60% of the cases. Substantial scatter is apparent, but there is some evidence of an inverse relationship between the magnitude of this ratio and mean annual runoff at the global scale. This conforms to the findings of Langbein and Dawdy (1964) and may be explained by the clear positive relationship between annual dissolved load and annual runoff, and the lack of any well-defined simple relationship between suspended sediment load and runoff, or perhaps a tendency for sediment loads to decrease with increasing runoff (cf. Figure 7.19(b)).

It is important to recognize that the data presented in Figures 7.19 and 7.20 and in Tables 7.7 and 7.8 do not provide a meaningful basis for evaluating the relative efficacy of mechanical and chemical denudation. In the first place it has already been demonstrated that only a proportion of the total dissolved solids load of a river will represent the products of chemical weathering. Secondly, there are many problems in attempting to relate downstream sediment yields to local upstream erosion rates (cf. Trimble, 1981; Walling, 1983). Available evidence suggests that in many locations as little as 10% of the total particulate material eroded within a drainage basin may find its way to the outlet. The values of the suspended-sediment: dissolved-load ratio for global, continental and individual rivers introduced earlier may therefore need to be increased by an order of magnitude in order to produce a

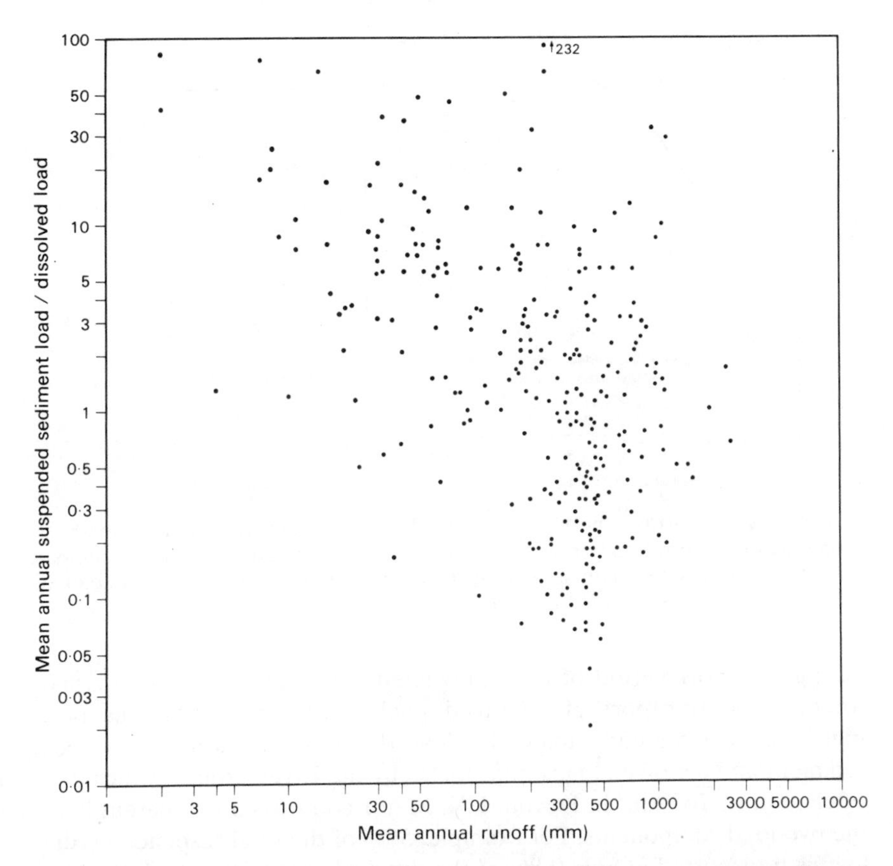

Figure 7.20. Relationship between the magnitude of the suspended sediment/dissolved load ratio and mean annual runoff for a sample of 302 world rivers

meaningful estimate of the relative importance of mechanical and chemical erosion.

Several significant contrasts also exist between the temporal behaviour of dissolved and suspended sediment loads. Figure 7.9 demonstrates that total solute concentrations exhibit an inverse relationship with discharge in most rivers and that the range of concentrations involved is generally considerably less than an order of magnitude. The relationship between suspended sediment concentration and discharge is, however, in nearly all cases positive and concentrations may range over several orders of magnitude. As a result, there are very important differences in the temporal distribution of the transport associated with the two load components (cf. Webb and Walling, 1982b). The majority of the suspended sediment load of a river will be moved

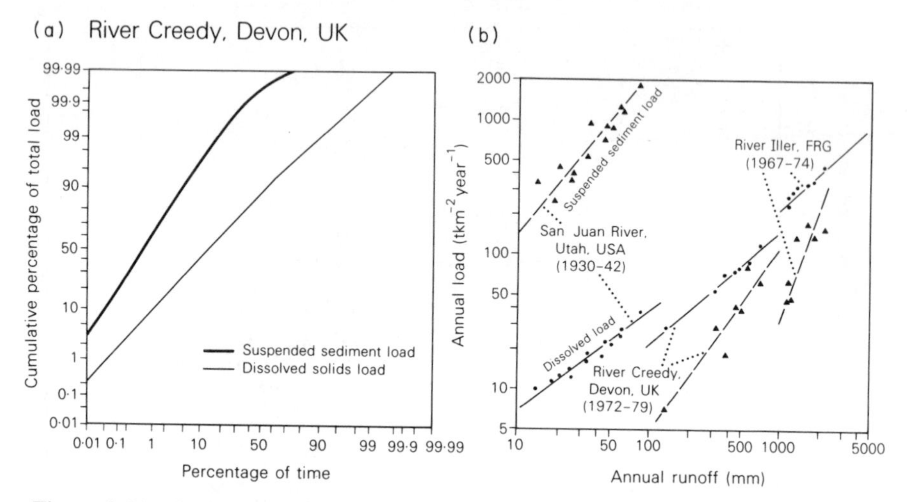

Figure 7.21. Contrasts between dissolved and suspended sediment loads. (a) compares load duration curves for the River Creedy in Devon, UK. (b) compares the relationships between annual suspended sediment and dissolved load and annual runoff for three rivers in different climatic zones

during the short period of time associated with high magnitude discharges, whereas the transport of dissolved load is a more continuous process. Figure 7.21(a) presents cumulative load duration curves for the suspended sediment and dissolved loads transported by the River Creedy, which drains a catchment of 262 km^2 in Devon, UK. Major contrasts are apparent between the two load components. For example, 50% of the total suspended sediment load is transported in only 0.8% of the time, or approximately three days per year, whereas a period of 44 days per year, or 12% of the time is required to move 50% of the dissolved load. Equivalent duration figures for 90% of the suspended and dissolved loads are 22 days per year or 6% of the time, and 204 days per year or 56% of the time respectively.

Contrasts are also apparent in the degree of variability of annual dissolved and suspended sediment loads and in the relationship between load and annual runoff. Figure 7.21(b) compares the relationships between annual suspended sediment and dissolved load and annual runoff for three rivers in different climatic zones. In the Rivers Creedy and Iller, dissolved loads exceed particulate loads, but the particulate component is dominant in the San Juan River. In all three rivers, the exponent of the relationship between sediment load and annual runoff is greater than 1.0, whereas values below 1.0 are characteristic of the dissolved load relationship. Dissolved loads therefore exhibit less variability than suspended sediment loads on an annual basis, and are less responsive to variations in annual runoff and to extreme events owing to the inverse relationship between total solute concentration and discharge.

7.5 CHALLENGES

There are many themes relating to solute transport by rivers which merit further study and development, but two in particular will be highlighted in this chapter. The first involves the need to develop a more rigorous approach to the design of sampling strategies and to the selection of load calculation procedures. The second relates to the potential for modelling the solute behaviour of rivers.

7.5.1 Sampling strategies and load calculation

Most texts dealing with methods for studying solute transport by streams (e.g. Hem, 1970) place emphasis on laboratory techniques for analysing individual water samples. Little is said about sampling frequency requirements or the procedure to be employed to estimate the load for a given period from a series of infrequent instantaneous samples. These considerations are of minimal importance when continuous monitoring apparatus or automatic sampling equipment are available, particularly when the latter incorporates a discharge-integrating facility (e.g. Claridge, 1970; Nelson, 1970). However, manual sampling programmes provide the basis for many studies of solute transport and it is important to recognize the potential problems associated with sampling frequency and load calculation procedures (e.g. Walling, 1975, 1978b; Ongley *et al*., 1977; Smith and Stewart, 1977; Dupraz *et al.*, 1982).

The discussion on temporal variations in solute transport presented previously has clearly demonstrated that solute concentrations may evidence considerable variation through time. Infrequent samples are unlikely to provide a meaningful representation of the pattern of variation, and estimates of such simple statistics as the annual mean concentration based on a small number of samples may involve significant errors. Walling and Webb (1982) have used a detailed 5-year record of solute transport by the River Dart, a 46 km^2 drainage basin in Devon, UK, to estimate the levels of precision associated with estimates of the annual mean concentration of Mg^{2+} and NO_3^- based on specified sampling frequencies. The results presented in Figure 7.22(a) suggest that these estimates may frequently be only within ±10% or ±20% of the true mean for Mg^{2+} and NO_3^- respectively. Similarly they indicate that, in order to obtain an estimate within ±10% at the 95% level of confidence, sampling frequencies of 7 days and 2 days would be required for Mg^{2+} and NO_3^-.

Considering methods which have commonly been used to estimate annual solute loads from infrequent grab samples, it is possible to distinguish the two basic approaches of interpolation and extrapolation. Interpolation procedures (e.g. Table 7.9) effectively involve the assumption that the concentration or load associated with an instantaneous sample is representative of the time

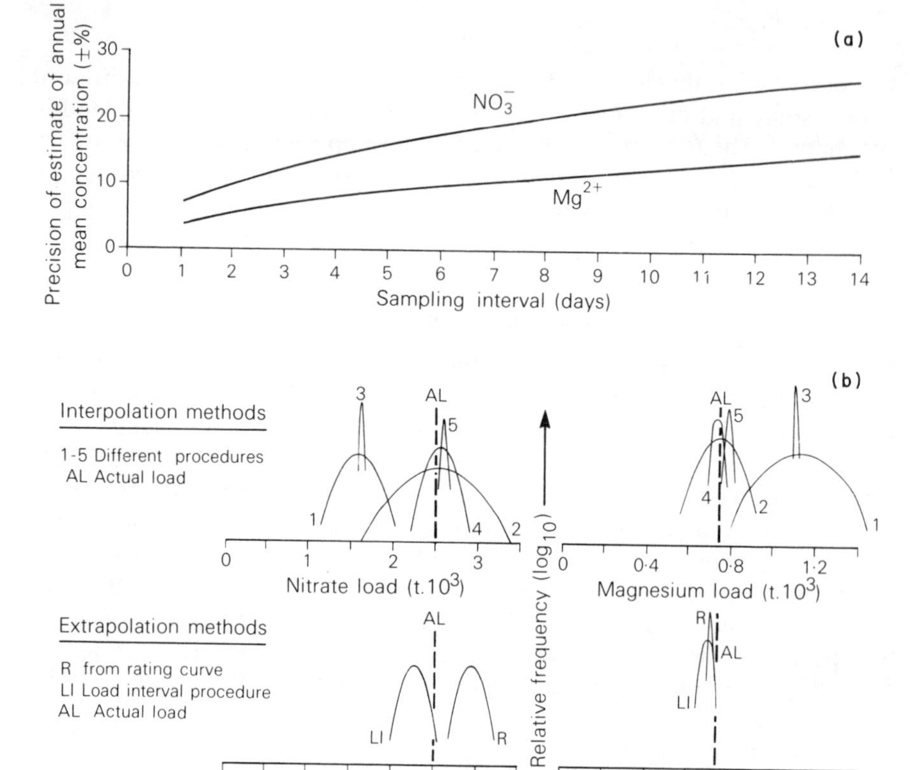

Figure 7.22. Assessing sampling frequency requirements and the accuracy and precision of load calculation procedures, using data collected from the River Dart, Devon, UK. (a) depicts the relationship between the precision of estimates of the annual mean concentration of Mg^{2+} and NO_3^- and sampling frequency at the 95% level of confidence. (b) provides a diagrammatic representation of the accuracy and precision of estimates of nutrient load for the 5-year study period obtained using regular weekly sampling and various load calculation procedures

period between samples, whereas rating curve techniques and the load interval procedure described by Verhoff *et al*. (1980) may be viewed as extrapolation techniques. Walling and Webb (1982) were able to use the same data set from the River Dart to estimate the accuracy of these load calculating procedures by generating representative data sets for a specified sampling frequency and comparing the resultant load estimates with loads calculated using the detailed records of solute concentration. Precision was also assessed by using replicate data sets for a given sampling strategy or

Table 7.9. Load interpolation procedures

Method	Numerical procedure
1	$\text{Total load} = K\left(\sum_{i=1}^{n}\frac{C_i}{n}\right)\left(\sum_{i=1}^{n}\frac{Q_i}{n}\right)$
2	$\text{Total load} = K\sum_{i=1}^{n}\left(\frac{C_iQ_i}{n}\right)$
3	$\text{Total load} = K\bar{Q}_r\left(\sum_{i=1}^{n}\frac{C_i}{n}\right)$
4	$\text{Total load} = \frac{K\sum_{i=1}^{n}(C_iQ_i)}{\sum_{i=1}^{n}Q_i}\bar{Q}_r$
5	$\text{Total load} = K\sum_{i=1}^{n}(C_i\bar{Q}_{pi})$

K = conversion factor to take account of period of record
C_i = instantaneous concentration associated with individual samples (mg l^{-1})
Q_i = instantaneous discharge at time of sampling ($m^3\ s^{-1}$)
$\bar{Q}_r$ = mean discharge for period or record ($m^3\ s^{-1}$)
$\bar{Q}_{pi}$ = mean discharge for interval between samples ($m^3\ s^{-1}$)
n = number of samples

frequency. High variability of the load estimates reflects a low degree of precision.

Figure 7.22(b) demonstrates the considerable variation in the accuracy and precision of various estimates of Mg^{2+} and NO_3^- load for the 5-year study based on regular weekly sampling. The individual interpolation procedures are detailed in Table 7.9, and the extrapolation procedures employed involve a simple logarithmic rating between concentration and flow and the load interval method. The variability of the estimates produced by the replicate data sets has been portrayed in an idealized form by plotting the normal distributions characterizing the appropriate values of mean and standard deviation. The ordinate scale has been logarithmically transformed, and the distributions, which have been truncated at two standard deviations, afford an indication of the 95% confidence limits of the replicate load estimates. In the case of the interpolation procedures, method 5, which assumes that the sampled concentration is representative of the sampling interval and calculates the total load as the sum of the products of sampled concentration and mean discharge for individual intervals, produces the most accurate and precise estimates of Mg^{2+} and NO_3^- load. Estimates of Mg^{2+} load obtained using rating curves and the load interval technique are notable for their high degree of precision, although they underestimate the actual load by about 5%. The equivalent estimates of NO_3^- load exhibit appreciably lower levels of accuracy and precision than those obtained using interpolation method 5.

The design of effective sampling programmes for load estimation must therefore involve detailed consideration of the most appropriate load calculation procedures. Extension of work of the type reported here to rivers in other environments could generate useful guidelines for the sampling frequencies required in order to attain a specified degree of accuracy and precision in estimating various characteristics of solute transport. Furthermore, it could also provide useful assessments of the potential reliability of existing load estimates which are frequently viewed as being absolute values rather than estimates with an associated error distribution.

7.5.2 Modelling strategies

Since the discussion of solute behaviour in rivers presented in the earlier part of this chapter focused on spatial and temporal variation, it is appropriate to adopt a similar structure in considering modelling strategies. The importance of climate, parent material, topography and biotic factors in controlling spatial variations in solute concentrations and dissolved loads has long been recognized by Gorham (1961) and subsequent writers (e.g. Douglas, 1972; Walling and Webb, 1975) and has been further emphasized in this chapter. However, there have been surprisingly few successful attempts to develop models capable of predicting such spatial variations over sizeable areas. To some extent this situation is a response to the difficulties involved in quantifying important physiographic factors such as geology, and the complex interactions between catchment characteristics and the processes governing solute transport in rivers depicted in Figure 7.1. Equally, however, it may reflect the lack of a pressing need for these models, since the presence of solutes in riverwater pose relatively few environmental problems. Some progress is, however, apparent in the development of multivariate statistical models relating solute concentrations and loads to a range of climatic and physiographic variables.

This approach was successfully employed in Australia by Douglas (1973), who found that contrasts in the solute loads from a number of small catchments in Queensland and New South Wales were significantly related to four variables representing a precipitation seasonality and volume index, mean annual precipitation, mean annual runoff and the percentage of catchment occupied by rain forest. A similar approach was used by Steele and

Jennings (1972) in Texas, where significant relationships were established between a number of solute parameters and variables indexing catchment area, mean annual precipitation, average number of thunderstorm days in the year, mean annual evapotranspiration and annual frequency values of streamflow at the sampling point. A further refinement was also introduced into this study by using the residuals from general trends to define regions with their own individual relationships. Other examples include the work of Reinson (1976) in the Genoa River basin in south-east Australia which demonstrated a significant correlation between the total dissolved solids concentration encountered in ten tributary basins and a geological index (percentage of catchment occupied by outcrops of quartz-diorite- granodiorite), and of Hill (1978, 1980) in relating nitrate-nitrogen and cation concentrations and loads measured in sub-watersheds of the 262 km^2 Duffin Creek drainage basin in Ontario, Canada, to several indices of land-use and a number of soil and topographic variables.

Webb (1983b) also reports a detailed stepwise multiple regression analysis of the relationship between several measures of baseflow solute concentration obtained for samples collected from 260 small catchments within the 601 km^2 basin of the Upper Exe in Devon, UK (cf. Figures 7.7 and 7.8) and a variety of physiographic variables. Final regression equations accounted for between 60% and80% of the spatial variation in individual solute levels. However, the summary of the results presented in Table 7.10 clearly demonstrates that both the composition of the sets of independent variables and the rank order of the variables within these sets differed considerably for the individual solute concentrations. Any attempt to develop multivariate statistical relationships between solute parameters and controlling factors must therefore expect the variables included in these relationships to vary according to the ion or constituent involved.

Although they are of value, the empirical and area-specific nature of the multiple regression equations described above limit their general applicability. Furthermore, problems of multicollinearity may hamper attempts to decipher the precise influence of individual controls. A more physically based approach to developing relationships for predicting variations in solute transport, which takes account of the sources of the dissolved material and their governing mechanisms, is required. In addition potential exists for use of rapid reconnaissance surveys to collect single water samples from a number of locations across an area. These data could be used to 'tune' prediction techniques which possess rather coarse resolution.

The use of solute concentration or dissolved load rating curves in conjunction with a continuous streamflow record provides a simple approach to modelling temporal variations in solute loads and concentrations that has been widely employed. Steele (1973), for example, advocates the use of this strategy to simulate specific conductance records which may in turn be used to

Table 7.10. Factors influencing spatial variations in baseflow solute concentrations from 260 small catchments within the Upper Exe Basin, Devon, UK (Webb, 1983b)

Significant independent controls[b]	Solute species					
	TDS[a]	Ca^{2+}	Mg^{2+}	Na^{+}	K^{+}	Cl^{-}
1	A−	C−	G−	G−	A−	G−
2	C−	B+	E+	D+	C−	C−
3	B+	E+	C−	C−	E+	D+
4	E+	G−	B+	F+	G−	B+
5	G−		F+	E+		F+
6	F+		A−	A−		E+
7	D+			B−		
Explained variance (%)	77.9	61.0	76.0	78.1	72.5	70.0

[a]Total dissolved solids indexed by specific conductance.
[b]Variables ranked according to order of entry in stepwise procedure.

Variables

A Percentage of catchment underlain by Devonian rock types.
B Percentage of catchment underlain by Pilton, Lower Carboniferous and Permian rock types.
C Percentage of catchment underlain by moorland and woodland.
D Significant settlement in catchment (dummy variable).
E Catchment area.
F Net catchment slope.
G Mean annual run-off percentage in catchment.
\+ Variable having direct relationship with solute parameter.
− Variable having inverse relationship with solute parameter.

predict the total solute concentration and individual ion concentrations. Inherent in this approach, however, are all the limitations of simple concentration/discharge relationships in describing the solute response of a river, many of which have already been discussed. In a survey of 88 river sampling stations located throughout the USA, Steele (1970) found that in more than 50% of the cases, the simple relationship between total dissolved solids concentration and discharge accounted for less than 49% of the variance of the sampled concentrations. Improvements have been introduced by using more complex rating relationships. For example, Lane (1975) demonstrated that the use of a simple rating relationship with periodic slope and intercept terms improved the degree of explanation of daily specific conductance records from five streams in the western USA from 54–83% to 81–87% of the variance. Attempts have also been made to include other variables besides discharge in a multivariate prediction equation. These include measures of precipitation, water temperature and catchment moisture status (e.g. Pionke *et al.*, 1972; Foster, 1978b; Fehér, 1983).

Simple mass-balance mixing models involving separation of the discharge record into a series of runoff components, each having a characteristic solute concentration, provide a further development of this approach (e.g. O'Connor, 1976). This is the method that has been used quite successfully by Birtles (1977) and by Oborne (1981) to model the behaviour of the concentrations of several solute components in the Rivers Severn and Wye in the UK. A flow model was employed to separate the discharge record at the measuring site into a storm runoff component, one or more baseflow components and effluent return flow, and the concentrations associated with these components were assumed either to be constant or to vary according to flow rate or time of year. Figure 7.23(a) illustrates the results obtained by Birtles (1977) in applying the flow separation procedure to the River Severn and the resultant simulation of Ca^{2+} concentrations over a two-year period. Model efficiencies in excess of 80% explained variance were reported for all the solute components considered, with the exception of total nitrogen. Another model involving a simple mixing concept is that developed by Johnson *et al*. (1969) for application to the Hubbard Brook catchment in New Hampshire, USA. This was based on the mixing of rainwater with soilwater in a catchment store, rather than mixing of flow components, but the volumes involved were similarly estimated from the discharge record.

The use of streamflow data, either in a lumped form or subdivided into its constituent components, can provide a worthwhile basis for simulating solute transport in many rivers, but its scope is necessarily limited. The runoff processes implicit in the streamflow record cannot provide an entirely effective surrogate for the processes involved in solute generation. Furthermore, the approach is essentially empirical and 'black box' and there is little scope for using models of this type to predict changes in solute responses associated with a changing catchment condition. Process-orientated models which take account of the methods of solute production illustrated in Figure 7.1 are required for this purpose and to provide a general improvement in model effectiveness.

Some attempts to develop models of this latter type have involved the 'pick-a-backing' of solute generation subroutines onto existing runoff simulation models. As an early example, Nakamura (1971) describes how specific conductance simulation was incorporated into a 'Tank' runoff model (Sugawara, 1961). There are also several examples of models developed in South Africa to simulate the salinity of river inputs to reservoirs (e.g. Hall and Görgens, 1978, 1979; Herold, 1980, 1981) in which solute pickup subroutines have been added to the Pitman (1976) runoff model. Under these local conditions, emphasis is placed on surface 'washoff' of stored salts and associated inputs via soil moisture to interflow and groundwater percolation. Simulation of total dissolved solids and Cl^- concentrations has been undertaken on both a monthly and daily time basis with encouraging results.

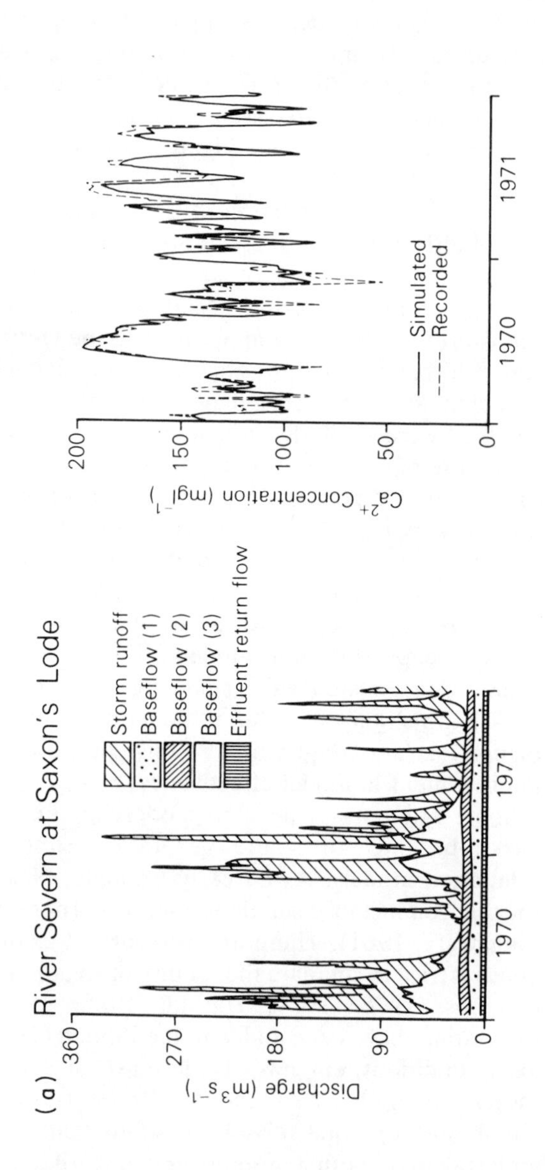
(a)
River Severn at Saxon's Lode
Discharge ($m^3 s^{-1}$)
360
270
180
90
0
1970
1971
Storm runoff
Baseflow (1)
Baseflow (2)
Baseflow (3)
Effluent return flow
Ca^{2+} Concentration ($mg l^{-1}$)
200
150
100
50
0
Simulated
Recorded

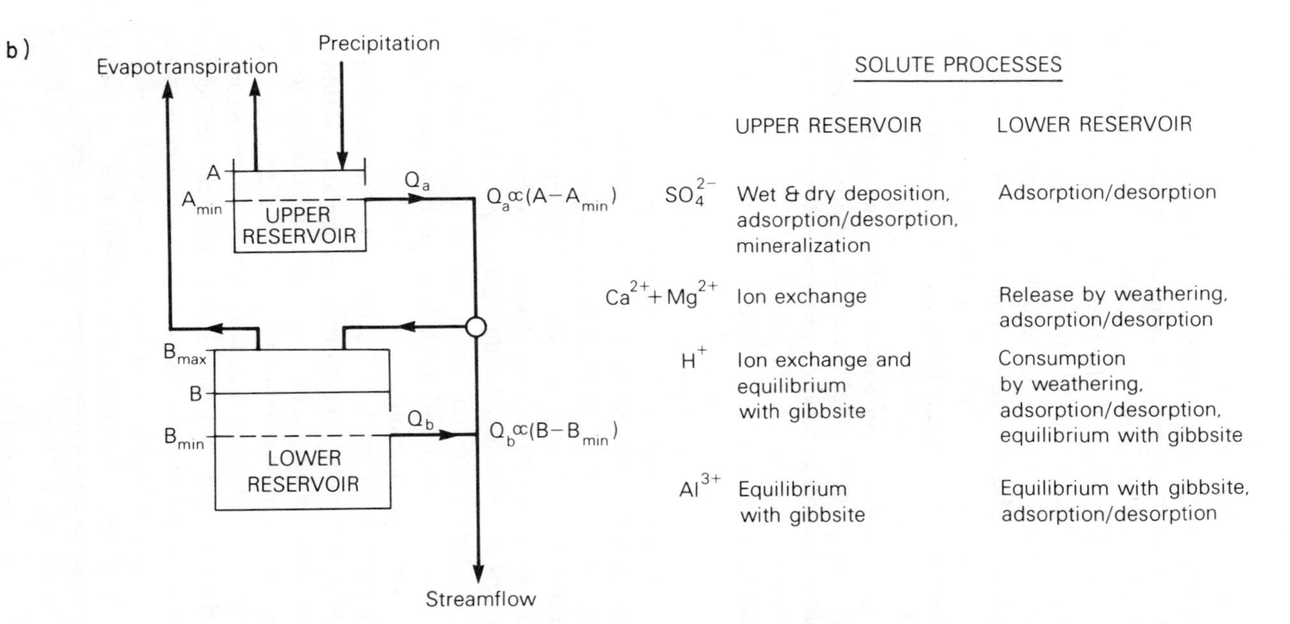

Figure 7.23. Some modelling strategies. (a) illustrates some of the results obtained by Birtles (1977) in applying a flow separation model and an associated solute mixing model to the River Severn. (b) depicts the basic structure of the runoff simulation model employed by Christophersen *et al*. (1982) and a summary of the processes included in the model for simulating major ion chemistry that was coupled with this

A further example of the addition of solute subroutines to a runoff model which attempts to represent many of the processes depicted in Figure 7.1 is provided by the work of Christophersen *et al.* (1982) and Grip (1982) in developing the simple two-reservoir hydrologic model proposed by Lundquist (1976) to simulate solute transport by small streams in Norway and Sweden. In this environment, considerable emphasis is placed on atmospheric contributions and their movement through the basin system. Sulphate is the dominant anion, and the sulphate submodel provides the key to simulating cation concentrations and release of Ca^{2+}, Mg^{2+} and Al^{3+} by weathering, since the sum of these three cations plus H^+ must balance the anion sulphate in each reservoir at a given time. Figure 7.23(b) illustrates the structure of the runoff model and the processes governing major ion chemistry represented in the model.

With models involving the addition of solute subroutines to an existing runoff simulation model, it is important that the latter component provides a meaningful representation of runoff generation processes. Recent years have seen considerable advances in the development of physically based distributed models of runoff production (e.g. Beven *et al.*, 1980), and scope clearly exists for extending these to include solute generation. The very substantial increase in computational complexity involved in moving from a lumped runoff model of the type employed in the previous examples, to a distributed model will necessarily introduce problems, but this step must be seen as highly desirable. The coupling of such models with an improved understanding and representation of the processes of solute generation could produce significant advances in the modelling of solute transport by rivers. An alternative and perhaps less mathematically demanding strategy could involve the extension of the distribution function approach, applied by Moore and Clarke (1981, 1983) to rainfall–runoff modelling and to sediment yield modelling, to simulating solute production.

When models are applied to very large catchments, attention must inevitably turn from detailed representation of processes operating within the basin to the routing of contributions from individual tributaries to the catchment outlet. In this case, a mass balance may be applied to individual reaches moving in a downstream direction (e.g. Dixon *et al.*, 1970; Dixon and Hendricks, 1970; Farrimond and Nelson, 1980), but the effects of sediment/solute interactions and other channel processes influencing solute behaviour should also be considered. Simulation of detailed temporal variations in solute transport at a downstream point will also necessitate attention to flow routing mechanisms and to dispersion processes.

7.6 CONCLUSIONS

This chapter has attempted to provide a general review of current knowledge concerning solute transport by rivers and its spatial and temporal variation. It has stressed that any attempt to understand these patterns must take account of basin-wide processes and the various sources and pathways involved in the movement of solutes through the drainage basin system (Figure 7.1). As such the response of the river can be seen to reflect the spatial and temporal integration of the various processes of solute mobilization and movement discussed in Chapters 2–6. As our comprehension of these processes expands and improves, so our understanding of the solute behaviour of rivers should develop. Inevitably, there will be a lag before many of the concepts and theories advanced in these chapters are fully exploited in terms of their implications and significance for solute transport by rivers. Nevertheless, considerable progress is evident and current knowledge is now providing the basis for the development of worthwhile models. One of the greatest needs must, however, be to more closely couple recent advances in our understanding of hillslope hydrology and runoff processes (e.g. Kirkby, 1978) with studies of solute behaviour in rivers, since much existing work on solute transport has been approached within a more traditional hydrological framework.

REFERENCES

Afifi, A. A., and Bricker, O. P. (1983). Weathering reactions, water chemistry and denudation rates in drainage basins of different bedrock types: I—sandstone and shale. In: *Dissolved Loads of Rivers and Surface Water Quantity/Quality Relationships*, International Association of Hydrological Sciences Publication No. 141, pp. 193–203.

Alekin, O. A., and Brazhnikova, L. V. (1962). The correlation between ionic transport and suspended sediment. *Dokl. Akad. Nauk. SSSR,* **146**, 203–206.

Alekin, O. A., and Brazhnikova, L. V. (1968). Dissolved matter discharge and mechanical and chemical erosion. *International Association of Hydrological Sciences Publication* No. 78, 35–41.

Al-Jabbari, M. H., Al-Ansari, N. A., and McManus, J. (1983). Variation in solute concentration within the River Almond and its effect on the estimated dissolved load. In: *Dissolved Loads of Rivers and Surface Water Quantity/Quality Relationships*, International Association of Hydrological Sciences Publication No. 141, pp. 21–29.

Anderson, M. G., and Burt, T. P. (1982). The contribution of throughflow to storm runoff: an evaluation of a chemical mixing model. *Earth Surface Processes and Landforms*, **7**, 565–574.

Andrews, E. D. (1983). Denudation of the Piceance Creek Basin, Colorado. In: *Dissolved Loads of Rivers and Surface Water Quantity/Quality Relationships*, International Association of Hydrological Sciences Publication No. 141, pp. 205–215.

Arnett, R. R. (1978). Regional disparities in the denudation rate of organic sediments. *Zeitschrift für Geomorphologie Neue Folge*, Supp. Band **29**, 169–179.

Beckinsale, R. P. (1972). The limestone bugaboo: surface lowering or denudation or amount of solution? *Transactions Cave Research Group of Great Britain*, **14** (2), 55–58.

Beven, K., Warren, R., and Zaoui, J. (1980). SHE: towards a methodology for physically-based distributed forecasting in hydrology. In: *Proceedings of the Symposium on Hydrological Forecasting*, International Association of Hydrological Sciences Publication No. 129, pp. 133–137.

Biesecker, J. E., and Leifeste, D. K. (1975). Water quality of hydrologic bench marks—an indicator of water quality in the natural environment. *US Geological Survey Circular* No. 460–E.

Birtles, A. B. (1977). River water quality models based on stream hydrograph components. Central Water Planning Unit, Reading, UK, Technical Note No. 23.

Blanc, P., and Conrad, G. (1968). Evolution géochimique des eaux de l'Oued Saoura (Sahara Nord-occidental). *Révue de Géologie Dynamique et de Géographie Physique*, **10**, 415–428.

Bond, H. W. (1979). Nutrient concentration patters in a stream draining a montane ecosystem in Utah. *Ecology*, **60** (6), 1184–1196.

Brady, J. A., and Johnson, P. (1981). Predicting times of travel, dispersion and peak concentrations of pollution incidents in streams. *Journal of Hydrology*, **53**, 135–150.

Brinson, M. M. (1976). Organic matter losses from four watersheds in the humid tropics. *Limnology and Oceanography*, **21**, 572–582.

Bruijnzeel, L. A. (1983). The chemical mass balance of a small basin in a wet monsoonal environment and the effect of fast-growing plantation forest. In: *Dissolved Loads of Rivers and Surface Water Quantity/Quality Relationships*, International Association of Hydrological Sciences Publication No. 141, pp. 229–239.

Buckney, R. T. (1977). Chemical dynamics in a Tasmanian river. *Australian Journal of Marine and Freshwater Research*, **28**, 261–268.

Burt, T. P. (1979). The relationship between throughflow generation and the solute concentration of soil and stream water. *Earth Surface Processes*, **4**, 257–266.

Butcher, R. W., Pentelow, K. F., and Woodley, J. W. A. (1939). Variations in composition of river waters. *Int. Rev. Gesamten Hydrobiol.*, **24**, 47–80.

Carbonnel, J. P., and Meybeck, M. (1975). Quality variations of the Mekong River at Phnom Penh, Cambodia, and chemical transport in the Mekong Basin. *Journal of Hydrology*, **27**, 249–265.

Carling, P. A. (1983). Particulate dynamics, dissolved and total load, in two small basins, northern Pennines, UK. *Hydrological Sciences Journal*, **28**, 355–375.

Casey, H. (1969). The chemical composition of some Southern English chalk streams and its relation to discharge. *Association of River Authorities Yearbook*, 1969, 100–113.

Casey, H. (1975). Variation in chemical composition of the River Frome, England, from 1965 to 1972. *Freshwater Biology*, **5**, 507–514.

Casey, H., and Clarke, R. T. (1979). Statistical analysis of nitrate concentrations from the River Frome (Dorset) for the period 1965–76. *Freshwater Biology*, **9**, 91–97.

Casey, H., Clarke, R. T., and Marker, A. F. H. (1981). The seasonal variation in silicon concentration in chalk-streams in relation to diatom growth. *Freshwater Biology*, **11**, 335–344.

Casey, H., and Farr, I. S. (1982). The influence of within-stream disturbance on dissolved nutrient levels during spates. *Hydrobiologia*, **91/92**, 447–462.

Casey, H., and Ladle, M. (1976). Chemistry and biology of the South Winterbourne, Dorset, England. *Freshwater Biology*, **6**, 1–12.

Casey, H., and Newton, P. V. R. (1972). The chemical composition and flow of the South Winterbourne in Dorset. *Freshwater Biology*, **2**, 229–234.

Casey, H., and Newton, P. V. R. (1973). The chemical composition and flow of the River Frome and its main tributaries. *Freshwater Biology*, **3**, 317–333.

Christophersen, N., Seip, H. M., and Wright, R. F. (1982). A model for streamwater chemistry at Birkenes, Norway. *Water Resources Research*, **18**, 977–996.

Claridge, G. G. C. (1970). Studies in element balances in a small catchment at Taita,

New Zealand. *International Association of Hydrological Sciences Publication* No. 96, 523–540.

Clarke, F. W. (1924). Data of geochemistry. US Geological Survey Bulletin No. 770.

Cleaves, E. T., Godfrey, A. E., and Bricker, O. P. (1970). Geochemical balance of a small watershed and its geomorphic implications. *Geological Society of America Bulletin*, **81**, 3015–3032.

Collins, D. N. (1979). Hydrochemistry of meltwaters draining from an alpine glacier. *Arctic and Alpine Research*, **11**, 307–324.

Colombani, J. (1983). Evolution de la concentration en matières dissoutes en Afrique. Deux exemples opposés: les fleuves de Togo et la Medjerdah en Tunisie. In: *Dissolved Loads of Rivers and Surface Water Quantity/Quality Relationships*, International Association of Hydrological Sciences Publication No. 141, pp. 51–69.

Cornish, P. M. (1982). The variations of dissolved ion concentration with discharge in some New South Wales streams. In: *The First National Symposium on Forest Hydrology, 1982, Melbourne, 11–13 May* (E. M. O'Loughlin and L. J. Bren, Eds.), pp. 67–71.

Cryer, R. (1976). The significance and variation of atmospheric nutrient inputs in a small catchment system. *Journal of Hydrology*, **29**, 121–137.

Cryer, R. (1980). The chemical quality of some pipeflow waters in upland Mid-Wales and its implications. *Cambria*, **6**, 28–46.

Cullen, P., and Rosich, R. S. (1979). Effects of rural and urban sources of phosphorus on Lake Burley Griffin. *Progress in Water Technology*, **11**, 219–230.

Davis, A. L., and Slack, J. G. (1964). The Rivers Blackwater and Chelmer—hardness, sulphate, chloride and nitrate content. *Proceedings Society of Water Treatment and Examination*, **13**, 12–19.

Davis, J. S., and Keller, H. M. (1983). Dissolved loads in streams and rivers—discharge and seasonally related variations. In: *Dissolved Loads of Rivers and Surface Water Quantity/Quality Relationships*, International Association of Hydrological Sciences Publication No. 141, pp. 79–89.

Dixon, N., and Hendricks, D. W. (1970). Simulation of spatial and temporal changes in water quality within a hydrologic unit. *Water Resources Bulletin*, **6**, 483–497.

Dixon, N., Hendricks, D. W., Huber, A. L., and Bagley, J. M. (1970). *Developing a Hydro-quality simulation model*, Technical Report, PRWG 67–1, Utah State University.

Douglas, I. (1964). Intensity and periodicity in denudation processes with special reference to the removal of material in solution by rivers. *Zeitschrift für Geomorphologie Neue Folge*, **8**, 453–473.

Douglas, I. (1972). The geographical interpretation of river water quality data. *Progress in Geography*, **4**, 1–81.

Douglas, I. (1973). Rates of denudation in selected small catchments in Eastern Australia. University of Hull Occasional Papers in Geography, No. 21.

Dupraz, C., Lelong, F., Trop, J. P., and Dumazet, B. (1982). Comparative study of the effects of vegetation on the hydrological and hydrochemical flows in three minor catchments of Mount Lozère (France)—methodological aspects and first results. In: *Hydrological Research Basins and their Use in Water Resources Planning*, Landeshydrologie, Berne, pp. 671–682.

Durum, W. H., Heidel, G., and Tison, L. J. (1960). Worldwide runoff of dissolved solids. *International Association of Hydrological Sciences Publication* No. 51, 618–628.

Edwards, A. M. C. (1973). The variation of dissolved constituents with discharge in some Norfolk rivers. *Journal of Hydrology*, **18**, 219–242.

Edwards, A. M. C. (1974). Silicon depletions in some Norfolk rivers. *Freshwater Biology*, **4**, 267–274.
Edwards, A. M. C. (1975). Long term changes in the water quality of agricultural catchments. In: *Science, Technology and Environmental Management* (R. D. Hey and T. D. Davies, Eds.), Saxon House, Farnborough, pp. 111–122.
Edwards, A. M. C., and Thornes, J. B. (1973). Annual cycle in river water quality: a time series approach. *Water Resources Research*, **9**, 1286–1295.
Eriksson, E., (1981). Hydrochemistry: chemical processes in the water cycle. UNESCO Technical Documents in Hydrology No. SC–81/WS/1.
Farrimond, M. S., and Nelson, J. A. R. (1980). Flow-driven water quality simulation models. *Water Research*, **14**, 1157–1168.
Fehér, J. (1983). Multivariate analysis of quality parameters to determine the chemical transport in rivers. In: *Dissolved Loads of Rivers and Surface Water Quantity/Quality Relationships*, International Association of Hydrological Sciences Publication No. 141, pp. 91–98.
Feller, M. C., and Kimmins, J. P. (1979). Chemical characteristics of small streams near Haney in Southwestern British Columbia. *Water Resources Research*, **15**, 247–258.
Finlayson, B. (1977). Runoff contributing areas and erosion. Research Papers, School of Geography, University of Oxford, No. 18.
Fischer, H. B. (1973). Longtitudinal dispersion and turbulent mixing in open-channel flow. *Annual Review of Fluid Mechanics*, **5**, 59–78.
Foggin, G. T., and Forcier, L. K. (1977). Using topographic characteristics to predict total solute concentrations in streams draining small forested watersheds in Western Montana. University of Montana Joint Water Resources Research Center Rept. No. 89.
Foster, I. D. L. (1978a). Seasonal solute behaviour of stormflow in a small agricultural catchment. *Catena*, **5**, 151–163.
Foster, I. D. L. (1978b). A multivariate model of storm-period solute behaviour. *Journal of Hydrology*, **39**, 339–353.
Foster, I. D. L. (1979a). Chemistry of bulk precipitation, throughfall, soil water and stream water in a small catchment in Devon, England. *Catena*, **6**, 145–155.
Foster, I. D. L. (1979b). Intra-catchment variability in solute response, an East Devon example. *Earth Surface Processes*, **4**, 381–394.
Foster, I. D. L. (1980). Chemical yields in runoff, and denudation in a small arable catchment, East Devon, England. *Journal of Hydrology*, **47**, 349–368.
Foster, I. D. L., Carter, A. D., and Grieve, I. C. (1983). Biogeochemical controls on river water quality in a forested drainage basin, Warwickshire, UK. In: *Dissolved Loads of Rivers and Surface Water Quantity/Quality Relationships*, International Association of Hydrological Sciences Publication No. 141, pp. 241–253.
Foster, I. D. L., and Walling, D. E. (1978). The effects of the 1976 drought and autumn rainfall on stream solute levels. *Earth Surface Processes*, **3**, 393–406.
Froehlich, W. (1983). The mechanisms of dissolved solids transport in flysch drainage basins. In: *Dissolved Loads of Rivers and Surface Water Quantity/Quality Relationships*, International Association of Hydrological Sciences Publication No. 141, pp. 99–108.
Geary, P. M. (1981). Sediment and solute transport in a representative basin. *Aust Geogr. Studies*., **19**, (2), 161–175.
Gibbs, R. J. (1967). Geochemistry of the Amazon river system, part I. The factors that control the salinity and the composition and concentration of the suspended solids. *Geological Society of America Bulletin*, **78**, 1203–1232.

Gibbs, R. (1970). Mechanisms controlling world water chemistry. *Science*, **170**, 1088–1090.
Glover, B. J., and Johnson, P. (1974). Variations in the natural chemical concentrations of river water during flood flows, and the lag effect. *Journal of Hydrology*, **22**, 303–316.
Gorham, E. (1961). Factors influencing supply of major ions to inland waters, with special reference to the atmosphere. *Geological Society of America Bulletin*, **72**, 795–840.
Goudie, A. (1970). Input and output considerations in estimating rates of chemical denudation. *Earth Science Journal*, **4**, 59–65.
Green, D. B., Logan, T. J., and Smeck, N. E. (1978). Phosphate adsorption–desorption characteristics of suspended sediment in the Maumee River Basin of Ohio. *Journal of Environmental Quality*, **7**, 208–212.
Gregory, K. J., and Walling, D. E. (1973). *Drainage Basin Form and Process: A Geomorphological Approach*, Arnold, London.
Grimshaw, D. L., Lewin, J., and Fuge, R. (1976). Seasonal and short-term variations in the concentration and supply of dissolved zinc to polluted aquatic environments. *Environmental Pollution*, **11**, 1–7.
Grip, H. (1982). *Water Chemistry and Runoff in Forest Streams at Kloten*, Report No. 58, Uppsala Universitet, Naturgeografiska Institutionen.
Gunnerson, C. G. (1967). Streamflow and quality in the Columbia River Basin. *Proceedings ASCE, Journal of the Sanitary Engineering Division*, **93**, SA6, 1–16.
Haith, D. A. (1976). Land use and water quality in New York rivers. *Proceedings ASCE, Journal of the Environmental Engineering Division*, **102**, EE1, 1–15.
Hall, F. R. (1970). Dissolved solids-discharge relationships. 1: Mixing models. *Water Resources Research*, **6**, 845–850.
Hall, F. R. (1971). Dissolved solids-discharge relationships. 2: Applications to field data. *Water Resources Research*, **7**, 591–601.
Hall, G. C., and Görgens, A. H. M. (Eds.) (1978) Studies of mineralization in South African Rivers. South African National Scientific Programmes Rept. No. 26.
Hall, G. C., and Görgens, A. H. M. (1979). Modelling runoff and salinity in the Sundays River, Republic of South Africa. In: *The Hydrology of Areas of Low Precipitation*, International Association of Hydrological Sciences Publication No. 128, pp. 323–330.
Hart, F. C., King, P. H., and Tchobanoglous, G. (1964). Predictive techniques for water quality inorganics. Discussion. *Proceedings ASCE, Journal of the Sanitary Engineering Division*, **90**, SA5, 63–64.
Hem, J. D. (1948). Fluctuations in concentrations of dissolved solids in some southwestern streams. *Transactions American Geophysical Union*, **29**, 80–84.
Hem, J. D. (1970). Study and interpretation of the chemical characteristics of natural water. US Geological Survey Water Supply Paper No. 1473.
Hendrickson, G. E., and Krieger, R. A. (1960). Relationship of chemical quality of water to stream discharge in Kentucky. *Report of 21st International Geological Congress, Copenhagen*, **1**, 66–75.
Hendrickson, G. E., and Krieger, R. A. (1964). Geochemistry of natural waters of the Blue Grass Region, Kentucky. US Geological Survey Water Supply Paper No. 1700.
Herold, C. E. (1980). *A Model to Compute on a Monthly Basis Diffuse Salt Loads Associated with Runoff*, Report No. 1/80, Hydrological Research Unit, University of the Witwatersrand, Johannesburg.

Herold, C. E. (1981). *A Model to Simulate Daily River Flows and Associated Diffuse-source Conservative Pollutants*, Report No. 3/81, Hydrological Research Unit, University of the Witwatersrand, Johannesburg.

Hill, A. R. (1978). Factors affecting the export of nitrate-nitrogen from drainage basins in southern Ontario. *Water Research*, **12**, 1045–1057.

Hill, A. R. (1980). Stream cation concentrations and losses from drainage basins with contrasting land uses in southern Ontario. *Water Research*, **14**, 1295–1305.

Holland, H. D. (1978). *The Chemistry of the Atmosphere and Oceans*, Wiley, New York.

Hotes, F. L., and Pearson, F. A. (1977). Effects of Irrigation on water quality. In: *Arid Land Irrigation in Developing Countries: Environmental Problems and Effects* (E. B. Worthington, Ed.), Pergamon, Oxford, pp. 127–158.

Houston, J. A., and Brooker, M. P. (1981). A comparison of nutrient sources and behaviour in two lowland subcatchments of the River Wye. *Water Research*, **15**, 49–57.

Hughes, B. D., and Edwards, R. W. (1977). Flows of sodium, potassium, magnesium and calcium in the R. Cynon, S. Wales. *Water Research*, **11**, 536–566.

Imeson, A. C. (1973). Solute variations in small catchment streams. *Transactions Instititue of British Geographers*, **60**, 87–99.

Iorns, W. V., Hembree, C. H., and Oakland, G. L. (1965). Water resources of the Upper Colorado basin. US Geological Survey Professional Paper No. 441.

Janda, R. J. (1971). An evaluation of procedures used in computing chemical denudation rates. *Geological Society of America Bulletin*, **82**, 67–80.

Johnson, F. A., and East, J. W. (1982). Cyclical relationships between river discharge and chemical concentration during flood events. *Journal of Hydrology*, **57**, 93–106.

Johnson, N. M., Likens, G. E., Bormann, F. H., Fisher, D. W., and Pierce, R. S. (1969). A working model for the variation in stream water chemistry at the Hubbard Brook Experimental Forest, New Hampshire. *Water Resources Research*, **5**, 1353–1363.

Judson, S., and Ritter, D. F. (1964). Rates of regional denudation in the United States. *Journal of Geophysical Research*, **69**, 3395–3401.

Keefer, T. N., and McQuivey, R. S. (1974). Investigation of diffusion in open-channel flows. *Journal of Research of the US Geological Survey*, **2**, 501–509.

Keller, H. M. (1970). Der Chemismus kleiner Bäche in teilweise bewaldeten Einzugsgebieten in der Flyschzone eines Voralpentales. *Schweiz. Anstalt für das forstl. Versuchsw. Mitt.*, **46**, 114–155.

Kemp, P. H. (1971). Chemistry of natural waters—VI: Classification of waters. *Water Research*, **5**, 943–956.

Kirkby, M. J. (Ed.) (1978). *Hillslope Hydrology*, Wiley, Chichester.

Kirschner, W. B. (1975). An examination of the relationship between drainage basin morphology and the export of phosphorus. *Limnology and Oceanography*, **20**, 267–270.

Klein, M. (1981). Dissolved material transport—the flushing effect in surface and subsurface flow. *Earth Surface Processes and Landforms*, **6**, 173–178.

Klepper, R. (1978). Nitrogen fertilizer and nitrate concentrations in tributaries of the upper Sangamon River in Illinois. *Journal of Environmental Quality*, **7**, 13–22.

Lane, W. L. (1975). Extraction of information on inorganic water quality. Colorado State University Hydrology Papers No. 73.

Langbein, W. B., and Dawdy, D. R. (1964). Occurrence of dissolved solids in surface waters of the United States. *US Geological Survey Professional Paper* No. 501D.

Leak, W. B., and Martin, C. W. (1975). Relationship of stand age to streamwater nitrate in New Hampshire. US Department of Agriculture Forest Service Research Note NE–211.

Ledbetter, J. O., and Gloyna, E. F. (1964). Predictive techniques for water quality inorganics. *Proceedings ASCE, Journal of the Sanitary Engineering Divison*, **90**, SA1, 127–151.

Lenz, A. T., and Sawyer, C. N. (1944). Estimation of stream-flow from alkalinity determination. *Transactions American Geophysical Union*, **25**, 1005–1010.

Likens, G. E., Bormann, F. H., Pierce, R. S., Eaton, J. S., and Johnson, N. M. (1977). *Biogeochemistry of a Forested Ecosystem*, Springer-Verlag, New York.

Livingstone, D. A. (1963a). Chemical composition of rivers and lakes: data of Geochemistry, Chapter G. *US Geological Survey Professional Paper* No. 440G.

Livingstone, D. A. (1963b). The sodium cycle and the age of the ocean. *Geochimica Cosmochimica Acta*, **27**, 1655–1669.

Loughran, R. J., and Malone, K. J. (1976). Variations in some stream solutes in a Hunter Valley catchment. Research Papers in Geography, University of Newcastle, N.S.W., No. 8.

Lundquist, D. (1976). Simulation of the hydrologic cycle. Report IR 23/76, SNSF Project, Norwegian Institute for Water Resources, Oslo.

McCann, S. B., and Cogley, J. B. (1971). Observations of water hardness on Southwest Devon Island, Northwest Territories. *Canadian Geographer*, **15**, 173–180.

McPherson, H. J. (1975). Sediment yields from intermediate-sized stream basins in southern Alberta. *Journal of Hydrology*, **25**, 243–257.

Meade, R. H. (1969). Errors in using modern stream-load to estimate natural rates of denudation. *Geological Society of America Bulletin*, **80**, 1265–1274.

Meybeck, M., (1976). Total dissolved transport by world major rivers. *Hydrological Sciences Bulletin*, **21**, 265–284.

Meybeck, M. (1977). Dissolved and suspended matter carried by rivers: composition, time and space variations and world balance. In: *Interactions between Sediments and Fresh Water* (H. L. Golterman, Ed.), Dr. W. Junk, B. V., The Hague, pp. 25–32.

Meybeck, M. (1979). Concentrations des eaux fluviales en éléments majeurs et apports en solution aux océans. *Révue de Géologie Dynamique et de Géographie Physique*, **21**, 215–246.

Meybeck, M. (1981). Pathways of major elements from land to ocean through rivers. In: *River Inputs to Ocean Systems*, UNEP/UNESCO Rept., pp. 18–30.

Meybeck, M. (1982). Carbon, nitrogen, and phosphorus transport by world rivers. *American Journal of Science*, **282**, 401–450.

Meybeck, M. (1983). Atmospheric inputs and river transport of dissolved substances. In: *Dissolved Loads of Rivers and Surface Water Quantity/Quality Relationships*, International Association of Hydrological Sciences Publication No. 141, pp. 173–192.

Miller, J. P. (1961). Solutes in small streams draining single rock types, Sangre de Cristo Range, New Mexico. *US Geological Survey Water Supply Paper* 1535–F.

Miller, W. R., and Drever, J. I. (1977). Water chemistry of a stream following a storm, Absaroka Mountains, Wyoming. Geological Society of America Bulletin, 88, 286–290.

Milliman, J. D., and Meade, R. H. (1983). World-wide delivery of river sediment to the oceans. *Journal of Geology*, **91**, 1–21.

Moore, R. J., and Clarke, R. T. (1981). A distribution function approach to rainfall-runoff modelling. *Water Resources Research*, **17**, 1367–1382.

Moore, R. J., and Clarke, R. T. (1983). A distribution function approach to modelling basin sediment yield. *Journal of Hydrology*, **65**, 239–257.

Nakamura, R. (1971). Runoff analysis by electrical conductance of water. *Journal of Hydrology*, **14**, 197–212.

Nelson, D. J. (1970). Measurement and sampling of outputs from watersheds. In: *Analysis of Temperate Forest Ecosystems* (D. E. Reichle, Ed.), pp. 242–285.

Newbury, R. W., Cherry, J. A., and Cox, R. A. (1969). Groundwater-streamflow systems in Wilson Creek Experimental Watershed, Manitoba. *Canadian Journal of Earth Science*, **6**, 613–623.

Oborne, A. C. (1981). The application of a water-quality model to the River Wye, Wales. *Journal of Hydrology*, **52**, 59–70.

Oborne, A. C., Brooker, M. P., and Edwards, R. W. (1980). The chemistry of the River Wye. *Journal of Hydrology*, **45**, 233–252.

O'Connor, D. J. (1976). The concentration of dissolved solids and river flow. *Water Resources Research*, **12**, 279–294.

Omernik, J. M. (1976). The influence of land use on stream nutrient levels. US Environmental Protection Agency, Ecological Research Series Report No. EPA–600/3–76–014.

Ongley, E. D., Ralston, J. G., and Thomas, R. L. (1977). Sediment and nutrient loadings to Lake Ontario: methodological arguments. *Canadian Journal of Earth Science*, **14**, 1555–1565.

Oxley, N. C. (1974). Suspended sediment delivery rates and the solute concentration of stream discharge in two Welsh catchments. In: *Fluvial Processes in Instrumented Watersheds* (K. J. Gregory and D. E. Walling, Eds.), Institute of British Geographers Special Publication No. 6, pp. 141–154.

Penck, A. (1894). *Morphologie der Erdoberfläche*, Verlag Von Engelhorn, Stuttgart.

Pilgrim, D. H., Huff, D. D., and Steele, T. D. (1978). A field evaluation of subsurface and surface runoff. II: Runoff. *Journal of Hydrology*, **38**, 299–318.

Pilgrim, D. H., Huff, D. D., and Steele, T. D. (1979). Use of specific conductance and contact time relations for separating flow components in storm runoff. *Water Resources Research*, **15**, 329–339.

Pinder, G. F., and Jones, J. F. (1969). Determination of the ground-water component of peak discharge from the chemistry of total runoff. *Water Resources Research*, **5**, 438–445.

Pionke, H. B., Nicks, A. D., and Schoof, R. R. (1972). Estimating salinity of streams in the Southwestern United States. *Water Resources Research*, **8**, 1597–1604.

Pitman, W. V. (1976). *A Mathematical Model for Generating Daily River Flows from Meteorological Data in South Africa*, Report No. 2/76, Hydrological Research Unit, University of the Witwatersrand, Johannesburg.

Popp, O. (1875). Ueber das Nilwasser. *Liebig's Annalen*, **155**, 334–348.

Reade, T. M. (1876). President's address. *Proceedings of the Liverpool Geological Society*, **3**, 211–235.

Reid, J. M., MacLeod, D. A., and Cresser, M. S. (1981a) Factors affecting the chemistry of precipitation and river water in an upland catchment. *Journal of Hydrology*, **50**, 129–145.

Reid, J. M., MacLeod, D. A., and Cresser, M. S. (1981b). The assessment of chemical weathering rates within an upland catchment in North-east Scotland. *Earth Surface Processes and Landforms*, **6**, 447–457.

Reinson, G. E. (1976). Hydrogeochemistry of the Genoa River Basin, New South Wales, Victoria. *Australian Journal of Marine and Freshwater Research*, **27**, 165–186.

Schwartz, F. W., and Milne-Home, W. A. (1982). Watersheds in muskeg terrain. 1: The chemistry of water systems. *Journal of Hydrology*, **57**, 267–290.

Sharp, J. V. A. (1969). Time-dependent behaviour of water chemistry in hydrologic systems. *Transactions American Geophysical Union*, **50**, 141.
Skakalskiy, B. G. (1966). Basic geographical and hydrochemical characteristics of the local runoff of natural zones in the European territory of the USSR. *Transactions State Hydrological Institute*, (Trudy GGI), **137**, 125–180.
Slack, J. G. (1977). River water quality in Essex during and after the 1976 drought. *Effluent and Water Treatment Journal*, **17**, 575–578.
Slack, K. V., and Feltz, H. R. (1968). Tree leaf control on lowflow water quality in a small Virginia stream. *Environmental Science and Technology*, **2**, 126–131.
Smith, R. V., and Stewart, D. A. (1977). Statistical models of river loadings of nitrogen and phosphorus in the Lough Neagh system. *Water Research*, **11**, 611–636.
Spraggs, G. (1976). Solute variations in a local catchment. *The Southern Hampshire Geographer*, **8**, 1–14.
Steele, T. D. (1968). Digital-computer applications in chemical quality studies of surface water in a small watershed. *Publications de l'Association Internationale d'Hydrologie Scientifique*, **80**, 203–214.
Steele, T. D. (1970). Beneficial uses and pitfalls of historical water quality data. In: *Proceedings National Symposium on Data and Instrumentation for Water Quality Management*, pp. 346–363.
Steele, T. D. (1973). Simulation of major inorganic chemical concentrations and loads in streamflow. US Geological Survey Computer Contribution.
Steele, T. D., and Gilroy, E. J. (1971). Statistical techniques for assessing long-term changes in streamflow salinity. *Transactions American Geophysical Union*, **52**, 846.
Steele, T. D., and Jennings, M. E. (1972). Regional analysis of streamflow chemical quality in Texas. *Water Resources Research*, *8*, 460–477.
Stevenson, C. M. (1968). An analysis of the chemical composition of rain-water and air over the British Isles and Eire for the years 1959–1964. *Royal Meteorological Society Quarterly Journal*, **94**, 56–70.
Strakhov, N. M. (1967). *Principles of Lithogenesis* (Vol. 1), Oliver and Boyd, London.
Stumm, W., and Morgan, J. J. (1970). *Aquatic Chemistry: an Introduction emphasizing Chemical Equilibria in Natural Waters*, Wiley, New York.
Sugawara, M. (1961). On the analysis of runoff structure about several Japanese rivers. *Japanese Journal of Geophysics*, **2**, 1–77.
Sutcliffe, D. W., and Carrick, T. R. (1973a). Studies on mountain streams in the English Lake District. I: pH, calcium and the distribution of invertebrates in the River Duddon. *Freshwater Biology*, **3**, 437–462.
Sutcliffe, D. W., and Carrick, T. R. (1973b). Studies in mountain streams in the English lake district. II: Aspects of water chemistry in the River Duddon. *Freshwater Biology*, **3**, 543–560.
Talsma, T., and Hallam, P. M. (1982). Stream water quality of forest catchments in the Cotter Valley, ACT. In: *The First National Symposium on Forest Hydrology, 1982, Melbourne, 11–13 May* (E. M. O'Loughlin and L. J. Bren, Eds.), pp. 50–59.
Ternan, J. L., and Williams, A. G. (1979). Hydrological pathways and granite weathering on Dartmoor. In: *Geographical Approaches to Fluvial Processes* (A. F. Pitty, Ed.), Geobooks, Norwich, pp. 5–30.
Thornes, J. B., and Clark, M. W. (1976). Modelling the Lea Marston water quality series. Non-sequential Water Quality Project Paper, Department of Geography, London School of Economics, No. 6.
Toler, L. G. (1965). Relation between chemical quality and water discharge in Spring Creek, Southwestern Georgia. *US Geological Survey Professional Paper* 525–C, C206–C208.

Tomlinson, T. E. (1970). Trends in nitrate concentrations in English rivers in relation to fertilizer use. *Water Treatment and Examination*, **19**, 277–289.
Tricker, A. S., and Scott, G. (1980). Spatial patterns of chemical denudation in the Eden Catchment, Fife. *Scottish Geographical Magazine*, **96**(2), 114–120.
Trimble, S. W. (1981). Changes in sediment storage in the Coon Creek basin, Driftless Area, Wisconsin, 1853–1975. *Science*, **214**, 181–183.
Troake, R. P., and Walling, D. E. (1973). The natural history of Slapton Ley Nature Reserve VII. The hydrology of the Slapton Wood Stream, a preliminary report. *Field Studies*, **3**, 719–740.
Troake, R. P., and Walling, D. E. (1975). Some observations on stream nitrate levels and fertiliser application at Slapton, South Devon. *Reports and Transactions Devonshire Association for the Advancement of Science*, **107**, 77–90.
Turvey, N. D. (1975). Water quality in a tropical rain forested catchment. *Journal of Hydrology*, **27**, 111–125.
Van Denburgh, A. S., and Feth, J. H. (1965). Solute erosion and chloride balance in selected river basins of the western conterminous United States. *Water Resources Research*, **1**, 537–541.
Verhoff, F. H., Yaksich, S. M., and Melfi, D. A. (1980). River nutrient and chemical transport estimation. *Proceedings ASCE, Journal of the Environmental Engineering Division*, **106**, 591–608.
Verstraten, J. M. (1977). Chemical erosion in a forested watershed in the Oesling, Luxembourg. *Earth Surface Processes*, **2**, 175–184.
Vitousek, P. M. (1977). The regulation of element concentrations in mountain streams in the northeastern United States. *Ecological Monographs*, **47**, 65–87.
Walling, D. E. (1971). Sediment dynamics of small instrumented catchments in south-east Devon. *Reports and Transactions Devonshire Association for the Advancement of Science*, **103**, 147–165.
Walling, D. E. (1974). Suspended sediment and solute yields from a small catchment prior to urbanization. In: *Fluvial Processes in Instrumented Watersheds* (K. J. Gregory and D. E. Walling, Eds.), Institute of British Geographers Special Publication No. 6, pp. 169–192.
Walling, D. E. (1975). Solute variations in small catchment streams: some comments. *Transactions Institute of British Geographers*, **64**, 141–147.
Walling, D. E. (1978a). Suspended sediment and solute response characteristics of the River Exe, Devon, England. In: *Research in Fluvial Geomorphology* R. Davison-Arnott and W. Nickling, Eds.), Geo Abstracts, Norwich, pp. 169–197.
Walling, D. E. (1978b). Reliability considerations in the evaluation and analysis of river loads. *Zeitschrift für Geomorphologie Neue Folge*, Supp. Band **29**, 29–42.
Walling, D. E. (1980a). Water in the catchment ecosystem. In: *Water Quality in Catchment Ecosystems* (A. M. Gower, Ed.), Wiley, Chichester, pp. 1–47.
Walling, D. E. (1980b). Solute levels in two Devon catchments. In: *Atlas of Drought in Britain* (J. C. Doornkamp, K. J. Gregory and A. S. Burn, Eds.), Institute of British Geographers Publication, p. 50.
Walling, D. E. (1983). The sediment delivery problem. *Journal of Hydrology*, **65**, 209–237.
Walling, D. E., and Foster, I. D. L. (1975). Variations in the natural chemical concentration of river water during flood flows, and the lag effect: some further comments. *Journal of Hydrology*, **26**, 237–244.
Walling, D. E., and Foster, I. D. L. (1978). The 1976 drought and nitrate levels in the River Exe Basin. *Journal of the Institution of Water Engineers and Scientists*, **32**, 341–352.

Walling, D. E., and Webb, B. W. (1975). Spatial variation of river water quality: a survey of the River Exe. *Transactions Institute of British Geographers*, **65**, 155–169.
Walling, D. E., and Webb, B. W. (1978). Mapping solute loadings in an area of Devon, England. *Earth Surface Processes*, **3**, 85–99.
Walling, D. E., and Webb, B. W. (1980). The spatial dimension in the interpretation of stream solute behaviour. *Journal of Hydrology*, **47**, 129–149.
Walling, D. E., and Webb, B. W. (1981). Water quality. In: *British Rivers* (J. Lewin, Ed.), George Allen and Unwin, London, pp. 126–169.
Walling, D. E., and Webb, B. W. (1982). The design of sampling programmes for studying catchment nutrient dynamics. In: *Hydrological Research Basins and their use in Water Resources Planning*, Landeshydrologie, Berne, pp. 747–758.
Walling, D. E., and Webb, B. W. (1983). The dissolved loads of rivers: a global overview. In: *Dissolved Loads of Rivers and Surface Water Quantity/Quality Relationships*, International Association of Hydrological Sciences Publication No. 141, pp. 3–20.
Ward, R. C. (1981). River systems and river regimes. In: *British Rivers* (J. Lewin, Ed.), George Allen and Unwin, London, pp. 1–33.
Waylen, M. J. (1979). Chemical weathering in a drainage basin underlain by Old Red Sandstone. *Earth Surface Processes*, **4**, 167–178.
Webb, B. W. (1976). Solute concentration in the baseflow of some Devon streams. *Reports and Transactions Devonshire Association for the Advancement of Science*, **108**, 127–145.
Webb, B. W. (1980). *Solute Levels in Streams of the Middle and Upper Exe Basin*. Unpublished Ph.D. thesis, University of Exeter.
Webb, B. W. (ed.) (1983a). *Dissolved Loads of Rivers and Surface Water Quantity/Quality Relationships*, International Association of Hydrological Sciences Publication No. 141.
Webb, B. W. (1983b). Factors influencing spatial variation of background solute levels in a Devon river system. *Reports and Transactions Devonshire Association for the Advancement of Science*, **115**, 51–69.
Webb, B. W., and Walling, D. E. (1980). Stream solute studies and geomorphological research: some examples from the Exe Basin, Devon, UK. *Zeitschrift für Geomorphologie Neue Folge*, Supp. Band **36**, 245–263.
Webb, B. W., and Walling, D. E. (1982a). Catchment scale and the interpretation of water quality behaviour. In: *Hydrological Research Basins and their use in Water Resources Planning*. Landeshydrologie, Berne, pp. 759–770.
Webb, B. W., and Walling, D. E. (1982b). The magnitude and frequency characteristics of fluvial transport in a Devon drainage basin and some geomorphological implications. *Catena*, **9**, 9–23.
Webb, B. W., and Walling, D. E. (1983). Stream solute behaviour in the River Exe basin, Devon, UK. In: *Dissolved Loads of Rivers and Surface Water Quantity/Quality Relationships*, International Association of Hydrological Sciences Publication No. 141, pp. 153–169.
White, R. J. (1983). Nitrate in British waters. *Aqua*, **2**, 51–57.
Williams, A. G., Ternan, J. L., and Kent, M. (1983). Stream solute sources and variations in a temperate granite drainage basin. In: *Dissolved Loads of Rivers and Surface Water Quantity/Quality Relationships*, International Association of Hydrological Sciences Publication No. 141, pp. 299–310.
Zeman, L. J., and Slaymaker, H. O. (1975). Hydrochemical analysis to discriminate variable runoff source areas in an alpine basin. *Arctic and Alpine Research*, **7**, 341–351.

Solute Processes
Edited by S. T. Trudgill

CHAPTER 8

Spatial distribution of solutional erosion

R. W. Crabtree

Department of Civil Engineering,
University of Birmingham

'In landform science many of the observations can only be interpreted.'

Professor J. MacDonald-Holmes

8.1 SOLUTE PROCESSES IN A GEOMORPHOLOGICAL CONTEXT

8.1.1 Introduction

This chapter aims to identify and quantify the geomorphological relationships between solute processes and landforms, in terms of solution erosion. Relatively few field studies, particularly outside 'karst' areas (see Chapter 9), have yet investigated this spatial variation in solutional erosion over the landscape.

Theoretical studies, modelling landform development by solutional processes, have been carried out at the regional and individual hillslope scales in relation to both climate and lithology. On small-scale process-based studies, however, concentration has been directed to process identification and rate measurement for the prediction of future landform development—though the actual steps of prediction and testing have rarely been made.

This chapter will consider the detailed development of hillslopes, with particular reference to spatial variation in the erosion of surfaces by solutional processes. A simple model will be proposed to predict such spatial patterns over a hillslope. This model can be applied to larger-scale catchments and regions, which are also reviewed, and will be discussed as a direction for future study. Recent trends have encouraged studies of increasing specialization and complexity in the understanding of solute processes, but have failed to consider the wider implications in a geomorphological sense. The present objective is to bridge the gap between solute processes and geomorphological effects. This objective, to use an analogy (Chorley, 1971), aims to contribute towards pulling together the two diverging tightropes of study along which a physical geographer may choose to tread, those of process–response and of temporal evolution. Therefore the hypothesis that spatial variation in solute processes will lead to spatial variations in solutional denudation rates over the landscape should be investigated. However, in interpreting the effects of solute processes, great care must be taken. In general, chemical and physical weathering mechanism are not independent and may be of equal importance. In addition, many landforms are not solely the products of present-day processes, but may be relict features which are being modified by present-day processes. It is extremely difficult to assess the morphological effects of current processes acting for short time periods, because of their generally small magnitude. Yet, despite this small magnitude, to quote from Carson and Kirkby (1972):

> Removal in solution is a process of the greatest importance to hillslope development, and its rate may sometimes exceed that of all mechanical processes combined. . . . The variations in the rate of chemical removal over a hillside are, therefore, of great importance in interpreting the evolution of hillside form.

Solution erosion and its effect on land surface lowering, that is solutional denudation, is the result of solute uptake. This is the chemical weathering of minerals to release ions into solution, and the transportation of those ions away from the locus of solution by water movement. These will now be considered.

8.1.2 Solute generation, chemical weathering and geochemical materials

Le Chatelier's principle governs the mechanisms of chemical weathering. Any system in equilibrium will tend to react to restore the equilibrium when an external force is applied. Minerals produced by geochemical processes (that is, formed geothermally in deep-seated crustal environments or in sedimentary basins) become unstable in a weathering environment, and chemical reactions take place. These reactions produce a new set of minerals in equilibrium with the weathering environment, which may release ions into solution. This decomposition of minerals to products that are more in equilibrium in the weathering environment is manifested as solute generation by chemical weathering. Details of these weathering processes can be found in standard texts (e.g. Ollier, 1969; Loughnan, 1969; Garrels and Christ, 1965) but some of the main principles are discussed below.

The main chemical process which releases ions into solution is hydrolysis. Hydrolysis is the chemical reaction that takes place between a mineral and a constituent of water, either H^+ or OH^-; most commonly it occurs in acidic water by the exchange of hydrogen ions with metal cations from minerals. The input of precipitation, acidified by the presence of carbon dioxide, sulphur and nitrogen compounds in the atmosphere, is universal over the earth's surface. Acidic precipitation infiltrates through the soil surface, and in the contact with soil and rock minerals hydrolysis takes place and solute uptake occurs. This releases mineral ions into the percolating water, which becomes progressively less acidic until a state of equilibrium is reached between minerals and solvent.

Water percolating through soils rich in carbon dioxide produced by the soil flora and fauna can become more acidic than waters initially at equilibrium with atmospheric carbon dioxide concentrations (Garrels, 1967; Trudgill, 1977) owing to the dissolution of carbon dioxide:

$$H_2O + CO_2 \rightleftharpoons H_2CO_3 \rightleftharpoons H^+ + HCO_3^- \quad (8.1)$$

Acidic soilwaters, therefore, are a more potent agent for solute production than is precipitation. Other chemical weathering processes, for example reduction, oxidation or chelation, may release ions into solution. The relative importance of each chemical weathering process will vary with the weathering material and the conditions of the weathering environment.

In carbonate terrains the process of neutralization of acidic water can be described by the reaction:

$$CaCO_3 + H^+ \rightleftharpoons Ca^{2+} + HCO_3^- \quad (8.2)$$

This is perhaps the simplest form of hydrolysis where the sole products are ions in solution. In non-carbonate rocks hydrolysis can produce a combination of insoluble residues, authigenic minerals (new minerals produced *in situ* and which may undergo further chemical weathering), and ions in solution (Stumm and Morgan, 1970). For example, the hydrolysis of the plagioclase feldspar, anorthite, will take the form:

$$CaAl_2\,Si_2O_8 + 2H^+ + H_2O \rightleftharpoons Al_2\,Si_2O_5(OH)_4 + Ca^{2+} \quad (8.3)$$

This reaction releases Ca^{2+} into solution and leaves the clay mineral, kaolinite, ($Al_2\,Si_2O_5(OH)_4$) as a weathering residue which may then be removed by mechanical processes.

A relatively small number of minerals are the dominant components of most rocks. Quartz is the most common mineral in sandstones and is present in most sedimentary rocks, acidic and intermediate igneous rocks, and most metamorphic rocks. Quartz is stable at the earth's surface and is therefore resistant to chemical weathering; it can persist in the environment through several cycles of deposition, diagenesis and erosion.

Table 8.1. Sample reactions for the weathering of some common minerals by hydrolysis

(i) Olivine: forsterite

$$Mg_2SiO_{4(s)} + 4H^+_{(aq)} \rightleftharpoons 2\ Mg^{2+}_{(aq)} + 2H_2O + SiO_2(s)$$

(ii) Pyroxene: diopside

$$CaMg(SiO_3)_{2(s)} + 4H^+_{(aq)} \rightleftharpoons Mg^{2+}_{(aq)} + 2H_2O + 2SiO_{2(s)}$$

(iii) Amphibole: tremolite

$$Ca_2\ Mg_5\ Si_{18}O_{22}\ (OH)_{2(s)} + 14H^+_{(aq)} \rightleftharpoons 5\ Mg^{2+}_{(aq)} + 2Ca^{2+}_{(aq)} + 8H_2O + 8SiO_{2(s)}$$

(iv) Plagioclase feldspar: anorthite

$$CaAl_2Si_2O_{8(s)} + 2H^+_{(aq)} + H_2O \rightleftharpoons Al_2Si_2O_5\ (OH)_{4(s)} + Ca^{2+}_{(aq)}$$

(v) Plagioclase feldspar: albite

$$2NaAlSi_3O_{8(s)} + 2H^+_{(aq)} + H_2O \rightleftharpoons Al_2\ Si_2O_5\ (OH)_{4(s)} + 4SiO_{2(s)} + 2Na^{2+}_{(aq)}$$

(vi) Orthoclase feldspar: microcline

$$2KAlSi_3O_{8(s)} + 2H^+_{(aq)} + H_2O \rightleftharpoons Al_2Si_2O_5\ (OH)_{4(s)} + 4SiO_{2(s)} + 2K^{2+}_{(aq)}$$

(vii) Mica: muscovite

$$2KAl_3Si_3O_{10}\ (OH)_{2(s)} + 2H^+_{(aq)} + 3H_2O \rightleftharpoons 3Al_2Si_2O_5\ (OH)_{4(s)} + 2K^{2+}_{(aq)}$$

(viii) Calcium silicate: wollastonite

$$CaSiO_{3(s)} + 2H^{2+}_{(aq)} \rightleftharpoons Ca^{2+}_{(aq)} + H_2O + SiO_{2(s)}$$

(ix) Garnet: grossular

$$Ca_3Al_2Si_3O_{12(s)} + 6H^+_{(aq)} \rightleftharpoons 3Ca^{2+}_{(aq)} + Al_2Si_2O_5\ (OH)_{4(s)} + H_2O + SiO_{2(s)}$$

(x) Magnesium aluminate: spinel

$$MgAl_2O_{4(s)} + 2H^+_{(aq)} + H_2O \rightleftharpoons Al_2O_3\ 3H_2O(s) + Mg^{2+}_{(aq)}$$

(s) = solid phase; (aq) = aqueous phase.

Table 8.1 illustrates hydrolytic reactions of some other common minerals. The main points concerning the hydrolysis of some of these minerals are as follows. Feldspar is the generic name for the alumino-silicate minerals. There are two main groups, orthoclase (K and Na rich, e.g. microcline (vi) on Table 8.1) and plagioclase (Na and Ca rich, e.g. anorthite (iv) and albite (v)). The plagioclase group is more readily weathered than orthoclases, but all feldspars are rapidly weathered by hydrolysis, usually producing the residual clay mineral, kaolinite. Feldspars occur as primary minerals in most igneous and metamorphic rocks but, owing to the ease of weathering, they are not usually found in sedimentary rocks. Feldspars are, however, common in arkoses, indicative of an arid environment of deposition, and in greywackes produced by rapid mechanical erosion and deposition.

Olivine (for example, forsterite (i) on Table 8.1) is a magnesium iron silicate found in basic and ultrabasic igneous rocks and is rapidly weathered by water to leave chlorite-type residual clay minerals.

Pyroxenes (for example, diopside (ii) are a group of Ca–Al silicates found in intermediate and basic igneous rocks such as dolerite, and metamorphic rocks. They weather rapidly to clays and are rarely found in sedimentary rocks.

Micas are basic K–Al silicates and biotite and muscovite ((vii) on Table 8.1) are the main forms. They occur mainly in igneous and metamorphic rocks and in some sedimentary rocks. Micas are easily weathered by hydrolysis to clay minerals. Biotite is more rapidly weathered than muscovite.

Carbonates are the dominant minerals in limestones and may also be present in marls, and calcareous sandstones and as metamorphosed limestone, marble. The two common forms are calcite ($CaCO_3$) and dolomite $(Ca,Mg)CO_3$. Solution of limestones by hydrolysis is rapid, and calcite is more readily dissolved than dolomite. However, magnesium impurities can increase the rate of solution of calcite (Picknett, 1964).

Clay minerals are the residual product of the weathering of primary materials and are relatively stable; therefore, they contribute little to solute production.

8.1.3 Weathering rates

Thermodynamic equilibrium conditions determine the maximum amount of a mineral that can be dissolved for a given set of environmental conditions. However, the actual magnitude of solute loss depends on the rates of reaction or dissolution. Simple equilibrium chemical reactions such as those in Table 8.1 only serve to illustrate the reaction, not its speed or direction. Hence they do not show whether or not the reaction will be an important solute-generating process. The importance of a reaction is, in fact, dependent on two factors: firstly, the rate of reaction (considered below), and secondly the rate of removal of material in solution (see §8.1.4). Rates of reactions are controlled by the energy involved in the reaction. The relative stabilities of different minerals can be considered in terms of the physical characteristics of the minerals, which imply weaknesses for the penetration of water, such as grain size, surface area, cleavage, hardness, solubility, and crystal and atomic structure (Loughnan, 1969; Curtis, 1976). The common primary minerals are often ranked in order of descending stability to weathering (Goldich, 1938), as shown in Figure 8.1. This order is the inverse of the Bowen reaction series for the order of crystallization of igneous minerals from molten magma. In essence, the first minerals to crystallize are the most rapidly weathered because there is the greatest difference in conditions between the environments of mineral formation and of weathering. The series thus implies that, for a given weathering environment, calcic plagioclase will weather more quickly than biotite. Quartz is the most stable and is persistent in the

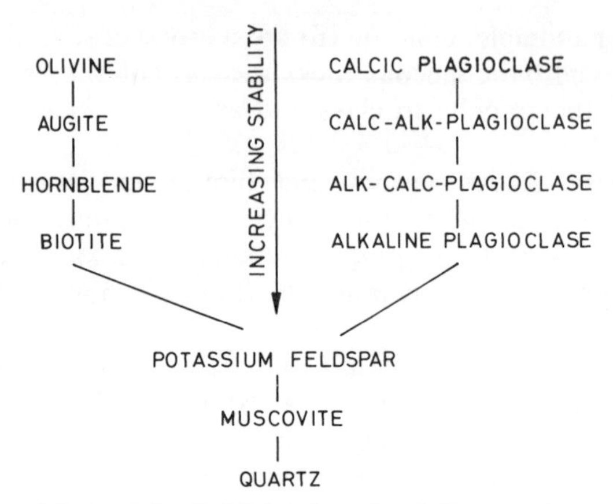

Figure 8.1. Goldich mineral stability ranking

weathering environment. A thermochemical approach can be used to explain the empirical Goldich weathering series.

Energy changes occur when a chemical reaction takes place, and the nature of these changes has been used by Curtis (1976) to rank minerals in order of stability (as described in Chapter 1.). Curtis demonstrates that the rate of weathering, and hence persistence, of a mineral in the weathering environment can be correlated with the total energy released by its decomposition into weathering products. The relative mobility, or degree of persistence of different minerals, can be used to develop weathering sequences of solute loss and residual production (Chesworth, 1973) or to assess the degree of chemical weathering at a regional scale (Kronberg and Nesbitt, 1981).

8.1.4 Solute uptake and transport

Once solutes have been released by chemical weathering, the removal of solutes then determines the subsequent rate of weathering. If solutes remain at the locus of solute production, a closed system equilibrium condition will develop and the rate of dissolution will decrease towards the equilibrium state. This follows a Nernst diffusion mechanism, as discussed in Chapter 1.

Closed system conditions will lead to a thermodynamic equilibrium when the mineral is in contact with static water, for example in a static soilwater-aquifer regime between rainfall events. Rainfall inputs to such systems are intermittent, and while water movement rates will be rapid during and immediately after rainfall, static conditions may develop in the intervening periods.

If equilibrium develops in a closed system, solute uptake and transport will be reduced by lack of transport of weathering products: dissolution will thus be transport limited. In this case the total amount of dissolution, assuming a constant rate of reaction of the mineral, will be proportional to the amount of transport. This assumption has been used (Carson and Kirkby, 1972) to produce a model for solute removal. The fundamental principle of this model is that the flow of water (solvent) governs the amount of solute removed and that the time needed to reach equilibrium (TC_s) is small compared with the residence time of the water in contact with the mineral phase, as described in Chapter 1.

In a strict sense, closed system conditions will never develop because the weathering system is in an open state of flux, with water entering and leaving the locus of solute production. Water in the weathering environment is thus not totally static, and therefore the residence time of the water in contact with the mineral must be considered. The application of closed system concepts is therefore limited in its interpretation of natural solute production, uptake and transport (Morgan, 1967). Also, different minerals will have different rates of reaction and different solubilities. The actual weathering mechanisms and important solute producing minerals in a complex mineral assemblage will be dependent on the residence time of the solvent (Trudgill, 1977; Deju and Bhappu, 1965). This may cause preferential dissolution of some minerals (Priesnitz, 1972).

The residence time of a solvent in the locus of solute production will be dependent on the prevailing hydrological regime and solvent flow processes; for example, overland flow, throughflow or groundwater percolation (Kirkby, 1978). Topography, infiltration, and the porosity and permeability of the solute producing medium must also be considered. The influence of hydrological regime, the interaction of rainfall and evaporation, must be taken into consideration as solutes will be produced and reprecipitated where potential evaporation losses are greater than actual rainfall inputs. In such a situation, chemical weathering will produce reprecipitated mineral crusts, such as silcretes and calcretes (Ollier, 1969), without any net chemical removal from the system.

There are three important parameters governing rates of solute removal. These are:

1. reaction rates of chemical weathering—releasing solutes;
2. solvent–mineral-phase contact;
3. solvent resident time.

These are controlled by:

(a) the solubility of the mineral phase in the solvent;
(b) the rate of transport of solvent to, and solvent plus solutes from, the locus of solute production;
(c) topography;

(d) hydraulic characteristics of the mineral-phase medium;
(e) hydrological regime;
(f) rainfall intensity and frequency.

Recent laboratory and field studies concerning dissolution of bedrock, mineral equilibrium and solvent residence times (Priesnitz, 1972; Trudgill *et al.*, 1980) suggest that for soil–rock–water weathering systems, residence times can be far shorter than those which would enable a state of chemical equilibrium to be achieved. Short-term kinetic behaviour would appear to be more important in controlling solute uptake. In such a dynamic system, closed conditions necessary for chemical equilibrium to develop do not exist, as the system is in a state of flux (Johnson, 1971). However, the system may be in a steady state with a constant mass balance of dissolution and solute output controlled by constant rate processes. This involves an open system of fluxes and not a closed system in equilibrium. The net load of solute transport will be constant over time, but solute concentrations will be dependent on the rainfall regime (Trudgill, 1977).

In summary, water flow is required for solute transport and solutional removal or erosion. This may be weathering limited, if solute transport rates are higher than weathering solute production rates or the amount of solute producing material is restricted; or solutional removal is transport limited because of the hydrological regime or flow processes.

8.1.5 The concept of solutional erosion and solutional denudation

Chemical weathering reactions and solute production are generally considered in the context of the chemical alteration of minerals to produce residual materials and ions in solution. The residual component is not strictly involved in a scheme of solutional erosion, which is the removal of material, ultimately by rivers draining to the sea. However, residual material such as clay minerals may be more susceptible to mechanical transport and erosion processes. Chemical weathering and solutional erosion should not be considered in isolation. If the solute load of a large river is calculated, then the net removal of material in solution, in excess of that input to the area from atmospheric sources, is a measure of the rate of solutional erosion. This can be expressed as:

$$S_{cw} = E_s - S_A \pm \Delta_S \qquad (8.4)$$

where E_s is net solutional transport, S_{cw} are solutes produced by chemical weathering, S_A are solutes from atmospheric sources, and Δ_S are changes in solutes stored in the ecosystem.

Such a simple budget equation usually assumes that solutes stored in the ecosystem as vegetation are involved in a quantitatively closed loop of nutrient cycling (Trudgill, 1977). However, depending on the vegetation status, this may not necessarily be the case, and the role of nutrient cycling should be considered (Likens *et al*., 1977). Solute budgets are lumped blackbox models of solutional erosion over an entire catchment. The results may be expressed as units of loss of mass (for example, kg km^{-2} yr^{-1}) or, by using an assumed rock density, a volumetric loss can be calculated (for example, m^3 km^{-2} yr^{-1}). By considering the catchment area, an annual rate of surface lowering can be estimated. Rates and formulae of calculating solutional erosion have been given for limestone in the literature for some years (Corbel, 1957; Williams, 1963; and Chapter 9); for example, the rate for the Mendip Hills, England (Corbel, 1957) was estimated as 40 m^3 km^{-2} yr^{-1}, equivalent to 40 mm surface lowering per thousand years. Comparison of surface lowering rates worldwide shows a great variation both between and within similar lithologies and climates (Waylen, 1979). Use of this type of unit is misleading and inappropriate. Values for surface lowering equivalent should only be used in a comparative sense and should not be considered as a direct measurement of surface lowering. Rates of weathering and erosion will not be uniform over a catchment. In a geomorphic context, the importance of solutional erosion in modelling the landscape lies in the spatial variation of solutional erosion. In particular the spatial variation in landscape lowering by solute processes must be considered and can be investigated by small-scale process studies.

In order to consider the solutional denudation component of solutional erosion, a conceptual framework for the system of solutional erosion needs to be proposed. This can be considered to have three main points:

1. Solutional erosion takes place when hydrological processes move water through a mineral phase medium.
2. Solute uptake occurs until the chemical composition of the moving water reaches a state of equilibrium with the mineral phase.
3. Geomorphologically effective solutional denudation involves the lowering of the bedrock surface by solute processes.

The first two points have already been considered. Acidic water moving through rock and soil material is neutralized by hydrolysis exchanging H^+ ions for solutes from a mineral. When the acidic water is neutralized, erosion ceases. The question of importance is, where does this neutralization take place? Obviously this will depend on water flow rates, flow routes, the availability of mineral material, contact times and reaction rates. It must not be assumed that neutralization will take place in the soil, at the soil–bedrock interface, or below the water table in the bedrock. What must be considered is a zone of weathering where reactions take place, or a weathering front, with its leading edge between weathered and unweathered material (Curtis *et al*., 1976). The locus of weathering will depend on the relative chemical composition of the soil material overlying the rock; if solute uptake can take place in the soil first, then bedrock dissolution at the soil–bedrock interface will be reduced. Alternatively, if the soil is deficient in easily weatherable minerals, acid water will pass relatively unaltered through the soil and the weathering front will be at the soil–bedrock interface. This is a simplistic but practical approach to understanding spatial variations in solution processes and solutional denudation. Few studies have actually examined the variation in erosion rates through a soil–bedrock weathering scheme. A study of limestone dissolution on the Mendip Hills (Atkinson and Smith, 1976) found that 10.1% of the total erosion took place in the soil, 57% in the uppermost 10 m of the limestone, 31.4% within the rock and only 1.5% in streams and cave passages.

The approach can be illustrated by considering the association of limestone bedrock erosion by solution and the nature of the overlying soil material (Trudgill, 1973, 1976). Weathering at the soil–limestone interface depends on how far the percolating acid soilwaters have been equilibriated with carbonates present in the soil. Limestone bedrock is protected under calcareous soil cover but is eroded under acid soil cover to produce a subsoil morphology of clints and grikes.

This simple example also serves to stress the importance of the nature and origin of the soil mantle overlying the rock. For this purpose a soil cover can be considered to have one of two origins.

In a 'classical' sense, the geomorphological interpretation of a soil mantle is as a residue produced by the *in situ* weathering of the bedrock. Soils contain the quartz, clay minerals and secondary silicates left after the solute component has been removed. The soil is chemically inert and weathering of the rock takes place beneath it. This is clearly the case in tropical regions where chemical weathering is rapid, and may have proceeded for a long period. This scheme of weathering follows the lines of the 'Aufbereitung' concept (Penck, 1924) of soil formation and slope development (Holmes, 1937; Beckett, 1968).

In temperate latitudes, where environmental conditions have changed dramatically within a short geological timespan, landforms are due to a range of processes and may exhibit features inherited from previous conditions. Soils may be depositional forms, not *in situ* weathering residuals. Where the timespan has been relatively short for soil development from deposited parent materials, the soils will not be in equilibrium with the environmental conditions. Then, depending on the chemical composition of the original deposits, the soil mantle may act to protect the underlying rock. In this case solutional erosion will produce soil mantle reduction and soil surface lowering until the soil minerals have been weathered and the solutes removed.

The nature and geomorphological importance of solutional erosion is in the spatial variations in solutional denudation over the landscape. These variations, for an area of a given climate, will be related to the distribution of soil and rock types and to variations in hydrological processes.

8.2 SOLUTIONAL EROSION AND CLIMATE: A GLOBAL PERSPECTIVE

8.2.1 Climate and solutional erosion

Climate can influence solute production by chemical weathering (Keller, 1957). The morphometric approach to climatic geomorphology (Peltier, 1950) has a major weakness in that it cannot be assumed that different zones produce different landforms. To investigate the relationship between solution and climate, the relationship between solution and other process factors must be considered first. Only then can the effect of climate and solution be considered.

It can be difficult to separate the effects of climate, lithology and topography because these may vary simultaneously. The purpose of this section is to compare the importance of solutional erosion on a global scale in terms of climate zones.

At the global scale it is difficult to isolate present influences from the effects of climatic change. Analysis of solutional erosion data (Harmon *et al*., 1972) suggests that regional climatic factors are relatively unimportant compared with local factors, such as lithology and hydrology. The effect of vegetation on chemical and hydrological processes must also be considered.

The role of morphoclimatic control has been exaggerated by many workers (Ollier, 1969). However, the interaction of lithology, solution processes and time are complex and often obscure. Also it is uncertain whether process intensity and time can compensate for one another. Therefore a morphometric or climatic classification is a useful but general approach for comparing the importance of solutional erosion on a global scale. Within a climate zone the effect of topography on mechanical erosion processes must be considered. Regions with a high degree of relief generally have a proportionally lower solutional erosion load (Leopold *et al*., 1964), as gradient is the major controlling factor in clastic load transport.

8.2.2 Solute load discharge in rivers

A great deal of work has been carried out to investigate the relationship between solute concentration, runoff and fluvial solute load transport.

For example, the solute content of rivers draining single rock types is essentially uniform (Miller, 1961). This implies a steady-state relationship between solute production and solute transport through a range of environmental conditions.

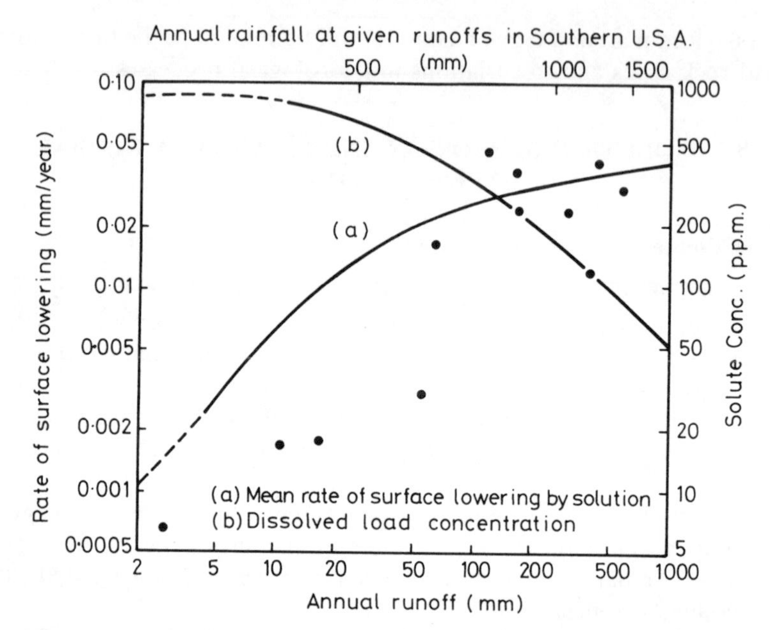

Figure 8.2. (a) Mean rate of surface lowering by dissolved load in USA (line a) and for a sample of streams in different climatic regions in USA (dots). (b) The same data expressed in terms of solute concentrations in streamwater (Carson and Kirkby, 1972) (Lines adopted from Langbein and Dawdy, 1964; sample dots from Leopold *et al.*, 1964, Table 3.10, p. 64). *Reproduced by permission of Cambridge University Press*

On a global scale, solutional erosion increases with runoff (Corbel, 1959) until the rate of dissolution becomes a controlling factor and solutional erosion is weathering-limited. If discharge increases beyond this point, transport is less effective in solute removal. Ultimately, if residence times are very short, solute concentrations will decrease. A typology of solute transport by major rivers (Maybech, 1976) can be developed on a morphoclimatic basis. Figure 8.2 shows the general relationships between annual runoff, solutional denudation and solute concentration in the southern USA. This illustrates that while generalized relationships can be proposed, field data (Leopold *et al.*, 1964) show great variability.

8.2.3 Climatic effects and morphoclimatic regions

The principal effects of climate on solutional erosion can be expressed in terms of temperature and hydrology. These are interrelated as the hydrological effects are related to the interaction of precipitation,

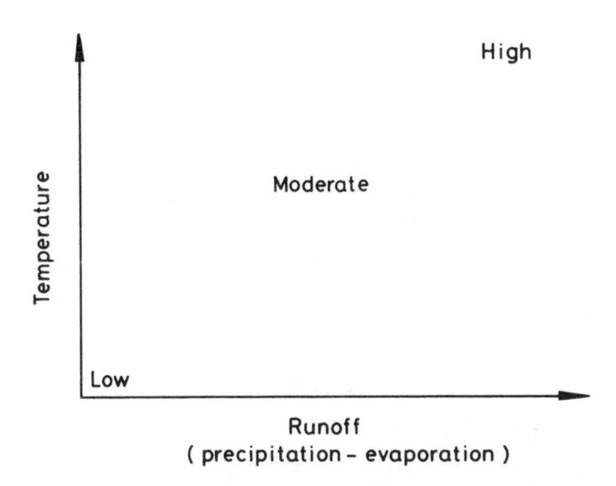

Figure 8.3. Relative importance of potential solutional erosion in different climates

evaporation, vegetation and temperature. The effect of climate on hydrology has been adequately discussed elsewhere (Kirkby, 1976). Rates of chemical reactions increase with temperature, and solutional erosion also increases with runoff. It can therefore be predicted that rates of solutional erosion will be high when runoff and temperature are high, as illustrated in Figure 8.3. This is a purely relative designation, expressed as 'high', 'moderate' and 'low' for comparison, assuming an adequate supply of mineral matter and contact time. This designation forms the basis for a morphoclimatic comparison of solutional erosion; for example, potential solutional erosion would be highest in the humid tropics and lowest in arid regions.

Peltier diagrams (Peltier, 1950) show a more formalized expression of this relationship, using rainfall and temperature to propose seven morphogenetic regions. These regions are hypothetical zones to illustrate the concept. A simpler expression of morphogenetic regions is shown in Figure 8.4. The proposed relative importance of mechanical and chemical processes is initiated by the order of listing.

An alternative approach is to use the continuity equation (Carson and Kirkby, 1972). A simple model can be proposed if it is assumed that:

1. the initial quantity of water involved in solute uptake is equal to the amount which infiltrates the soil and is a large proportion of the rainfall;
2. soilwater and solutes rapidly come to equilibrium;
3. soilwater and solutes are concentrated by evaporation until discharged from the catchment as runoff.

If the solute concentration rises above the saturation concentration, redeposition will occur within the soil. The model is a two-stage process involving deposition and removal. Figure 8.5 shows the calculated rates of solutional erosion for an igneous rock soil of constant composition with

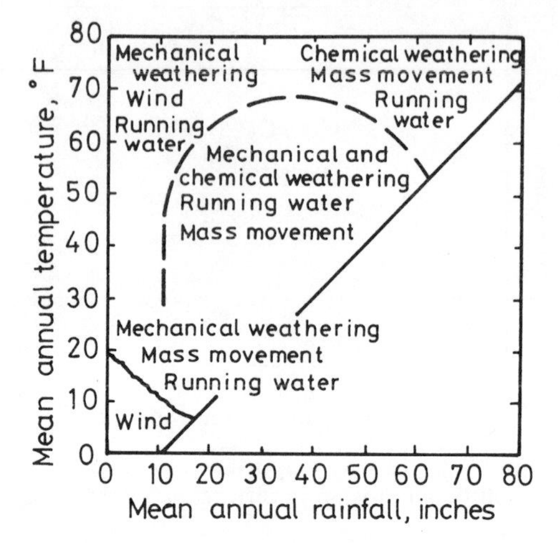

Figure 8.4. Hypothetical morphogenetic regions. The possible order of importance of specific processes in indicated, as listed, in decreasing order. *From Leopold et al.*, Fluvial Processes in Geomorphology, *W. H. Freeman & Company. Copyright © 1964*

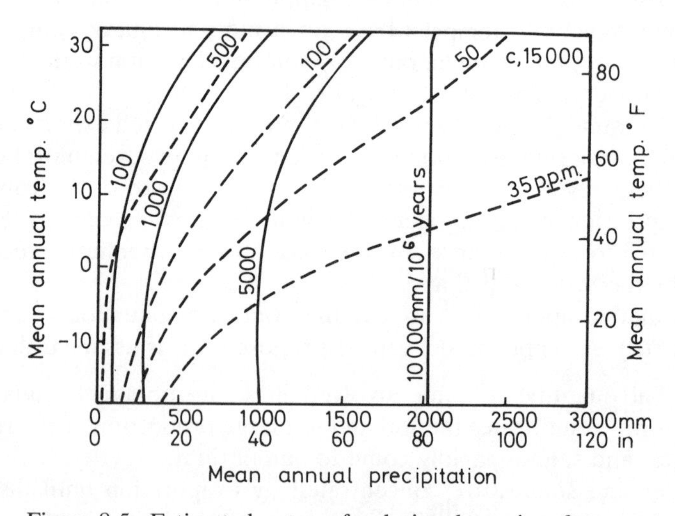

Figure 8.5. Estimated rates of solutional erosion from an igneous rock soil in different climatic conditions. Solid lines show total removal rates, for cations only (mm/10^6 years). Broken lines show predicted solute concentrations (cations and anions, parts in 10^6) (Carson and Kirkby, 1972). *Reproduced by permission of Cambridge University Press*

different climates, based on the continuity equation (Carson and Kirkby, 1972). The highest rates of solutional erosion are related to high precipitation, irrespective of temperature. The lowest rates are related to hot, dry climates where soilwater is saturated with solutes, leading to the formation of mineral crusts.

It should be remembered that it is the ratio of solution rates of different rock materials that produces the landforms of an area. For example, in arid regions granite weathers faster than limestone, so that in arid and semi-arid areas limestone is a relief former. In other climates granite is more resistant to solutional erosion and forms upland areas.

8.2.4 Solutional erosion rates in climatic regions

The relative importance of solutional erosion rates in the major climatic regions can be predicted on theoretical grounds. However, it must be stressed that these are only generalizations and take no account of local topography and lithology. Climatic influences on solute processes have not yet been investigated in any great detail.

In the rainforest zone (hot and wet), chemical weathering is rapid and even quartz may dissolve. Consequently solutional erosion may be the most important process of mineral removal, if minerals are available. Weathering may proceed at depth beneath a thick weathered mantle of regolith, or weathered residue (Berry and Ruxton, 1960; Ollier, 1969). Recent work suggests that solutional erosion may be limited by short residence times and high throughflow rates due to high rainfall intensities (Nortcliff *et al*., 1979).

In the Savanna zone, where rainfall is intermittent, rapid chemical weathering of crystalline rocks occurs at a depth below the water table, with a high irregular regolith of *in situ* residues (Thomas, 1966). In arid and semi-arid zones, solutional erosion is restricted by the lack of water for solute removal. Chemical weathering rates may be high and can produce depositional mineral crusts.

Solutional erosion may be the most important erosion process in the humid temperate zone. Chemical weathering is not restricted by weathering rates or availability of water for transport. The next section will review solutional erosion in this zone where most studies have been carried out.

In the polar and montane zones, solutional erosion is generally not important, compared with mechanical processes. Solutional erosion may be weathering-limited or transport-limited (Smith, 1969; Ollier, 1969).

Few studies have tried to investigate the effects of climate on solute processes by field comparisons of different regions. A study of this nature (Johnson *et al*., 1977) compared rates of hydrolysis in terms of soil leaching by carbonic acid in tropical rainforest, temperate and polar forest field sites. The effectiveness of leaching was found to decrease with temperature.

A similar result was found by comparing the weight loss of rock tablets from the same source emplaced in temperate woodland and tropical forest soils (Day *et al*., 1980). Weathering rates were found to be 3.5 times faster in the tropical environment. The difference in weathering rate was thought to be due to temperature differences influencing rates of reactions.

Detailed morphoclimatic models of solutional denudation at a global scale can only be related to an understanding of landform development by solute processes. This is an enormous task and requires a great deal more process-based studies; however, a theoretical foundation has been laid (Derbyshire, 1976). The theoretical assessment of climatic control on solutional denudation (Carson and Kirkby, 1972) in terms of process–response modelling has begun to be evaluated by field study (Saunders and Young, 1983).

8.3 CATCHMENT SCALE STUDIES OF SOLUTIONAL EROSION

8.3.1 Techniques

The main technique used in assessing rates of solutional erosion, at the scale of an individual river catchment, is solute load budgeting. This technique can be used to estimate relative rates of solution between catchments. However, quantitative estimates of surface lowering should be used with care, as they are catchment averaged rates and do not take into account any intracatchment variability produced by small-scale spatial variations in lithology or hydrological flow processes. In general, at a regional scale, catchments with a similar climate exhibit variations in solutional erosion in relation to the rates of weathering of the soil and rock materials producing solutes. This suggests that at the regional scale, for a fixed hydrological regime, solutional erosion is dominantly controlled by weathering rates, not transport of solutes. Individual examples may be transport limited or weathering limited, depending on the general climatic conditions.

Problems of solute budgeting techniques will be reviewed in terms of the actual calculation of fluvial discharge solute load. Other problems such as solute input–output relationships, nutrient cycling and presentation of results have been considered in §8.1.

A large number of solute budget studies have been carried out particularly for catchments with carbonate lithologies. To illustrate the use of solute budgeting techniques, and how they can be expanded, examples from the British Isles will be used. The purpose of this is to illustrate that, for the given climatic regime, humid temperate, there will be large variations in hydrological regime and lithology over for a comparatively small area. Other examples will be considered, but it must be stressed that generalizations cannot be made, and individual catchment solute processes may be best considered in isolation.

The major problem with a solute budgeting study lies in data collection. The technique involves the calculation of load by multiplying discharge and solute concentration. Ideally, to obtain an absolute budget, discharge and concentration should be monitored continuously. This is rarely practical, although continuous monitoring of discharge is often carried out. Solute sampling and analysis is a logistical drawback in obtaining a solute budget. One approach is to produce a summation of load based on actual discharge and one sample of solute concentration for a given time period. For example, for a weekly sampling scheme:

$$SL = \sum_{n=1}^{n=52} Q_n C_n \tag{8.8}$$

where SL is the annual solute load, Q_n is the stream discharge for week n, and C_n is the solute concentration for sample for week n. Obviously this procedure will produce errors as solute concentration will not necessarily be constant with time. The longer the sampling interval, the larger the error is likely to be.

While lithology can be the dominant controlling factor in spatial patterns of solute concentration (Webb and Walling, 1974; and see Chapter 7 for further discussion), hydrological regime, in terms of discharge variations, is the dominant controlling factor in temporal variations in solute concentration (Imeson, 1973; Walling, 1975). Solute concentrations can exhibit marked seasonal (long-term) and storm-based (short-term) temporal variations. Many workers have investigated temporal effects, such as dilution during periods of seasonally high flow. Hysteretic effects in the relationship between solute concentration and discharge during storm events have been shown to be positive (Edwards, 1973a) or negative (Toler, 1965). The decrease in solute concentration in storms can be attributed to the dilution of baseflow by short-residence-time storm runoff (overflow model), or increased solute concentrations can be attributed to displacement of long-residence-time water (plugflow model). While multivariate models can be used to explain long- and short-term temporal solute behaviour (Foster, 1978), hydrological flow mechanisms will be the most important factor (see Chapter 7).

An examination of the solute budget equation shows that the estimate of discharge will be the controlling factor in the actual load estimate. This will normally be several orders of magnitude greater than the measured concentration and may show proportionately greater magnitude temporal variations. Solute load estimates are therefore discharge-dependent (Crabtree, 1981), and the continuous measurement of discharge should be as accurate as possible.

The reliability of solute load budgets for different solute sampling schemes have been estimated for rivers in Devon, England (Walling, 1978). Weekly sampling, unless carried out for a period of several years, is insufficient to determine the extent of temporal variations in solutes in small streams. More frequent sampling such as daily or continuous sampling is required. Compared with continuous monitoring the results of daily, weekly and monthly sampling schemes produced errors of between −1% and +4%; −5% and +10%; and ±15% and ±20% respectively.

Load transport estimation errors can be minimized by using a statistically based sampling strategy incorporating random time intervals, (Frere, 1971; Yaksich and Verhoff, 1983) rather than a fixed time interval and fixed length of study for sampling. Finally, temporal land-use changes should be considered (Foster, 1979). A solute budget can only be considered as being representative of the period of study and it may not be valid to extrapolate such results over a long time period.

8.3.2 Examples of solute budget studies in the UK

Five studies will be used to illustrate the effects of lithology and hydrological regime in an area of generally similar climate.

Linear regression models of discharge and solute concentration were used to produce tentative solute budgets for two catchments in Norfolk, in the south-east of England (Edwards, 1973b). The results of this study were inconclusive because of the effects of intensive land-use, fertilizer applications and the input of solutes to rivers from sewage sources. Calculated average rates of erosion of silicon were 0.94×10^3 kg km^{-2} yr^{-1}. Average rates of erosion for calcium carbonate and calcium sulphate were 48.6 and 13.1×10^3 kg km^{-2} yr^{-1} respectively. For a small arable catchment in East Devon underlain by siliceous Permian sandstones and conglomerates (Foster, 1980) rates of denudation based on solute rating curves gave a net denudation rate of 22.8 ± 13.2 m^3 km^{-2} yr^{-1} for a three-year period.

A solute budget study based on weekly sampling, for a catchment underlain by a dolomite aquifer in eastern England (Crabtree and Trudgill, 1984) showed that dissolution of the dolomite bedrock produced Ca^{2+}, Mg^{2+} and HCO_3^- in solution. The solutional denudation component of the net output was far greater than atmospheric inputs. The output rates (in 10^3 kg km^{-2} yr^{-1}) for a one-year budget period were:

Ca^{2+}	18.0
Mg^{2+}	10.1
HCO_3^-	48.6
Si	0.5

In an upland catchment in the Lake District, underlain by andesites and granites, rates of solutional erosion were found to be negligible compared with the input of solutes from atmospheric sources. Estimated solutional erosion rates were (White *et al*., 1971):

Ca^{2+}	2.11–2.57
Mg^{2+}	0.58–0.72
Na^{+}	<0.8
Cl^{-}	<0.12
Si	1.35

A study of a catchment underlain by granite and gneiss in north-east Scotland (Reid *et al*., 1981) indicated considerable solutional erosion, yet atmospheric inputs were more important than chemical weathering in the discharge of solutes from the catchment. The loss of solutes produced by chemical weathering decreased in the order:

$$Si > Ca > Mg > K > Fe > Al$$

The weathering of plagioclase was considered to account for 75% of the chemical weathering output. Calculated solute output rates from solutional erosion were:

Ca^{2+}	1.70
Mg^{2+}	0.50
K^{+}	0.29
Na^{+}	0.90
Si	3.85
Fe	0.27
Al	0.18

8.3.3 Other selected examples

This is not intended to be an exhaustive review but to give an indication of some of the studies that have been undertaken, particularly in climatically different parts of North America. The most famous of these is the Hubbard Brook study (Likens *et al*., 1977) which investigated a temperate forest catchment. A study of a dolomite catchment, Walker Branch watershed (Henderson *et al*., 1977) yielded similar results to a UK example (Crabtree and Trudgill, 1984).

In northern Canada, a muskeg terrain catchment has been studied (Schwartz and Milne-Home, 1982). In the arid area of Colorado, solute uptake and transport from shales has been investigated (Laronne and Shen, 1982). Seasonal patterns of solutional erosion related to river discharge have been identified by solute studies in Newfoundland and Nova Scotia (Thompson, 1982).

Solute budget studies used to investigate solutional erosion and denudation are few outside the UK and USA. For example, the distribution of solutional erosion over the Cooleman Plains, New South Wales, Australia (Jennings, 1972) was found to be similar to the pattern found on the Mendip Hills, UK (Atkinson and Smith, 1976). Denudation rates per 1000 years of 11.3 mm have been calculated for a dolomitic limestone catchment in West Spitzbergen (Hellden, 1973), while a figure of 2.05 mm has been calculated for a limestone area of Somerset Island, northern Canada (Smith, 1969).

8.3.4 Extension of solute budgeting techniques

The problems and limitations of the conventional use of solute budgeting results, with the production of 'lumped' catchment solutional erosion rates, are such that the technique is at best a means of relative comparison between similar catchments. No interpretation can be made of intracatchment variability or the nature of the solute mechanisms involved. Two studies have shown how conventional solute budget data can be extended.

The first example (Waylen, 1979) illustrates the use of chemical thermodynamics to interpret solute budget data. By investigating the soil and rock chemistry for a siliceous catchment in south-west England, the solute output from predicted weathering reactions could be compared with the actual solutional denudation solute output. From this comparison tentative estimates of primary mineral alteration and the nature of the weathering processes were made.

The second example (Walling and Webb, 1978) illustrates a different approach to solute budget data presentation, which investigates and explains intracatchment variability of tributary streams (Walling and Webb, 1980). This approach produces solute load maps by spatial extrapolation of solutional rates, lithological groupings and hydrological schemes. Similar studies have been carried out a regional scale, for example, for Poland (Pulina, 1972).

8.4 SOLUTIONAL EROSION ON HILLSLOPES

8.4.1 Qualitative considerations

Solute process studies are best carried out at the scale of the individual hillslope, where studies can be carried out with fixed climatic and lithological controls. For this reason, most work at this scale has concentrated on hillslopes underlain by single rock types. Such studies have emphasized the role of hydrological processes on the spatial distribution of solutional erosion. However, few have actually measured spatial differences in solutional erosion and denudation or have considered the influence of the overlying soil mantle.

The investigation, identification and quantification of spatial variations in solutional denudation over individual hillslopes can be extended to solutional denudation over the landscape. This is essential to a geomorphological interpretation of solute processes.

Early theoretical considerations of the distribution of solutional erosion, and hence denudation over hillslopes, suggested either than rates would be uniform over a slope (Scheidegger, 1961) or that rates would increase downslope (Young, 1972). More complicated process–response models of slope development sequences for generalized lithologies and climates have been produced (Carson and Kirkby, 1972). These models are based on hydrological flow process models (Kirkby and Chorley, 1967; Kirkby, 1976) and solute transport continuity (Equation (8.6)). The models incorporate a soil development component (Kirkby, 1977) but relate to bedrock lowering rates at the soil–bedrock surface. Studies of direct bedrock lowering using micro-erosion meters (Trudgill, 1979; Spencer, 1981) are outside the scope of this chapter as their main interest is not solute processes but micromorphological development.

8.4.2 Process–response models of hillslope development by solutional erosion

Process–response models (Carson and Kirkby, 1972) can indicate the direction of hillslope development by solutional denudation for different climatological conditions and parent materials. The models make two important assumptions. First, the amount of material removed in solution is proportional to the flow of water. This suggests that solute removal is not weathering limited, and that solute concentrations reach equilibrium in a time shorter than the residence time of the water. Second, the potential for solution is constant over the slope. This implies that the distribution of weatherable minerals is uniform.

In general over a hillslope, soil moisture levels tend to increase downslope, owing to downslope water movement. Usually evapotranspiration does not proceed to the maximum potential rate (illustrated in Figure 6.13(a)). Figure 6.13(b) shows the change in downslope runoff contribution with rainfall multiplied by the proportion p of an oxide in the rock. The quantity of rock removed in solution is proportional to the lower value of either $p \times$ rainfall or runoff $\times$ 1.0 (Carson and Kirkby, 1972, p. 251). In Figure 6.13(b) the rate of solutional denudation is initially constant over the slope and then decreases downslope; the difference between the two lines indicates the amount of deposition. Three types of downslope variation in solutional denudation can be identified where:

1. rainfall $\times p <$ runoff over slope—solutional denudation is constant;
2. rainfall $- p >$ runoff over slope—solutional denudation decreases downslope;

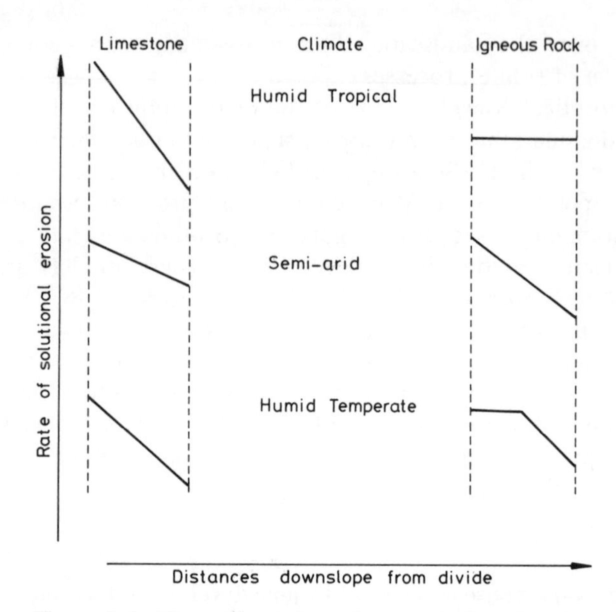

Figure 8.6. Generalized downslope variations in solutional erosion over igneous and limestone bedrock slopes for different climates. Adapted from process–response models (Carson and Kirkby, 1972)

3. intermediate cases—solutional denudation is initially constant, then decreases downslope.

Figure 8.6 shows the generalized downslope variation produced by this scheme for limestones and igneous rocks in different climates. Rates of solutional denudation for limestones will always decrease downslope, whereas for an igneous rock the pattern will change as the ratio of runoff to rainfall changes with increased runoff in more humid climates.

These simple variations of solutional denudation can be converted to hillslope development sequences, by solution alone, assuming an initially straight slope, fixed stream divide and no stream downcutting. These are shown in Figure 8.7. Taking the case of a limestone hillslope under temperate conditions, the model predicts that solutional denudation will increase upslope. With time this mode of development will lead to a slope decline. The basis of this is that soilwater content and solute concentration will increase downslope. While the potential evapotranspiration rate will remain constant downslope, actual evapotranspiration will increase downslope. This will lead to soilwater solute concentration and eventual redeposition. Solutional denudation rates will decrease downslope as the water moving through the soil becomes less aggressive.

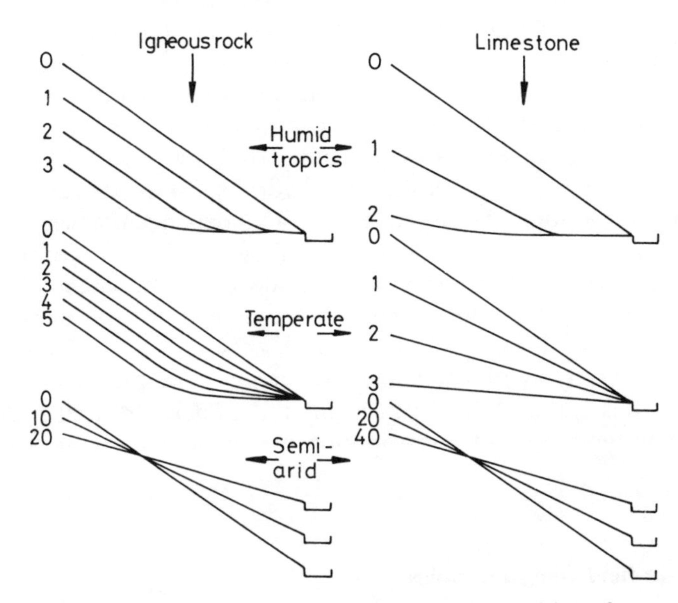

Figure 8.7. Slope development sequences resulting from solutional denudation only. Slopes are initially of uniform gradient with fixed divides and fixed stream position with no downcutting. Numbers indicate relative time periods to development stage (Carson and Kirkby, 1972). *Reproduced by permission of Cambridge University Press*

The simple process–response models show how the effects of climate and lithology influence the slope forms resulting from solutional denudation. For example, limestones will undergo slope decline in temperate climates, whereas other lithologies produce lower slope concavities. In hot, humid climates, high but uniform rates of solutional erosion lead to parallel slope retreat.

Lateral variations in soil moisture movement processes may also influence solutional denudation rates. Increased solutional erosion occurs in hillslope hollows (Burt, 1979) where water flow is concentrated and soils remain saturated.

The two main modelling assumptions of these process–response models are open to question and the models should be applied with care to individual hillslopes. Recent dissolution studies suggest that for a hillslope soil–rock–water system, residence times can be shorter than those which would enable a state of equilibrium to be achieved, and that short-term kinetic behaviour is more important in controlling solute uptake. The dynamic hillslope solute system is in a state of flux, and closed system equilibrium conditions will not develop. The first assumption may therefore be invalid in many cases.

The second assumption, of uniform potential solutional rates over a hillslope, must be examined empirically. The purpose of this is to investigate whether or not the distribution of potential solution rates as related to changes in soil type and solute uptake over a slope may be more important than the hydrological processes operating over the slope. This empirical approach involves the monitoring of hydrological processes and solute behaviour over a slope. At the same time, an independent assessment of the spatial and vertical distribution of solutional denudation rates over the slope should be undertaken. These two lines of evidence—hydrochemical processes and geomorphological effect—must be combined to assess the relative importance of hydrological and solutional processes on the development of an existing hillslope by solutional denudation under present-day conditions. The problem of this approach lies in assessing the influence of past hydrological processes on the present-day distribution of soluble material in the soil and rock.

8.4.3 Two field study examples

Patterns of solute removal from hillslopes have been inferred from the study of solute concentrations in streamwater, and yet the literature contains few measurements specifically related to the problem of subsurface erosion and slope development. Micro-weight-loss bedrock tablets (Trudgill, 1975) have been used to provide a direct measurement of solutional rates on hillslopes in relation to hydrological processes (Finlayson, 1977). There are limitations on the use of these micro-weight-loss techniques when measuring absolute erosion rates; but they can be adapted, by using standardized emplacement methods, to produce a spatial pattern of relative solutional denudation over a hillslope (Crabtree, 1981). Studies on two magnesian limestone hillslopes in eastern England showed that weight loss of tablets emplaced at the soil–bedrock interface, used as a surrogate for relative solutional denudation, was related to soil chemistry. At one slope, relative denudation rates were uniform over the slope, and soil chemistry (pH and carbonate content) showed little variation. Over a similar slope, where soil pH and carbonate content decreased upslope, relative solutional denudation rates increased upslope. This increase could not be attributed to soilwater movement processes as water movement over the slope was by vertical percolation. Statistically significant relationships linked tablet weight loss, pH and distance upslope. Table 8.2 shows the results from these slopes. Lower erosion rates at the slope bases were due to lateral movement of alkaline groundwater (Crabtree, 1981). These two slopes will develop by slope decline, owing to soil conditions not hydrological processes, because soils over the slope are

Table 8.2. Relative solutional denudation rates and changes in soil chemistry over two magnesian limestone hillslopes (Crabtree, 1981)

Distance upslope from stream (m)	Soil pH	Soil carbonate content (%)	Wt. loss (%)
SLOPE 1			
2.0	7.6	19.0	0.15
4.8	7.4	16.5	0.29
8.1	7.5	3.6	0.25
11.7	7.4	2.3	0.22
17.9	5.2	1.4	0.19
24.2	5.9	1.4	0.22
28.9	6.6	2.3	0.22
SLOPE 2			
1	7.9	18.9	0.10
10	6.7	3.0	0.27
20	6.7	5.2	0.21
30	6.6	2.1	0.31
40	5.4	2.6	0.31
40	5.4	2.6	0.31
50	5.6	3.7	0.37
60	4.9	1.8	0.62
70	5.2	0.0	0.39
80	4.6	2.8	0.48

depositional, not *in situ* weathering residues (Reeve, 1976). The process–response models are therefore inapplicable but suggest similar lines of slope development. A micro-weight-loss tablet study over a hillslope hollow, underlain by siliceous rocks (Crabtree and Burt, 1983) indicated an upslope increase in solutional denudation rates in relation to upslope changes in soil acidity. Solutional denudation rates in the hollow were greater than on the adjacent spurs, owing to the effects of soilwater flushing in the hollow (Burt, 1979). The base of the hollow had a reduced rate of solutional denudation due to the movement of solute-rich throughflow. The results of this study suggest that the topographic contrast between the spurs and hollow will continue to develop owing to hydrological processes, although the slope will decline because of the upslope soil changes.

The results of these two studies show the limitations of hydrological process–response models in explaining present-day spatial distributions of solutional erosion and denudation. More simple soil chemistry-based models are discussed below (and in Chapter 11) to predict the locus and relative rates of solutional erosion in a soil–rock–water weathering system.

8.5 GEOMORPHOLOGICAL SOIL CHEMISTRY-BASED MODEL

8.5.1 General considerations and model development

A geomorphological model of slope development by solutional erosion and denudation is proposed. This model relates to changes in soil chemical conditions and the resulting pattern of soil and rock solute uptake rather than hydrological conditions. The major assumption, which may not be generally applicable, is that the process of solutional denudation is weathering limited. The rate of solute production is controlled by rates of hydrolysis and the availability of weatherable minerals. Hydrological processes only act to transfer solutes from the locus of solute uptake. Hydrological conditions are relatively unimportant (except for slope base groundwater flushing) to the pattern of solutional denduation. However, the presence of water is necessary for mineral dissolution by hydrolysis. The results of the two field studies, described in §8.4, suggest that hydrological processes do not play a major role in controlling the spatial distribution of solutional denudation rates.

The model has two components. The first predicts the locus of solute uptake and solutional denudation. The second shows that following the first component, the pattern of relative solutional denudation over an individual hillslope will be related to the changes in soil chemistry over the hillslope, assuming a uniform bedrock lithology. The model of hillslope solutional denudation can be deduced from a knowledge of the interaction of pedological, lithological, chemical and hydrological processes over the hillslope. For example, rates of solutional denudation will increase upslope if soil acidity increases upslope. Assuming a uniform input of acidic

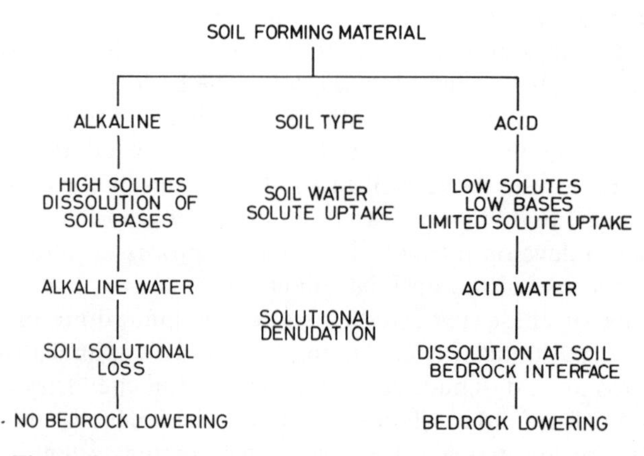

Figure 8.8. General solutional denudation model based on soil chemistry

precipitation to the slope, solute uptake and acid neutralization in the soil will decrease upslope and bedrock dissolution will increase upslope.

The first component of the model is illustrated in Figure 8.8. Under acid soil conditions, such as weathering residual soils, soilwater solute uptake will be limited by the lack of available bases, and acid water will move through the soil to bedrock. Dissolution and solutional erosion will then occur at and below the soil–bedrock interface, and thus lead to bedrock lowering. Under alkaline soils, for example depositional soils, acidic soilwater will lead to dissolution of soil bases and soil solute loss, but no bedrock lowering. It must be realized that Figure 8.8. illustrates the two extreme cases. Actual weathering patterns will be influenced by soil chemistry and soilwater residence times.

8.5.2 General applicability of a solutional denudation model based on soil chemistry

The model illustrated in Figure 8.8 can be used to predict that, if soil acidity increases upslope, rates of solutional denudation will also increase upslope. Two points must be considered to test the general application of the model:

1. Is dissolution by hydrolysis the dominant solutional erosion process?
2. Does the soil pH decrease upslope over all slopes?

If these two points are generally, rather than universally, applicable the simple soil chemistry-based model will predict that the solutional development of all hillslopes will lead to eventual slope decline.

If the model is generally applicable, it is nevertheless only related to slope development by solutional processes acting alone. In some situations mechanical processes may be more important as erosional agents. The model takes no consideration of lithological changes over the slope or lithologically-based pedological changes.

The first point to consider is the nature and influence of the solution process and the effect of soil type. The input of acid precipitation is a universal process and its effect is solute uptake. This is achieved, as previously discussed, by hydrolysis, in carbonate and non-carbonate terrains. Hydrolysis can, therefore, be considered to be universally important, although other chemical weathering processes may also contribute to the solute load.

Soils may be weathering residuals or mechanical deposits and may or may not be in equilibrium with present-day conditions. The influence of the original soil parent materials may be greater than present-day conditions of climate and hydrology. On theoretical geochemical grounds the relationship between erosion rates and soil types, especially with respect to soil pH, will hold for any type of material being weathered by hydrolysis, as illustrated in Figure 8.8.

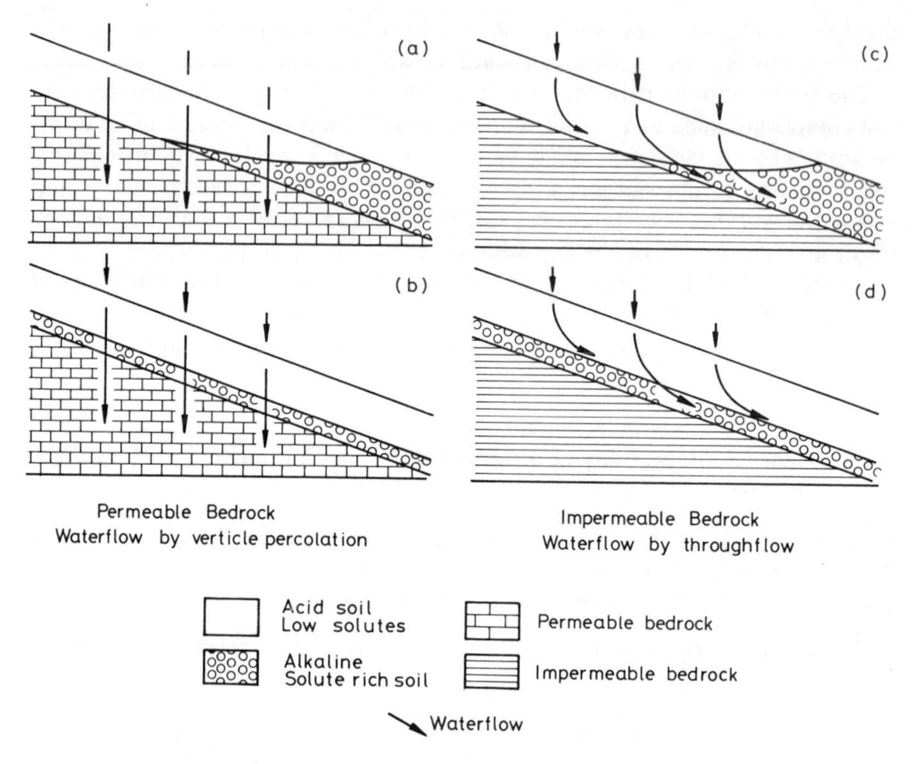

Figure 8.9. Theoretical combinations of soil and hydrological conditions over hillslopes

The second point to consider is upslope changes in soil chemistry. It is common for soil types and soil chemistry on slopes underlain by single rock types to change upslope. In general these upslope soil changes involve an increase in acidity and a decrease in base status. Numerous studies have investigated soil changes over slopes (Aandahl, 1948; Beckett, 1968; Furley, 1968, 1971). On chalk hillslopes soil chemistry has been related to distance downslope (Anderson and Furley, 1975). It was considered that the slope underwent solutional erosion while the valley bottom was an area of net deposition. Soil pH increased upslope.

Pedogenesis tends to produce a soil chemistry sequence whereby the acidity of the soil decreases downslope to a point near the slope base where conditions of acidity change to conditions of alkalinity. It cannot be determined whether these changes are due to initial pedological conditions or post soil alteration by hydrological and hydrochemical processes acting on the slope. Soil development will have been influenced by both. A general principle seems to operate, in that the upper parts of hillslopes undergo solutional denudation, solute uptake and transport, whereas the base of the slope is an area protected from solutional denudation by the transport and deposition of solutes under alkaline conditions. This model will hold for the general slope case with the expected soil chemical sequence over the slope. This may not be valid in specific conditions—for example, where soils are formed from calcareous deposits over slopes and acidic deposits in valley bottoms.

The theoretical combinations of the downslope distribution of hydrological and soil conditions are illustrated in Figure 8.9. This shows a permeable bedrock with and without downslope chemical change (a,b), and an impermeable bedrock (c,d). Combination (d) is unlikely to occur because a downslope increase in solute concentrations will be caused by throughflow. If (b) arises, solutional erosion will be uniform over the slope; but this situation is thought to be unlikely because soil mass movement processes lead to the accumulation of less weathered materials at the slope foot. This leaves (a) and (c) as likely combinations of soil and hydrological conditions, and such patterns of solutional erosion will be found on slopes (a) in relation to soil chemistry and (c) in relation to either the parent material distribution (as in (a)) or to throughflow solute processes.

It can therefore be proposed that the geomorphological model of hillslope solutional denudation and solute uptake in relation to the spatial variations in soil acidity can be applied to hillslopes under different climatic, hydrological. lithological, pedological and geochemical conditions.

This simple model has been proved to be more valid and applicable than previous models of hillslope development by solutional denudation based on assumptions of hydrological and hydrochemical processes.

8.5.3 Scaling-up hillslope solute processes to landform development

Spatial variations in hillslope solute processes, identified by detailed process investigation, can readily be scaled up to predict catchment and regional scale spatial variations in solutional denudation over the landscape. At the level of an individual catchment or hillslope, spatial variations in soil chemistry and hydrological process will be the dominant controlling factors. However, at a regional scale the effects of climate, past processes and especially lithology must be considered. Small-scale solute processes are insufficiently understood to be able to make a detailed evaluation of the role of solutional erosion and denudation on landform development, especially outside areas of temperate climate and carbonate lithology. Future studies must investigate the role of solute uptake and erosion over the landscape in relation to other mechanical processes. Finally the role of solutional denudation in landform development must be investigated empirically to obtain a greater understanding of the spatial distribution of solute processes.

ACKNOWLEDGEMENTS

Many thanks to Sîan for making English out of my English, to Sylvia for typing the manuscript and to Mary for producing the illustrations.

REFERENCES

Aandahl, A. R. (1948). The characterisation of slope positions and their influence on the total nitrogen content on a few virgin soils of Western Iowa. *Proc. Soil Sci. Soc. America*, **12**, 449–454.

Anderson, K. E., and Furley, P. A. (1975). An assessment of the relationship between the surface properties of chalk soils and slope form using principal components analysis. *Journal of Soil Science*, **26**, 130–143.

Atkinson, J. C., and Smith, D. I. (1976). The erosion of limestones. In: *The Science of Speleology* (T. C. Ford and C. H. Cullingford, Eds.), Academic Press, London, pp. 151–177.

Beckett, P. H. T. (1968). Soil formation and slope development. I: A new look at Walther Penck's Aufbereitung concept. *Zeitschrift für Geomorphologie, N.F.*, **12**, 1–24.

Berry, L., and Ruxton, B. P. (1960). The evolution of Hong Kong harbour basin. *Zeitschrift für Geomorphologie, N.F.*, **4**, 97–115.

Burt, T. P. (1979). The relationship between throughflow generation and the solute concentration of soil and streamwater. *Earth Surface Processes*, **4**, 257–266.

Carson, M. A., and Kirkby, M. J. (1972). *Hillslope Form and Process*, Cambridge geographical studies No. 3, Cambridge University Press.

Chesworth, W. (1973). The residual system of chemical weathering: a model for the chemical breakdown of silicate rocks at the surface of the earth. *Journal of Soil Science*, **24**, 69–81.

Chorley, R. J. (1971). The role and relations of physical geography. *Progress in Geography*, **3**, 87–109.

Corbel, J. (1957). Les karst du nord-ouest de l'Europe. Mem. Institut Etudes Rhodaniennes de l'Université de Lyon, No. 12, 544 pp.

Corbel, J. (1959). Vitesse de l'erosion. *Zeitschrift für Geomorphologie, N.F.*, **3**, 1–28.

Crabtree, R. W. (1981). *Hillslope Solute Sources and Solutional Denudation on agnesian limestone*. Unpublished PhD thesis, University of Sheffield.

Crabtree, R. W., and Burt, T. P. (1983). Spatial variation in solutional denudation and soil moisture over a hillslope hollow. *Earth Surface Processes and Landforms*, **8**, 151–160.

Crabtree, R. W., and Trudgill, S. T. (1984). Hydrochemical budgets for a Magnesium Limestone catchment in lowland England. *Journal of Hydrology*, **74**, 67–79.

Curtis, C. D. (1976). Stability of minerals in surface weathering reactions: a general thermochemical approach. *Earth Surface Processes*, **1**, 63–70.

Curtis, L. F., Courtney, F. M., and Trudgill, S. T. (1976). *Soils in the British Isles*, Longman, London.

Day, M. J., Leigh, C., and Young, A. (1980). Weathering of rock discs in temperate and tropical soils. *Zeitschrift für Geomorphologie, N.F.*, Supp. Band. **35**, 11–15.

Deju, R. A., and Bhappu, R. B. (1965). Surface properties of silicate minerals. New Mexico Institute of Mining and Technology, State Bureau of Mines and Technology, Circular **82**, pp.67–70.

Derbyshire, E. (1976). *Geomorphology and Climate*, Wiley, London.

Edwards, A. M. (1973a). The variation of dissolved constituents with discharge in some Norfolk rivers. *Journal of Hydrology*, **18**, 219–242.

Edwards, A. M. (1973b). Dissolved load and tentative solute budgets of some Norfolk catchments. *Journal of Hydrology*, **18**, 210–217.

Finlayson, B. (1977). Run off contributing areas and erosion. School of Geography, University of Oxford, Research Paper 18.

Foster, I. D. (1978). A multivariate model of storm-period solute behaviour. *Journal of Hydrology*, **39**, 339–353.

Foster, I. D. (1979). Intra-catchment variability in solute response: an east Devon example. *Earth Surface Processes*, **4**, 381–394.

Foster, I. D. (1980). Chemical yields in runoff, and denudation in a small arable catchment: east Devon, England. *Journal of Hydrology*, **47**, 349–368.

Frere, M. H. (1971). Requisite sampling frequency for measuring nutrient and pesticide movement in runoff waters. *Journal of Agricultural and Food Chemistry*, **19**, 837–839.

Furley, P. A. (1968). Soil formation and slope development. 2: The relationship between soil formation and gradient angle in the Oxford area. *Zeitschrift für Geomorphologie, N.F.*, **12**, 25–42.

Furley, P. A. (1971). Relationships between slope form and soil properties developed over chalk parent materials. In: *Slope Form and Process* (D. Brunsden, Ed.), Institute of British Geographers, Special Publication 3, pp. 141–163.

Garrels, R. M. (1967). Genesis of some groundwaters from igneous rocks. In: *Research in Geochemistry* (Vol. 2) (P. H. Abelson, Ed.), Wiley, New York, pp. 405–420.

Garrels, R. M., and Christ, C. L. (1965). *Solutions, Minerals and Equilibria*, Harper and Row, New York.

Goldich, S. S. (1938). A study in rock weathering. *Journal of Geology*, **46**, 17–58.

Harmon, R. S., Hess, J. W., Jacobson, R. W., Shuster, E. J., Haygood, C., and White, W. B. (1972). The chemistry of carbonate denudation in northern America. *Transactions Cave Research Group, Great Britain*, **14**, 96–103.

Hellden, U. (1973). Some calculations of the denudation rate in a dolomitic limestone area at Isfjord-Radio, West-Spitzbergen. *Transactions Cave Research Group, Great Britain*, **15**, 81–87.

Henderson, G. S., Hunley, A., and Selvidge, W. (1977). Nutrient discharge from Walker Branch Watershed. *Watershed Research in North America*, **1**, 307–322.

Holmes, J. M. (1937). The growth of soils on slopes. *Proc. Linn. Soc. NSW.*, **62**, 230–242.

Imeson, A. C. (1973). Solute variations in small catchment streams. *Transactions Institute of British Geographers*, **60**, 87–99.

Jennings, J. N. (1972). The Blue Waterholes, Cooleman Plain, N.S.W., and the problem of Karst denudation rate determination. *Transactions Cave Research Group, Great Britain*, **14**, 109–117.

Johnson, D. W., Cole, D. W., Gressel, S. P., Singer, M. J., and Minden, R. V. (1977). Carbonic acid leaching in a Tropical, Temperate, Subalpine and Northern Forest Soil. *Arctic and Alpine Research*, **9**, 329–343.

Johnson, N. M. (1971). Mineral equilibria in ecosystem geochemistry. *Ecology*, **52**, 529–531.

Keller, F. (1957). *Principles of Chemical Weathering*, Lucas Bros., Columbia, Missouri.

Kirkby, M. J. (1976). Deterministic continuous slope models. *Zeitschrift für Geomorphologie, N.F.*, Supp. Band. **25**, 1–19.

Kirkby, M. J. (1977). Soil development models as a component of slope models. *Earth Surface Processes*, **2**, 203–230.
Kirkby, M. J. (1978). *Hillslope Hydrology*, Wiley, Chichester.
Kirkby, M. J., and Chorley, R. J. (1967). Throughflow, overland flow and erosion. *Int. Assoc. Sci. Hydrol. Bull.*, **12**, 5–12.
Kronberg, B. I., and Nesbitt, H. W. (1981). Quantification of weathering, soil geochemistry and soil fertitily. *Journal of Soil Science*, **32**, 453–459.
Langbein, W. B., and Dawdy, D. R. (1964). Occurrence of dissolved solids in surface waters in the United States. United States Geological Survey Professional Paper No. 501 D, pp. 115–117.
Laronne, J. B., and Shen, H. W. (1982). The effect of erosion on solute pickup from Mancos Shale hillslopes, Colorado, U.S.A. *Journal of Hydrology*, **59**, 189–207.
Leopold, L. B., Wolman, M. G., and Miller, J. P. (1964). *Fluvial Processes in Geomorphology*, W. H. Freeman & Co., New York.
Likens, G. E., Borman, F. H., Pierce, R. S., Eaton, J. S., and Johnson, N. M. (1977). *Biogeochemistry of a Forested Ecosystem*, Springer-Verlag, New York.
Loughnan, F. C. (1969). *Chemical weathering of silicate minerals*, American Elsevier, New York.
Maybech, M. (1976). Total mineral dissolved transport by world major rivers. *Hydrological Sciences Bulletin*, **21**, 265–284.
Miller, J. P. (1961). Solutes in small streams draining single rock types—Sangre de Cristo range, New Mexico. United States Geological Survey Water Supply Paper No. 1535F.
Morgan, J. J. (1967). Applications and limitations of chemical thermodynamics in natural water systems. In: *Equilibrium Concepts in Natural Water Systems* (R. F. Gould, Ed.), American Chemical Society, Advances in Chemistry Series No. 67, pp. 1–29.
Nortcliff, S., Thornes, J. B., and Waylen, M. J. (1979). Tropical forest systems: a hydrological approach. *Amazoniana*, **4**, 557–568.
Ollier, C. D. (1969). *Weathering*, Geomorphology Texts No. 2, Longman, London.
Peltier, L. (1950). The geographic cycle in periglacial regions as it is related to climatic geomorphology. *Ann. Assoc. Amer. Geog.*, **40**, 214–236.
Penck, W. (1924). *Die morphologische Analyse. Ein Kapitel der physikalischen Geologie*, Engelhorns, Stuttgart.
Picknett, R. G. (1964). A study of calcite solution at 10°C. *Transactions Cave Research Group, Great Britain*, **7**, 39–62.
Priesnitz, K. (1972). Methods of isolating and quantifying solution factors in the laboratory. *Transactions Cave Research Group, Great Britain*, **14**, 153–158.
Pulina, A. (1972). A comment on present day chemical denudation in Poland. *Georgr. Pol.*, **23**, 45–62.
Reeve, M. J. (1976). Soils in Nottingham III. Sheet SK57 (Worksop). *Soil Survey Record*, 33. Soil Survey of England and Wales, Harpenden.
Reid, J. M., Macleod, D. A., and Cresser, M. S. (1981). The assessment of chemical weathering rates within an upland catchment in north-east Scotland. *Earth Surface Processes and Landforms*, **6**, 447–457.
Saunders, I., and Young, A. (1983). Rates of surface processes on slopes, slope retreat and denudation. *Earth Surface Processes and Landforms*, **8**, 473–501.
Scheidegger, A. E. (1961). *Theoretical Geomorphology*, Springer-Verlag, Berlin.
Schwartz, F. W., and Milne-Home, W. A. (1982). Watersheds in muskeg terrain. I: The chemistry of water systems. *Journal of Hydrology*, **57**, 267–290.
Smith, D. I. (1969). The solution of limestones in an arctic morphogenetic region. *Studia Geographica*, **5**, 100–107.

Spencer, T. (1981). Microtopographic change on calcarenites, Grand Cayman Island, West Indies. *Earth Surface Processes and Landforms*, **6**, 85–94.

Stumm, W., and Morgan (1970). *Aquatic Chemistry*, Interscience, New York.

Thomas, M. F. (1966). Some geomorphological implications of deep weathering patterns in crystalline rocks in Nigeria. *Transactions Institute of British Geographers*, **40**, 173–193.

Thompson, M. E. (1982). The cation denudation rate as a quantitative index of sensitivity of eastern Canadian rivers to acidic atmospheric precipitation. *Water, Air and Soil Pollution*, **18**, 215–226.

Toler, L. G. (1965). Relation between chemical quality and water discharge in Spring Creek, southwestern Georgia. United States Geological Survey Professional Paper No. 525C, pp. 209–213.

Trudgill, S. T. (1973). Limestone erosion under soil. *Proc. 6th Int. Cong. of Speleology, Olomouc, Czechoslovakia*, pp. 409–422.

Trudgill, S. T. (1975). Measurement of erosional weightloss of rock tablets. British Geomorphological Research Group Technical Bulletin, No. 17.

Trudgill, S. T. (1976). The erosion of limestones under soil and the long term stability of soil-vegetation systems on limestone. *Earth Surface Processes*, **1**, 31–41.

Trudgill, S. T. (1977). *Soil and Vegetation Systems*, Oxford University Press.

Trudgill, S. T. (1979). Surface lowering and landform evolution on Aldabra. *Philosophical Transactions of the Royal Society*, B, **286**, 35–45.

Trudgill, S. T., Laidlaw, I. M. S., and Smart, P. L. (1980). Soil water residence times and solute uptake on a dolomite bedrock—preliminary results. *Earth Surface Processes*, **5**, 91–100.

Walling, D. E. (1975). Solute variations in small catchment streams: some comments. *Transactions Institute of British Geographers*, **64**, 141–147.

Walling, D. E. (1978). Realiability considerations in the evaluation and analysis of river loads. *Zeitschrift für Geomorphologie, N.F.*, Suppl. Band. **29**, 29–42.

Walling, D. E., and Webb, B. W. (1978). Mapping solute loadings in an area of Devon, England. *Earth Surface Processes*, **3**, 85–99.

Walling, D. E., and Webb, B. W. (1980). The spatial dimension in the interpretation of stream solute behaviour. *Journal of Hydrology*, **47**, 129–149.

Waylen, M. J. (1979). Chemical weathering in a drainage basin underlain by old red sandstone. *Earth Surface Processes*, **4**, 167–178.

Webb, B. W., and Walling, D. E. (1974). Local variation in background water quality. *The Science of the Total Environment*, **3**, 141–153.

White, E., Starkey, R. S., and Saunders, M. J. (1971). An assessment of the relative importance of several chemical sources to the waters of a small upland catchment. *Journal of Applied Ecology*, **8**, 743–749.

Williams, P. J. (1963). An initial estimate of the speed of limestone solution in County Clare, Ireland. *Irish Geography*, **4**, 432–441.

Yaksich, S. M., and Verhoff, F. H. (1983). Sampling strategy for river pollutant transport. *Journal of Environmental Engineering*, **109**, 219–231.

Young, A. (1972). *Slopes*, Longman, London.

Solute Processes
Edited by S. T. Trudgill

CHAPTER 9

Solute processes and karst landforms

John Gunn

Department of Environmental and Geographical Studies,
Manchester Polytechnic

'In some ways geomorphologists seem to have been left on their own, and to have made most progress, in the area of limestone studies'

(Clayton, 1980, p. 173)

In reviewing the contribution of British geomorphologists to *Geography Yesterday and Tomorrow*, Clayton has been particularly generous to karst researchers. Whether or not this is entirely deserved is arguable, but it is certainly welcome as karst has to a large extent lain outside the mainstream of geomorphological (and hydrological/hydrogeological) research in Britain. The same would also appear to be true of many other parts of the English-speaking world, although in the USA there has been considerable interest in karst hydrology and water resources and this is particularly evident in the volumes edited by Yevjevich (1976, 1981) and Dilamarter and Csallany (1977). This theme is taken up in §9.4 of the present review, where progress in karst hydrology and water tracing is discussed and a revised conceptual model of limestone drainage is presented. During the late 1970s and early 1980s advances in the quantitative analysis of cave sediments, and particularly the absolute dating of speleothems, achieved a new prominence for karst. Data from caves have now been used to establish and revise Pleistocene chronologies, to infer palaeoclimates, to determine rates of uplift and sea-level change, and to reconstruct the evolution of both specific landforms and large tracts of country (§9.5). Exciting though these advances are, it must be remembered that the prime reason for identifying karst as a specific branch of geomorphology is the dominance of one erosive process in landform evolution—the aqueous solution of minerals. Hence, the present chapter includes discussion of both solution processes (§9.2) and solutional erosion rates (§9.3). Finally, it is necessary to consider just what is meant by karst. Clayton (1980) avoids the term, preferring instead 'limestone studies'. Others would agree with Bloom (1978, p. 136) that 'other soluble-rock terranes may be karst and limestone weathering in some regions is not characterized by solution'. As this question is fundamental it is considered in §9.1 and forms a precursor to more detailed discussion of processes and landforms.

9.1 THE NATURE OF KARST

The term 'karst' is derived from the Slavian word krš, which originally had a toponymic meaning, defining geographic areas with similar landforms and drainage to those in the limestone country of western Slovenia (Herak and Stringfield, 1972). The subsequent passage of karst into international usage has been somewhat imprecise, and although several definitions have been proposed none has achieved universal acceptance (Jakucs, 1977). There is general agreement that karst systems are distinguished from others by the predominance of one erosion process, the aqueous solution of minerals. Other processes may contribute to the final dimensions of karst landforms, but in all cases solution is an essential preliminary stage, for example in collapse (cavitation). Formation of karst landforms is therefore related to the occurrences of specific soluble rocks, although there is no consensus as to which rocks should be classed as karstic. Hence, the characteristics of karst rocks must be examined prior to discussion of solution processes and landforms.

The principal prerequisite for a rock to be classified as karstifiable is 'a higher degree of rock solubility in natural waters than is found elsewhere' (Jennings, 1971, p. 1). On these grounds three rock groups are potentially karstifiable: silicates, evaporites and carbonates. However, a further distinction is usually made between rocks in which the individual mineral components are dissolved leaving little residual matter (evaporites, carbonates), and those in which solutional weathering disintigrates the minerals, creating large quantities of insoluble residues (silicates). Only the former group are regarded as true karst rocks, although Löffler (1978) has disputed this distinction.

9.1.1 Silicates

Silicate rocks are quite soluble, particularly under humid tropical conditions (Douglas, 1978), but the weathering process produces insoluble clays and hydroxides. As these residues accumulate they block the solutionally widened interstices and inhibit the development of secondary permeability and underground drainage which is characteristic of most karst regions outside of the permafrost zone. Hence, the small solution pools and solution grooves which resemble karst forms and which form on bare silicate rock surfaces in both humid temperate areas (LeGrand, 1952; Reynolds, 1961; Reed *et al.*, 1963; Clayton, 1966) and humid tropical areas (Palmer, 1927; Tschang, 1961; White *et al.*, 1966; Wall and Wilford, 1966) are generally referred to as pseudokarst landforms, although more specific terms such as silikatkarren (Jennings, 1971) and pseudokarren (Tschang, 1961) are also used. However, Löffler (1978) has argued that since these features result from high rock solubility in natural waters they should be regarded as true karst landforms. Pseudokarst would then be restricted to those features bearing a superficial resemblance to karst landforms but formed by processes other than solution, the main categories being piping, in which clastic sediment is removed by subsurface water (e.g. Parker *et al.*, 1974; Löffler, 1974), and lava tube caves and related landforms in rocks of volcanic origin (e.g. Halliday, 1960; Wood, 1976).

In support of his argument Löffler (1978, p. 248) describes a suite of landforms 'similar to those usually found in limestone karst from temperate regions', but developed on dunite, a magnesium-rich silicate rock. The landforms include closed depressions, blind valleys with stream-sinks, and a few short caves but no evidence of integrated underground drainage or of a steady growth of secondary permeability is provided. Hence, even this exceptional area cannot be regarded as a true karst, since the landforms are essentially limited to surface features.

It is therefore apparent that a widening of the definition of karst to include all landforms resulting from solution, and in particular solutional landforms on silicate rocks, would result in an undesirable loss of precision (Sweeting, 1972). However, the morphological similarities between silicate pseudokarst landforms and true karst landforms suggest that karst processes form part of a continuum of processes operating in natural landscapes, and this is considered further in §9.3.

9.1.2 Evaporites

The principal evaporite rocks are halite (rock salt), anhydrite and gypsum. All are soluble in pure water, leaving little or no residue, and they are therefore regarded as true karst rocks. However, their solubility products are so high that they are effectively infinite under most climatic regimes, and as a result evaporite karsts can only survive in areas of relatively low rainfall (Nicod, 1976).

Halite (NaCl) is the most soluble evaporite rock and is therefore rarely exposed at the surface in pluvial climates. However, karstification can take place at depth, forming salt caverns which may in turn collapse and give rise to closed depressions on the surface (Pfeiffer and Hahn, 1972; Gustavson *et al.*, 1982). Gypsum ($CaSO_4.2H_2O$) is less soluble than halite and forms surface outcrops in semi-arid areas. Karstification is by simple solution, and characteristic landforms such as karren, dolines, stream-sinks and risings, and caves may be formed (e.g. Wigley *et al.*, 1973). Anhydrite ($CaSO_4$) is the most common evaporite rocks as it is preferentially deposited when seawater is evaporated. On contact with freshwater it is converted into gypsum by hydration, and this process is accompanied by an increase in volume of about 36.5% (Jakucs, 1977, p. 79). Where the overburden is sufficiently thick and/or strong to resist the stresses generated, hydration is prevented and anhydrite may remain stable over long geological periods. Jakucs (1977, p. 79) has suggested that as little as 5 m of gypsum overburden may forestall further hydration, although Pfeiffer and Hahn (1972, p. 212) suggest that conversion of anhydrite to gypsum may occur at depths of up to 30–40 m.

While the depth at which cessation of hydration occurs may be debatable, it is clear that any joints of fissures which penetrate anhydrite are likely to close up after a short time as a result of volumetric expansion, and that this will inhibit development of secondary permeability and karstic underground drainage (Fulda, 1912; cited by Jakucs, 1977). Hence, full development of evaporite karst is only likely in semi-arid areas where primary gypsum beds (deposited by freshwater), secondary gypsum beds (developed by hydration of anhydrite) or mixed gypsum–halite beds (Baker, 1977; Gustavson *et al.*, 1982) outcrop. These conditions are very restrictive, and as a result evaporite karsts occupy a very small proportion of the total world area of karst (Sweeting, 1981, p. xiii) and will not be considered further in this chapter.

9.1.3 Carbonates

Carbonates occupy about 15% of the earth's land surface and are the most commonly occurring rocks of high solubility in natural waters. They are the dominant karst rocks, although karst development does not take place on all carbonates. For example, the Cretaceous chalks and Jurrassic limestones of Great Britain are largely devoid of karst landforms. Although this may in part be due to a low mechanical strength (Sweeting, 1972), it seems that the high primary porosity of these rocks is of prime importance. This results in solvent attack being expended on the abundant pores at the soil–rock interface, with the consequent development of diffuse flow aquifers rather than the conduit flow systems with point-centred forms (depressions, caves) which are typical of karst terrains (Ford, 1980a).

By definition all rocks containing 50% or more by weight of carbonate minerals are classed as limestones, although the term is ambiguous since it covers a polygenetic group of rocks with a wide variety of properties. In the present account the broad usage will be followed except where more specific mention is required. However, a brief account of the nature of limestones is necessary as bulk chemical purity and 'petrovariance' (Jakucs, 1977) exert important controls on rock solubility and karst landform development (Sweeting, 1976, 1979; Ford, 1980a).

The main chemical constituent of most limestones, calcium carbonate, exists in three crystalline polymorphs—calcite, aragonite and vaterite—which differ from one another in the spatial arrangement of calcium and carbonate ions in the crystal structure. Vaterite is the least stable and changes to calcite or aragonite very rapidly in the presence of water. Aragonite (orthorhombic crystals) is often a major constituent of limestones deposited in warm seas, and particularly of coral reef limestones; but it is metastable and 16% more soluble than calcite, so that it normally recrystallizes to calcite over the course of time (diagenesis). Hence, calcium carbonate is almost always present in limestones as the mineral calcite (hexagonal-rhombohedral crystals), although both calcite and aragonite may be precipitated from supersaturated solutions in caves (Picknett *et al.*, 1976).

Calcium carbonate may be extremely pure but it is often associated with magnesium carbonate as part of the isomorphous series $CaCO_3$–$MgCO_3$. The end members of this series are calcite and dolomite ($CaMg(CO_3)_2$), a double carbonate which contains the equivalent molecular proportions of 54.4% $CaCO_3$ and 45.6% $MgCO_3$. Primary dolomite is rare, and dolomitization is usually a secondary process (diagenesis) in which magnesium is introduced into the calcite molecule. The process results in a theoretical reduction of molecular volume of 12% (Ford, 1976), and as a result the primary permeability of carbonate rocks tends to increase with their content of dolomite mineral. The relative proportions of calcite mineral, dolomite mineral and impurities were used by Leighton and Pendexter (1962) to provide more detailed definitions for mixed carbonate rocks (Figure 9.1).

Several other systems for the classification of limestones have been proposed (e.g. Folk, 1959; Chilingar *et al.*, 1967; Hatch *et al.*, 1971; Bathurst, 1972), and Ford (1976) has provided a useful summary of the various classes. High calcium limestones (limestone *sensu stricto*) are the most abundantly occurring and most widely studied carbonate rocks. They are also the rocks on which karst landforms are best developed since the presence of Mg as part of the calcite lattice, as $MgCO_3$, or as dolomite mineral usually decreases rock solubility (Picknett *et al.*, 1976; Sweeting, 1979). Rocks containing dolomite mineral are generally much less soluble than high-calcium limestones, although it has been argued that dolomite solution is temperature-dependent so that striking karst landforms may develop on some tropical dolomites (Marker, 1970; Sweeting, 1979). Under temperate conditions weathering of dolomitic limestones and dolomites often leaves angular grains of residual dolomite 'sands' which may inhibit karstification (Bogli, 1980). However,

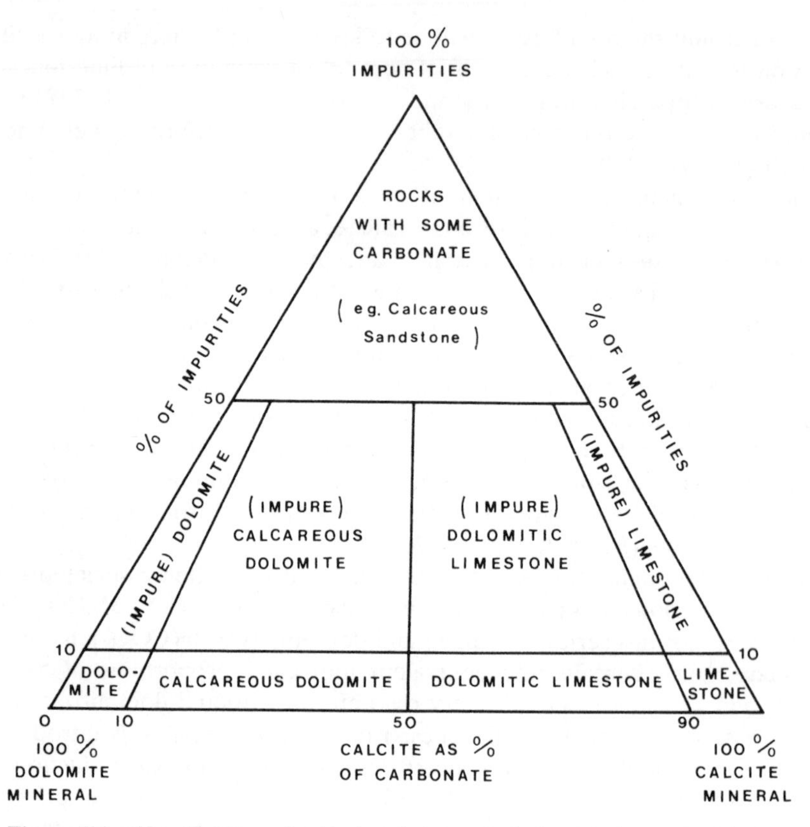

Figure 9.1. Classification of mixed carbonate rocks by relative proportions of calcite mineral, dolomite mineral and impurites (Leighton and Pendexter, 1962)

dolomites on the Bruce Peninsula of Ontario, Canada, display 'one of the most mature assemblages of karst landforms of postglacial age in eastern Canada, and well developed groundwater circulation' (Cowell and Ford, 1980, p. 525), and this illustrates the complexity of rock control on karstification.

Insoluble impurities in the limestone may also inhibit the karstification process, although their role is somewhat equivocal. Ford (1980a) suggests that a fully karstic landform assemblage (i.e. a holokarst) is unlikely to develop where the limestone comprises more than 20–30% insolubles, although a parakarst (Birot, 1954) or fluviokarst (Roglic, 1960) with a low density of small dolines in an otherwise fluvial landscape may form on limestones of low bulk purity (e.g. arenaceous limestones) and on some non-limestones which have a relatively high carbonate content (e.g. calcareous sandstones). However, Sweeting (1979, p. 107) cites the widespread occurrence of karstic features in limestones containing up to 40% quartz as evidence that 'the presence of quartz does not reduce the actual solubility of limestone significantly', although its presence controls other types of erosion, particularly abrasion. Jakucs (1977) also concluded that the concentration of non-carbonate constituents is not a primary control on limestone solubility.

Hence it would seem that impure limestones, like dolomitic limestones and dolomites, have a potential to develop a holokarst assemblage. Whether that potential is realized is largely controlled by other factors such as structure, porosity and permeability which influence the movement of water, and hence of insoluble residues. Where the residues can be removed (as in the Gordon Limestone, Tasmania, cited by Sweeting (1976)), caves and a full range of karst landforms may develop. By way of contrast, accumulation of residues will choke development of a karstic groundwater system and prevent the growth of a full holokarst.

'Rock control' (Ford, 1980a) on karstification and karst landform variability extends beyond carbonate geochemistry to include petrology, lithology, texture, structural features, and position of the carbonates in the gross stratigraphy. These factors are considered in some detail by Sweeting (1976, 1979), Jakucs (1977) and Ford (1980a) but are beyond the scope of the present discussion. However, it can be stated with some certainty that the 'ideal' karst rocks are massive, well-jointed, high-calcium limestones, which, fortuitously, are the most common of the carbonate rocks.

9.2 SOLUTION PROCESSES

The solution chemistry of limestone is relatively simple as only two major minerals, calcite and dolomite, are involved. Both are only slightly soluble in pure water and the solvent action of natural waters depends on their acid content. Although organic and mineral acids may be important in some localities (e.g. Williams, 1963; Corbel, 1968), dissolution of calcite and dolomite is dominated by carbonic acid from dissolved carbon dioxide. It is frequently stated that the reaction between carbonic acid and 'insoluble' limestone produces calcium bicarbonate ($(CaCO_3)_2$) which is a soluble. However, this is incorrect as 'there is no evidence for the existence of calcium bicarbonate molecules in solution (Nakayama, 1968), and the ratio of $CaCO_3$ molecules dissolving to molecules of CO_2 in solution is not 1 : 1, as implied' (Picknett *et al.*, 1976). In addition, the HCO_3^- ion has often been referred to as the bicarbonate ion as two are formed in the overall reaction of CO_2, water and limestone (see Equation (9.6) below, and Chapter 5). It is now more usefully and more accurately termed the hydrogen carbonate ion. Picknett *et al.* (1976), Bogli (1980) and Spears (Chapter 5, this volume) have provided more accurate accounts of the sequence of reactions which may be summarized as:

Process equation	*Kinetics*	*Description*
$CO_2 \underset{}{\overset{H_2O}{\rightleftharpoons}} CO_2^0$ (air) (physically dissolved)	Slow	Diffusion of CO_2 into water (9.1)
$CO_2^0 + H_2O \rightleftharpoons H_2CO_3$	Slow	Hydration of physically dissolved CO_2 to form carbonic acid (9.2)
$H_2CO_3 \rightleftharpoons H^+ + HCO_3^-$	Fast	Dissociation of carbonic acid into hydrogen and hydrogen carbonate ions (9.3)
$CaCO_3 \overset{H_2O}{\rightleftharpoons} Ca^{2+} + CO_3^{2-}$	Slow	Dissociation of calcite—ions freed from crystal lattice (9.4)
$H^+ + CO_3^{2-} \rightleftharpoons HCO_3^-$	Fast	Association of CO_3 from (9.4) with H from (9.3) to form hydrogen carbonate (9.5)

As a result of (9.5) the previous four processes are all reactivated as follows:

1. Removal of CO_3^{2-} from the solution–mineral interface disturbs the equilibrium of Equation (9.4). Since the ion product is no longer equal to the solubility product K_s, further $CaCO_3$ must dissolve to replace CO_3^{2-} and restore the balance.
2. Combination of H^+ with CO_3^{2-} disturbs the equilibrium in (9.3) and further dissociation results. This in turn disturbs the equilibrium in (9.2) and causes further CO_2 to diffuse into the water by (9.1).

These dynamic processes continue until the forward and reverse reaction rates become equal (see Chapter 1), at which point the system is in equilibrium and the solution is said to be saturated with calcite (Picknett *et al*., 1976). Addition of any acid to the system will increase the concentration of hydrogen ions and displace the equilibria of (9.3) and (9.5) in a forward direction. This reduces the concentration of CO_3^{2-} and permits more $CaCO_3$ to dissolve as in (1) above, so that when equilibrium is re-established the saturated solution has a higher calcium concentration. Dissolved carbon dioxide is an acid and acts as described, but it also provides a supply of hydrogen carbonate (HCO_3^-) and carbonate ions ((9.3) and 9.5)) and involves gas–liquid exchanges (9.1) and (9.2).

Combination of Equations (9.1)–(9.5) gives the commonly quoted dissolution equation for calcite:

$$CaCO_3 + H_2CO_3 \rightleftharpoons Ca^{2+} + 2HCO_3^- \quad (9.6)$$

The presence of Mg^{2+} considerably complicates conditions at the solution–mineral interface (9.4), but the dissolution equation for dolomite is generally written as:

$$CaMg\,(CO_3)_2 + 2H_2CO_3 \rightleftharpoons Ca^{2+} + Mg^{2+} + 4HCO_3^- \quad (9.7)$$

As Ford (1980a) has noted, there are no chemical thresholds in these reactions, and as a result solution may occur in static water as well as through the known range of natural water velocities. This is the major difference between karst systems and those dominated by mechanical erosion in which little work is accomplished below a threshold discharge. These differences are considered further in §9.3. Although there are no chemical thresholds there are chemical limits, the most important being temperature, the nature of the evolution system under which the mineral is dissolved (open or closed system) and the concentration of carbon dioxide in the gas phase. Other limits include the minerals solubility product which is constant, and the reaction kinetics which are beyond the scope of this discussion but which have been considered by Plummer *et al*. (1978, 1979), Dreybrodt (1981a,b,c) and briefly in Chapters 1 and 5.

9.2.1 Temperature

For any fixed carbon dioxide concentration in a gas mixture in contact with water and rock, the calcite solubility decreases with increasing temperature at a rate of approximately 1.3% per degree Celsius (Figure 9.2; Picknett *et al*., 1976). This effect, together with some questionable field data, led Corbel (1957, 1959) to claim that the concentrations of dissolved $CaCO_3$ in natural waters are inversely proportional to temperature, with higher values in the arctic than in the tropics. By extension he claimed that erosion rates (essentially the product of solute concentration and discharge) would be greatest in cold humid regions and least under hot, arid climates. Subsequently, more detailed studies have shown that carbon dioxide concentrations in the gas phase and the nature of the evolution system are more important controls on solute concentrations (e.g. Smith and Atkinson, 1976; Drake and Ford, 1981). Thus, while CO_2 is more soluble under cold conditions, it is less *available* than under tropical conditions. In addition regional runoff variations account for a greater proportion of the observed variability in erosion rates than do solute concentration variations (§9.3).

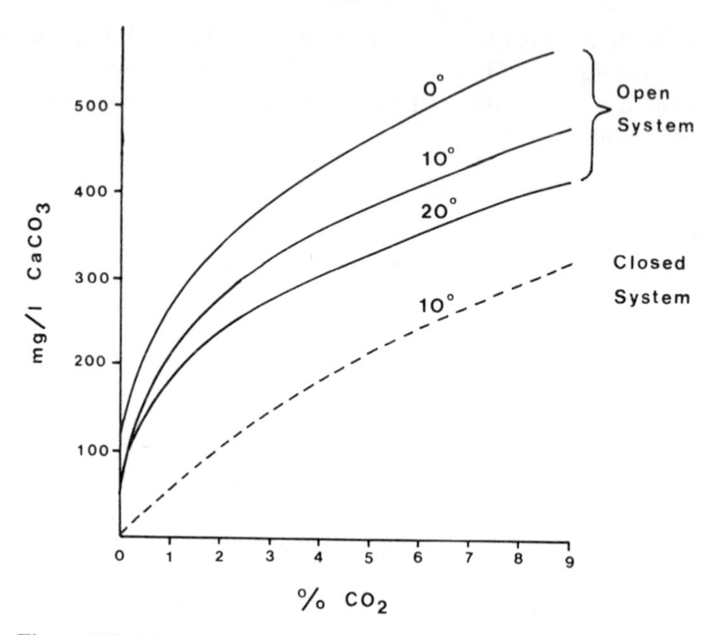

Figure 9.2. The amount of calcium carbonate which can dissolve in water which is in contact with air containing carbon dioxide under open and closed system equilibrium conditions at 0°C, 10°C and 20°C (Picknett *et al*., 1976)

9.2.2 Equilibrium conditions

There are two principal equilibrium conditions under which limestone may be dissolved; firstly, the 'open' system in which gas, water and rock are all in contact together such that CO_2 is available to replace that used up in the reaction of limestone and carbonic acid, and secondly, the 'closed' system in which gas and water come into equilibrium but the gas supply is cut off before contact with rock (Garrels and Christ, 1965). Since there is no replacement of CO_2 under closed system conditions, the amount of limestone which can be dissolved is less than under open-system conditions (Figure 9.2). The existence of 'pure' open or closed systems in real karst terrains has been questioned by Drake (1983), who suggested that these two geochemical terms could be respectively approximated by 'coincident' and 'sequential' (environmental) systems. Coincident systems are those in which soil atmosphere, water and limestone are spatially and temporally coincident—for example, in a carbonate-rich regolith or a carbonate bedrock with a large air volume due to a high primary or secondary permeability (Drake, 1983). In sequential systems the water first equilibrates with an atmosphere of CO_2 in the absence of limestone and then (sequentially) with limestone in the absence of an atmosphere—for example in an interstratal karst where a non-carbonate unit intervenes between soil and limestone (Drake, 1983). It would seem from these definitions that open and coincident systems are in fact identical, as are closed and sequential systems, and it is not readily apparent what benefits accrue from use of the new terms; the term sequential is perhaps more explicit of how a closed system operates in reality. However, there would appear to be a good case for establishing the range of equilibrium conditions between 'pure' open/coincident systems and 'pure' closed/sequential systems, and in this respect the identification by Drake (1983) of the geomorphic situations leading to different evolution systems provides a useful contribution (Figure 9.3).

9.2.3 Carbon dioxide concentration

The exchange between gas and solution is governed by Henry's law such that the concentration of carbon dioxide in solution is proportional to the concentration of CO_2 in the gas mixture. An increase in CO_2 in the mixture causes more CO_2 to dissolve (Equation (9.1)), more carbonic acid to form (9.2) and hence an increase in limestone solubility ((9.3)–(9.5); Figure 9.2).

The atmospheric concentration of CO_2 is close to 0.03%, which would yield a saturation value of about 70 mg per litre of $CaCO_3$ at 10°C under open-system conditions. Observed $CaCO_3$ concentrations in karst areas frequently exceed this value and may be as high as 500 mg/l^{-1} (Picknett *et al.*, 1976), although concentrations in excess of 300 mg/l^{-1} are uncommon. It is generally assumed that the source of the extra CO_2 required to attain these high figures is the soil atmosphere (Adams and Swinnerton, 1937; Smith and Mead, 1962). Prior to 1965 measurements of soil CO_2 concentrations involved withdrawing a large volume of gas from the soil and they are probably unreliable because of possible intake of atmospheric air (Russell, 1973, p. 411). Although accuracy has since been improved by a number of new laboratory and field techniques, notably that of Miotke (1972, 1974), there is still a paucity of field measurements. Moreover, many of the published results (summarized in Atkinson & Smith (1976) are from spot field measurements, and few studies have attempted to assess the extent of temporal and spatial variability. Hence, a recent attempt to produce a world model of soil carbon dioxide in which CO_2 concentrations are predicted from actual evapotranspiration (Brook *et al.*, 1983) is probably unrealistic and of little use for karst studies (Gunn, 1984a).

Those studies which have involved repeated measurements of soil CO_2 concentrations at single sites have shown considerable temporal variability (e.g. Gerstenhauer, 1972; Atkinson, 1977a; Gunn and Trudgill, 1982; Crowther, 1983a). As the CO_2 is derived by microbial activity in the soil, which is itself climatically controlled, this variability is to be expected,

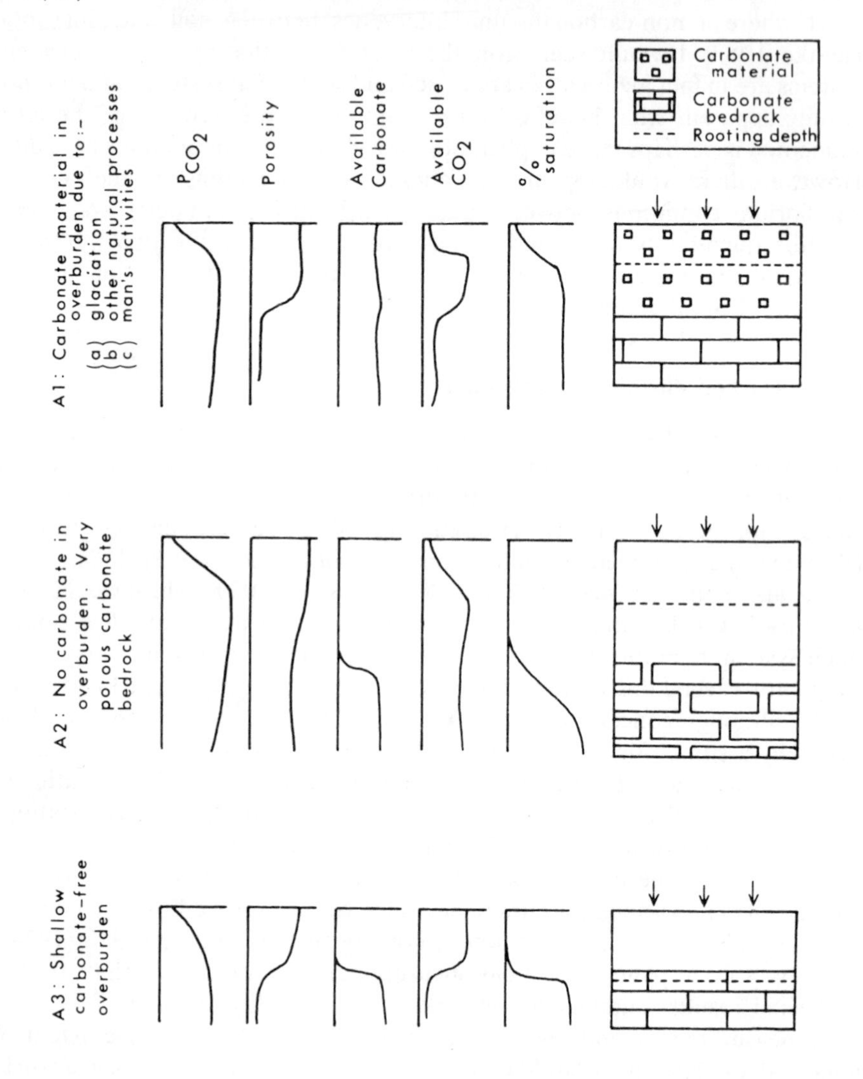

although it should be noted that soil gas diffusivity, which is a function of local factors, may also introduce significant variability (Miotke, 1974; Gunn and Trudgill, 1982). Despite this variability in CO_2 concentrations many karst risings, and particularly those from systems dominated by autogenic recharge, show relatively little fluctuation in total hardness (($CaCO_3 + MgCO_3$) measured, in fact, as Ca^{2+} and Mg^{2+} but often expressed as the carbonate). One such area is the Mendip Hills, Somerset, England, and this led Atkinson

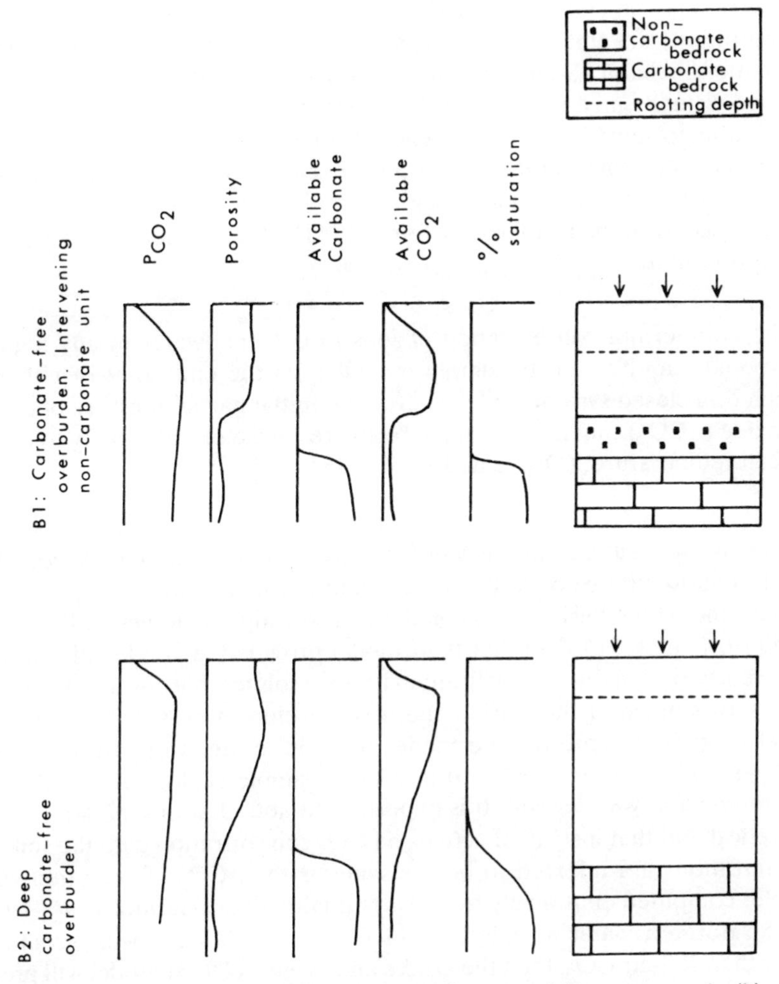

Figure 9.3. Geomorphic situations leading to (a) open- and (b) closed-system evolution (Drake, 1983)

(1977a), to suggest a source of CO_2 in the unsaturated epikarstic zone which he termed 'ground-air carbon dioxide'. The source of this ground-air CO_2 was thought to be microbial decay of organic matter in the joints and fissures of the unsaturated zone; and since conditions at depths of more than a few metres are almost isothermal, a more or less constant supply of debris would assure a high and more or less constant rate of CO_2 production (Atkinson, 1977a). Later work in the same area by Friederich (1981, p. 133) confirmed the presence of 'a buffer zone of moderate CO_2 concentrations at the top of the unsaturated zone'. Friederich (1981) suggested that although the influence of this zone of ground-air CO_2 may be minimal in summer, when concentrations of soil CO_2 are high, winter measurements of soil CO_2 concentrations are unlikely to be representative of the solutional power of percolation recharge. Somewhat surprisingly the ground-air CO_2 concept has received virtually no attention in subsequent studies, and the writer knows of no attempts to substantiate its presence in other areas. This is unfortunate as the concept is of considerable importance for karst studies, not least because of recent attempts to produce a global model for carbonate groundwater solute concentrations (Drake and Ford, 1981). This model is based on the assumption that:

> 'The equilibrium concentration of ions in a groundwater in any regional carbonate aquifer can be simply modelled as the end point of either an open or a closed system evolution from an initial partial pressure of carbon dioxide (pCO_2) that is determined by the response of soil activity to annual mean temperature' (Drake and Ford, 1981, p. 223).

Clearly, if the results from the Mendip Hills apply elsewhere and ground-air CO_2 is a major source of carbonic acid (and during certain seasons the main source), the whole basis for this model is undermined. However, Drake and Ford (1981, p. 229) found that their model provided 'a good explanation of observed variations in regional annual mean solute concentrations in karst areas across much of the world', and it is therefore necessary to subject the model to greater scrutiny. The model rests on earlier work by Drake and Wigley (1975) and Drake (1980) in which a simple ecological model of soil CO_2 *production* was derived. It is important to note that this model is largely theoretical and that instead of actual soil CO_2 concentration data the soil CO_2 concentrations are inferred to be the same as the pCO_2 of saturated water samples computed empirically from water quality data (Harmon *et al.*, 1975). If the saturated water samples are in fact in equilibrium with ground-air rather than soil-air CO_2, then the Drake and Wigley (1975) model will predict ground-air and not soil-air concentrations. This in turn would explain its success when applied by Drake and Ford (1981) to the prediction of solute concentrations. Moreover, this explanation would be conceptually more satisifying since the model suggests a high degree of temperature dependence and a relatively low seasonal variability of CO_2 concentrations. This is precisely what Atkinson (1977a) and Friederich (1981) have inferred for ground-air CO_2 concentrations, but it does not match observed soil CO_2 concentrations which show a greater seasonal variability and lower temperature dependence. It was this that led Gunn and Trudgill (1982, p. 92) to suggest that, contrary to the Drake and Wigley model:

> 'Prediction of CO_2 concentrations from climatic data is not generally possible as the statistical relationship between the weather parameters and *concentrations* are weaker than the relationships between these parameters and the rate of CO_2 *production*.'

Much of this discussion has been based on inference and conjecture and it is clear that detailed studies of the relative significance of soil-air and ground-air as sources of CO_2 for carbonate dissolution are urgently called for.

9.2.4 Mixing corrosion

Although there are no chemical thresholds in the limestone solution process, the mixing of waters which have followed very different evolutionary paths and which therefore have different solute concentrations may have a threshold result (Ford, 1980a). This results from the mischungskorrosion effect, first described by Bögli (1964), whereby the mixing of two saturated waters produces an unsaturated (aggressive) solution, and the mixing of a saturated and an aggressive solution, or of two aggressive solutions, may result in increased aggressiveness (Figure 9.4). In extreme cases the new

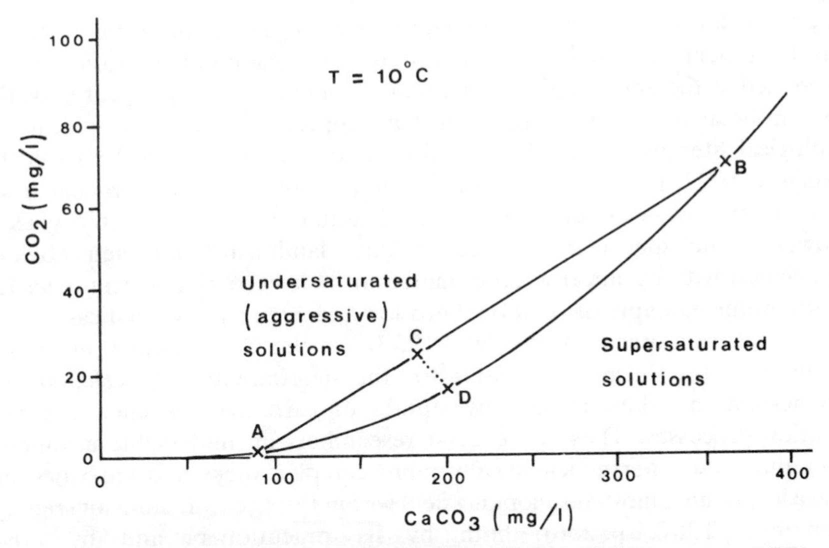

Figure 9.4. The 'mixing corrosion' effect. The equilibrium curve ADB (based on Dreybrodt (1981c)) shows the solubility of $CaCO_3$ at 10°C with respect to total CO_2 in solution. Mixing of the two saturated solutions, A and B, produces solution C which is undersaturated. Solution C then moves to saturation along the line CD

solution may be capable of dissolving 20% more calcite, but 1–2% is more usual in natural waters. Hence, the mixing effect is generally less effective than 'normal' solution, and its importance lies in its ability to operate in conditions under which normal solution is impossible such as narrow fissures and in the phreatic zone. In particular, recent work by Wigley and Plummer (1976) on the mixing of carbonate waters, together with an improved understanding of solution kinetics (Plummer *et al*., 1978, 1979) has been used by Dreybrodt (1981a) to calculate the development of solution conduits from the crossing of fine fractures (joints and/or bedding planes) with respect to time. He found that over a period of the order of 10^4 years a solution conduit of a few centimetres diameter could develop at the crossing of joints with original widths of between 2×10^{-3} and 10^{-2} cm, and that 'cave development to passage diameters of 1 m is possible in a few million years' (Dreybrodt, 1981a, p. 245). In a related paper Dreybrodt (1981c) calculated the saturation lengths and the dissolution rates for mixing corrosion in a cylindrical tube and hence the development of tube diameter with respect to time. His final mathematical model 'explains karstification exclusively from the solution kinetics of calcite and the flow properties of water in conduits.' (Dreybrodt, 1981c, p. 235) and is considered further in §9.6.

9.3 SOLUTIONAL EROSION RATES

The importance of chemical denudation as an agent of landscape evolution has long been accepted, and measurements of solutional erosion rates, expressed as the amount of material removed per unit area per year or as the amount of surface lowering per year, have appeared in the geographical and geological literature since the mid-nineteenth century (e.g. as reviewed by Douglas, 1972, p. 45). Early work was not confined to any particular rock type as the widespread significance of solute removal was recognized. However, the spectacular nature of karst landforms and their obvious association with chemical weathering led to their becoming widely regarded as the prime example of solutional erosion and even as a 'special case'. This trend was reinforced in the late 1950s by the development of simple complexometric titration methods for the determination of calcium and magnesium in solutions, and by studies of carbonate geochemistry and solution processes. These gave karst researchers an undeniable advantage over those on other geochemically more complex rocks and karst became regarded as 'an almost autonomous field within the scientific study of scenery' (Jennings, 1967, p. 256), both by its practitioners and by other geomorphologists. To a large extent this has been to the detriment both of karst studies which became somewhat insular and isolated from general geomorphological thought (Williams, 1978), and also of general geomorphology in which solute studies were neglected. Fortunately hydrologists have shown a greater interest in 'quality as well as quantity' since the early 1970s (Walling, 1979, p. 71), and this in turn has stimulated interest in solute studies by geomorphologists (e.g. Webb and Walling, 1980, and see Chapters I and II). Research in areas underlain by non carbonate lithologies have shown that removal of material in solution may be an important, and often the dominant, component of denudation, and that both the stream solute loads and net solutional erosion rates on these lithologies may be similar to or even exceed those of karst areas. For example, a detailed study of the River Exe drainage basin in which there is no significant karst development found that solutional erosion rates are in the range 10–70 $m^3km^{-2}yr^{-1}$ (Walling and Webb, 1978). These results overlap with those from karst basins in the Mendip Hills, some 100 km to the north-east and with a similar climate, where five separate studies obtained solutional erosion rates in the range 23–102 $m^3km^{-2}yr^{-1}$ (cited by Waylen, 1979, Table IX). Similarly Gunn (1982a) has shown that the magnitude and frequency properties of solute removal from 10 basins underlain by carbonates do not differ markedly from those of 14 non-carbonate basins, although medium to low flows tend to transport a greater proportion of the solute load in carbonate basins.

Catchment scale studies of solutional denudation on all lithologies and the contrasts between carbonate and non-carbonate terrains are discussed in Chapter 8, and they will not be considered further in this section. However, two important points which follow from the above account and from Chapter 8 must first be made in order to put the present discussion of karst erosion rates into its correct geomorphological context:

1. Karst is not an autonomous field. 'There is no such thing as a unique karst process, and clearly karst areas obey the normal laws of hydrology, chemistry and physics' (Sweeting, 1981, p. 2). Similarly, karst rocks are far from being the only lithologies on which solutional denudation is significant. However, they differ from other lithologies in that solutional erosion occurs beneath the surface and results in a marked enlargement of the secondary permeability.
2. Karst researchers would profit from consideration of solute studies on other lithologies which include hydrochemical budgets for small drainage basins (e.g. Verstraten, 1977; Waylen, 1979); detailed examinations of temporal variations in solute concentrations (e.g. Foster, 1978) and solute loads (e.g. Walling, 1978; Webb and Walling, 1982); and investigations of spatial variations in solute concentrations and loads at both the catchment scale (e.g. Walling and Webb, 1975, 1978; Foster, 1979) and on hillslopes (Crabtree and Burt, 1983). In addition there may be substantial benefits to be gained from extending the scope of karst solution studies to include a wider range of elements than has previously been the case. A study of the hydrogeochemistry of karst in North Derbyshire by Christopher (1980;

Christopher and Wilcock, 1981a,b) has clearly shown the value of this approach.

9.3.1 Net solutional erosion rates

Most investigations of denudation processes in karst areas have had as their aim the estimation of the net solutional erosion rate for a drainage basin or similar unit. Prior to 1970 computations were generally based on the formula of Corbel (1957) which has since been shown to give inaccurate results under most normal conditions (Hellden, 1973; Drew, 1974; Drake and Ford, 1976). Moreover, most of these early studies failed to take account of solute inputs in precipitation and from allogenic sources and their absolute values cannot be considered as in any way reliable. Nevertheless, Smith and Atkinson (1976) were able to use these data to demonstrate broad differences between the three main climatic zones (arctic/alpine; temperate; tropical). They also showed a general relationship between mean annual runoff and erosion rate, although the general unreliability of the data set and failure to take into account carbon dioxide concentrations and the evolution system mean that the equations have little predictive value (Gunn, 1981c). Similar attempts by Lang (1977) and Engh (1980) to predict erosion rates from rainfall are of even less value as it has been argued that the original work contains serious errors and has a dubious logical basis (Gunn, 1981c). However, there is no doubt that runoff variations account for the greatest part of the observed variability of global solutional erosion rates in karst areas. The remainder is due to variations in solute concentrations which are a function of temperature (through its control of CO_2 production, CO_2 solubility in water, and reaction kinetics) and the evolution system (open, closed). Drake and Ford (1981) have produced a global model for equilibrium groundwater concentrations based on these two factors, but its potential for prediction of solutional erosion rates is restricted since relatively few large karst risings discharge saturated water throughout the year.

In order to obtain a reliable estimate of net solutional erosion it is necessary to establish a hydrochemical budget for a drainage basin. The minimum requirements to compute solute load removal during a specific year are:

1. a complete discharge record;
2. water samples through the range of flows experienced to establish a rating curve for discharge and solute concentration (either total hardness or Ca and Mg ions separately);
3. the basin area.

The rating curve(s) may be applied either to the flow duration curve (Atkinson and Smith, 1976) or directly to the discharge data (Gunn, 1981b). If the drainage basin is entirely autogenic, then the only other requirments are precipitation inputs and the average solute concentration in precipitation. These are used to estimate solute accessions which are subtracted from the output to obtain the net solute load and, by dividing by basin area, the erosion rate. Unfortunately, most studies have taken place in mixed allogenic/autogenic catchments, and difficulties in accounting for the solute inputs from other lithologies may result in a serious overestimate of the rate of limestone solution. Where a long discharge record is available it may be possible to extend the erosion rates over a period of several years or even decades (Ford and Drake, 1982), although Douglas (1968) has sounded a cautionary note by showing that similar dissolved load/discharge rating curves may not be obtained for successive periods, either seasonally or from year to year. Moreover, the few studies in which estimates have been made of potential errors in the computed erosion rate for a specific year have shown that, even with high-quality data, these may be substantial (Williams and Dowling, 1979; Gunn, 1981b; and see Chapter 7). It is therefore suggested that extrapolations of erosion rates over the short period (10–100 years) should include some estimate of potential errors. Extrapolations over the longer period when climatic fluctuations must also be taken into account are extremely hazardous, and in this respect the expression of erosion rates as mm per thousand years is particularly unhelpful.

Finally, it should be noted that there are relatively few 'reliable' (i.e. based on hydrochemical budgeting) estimates for solutional erosion rates in karst areas, and although collection of this type of data is no longer regarded as being at the forefront of karst research it still represents an important contribution.

9.3.2 Spatial distribution of solutional erosion

Net solutional erosion rates provide a useful measure of the overall efficiency of solution processes in individual areas but are largely unrepresentative of solution at any specific location within those areas. In surface drainage basins the distribution of erosion is largely a two-dimensional problem, but in karst areas the third, vertical, dimension must also be considered in order to establish the significance of solutional erosion to landform evolution. For example, examination of 134 published erosion rates by Smith and Atkinson (1976, p. 388) demonstrated that 'there is no difference in "lumped" erosion rates between temperate and tropical areas, in spite of the fact that there are differences in landforms.' They then examined the hypothesis that landform differences could be accounted for by different distributions of solutional erosion within the landscape, but they found that it could not be maintained. However, it must be noted that only 8 of the 134 studies examined provided any indication of the spatial distribution of solution, and that of these 8 the only detailed breakdown is for the Mendip Hills. To the writer's knowledge there have been only two subsequent studies in which net erosion budgets have been dissected and the contribution of individual components assessed, those of Williams and Dowling (1979) and Gunn (1981b). Hence, there is an urgent need for more information, particularly for karst regions outside the temperate zone.

This information may best be obtained by an extension of the hydrochemical budgeting method discussed above. Water samples are collected from the full range of sites in the karst system—bare limestone surfaces, the soil zone, the subcutaneous zone, main body of bedrock (sampled as vadose flows and seepages), and cave streams in both vadose and phreatic zones. These, together with estimates of the proportion of water following the various pathways through the system (§9.4.5), permit the breaking down of the overall erosion budget (Atkinson and Smith, 1976; Gunn, 1981b). Those studies which have been made have shown that a high proportion of solution (50–85%) occurs within several metres of the surface in the soil (if present) and subcutaneous zone (uppermost bedrock). Caves account for very little of the erosion when averaged over the whole basin, but it is important to recognize that in reality erosion is concentrated in a very small area. For example, in a hypothetical drainage basin 1 km^2 and with no soil cover the erosion rate is estimated at 50 m^3 $km^{-2}yr^{-1}$, of which 80% takes place in the subcutaneous zone and contributes directly to surface lowering, 19.5% in the main mass of bedrock where it contributes to the enlargement of joints, fractures and bedding planes, and the remaining 0.5% is contributed by enlargement of cave passage. If the length of cave passage is taken to be 100 m and its average width 1 m, then this small proportion of the overall erosion budget will result in a passage floor lowering of 2.5 mm per year, which is over 60 times faster than the rate of surface lowering (0.04 mm per year).

The principal drawback of the hydrochemical approach is that it requires frequent sampling in order to establish the pattern and extent of variations in solute concentrations. This is not always possible, and alternative methods which integrate erosion over a longer time period have therefore been derived (Goudie, 1981). The two most commonly used in karst areas are the micro-erosion meter (High and Hanna, 1970; Trudgill *et al*., 1981) and rock tablets (Trudgill, 1975). In contrast to the hydrochemical method, these techniques are highly site-specific and may only be used to assess erosion rates on bare limestone surfaces, in the soil zone, at the soil–bedrock interface, and in cave streams. Trudgill (1977, p. 256) found that they could be used to detect seasonal differences in erosion rates and that 'further work is needed in order to evaluate whether reliable measurements over shorter time scales is possible'. However, this is at variance with recent work by Crowther (1983b), who compared the rock tablet and hydrochemical methods for determining solutional erosion rates on bare and soil covered limestone surfaces in West Malaysia. He found that the rock tablets gave estimates two orders of magnitude less than those calculated using water hardness data, the most likely explanation being that 'natural rock surfaces come in contact with larger volumes of water than do isolated rock tablets, simply because of their greater lateral flow component' (Crowther, 1983b, p. 62). If this is the case then it is clear that the two methods measure fundamentally different phenomena and that the hydrochemical method provides the only reliable means of estimating solutional erosion rates on limestone surfaces.

Different problems arise if tablets are placed in cave streams as they will project above the natural surface, and as a consequence are likely to erode more rapidly. They are also likely to suffer from abrasion as well as corrosion, although this can be exploited by placing the tablets in nylon cages with differing mesh sizes and comparing the erosional losses suffered (Newson, 1971; Trudgill, 1975). The Commission on Karst Denudation of the International Speleological Union (ISU) is at present conducting a comparison of the losses suffered by Yugoslavian limestone tablets exposed in standard ways in many different parts of the world, and it is clear that further comparisons of the tablet and hydrochemical methods will be necessary in order to interpret the results of this experiment. For the present it may be concluded that although the micro-erosion meter and tablet methods may provide useful data on erosion rates in certain parts of the karst system, the hydrochemical method permits a more comprehensive spatial coverage and provides better resolution for short (<1 year) time intervals.

Irrespective of the technique employed it is recommended that results be reported in mm per year and not mm per thousand years as this brings out the point that extrapolation of short-term estimates over a long time period is usually unwarranted (Trudgill *et al*., 1981).

However, it may be possible in some areas to estimate long-term average erosion rates for two elements in the karst system—bare limestone pavements and cave passages. In the first case surface lowering rates may be estimated from the height of limestone pedestals formed when part of a limestone pavement is protected from solution because it is capped with an erratic or a limestone block (e.g. Bogli, 1961; Sweeting, 1966; Williams, 1966; Peterson, 1982). The technique is simple and gives unequivocal results, but it is limited to areas which were ice-covered during the last glaciation.

Estimates of the rate of erosion in cave streams may be made using the novel approach proposed by Gascoyne (1981), which makes use of uranium series dating of speleothems (§9.5). If an old speleothem can be found in its growth position and near to an active streamway, then its height above present stream level divided by its basal age provides an estimate of the mean maximum rate of passage lowering. For example, wall flowstone 2.5 m above present stream level in Lost John's Cave, north-west England, was dated at

115,000 years, giving a mean maximum downcutting rate of 22 mm per 1000 years (Gascoyne, 1981, Table 1). The main shortcoming of this method is that the routes of cave streams may vary through time, and if the streamway at the measurement point has not been occupied continuously since speleothem deposition the calculated erosion rate is no longer a maxium value. Hence it is necessary to conduct a preliminary site investigation in order to establish that no alternative drainage routes exist, or have existed, for the chosen length of passage. With this one proviso the method would appear to provide an extremely powerful tool for estimating rates of cave stream lowering for periods of up to 350,000 years. However, some care is necessary in interpreting the results in terms of the overall distribution of solutional erosion, since both solution and mechanical erosion are likely to be active in cave floor lowering. In addition, it has already been noted that the lowering of a cave passage floor by, say, 10 mm requires the removal of a substantially smaller volume of limestone than does a similar rate of surface lowering. Direct comparison of long-term solutional erosion rates estimated from pedestal heights or speleothem ages with the results of modern process studies using hydrochemical or alternative methods is, strictly speaking, unsound. Nevertheless comparisons have been made and it is interesting to note that on the whole there is quite good agreement.

9.3.3 Temporal variability of solutional erosion

Variations in limestone solution rates over time have received greater attention than the spatial distribution of solution, and two main areas of interest may be identified—seasonal variations and magnitude/frequency relationships. Most studies have used the rating curve approach relating solute load (itself the product of discharge and solute concentration) to discharge, and applying this relationship to the continuous-flow record or flow duration curve. Major assumptions of this method are that the concentration–discharge relationship is linear and time-invariant. However, studies on other lithologies have shown that storm period solute concentration–discharge relationships are extremely complex and difficult to model owing to the occurrence of both hysteretic and flushing events (e.g. Foster, 1978). Little work of this kind has been undertaken in karst areas, and it could provide useful information particularly if combined with studies of natural flood pulses (Christopher, 1980; see also §9.4.4). A further complication is that the concentration–discharge relationship may show seasonal variations unrelated to flow if weathering rates are depressed by low levels of carbon dioxide production during colder months. Again there have been few attempts to consider seasonal variability of solute concentration or load rating curves in karst areas, although Douglas (1968, 1972) and Gams (1976) have provided some discussion of the problem. Fortunately, research on other lithologies suggests that the errors introduced into temporal studies of erosion rates by storm-period and seasonal variations in rating curves will be relatively small, although they are likely to increase as the time base decreases (Walling, 1978).

If the rating curve approach is used then the computed seasonal variability of solutional erosion rates will reflect the discharge regime and the slope of the rating curve. In most karst areas the decline of solute concentration with increasing discharge is relatively small, so that solute load increases with discharge and the months with the greatest runoff are also the months of greatest solute removal. Hence, studies in humid temperate areas have shown that solutional erosion is fairly evenly distributed throughout the year but with maximum rates during the winter months when discharges are highest (Drew, 1974; Gunn, 1981b). By way of contrast colder climates show a greater seasonal range in erosion rates, a large part of the solute load being evacuated during the snowmelt period (Hellden, 1973, 1976; Drake and Ford, 1976). Seasonally dry areas will also show a marked range in erosion rates, with maximum values during the months when recharge occurs. In summary seasonal variations in erosion rates would appear to be largely a function of local climate through its control on recharge, and the only scope for further study lies in possible seasonal changes in the discharge-solute concentration relationship discussed above and in the effects of seasonally variable solution on landform evolution (Jennings and Sweeting, 1963; Williams, 1978).

The magnitude and frequency properties of dissolved solids transport from both carbonate and non-carbonate basins have recently been reviewed by the writer (Gunn, 1982a). Data are available for only 24 drainage basins, of which 10 are in karst areas. The mean proportion of annual solute load transported by the highest flows which operate for only 5% of the time is 17% in the karst basins and 26% for other lithologies, although the ranges overlap. The corollary is that in karst areas lower flows, less than the median discharge, transport slightly more of the annual load and these differences may be accounted for by the tendency of karst basins to have less flashy discharge regimes than surface basins. However, it is noteworthy that in all basins at least half of the solute load is removed by high-to-medium flows operational for 30% of the time or less. This is contrary to the suggestion of Wolman and Miller (1960) that flows comparable to or less than the mean should be more important in dissolved solids transport, although it should be noted that the data set presently available is too small to permit firm conclusions to be drawn. Moreover it is clear that high-frequency/low-magnitude flows play a much greater role in dissolved solids transport than they do in the removal of clastic load, even though their overall role in solutional erosion may have been over-emphasized.

In conclusion it is apparent that despite 20 years of studies of solutional erosion in karst areas there is still a marked paucity of reliable data on net erosion rates and their spatial and temporal variability. Furthermore, those studies of temporal variability which have been undertaken have focused on the basin outlet (usually a rising), and very little is known of temporal variations in solution at different loci within karst systems. The hydrochemical method provides the best technique for obtaining this information, as all three aspects may be incorporated into a single study (Williams and Dowling, 1979; Gunn, 1981b). Until this information is available it is difficult to substantiate process–form relationships, although broad climatic controls of the type discussed by Sweeting (1980) are undoubtedly valid.

9.4 KARST HYDROLOGY

The primary difference between drainage basins on karst rocks and those on other rocks is that most karst areas are characterized by an absence of surface drainage. Four main possibilities exist for the investigation of underground water movement: direct exploration, theoretical models, mathematical statistical models, and water tracing. Each has its own strengths and weaknesses and each may both contribute to and benefit from the construction of a conceptual model of karst drainage systems.

9.4.1 Direct exploration

This is unique to karst areas as caves are infrequent and of short length on most other rocks, the only major exception being caves in rocks of volcanic origin (Wood, 1976) which are usually unrelated to underground hydrology. Over 8000 km of cave passages with vertical ranges of up to 1400 m have been surveyed in the world's limestones, and it is almost certain that at least as many systems still await discovery and exploration (Waltham, 1981a). Between them these caves carry an enormous amount of underground drainage.

Direct exploration by speleologists has been of considerable value both in the development of theory and in local studies of the nature and patterns of underground drainage. Of particular note is the 'water-table controversy' (reviewed in detail by Jennings, 1971; Sweeting, 1972; Smith *et al*., 1976) where one school of thought was influenced by experience of predominantly vadose caves with considerable vertical development (Katzer, 1909; Martel, 1910, 1921), while the opposing school were more familiar with caves which developed under phreatic conditions and which had a greater degree of horizontal development (Davis, 1930; Swinnerton, 1932). Resolution of the problem has been due in part to an appreciation of the wide variety of underground forms as well as to increased data collection, both surface and underground, and to other hydrological techniques (Smith *et al*., 1976).

The practical value of direct exploration was powerfully demonstrated by the recent Gunung Sewu (Java) Cave Survey (Waltham *et al*., 1983) in which 250 entrances were identified, 170 caves explored and 62, with a combined length of 28 km, were surveyed. The survey provided an assessment of the water resources of the Gunung Sewu karst aquifer and identified a number of economically usable sources of supply with both short and long-term potential. The branch of direct exploration which has seen the greatest advances during the last decade is cave diving (Farr, 1980), and it is hoped that studies by divers, and perhaps of diving hydrologists, will result in a better understanding of water movement in the phreatic zone (Fisk and Exley, 1977).

9.4.2 Theoretical models

The general principles of groundwater flow are used in these models to deduce patterns of water movement in karst systems. They form the basis of deterministic models in which physical laws are used to calculate runoff from input by determining the effect of system characteristics on the runoff process. The starting point for most discussions is the long established Darcy law which provides a general case for groundwater flow (Freeze and Cherry, 1979). More advanced treatments include the development of steady-state equations for flow in saturated and unsaturated media (Basmaci and Sendlein, 1977), and of equations for fissure flow in fractured aquifers (Boulton and Streltsova, 1977). In general these models are restricted to situations in which flow is laminar, and they are therefore not applicable to the majority of karst systems in which flow is turbulent in solutionally enlarged conduits. Those attempts which have been made to apply groundwater terminology and Darcian concepts to cavernous limestones have on the whole caused misunderstandings and created difficulties (Smith, 1977). However, some benefit may be gained from applying these principles to the development of underground drainage in karst rocks (White and Longyear, 1962; Atkinson, 1968; Thrailkill, 1968). No theoretical models are currently available for water movement in conduit flow systems, but the similarity between this type of system and surface drainage basins suggests that benefit could be gained from application of recent advances in surface water hydrology.

9.4.3 Mathematical statistical models

These are the main means by which outputs from groundwater systems may be predicted. In contrast to theoretical/deterministic models, they treat the system as a 'black box' and do not require any knowledge of physical conditions within the box.

The first attempt to apply these concepts to karst systems was undertaken by Knisel (1972), who fitted Fourier series to the mean and standard deviation of rainfall and runoff data from springs in the USA and obtained good prediction results for medium flows. Froidevaux and Krummenacher (1976) used similar spectral analysis techniques to study rainfall and runoff data from a karst basin in the French Jura. Although the rainfall and runoff series both had good internal coherency, the coherence between the two series was not significant, thereby limiting the ability for prediction.

A more advanced approach by Graupe *et al*. (1976) yielded a linear optimal model for the prediction of outflow from karstified drainage basins and karst aquifers, based on autoregressive moving-average methods. The model was calibrated using daily precipitation and runoff data and optimal prediction was obtained by the whitening of residuals and application of Kalman filtering of measurement errors. More recently Dreiss (1982) has suggested that the discharge response to recharge of karst systems may be estimated if single linear kernel functions are used to characterize the system. This approach achieves a greater degree of physical realism by using information on the hydrogeology of the system to determine the input series which is taken to be groundwater recharge rather than simple precipitation data. The model was tested by deriving an average kernel from all of the input and output series bar one for Big Spring, Missouri, and using this to predict the storm response of the omitted series. Spring flow responses were predicted with an accuracy of 10–20%, although the technique is somewhat limited by its inability to predict discharge response to storms of unknown volume (Dreiss, 1982). Other studies in which mathematical statistical models have been used to predict outputs from karst systems include those of Mangin (1974–75), Knezevic (1976), Ozis and Keloglu (1976), and Drogue and Guilbot (1977).

Mathematical models may also be used to obtain information on the physical properties of karst systems. Yevjevich (1981, p. 121) has suggested that:

> 'The black-box response hydrographs promise to yield, if properly developed and applied, knowledge on the water residence time in underground, the average water travel times as a function of hydrograph properties . . . [and] the storage effects of various porosities . . . as a function of hydrograph properties.'

To date these studies have generally been undertaken in conjunction with water-tracing experiments and they are therefore discussed in §9.4.4.

9.4.4 Water tracing

This is probably the area of karst research which has received greatest publicity in both the popular and scientific press, partly because of the sense of mystery which surrounds inaccessible places and partly because of the practical and economic value of knowing the destination of sinking water and the origins of water issuing from underground.

Most karst areas have an oral tradition concerning underground connections, often involving the descent of unfortunate animals into the bowels of the earth to emerge some time later at a distant location with scorched fur or feathers. A modern sequel is the use of marked eels to establish a 55 km connection in Yugoslavia (Wagner, 1954; cited by Bogli, 1980, p. 138). In Ireland the writer was given an account of an illegal still being dumped into a stream-sink when warning was given of the imminent arrival of revenue officers, and of its later recovery at a rising some 2 km away. This connection was later proved using fluorescent dyes (Gunn, 1982b). On a similar theme Bögli (1980, p. 138) records the inadvertent use of Pernod in establishing a French connection!

Deliberately designed tracing experiments date back to at least the seventeenth century and usually involved pouring pine needles, chaff or similar materials into sinking streams and visual observation of their emergence at rising. Like polypropolene floats and polyethylene powder, which are modern equivalents, they are likely to become trapped where cave passages become water-filled (Atkinson *et al.*, 1973), and hence early traces must be regarded with sceptisism. However, all of the chaff tests reported to the writer in Ireland have since been proven using fluorescent dyes (Gunn, 1982b).

Among the earliest scientific water tracing experiments where those undertaken in the late nineteenth century in the Malham area of Yorkshire, England, where chaff was replaced by magenta dye, Fluorescein dye, ammonium sulphate, sodium chloride and water pulses (reviewed by Smith and Atkinson, 1977). Similar, but often less detailed, experiments were undertaken in many other areas (e.g. Brodrick, 1908), but few advances were made prior to 1959 when Maurin and Zotl pioneered the use of dyed *Lycopodium* spores. Since then the number of tracing agents available for hydrological studies has multiplied. Those used in karst research are now divisible into six groups:

1. particulate tracers—principally *Lycopodium* spores (Drew and Smith, 1969);
2. chemical tracers—usually an inorganic salt, e.g. NaCl, KCl, LiCl;
3. radioisotopes—a wide variety have been used in groundwater tracing (White, 1974, 1977), but relatively few have been applied to karst areas.

Examples include tritium (Burdon *et al*., 1963), chromium-51 (Pirs *et al*., 1976) and scandium-46 (Ramljak *et al*., 1976);
4. non-fluorescent dyes—mainly Malachite Green and Rhodamine B (Drew and Smith, 1969) (now rarely used because of toxicity);
5. fluorescent dyes—a wide variety in three main groups: blue (e.g. amino G acid, Photine CU, Tinopal CBS-X), green (e.g. Fluorescein, Pyranine, Lissamine) and orange (e.g. Rhodamine WT; Eosine);
6. pulse-wave techniques—differ from the other groups in that no substance is injected into the water. Instead the water quantity of the sinking stream is changed by the release of a pulse of water. Associated risings are monitored and a rise in stage indicates a hydraulic connection (Smith and Atkinson, 1977).

Two other groups of tracing substances, bacterial tracers (e.g. *Serratia indica* and bacteriophages of *E. coli* (Wimpenny *et al*., 1972) and fluorocarbons (Thompson *et al*., 1974; Weeks and Thompson, 1982) have been used in surface and groundwater studies but have not been applied to karst systems.

In addition to the above references karst water-tracing techniques and their relative utility have been reviewed by several authors including Brown and Ford (1971), Atkinson *et al*. (1973), Back and Zötl (1975), Gospodarič and Habič (1976), Smart and Laidlaw (1977) and Atkinson and Smart (1981). Hence, the present account will consider only the applications of fluorescent dyes which are by far the most commonly used tracer in karst studies, and of pulse wave techniques which may be used with fluorescent dyes to obtain greater amounts of data on conduit systems. However, it should be noted that in some areas where water from karst risings contributes to domestic supply the use of dyes is not allowed owing to worries as to their toxicity. Smart (1982) has reviewed published toxicity information on 12 fluorescent dyes and concluded that fluorescein, fluorescent whitening agents and eosine are not hazardous but the rhodamine group of dyes are probably mutagens. Further screening of tracer dyes for mutagenicity is still needed, and until data are available water authorities may continue to insist on the use of *Lycopodium* spores which are demonstrably non-toxic.

The earliest, and still the most frequent, application of tracers in karst studies is to determine underground flow paths and average linear velocities (the straight-line distance between point of injection and point of recovery divided by travel time). Such experiments frequently have an applied value indicating the sources of groundwater pollution, (Aley, 1972; Bristol Water Company, 1972; Gunn, 1984b) or the potential for future pollution (Atkinson and Smith, 1974) and the time taken for a pollutant to reach the rising. When a number of underground connections have been established it may be possible to delimit the catchment boundaries of karst risings and to prepare maps of use in water resource planning. Among the best examples are the Mendip Hills, England, where over 90 underground connections have been established (Stanton and Smart, 1981), enabling the boundaries of 10 drainage basins to be defined, and the Central Kentucky karst, USA, where 28 groundwater basins have been delimited on the basis of over 400 dye tests, 1500 water-level measurements in wells, and 80 km of cave mapping (Quinlan, 1982).

Many water-tracing experiments are undertaken with the sole objective of establishing a hydrological connection between two points by means of a single unequivocal trace. While this provides basic data of use in pollution control and general studies, it is often useful to obtain further information on the nature of underground drainage—for example the relative contribution of particular input points to the overall system output or the volume of flooded conduits. The three main methods by which this type of information may be obtained are quantitative tracing techniques, pulse-wave techniques and repeated tracing experiments using fluorescent dyes.

Quantitative tracing techniques

These differ from simple tracing in that the amount of the injected tracer which is recovered at the output point or points is calculated. If the stream discharge at the input point and output point(s) is also measured, then water budgets may be drawn up showing the relative contribution of the input water to each output point (Brown *et al.*, 1969). Brown and Wigley (1969) and Brown and Ford (1971) have shown how these same data may also be used to deduce the topological form of the underground flow network (Figure 9.5), and their work has been extended by Atkinson *et al.* (1973). A major assumption of these methods is that the tracer substance is conservative, i.e. that none is lost within the system. In reality it is probable that no truly conservative tracer exists, and even tritium which forms part of the water molecule can be lost by adsorption (Kaufman and Orlob, 1956). Certain fluorescent dyes, notably amino G acid, Lissamine FF and Rhodamine WT, are generally considered to show relatively low adsorption losses (Smart and Laidlaw, 1977), and they have been used in the majority of quantitative experiments. However, it is becoming increasingly apparent that their use is prone to error, and even Rhodamine WT which was once regarded as ‘truly conservative’ and ‘the best readily available dye’ (Smith and Atkinson, 1977, p. 601) has been shown to suffer major losses which are generally progressive but occasionally apparently random. Of particular importance in this respect are the repeated tracing experiments carried out by Stanton and Smart (1981) in the Mendip Hills, England. They found that dye losses varied between systems and within systems with resurgence output. In the St. Cuthberts Swallet–Wookey Hole Rising system, dye recovery varied from 100% during

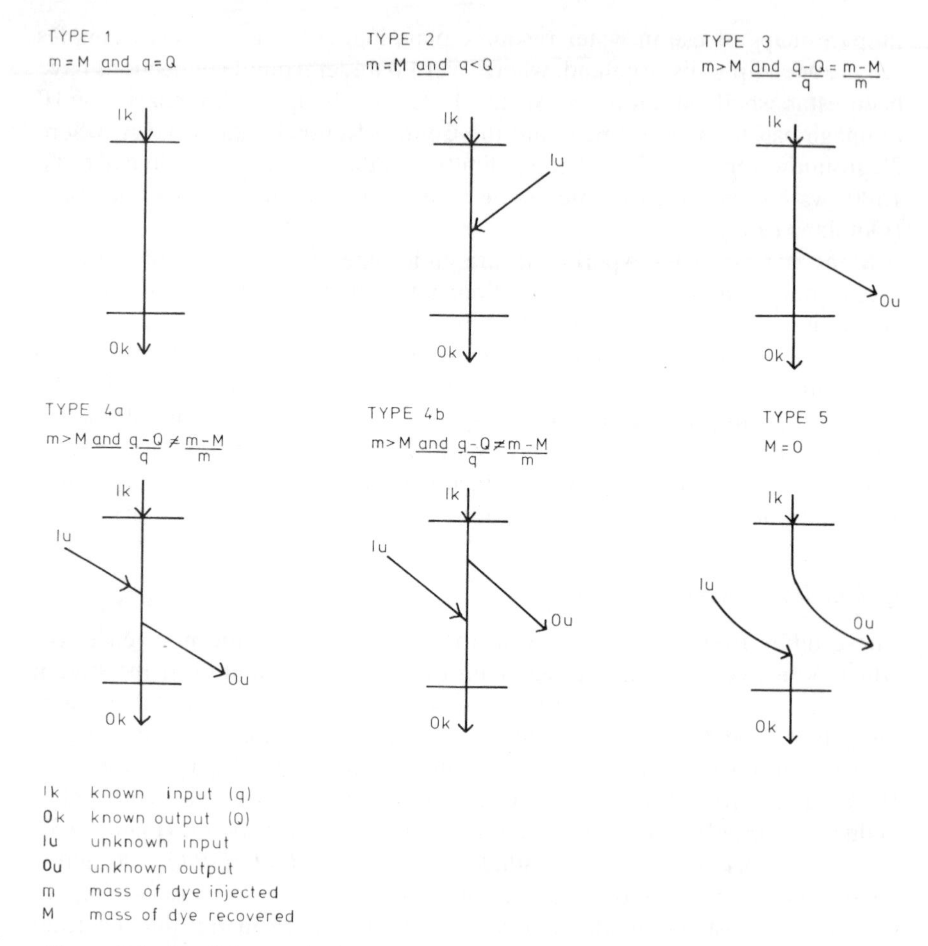

Figure 9.5. Possible topological types of conduit flow network in karst limestones (Brown and Ford, 1971)

a flood event (10 hours travel time) to 12% under base flow conditions (105 hours travel time), and two traces under similar flow conditions (23 and 25 hours travel times) yielded very different dye recoveries of 62% and 102%. Further problems result from the natural fluorescent background at the outlet point which is caused by the presence of dissolved organic material (Smart *et al.*, 1976) and, arguably, of fluorescing bacteria such as Chlorella (Brown and Ford, 1971). This background varies with time, particularly during storm events; and although Smart (1976) has shown that some correction may be possible, 'the technique is not infallible and background variations are still the most significant problem in quantitative fluorescent dye tracing work' (Smart, 1976, p. 293). More recently, Christopher (pers. comm.) has independently developed a similar method for detecting the presence of two fluorescent dyes simultaneously on activated charcoal detectors in the presence of a high concentration of dissolved organic material. By plotting the logarithm of absorption at a specific wavelength against wavelength in nanometres, a straight-line graph with negative slope is obtained. The presence of dye is characterized by deviations from this line at the dye absorption wavelength.

Despite these advances it must be concluded that quantitative tracing theory is presently in advance of the available tracer substances and detection techniques. In particular it is not possible to infer that flow networks of types 3 or 4 (Figure 9.5) exist solely on the basis of low dye recovery or even to infer a type 5 system on the basis of null recovery.

Pulse-wave experiments

Although the use of artificial flood pulses to establish a hydraulic connection between two points dates back over 100 years (Tate, 1879) the modern applications of the technique were first described by Ashton (1966), who based his analysis on the expected changes in water quantity and quality as it passes through a conduit system. In the vadose zone water pulses are transmitted in a similar way to surface streams, and the effects of tributary streams will be to modify the form of the hydrograph (Figure 9.6). The flood routing techniques developed for use in surface streams (e.g. Dunne and Leopold, 1978, p. 350) may prove to be of use in investigating the form of these hydrographs. If the pulse enters a flooded zone it will then be transmitted instantly owing to direct displacement of water. Analysis of water quality variables at the output point provides an indication of the arrival of actual floodwater, and the difference between this time and that of the arrival of the pulse may be used to estimate the volume of phreatic conduits. The method has since been extended by Wilcock (1968), who provided a form suitable for computer analysis, and by Brown (1973), who dispensed with water quality variables and made inferences as to the nature of the underground flow system by cross-spectral analysis of the input and output series of water quantity (discharge).

Although a potentially powerful tool, there have been very few studies of natural flood pulses, primarily because the multiple effect of several individual tributary streams is often difficult to dissentangle (Smith, 1977). Nevertheless, Christopher (1980) has shown that the detailed analysis of a natural flood pulse may provide a considerable amount of information on the hydrology of a complex karst system. To improve the practicability of the Ashton technique an artificial flood pulse may be used such that any complexity in the output hydrograph can be ascribed to bifurcation of passages rather than to the effect of additional pulses from tributaries. Such a study was undertaken by Williams (1977), who used pulses released from a

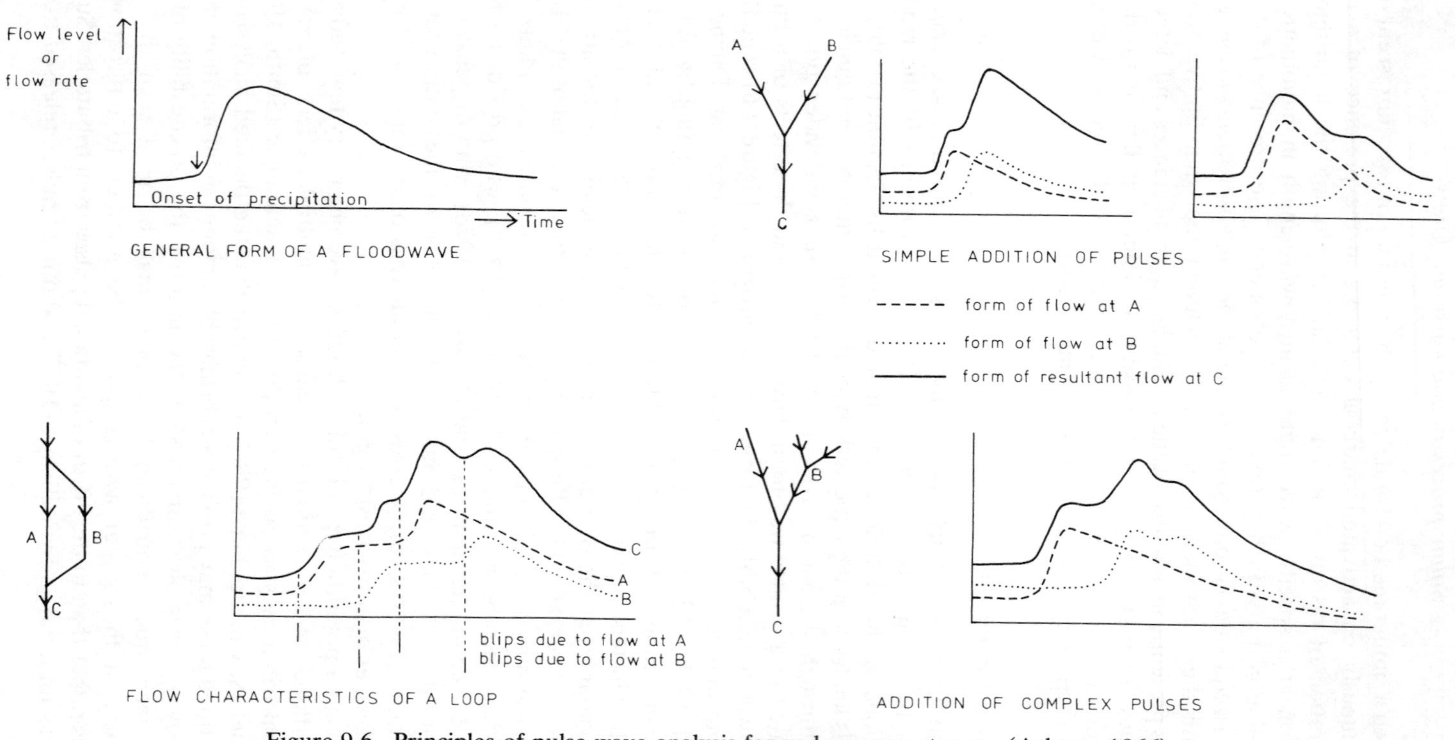

Figure 9.6. Principles of pulse-wave analysis for vadose zone streams (Ashton, 1966)

hydroelectric power plant to demonstrate a hydraulic connection between the Upper Takaka River in north-west Nelson, New Zealand, and the Waikoropupu Springs, a major tidal karst resurgence some 16–18 km to the north. Another improvement is to tag the natural or artificial flood pulse with a fluorescent dye, facilitating detection of the arrival of the actual floodwater and hence computation of the volume of flooded conduits (Atkinson *et al.*, 1973; Smith and Atkinson, 1977; Smart and Hodge, 1980; Christopher *et al.*, 1981).

Repeated tracing experiments

These should provide useful information on the nature of the underground drainage system since tracer travel times, and possibly also tracer losses, are a function of system form and hydrological state. Surprisingly few traces have been repeated, and the following account is based on the only systematic study known to the writer, that of Stanton and Smart (1981) who examined the relationship between tracer travel time and system output, the discharge at the rising. In theory a simple phreatic conduit should respond to increased inputs in the same way as a water pipe—the capacity is constant so that if the speed of water entry doubles it will flow through the system in half the time. Potential complications in real-world karst systems include branching of conduits of different bore sizes, the joining of phreatic streams of different lengths, and non-uniform rainfall inputs. Of these the first is considered to be unimportant as proportionality is likely to be maintained throughout the system; the second will produce a variable response and is perhaps best investigated using pulse-wave techniques; and the third will only be significant in large basins where the response will vary according to the relative positions of input site and storm centre.

As previously noted karst systems dominated by flow in vadose conduits represent underground extensions of surface drainage basins, and a similar response to increases in flow might therefore be expected. Initially water levels in the channel will rise, increasing the volume of in-channel storage. This increase in volume will partly offset the increased flow velocities, so that the decrease in travel time as system output increases will be less marked than in phreatic systems; and where there is a large available storage (e.g. overflow lakes) travel time may actually increase with discharge. In a surface stream bankfull stage forms a threshold of reversal and spillage of water on to the floodplain is likely to result in a much slower decrease or even an increase in travel times with a further increase in discharge. The filling to the roof of a normally vadose conduit is also a threshold state, but in this case the transformation is into a phreatic conduit so that travel times decrease more rapidly than had been the case. Hence a spectrum of karst systems may be envisaged from those with very large, entirely vadose conduits which are

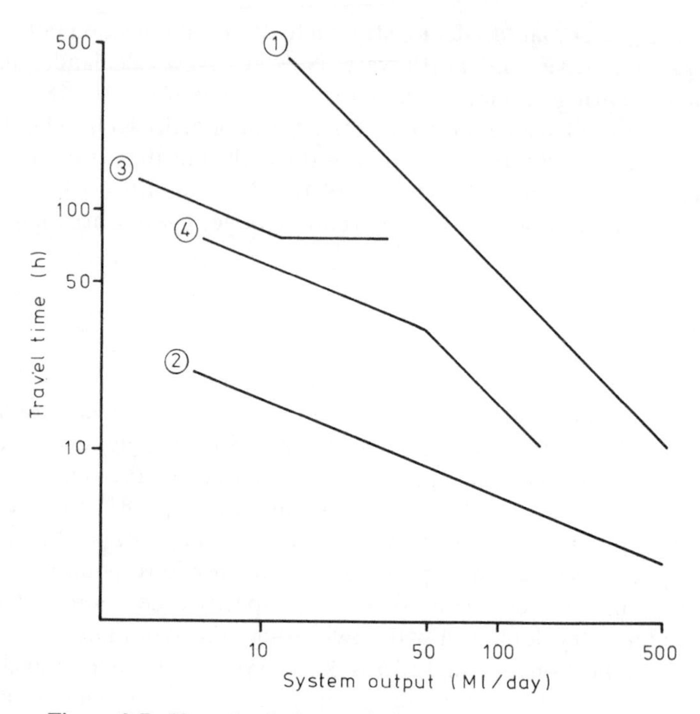

Figure 9.7. Hypothetical travel time relationships for (1) a simple phreatic system; (2) a simple vadose system (large passage relative to volume of water); (3) vadose system with overflow storage (e.g. large lakes); (4) vadose system which fills to roof under high flows and is thereby transformed to a phreatic conduit

physically incapable of flooding to the roof (continuous decrease in travel time unless lakes present), through systems with relatively small conduits which are normally vadose but rapidly fill to the roof under flood conditions (gradual decrease in travel times, then more rapid), to systems which are entirely or almost entirely phreatic (travel times inversely proportional to system output). On the basis of these differences Stanton and Smart (1981) suggested that log–log plots of travel time against system output (Figure 9.7) should provide a means for distinguishing between simple phreatic systems (straight line, gradient near unity), simple vadose systems (straight line, gradient less than unity) and complex systems (connected straight lines of various gradients). For simple systems the length of the straight line would also have hydrological significance, being limited at one end by the system's maximum flood capacity and at the other by the drying up of the input stream, thereby preventing further tracing. Stanton and Smart (1981) also carried out the first systematic study of travel time/discharge-relationships, tracing one connection 24 times, one 15 times, and a third 4 times. In all three systems the gradient of the best-fit straight line for plots of travel time against discharge was close to unity, indicating that the underground drainage systems are basically of the simple phreatic type. It is clear that this approach has considerable potential and should form an important part of future hydrological studies of karst systems. In particular it would be useful to carry out repeated traces of systems whose form is largely known from direct exploration in order to establish the degree to which system form can be inferred from travel-time/system-output relationships.

9.4.5 Conceptual models

These may be regarded as generalized pictures of water movement in karst systems. One of the earliest and best-known classifications is that of White (1969) who recognized that water movement in carbonate rocks spans a spectrum from slow laminar flow in 'diffuse flow aquifiers' through to the rapid, turbulent conduit flow of 'free-flow aquifers' in which 'solution has developed a subsurface drainage system logically regarded as an underground extension of surface streams' (White, 1969, p. 15). Confined flow aquifers, in which carbonate beds are either part of an artesian basin or form thin beds sandwiched between impervious rocks, complete the classification. In a later revision, White (1977, p. 187) suggested that 'the value of conceptual classifications is that they permit one to identify particular aquifer types and also indicate some of the variables that should be incorporated into quantitative models of aquifer behaviour'. While this is true it may be argued that neither diffuse flow nor confined flow systems are true karst forms, since a primary characteristic of karst aquifers is the development of secondary permeability over time and under normal conditions this would ultimately result in a conduit flow system. Indeed, White (1977, p. 182) has presented a useful evolutionary model for the transition of a diffuse flow system into one which is dominated by conduit flow, although 'the diffuse groundwater body remains and there is an interchange both between it and the underground drainage (the conduit system) and the surface drainage'. A somewhat similar concept was developed by Smith *et al*. (1976, p. 209), who suggested that 'limestone drainage occurs both by conduit and diffuse flow, which interact to form a single integrated system rather like a river basin on the surface'. Smith *et al*. (1976) further noted that pure diffuse flow and pure conduit flow systems are rare. Moreover, analysis of diffuse flow and confined flow systems is best undertaken using the traditional tools of the groundwater scientist, and in particular by theoretical models based on Darcy's law (e.g. Freeze and Cherry, 1979), while the application of Darcian theory and concepts to systems including a conduit flow component presents difficulties and may cause misunderstandings (Smith, 1977). The present account will therefore focus on 'free-flow' systems which have a well-developed conduit permeability, taking a conduit to have a diameter greater than 1 cm which will generally ensure that flow is turbulent. These systems may be regarded as truly karstic, and they are also the aquifers 'that pose the greatest problems in theoretical analysis, in water resource evaluation and development, in land-use hazards, and in pollution dispersion' (White, 1977, p. 176). As both conduit and diffuse flow components are present in most free-flow systems, it is important to identify their relative importance. In the first study which attempted to measure the relative contributions of conduit and diffuse flow, Atkinson (1977b) applied a qualitative hydrograph separation procedure to the outflow hydrograph from Cheddar Spring in the Mendip Hills, England. He found that 'quickflow' from sinking streams and closed depressions accounted for about 50% of recharge and slow percolation for the remainder. Quickflow was identified as the main source of conduit flow and slow percolation of diffuse flow, but Atkinson (1977b, p. 108) noted that 'the conduits also serve to drain the diffuse component of the aquifer and contain a baseflow derived from this component'. Recession curve analysis suggested that storage in flooded conduits was relatively small and that 'the majority of storage is in the form of true groundwater found in narrow fissures, where laminar flow prevails' (Atkinson, 1977b, p. 108). It is important to note that in arriving at this conclusion Atkinson assumed that 'drainage of the unsaturated zone has almost ceased by the end of one month of drought', so that all recession after 40 days is due to 'drainage of the unsaturated zone in the absence of recharge' (Atkinson, 1977b, p. 102). Later studies which have challenged this assumption will now be discussed.

Friederich (1981, p. 97) examined the recharge mechanisms of limestones in the Mendip Hills and found 'seepage flow' (which Atkinson called slow percolation) accounted for only 11% of unsaturated zone drainage and that a similar amount was contributed by leakage from Devonian sandstones. He also concluded that 'quickflow inlets' are responsible for most of the transport of percolation water, although 'a large proportion of the limestone drainage (i.e. baseflow) is stored in the unsaturated zone' (Friederich, 1981, p. 98). These conclusions, which are in direct contradiction with those of Atkinson (1977b), are similar to those obtained by the writer in the Waitomo District, New Zealand, where it was found that 'base flow' storage in the unsaturated zone formed 65–80% of available storage (Gunn, 1978, p. 150). Current research in Peak Cavern, Derbyshire, also suggests that the unsaturated zone forms an important store, although its relative importance cannot as yet be quantified. In their studies of unsaturated zone Gunn (1978) and Friederich (1981) independently identified two main stores—the soil zone and the uppermost 5–10 m of weathered limestone. Gunn (1978) followed Birot (1966) and Williams (1972) in referring to this as the 'subcutaneous' zone, while Friederich (1981) followed Mangin (1974–5) in using the term 'epikarstic' zone. For the present discussion subcutaneous will be employed as it has received wider usage in karst studies (e.g. Williams (1978, 1983); Day (1979); Gunn (1981a,b, 1983). The soil zone has previously been rather neglected by karst hydrologists as soils formed on bare limestone surfaces tend to be thin (e.g. Curtis *et al.*, 1976). However, many limestones are covered by superficial deposits (till, loess, volcanic ash), and under these conditions storage of water in and transmission of water through the soil zone is likely to be of considerable significance. For example, in the Waitomo area, throughflow discharges increase during the latter stages of prolonged, low-intensity rainfall events and following the cessation of short periods of high intensity rainfall, but a period of 5–15 weeks without rainfall is required before flow ceases (Gunn, 1981a). Similarly in the Mendip Hills, the bulk of soil water has a residence time of around 8 weeks (Friederich, 1981). The subcutaneous zone has also received scant attention from karst hydrologists, and a recent review has noted that its nature and role 'are not well or widely understood, despite its apparent importance in karst terrains' (Williams, 1983, p. 46). The development of a subcutaneous zone is a consequence of the concentration of limestone solution in the soil and uppermost bedrock (Williams, 1968; Jennings, 1972; Smith and Atkinson, 1976; Williams and Dowling, 1979; Gunn, 1981b). As a result of this solution, permeability is higher than in the underlying rock (Friederich (1981) suggests 2% for the Mendip Hills), and this encourages lateral movement of water often focusing on closed depressions (Gunn, 1981a, 1983). Williams (1983) has provided the first comprehensive review of subcutaneous hydrology, and it is clear from his work together with that of Friederich (1981) and Gunn (1978, 1981a, 1983) that, in common with the soil, it is an important zone of both storage and transmission and that this transmission may either be rapid, probably via a piston flow mechanism, or delayed. Gunn (1981a) and Williams (1983) have also argued that a subcutaneous zone will probably be present in most karst areas, and the presence of a soil cover is not thought to be a prerequisite since subcutaneous storage has been identified in the semi-arid Carlsbad Caverns region of New Mexico, where there is a minimal soil cover (Williams, 1983).

It is apparent from the field and theoretical evidence presented above, and from the further discussion in Williams (1983), that the current conceptual models of limestone drainage which emphasize saturated zone storage are inaccurate. Hence, a modified model based on Smith *et al.* (1976) but incorporating the new ideas on the soil and subcutaneous zones is proposed (Figure 9.8). Two main modes of recharge are recognized, allogenic inputs from non-carbonate rocks and autogenic inputs from precipitation falling on the limestone outcrop. Allogenic recharge is primarily via stream-sinks (point recharge), although more diffuse recharge is possible where permeable strata overlie carbonates. Autogenic recharge is frequently concentrated by closed

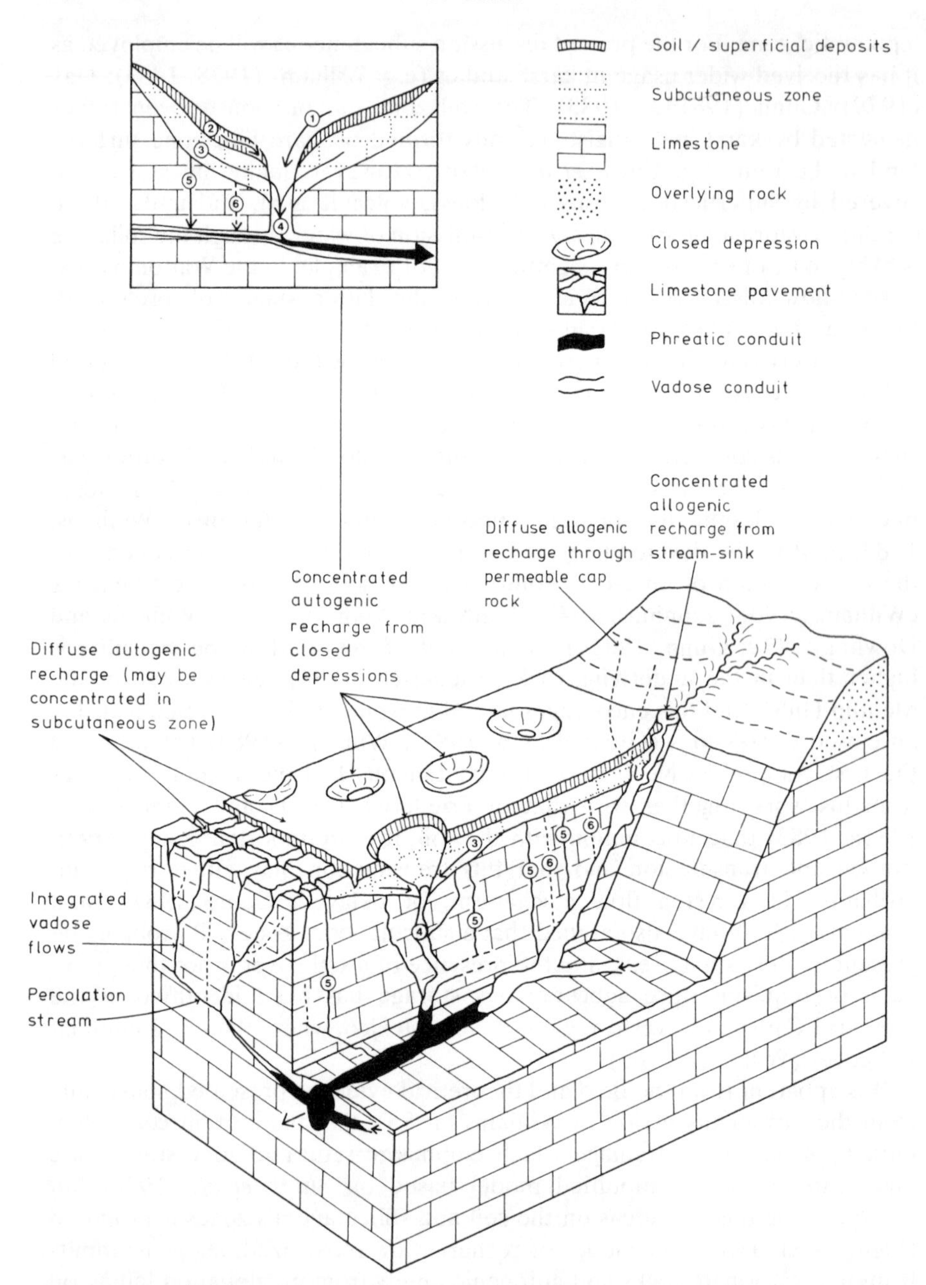

Figure 9.8. Conceptual model for the drainage of limestone areas. Some of the features shown may be absent in particularly areas. (1) Overland flow; (2) throughflow; (3) subcutaneous flow; (4) shaft flow; (5) vadose flow; (6) vadose seepage

depressions and may also be diffuse through a soil cover or direct through a bare bedrock surface. Stream-sinks feed directly into significant conduits, usually in the vadose zone and often but not always of accessible dimensions. Closed depressions generally feed via vertical shafts into more minor tributary conduits which may connect with larger conduits in either the vadose or the phreatic zone. Some depressions, particularly those of collapse rather than of solution origin, feed directly into larger conduits.

The hydrological function of depressions has been discussed by Gunn (1981a, 1983) who proposed a model incorporating three concentration mechanisms (overland flow, throughflow and subcutaneous flow) and three input mechanisms (shaft flow, vadose flow, and vadose seepage). Subcutaneous flow is possible in depressions without a soil cover (Williams, 1983), but overland flow will be extremely rare and throughflow, by definition, absent. In depressions with a soil cover and on soil-covered limestone surfaces where depressions are absent, the well established principles of hillslope hydrology will apply (Ward, 1975; Kirkby, 1978; Derbyshire *et al.*, 1979). In brief these suggest that overland flow will be relatively uncommon on vegetated slopes in temperate karst areas and that, where it does occur, return flow and precipitation on to saturated areas will be significant contributors. Greater volumes of overland flow might be expected in tropical karst areas under conditions of prolonged and/or intense rainfalls. Throughflow may be expected in both saturated soils where its velocities may be sufficiently fast to contribute to stormflows and also in unsaturated soils where velocities are slower and the contribution is to baseflow. Pipeflow is becoming increasingly recognized as an important contributor to stormflows, and there is no reason to suppose that it cannot occur on soils above limestones, although no examples are known to the writer. In view of the similarity between soil-covered slopes on limestones and those on other lithologies, it may be further argued (Gunn and Turnpenny, in press) that theories of stormflow generation from dynamic contributing areas within drainage basins may also be applied to stormflow generation in karst depressions, although subcutaneous zone processes would have to be taken into account. Further research on subcutaneous hydrology is required, but the present model envisages most flow being directed laterally to a depression outlet area. However, it is possible that in old karsts or karsts with no soil cover the vertical permeability may exceed the lateral permeability, thereby changing the principal direction of subcutaneous water movement.

The three concentrating mechanisms channel water towards the basal depression outlet area from where it is generally transmitted to a conduit as shaft flow. Shaft dimensions may vary from a few centimetres to several metres, and their hydrological significance lies in the rapid transmission of water through the limestones. Smaller vertical fissures may intercept water moving laterally through the subcutaneous zone, transmitting it underground as vadose flows which may combine to form percolation streams and/or be tributary to shafts, conduits, or a general groundwater body if one exists. The final transmission mechanism is vadose seepage, which is made up of flow through the rock's primary pores and through tight fissures. It may be tributary to flows or contribute directly to conduits or the general groundwater body, but its hydrological significance is generally small. Shaft flow, vadose flow and vadose seepage may be regarded as points on a continuum of flow paths characterized by diminishing physical size and decreasing flow velocity, and each point is capable of further subdivision. Shaft flow and those flows linked to significant conduits will contribute primarily to the system's stormflow response, while the remaining flows and the seeps contribute delayed flow. The vertical fissures which transmit vadose flows and seeps also form a static water store which forms about 10% of the baseflow store in the Mendip karst (Friederich, 1981, p. 98). Atkinson (1977b) estimated that about 3.5% of baseflow is stored in flooded conduits, and it would seem on the basis of work by Friederich (1981) that most of the remaining 86.5% is stored in the soil and subcutaneous zones. Water from these stores will be transmitted by shaft and vadose flows as well as by seeps, and it is therefore apparent that the routes which transmit stormflow may also transmit significant quantities of baseflow. On bare limestone surfaces vertical transmission mechanisms are of greater significance than concentration mechanisms; and although the subcutaneous zone may still be present (Williams, 1983), the absence of soil zone storage is likely to result in higher stormflows and lower baseflows than those karsts with a soil cover.

Flow in conduits may take place under vadose or phreatic conditions. Vadose streams respond to recharge in the same way as surface streams, and Gunn and Turnpenny (in press) have argued that their characteristics may be best analysed by the techniques of surface water hydrology. Application of these techniques to three karst streams in New Zealand showed that their storm hydrograph characteristics lay within the range of observed values from experimental basins on other lithologies in New Zealand, and that the characteristics provided useful insights into stormflow generation mechanisms in the karst basins. Many vadose streams feed into phreatic conduits before rising, and their discharge characteristics will therefore be modified according to the length and complexity of the phreatic zone. Water tracing experiments and flood pulse analyses (§9.4.4) suggest that in many karsts storage in phreatic conduits is relatively small and that flow velocities are slower than in vadose streams, although flood pulses are transmitted instantaneously by a piston effect. The final element in the model is diffuse flow (possibly laminar) and storage in flooded fissures. Its role is still a matter for debate, but the writer would contend that in most mature karsts it is of minor importance.

The model presented here should be applicable in a wide range of karst terrains, with the relative importance of the different stores and flow mechanisms varying according to local environmental factors. It should therefore contribute to, and be modified as a result of, more detailed hydrological studies.

9.5 KARST LANDFORM EVOLUTION

> 'The ultimate goal of karst and cavern research, so far as geomorphology is concerned, is to postulate, or perhaps even to prove, a scheme of landscape evolution.' (Powell, 1975, p. 234).

Two main types of evolutionary scheme may be envisaged; firstly, generalized models of the origin and sequence of development of karst landforms, and secondly, regional denudation chronologies in which the history of karst landform evolution is reconstructed. In both cases, the general principles are similar to those for other terrain (Jennings, 1982), but karst researchers have three important advantages over other evolutionary geomorphologists:

1. the relative simplicity of the dominat erosion process, aqueous solution;
2. the tendency of karst surface landforms to persist longer and with greater perfection than those on most other rocks ('karst immunity'), since water usually passes underground as karstification proceeds leaving an outer skin no longer subject to fluvial erosion;
3. the facility for absolute dating of certain cave deposits.

However, there are also special complexities which arise in karst development such as the problems of interpreting buried, exhumed and subjacent (interstratal) karsts (Jennings, 1982).

9.5.1 Sequences of karst landform evolution

Early theories of the development of karst topography which were largely based on the geographical cycle concept included consideration of both surface (exokarst) and underground (endokarst) landforms and their sequential development. The two most widely known and most frequently contrasted models are those of Grund (1914) and Cvijic (1918), although Powell (1975) has drawn attention to the pioneering work of Beede (1911) which unfortunately had little impact at the time. Several reviews of these and other early models have been published (e.g. Jennings, 1971; Powell, 1975; Sweeting, 1972, 1981). More recently there has been a tendency to consider the development of exokarst and endokarst landforms separately. For example, Ford and Ewers (1978) have provided a set of principles of cavern genesis which resolve most of the contradictions in earlier papers, and these are further developed by Waltham (1981a) in his review of the origins and development of limestone caves. However, in both cases cave development is

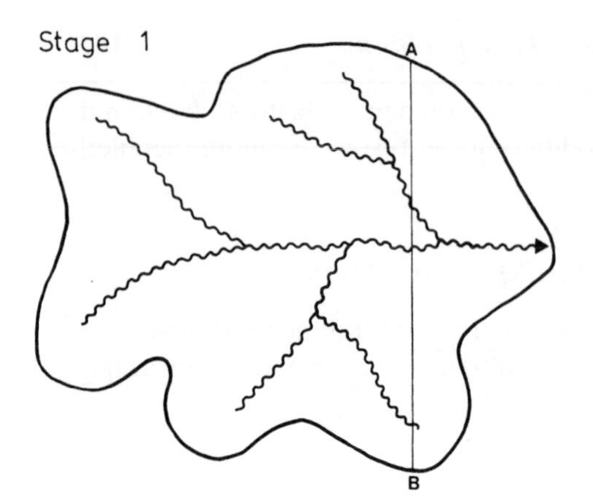

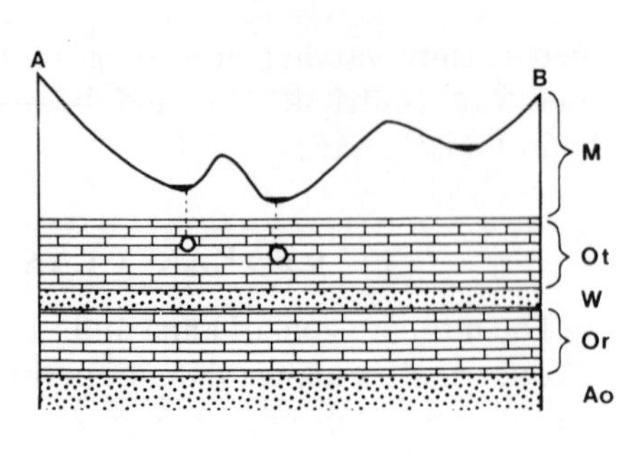

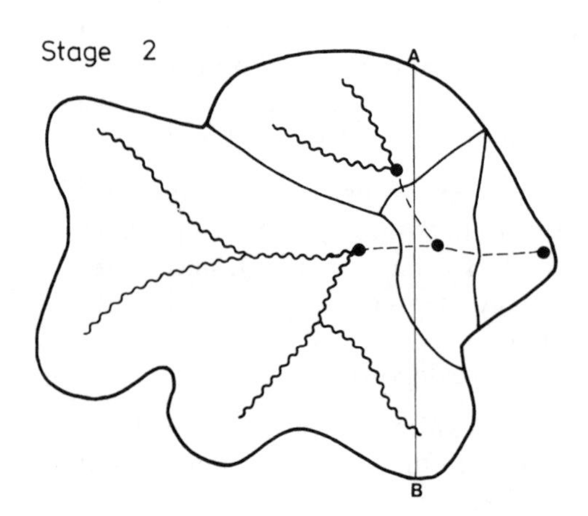

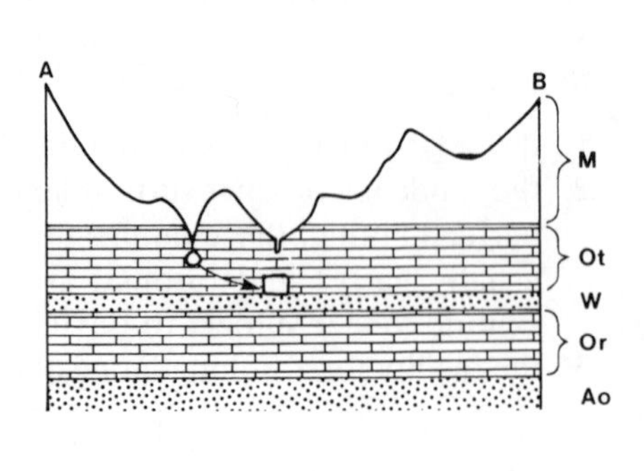

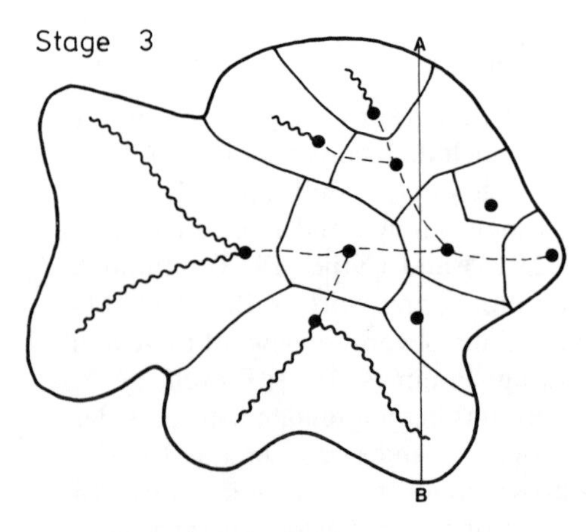

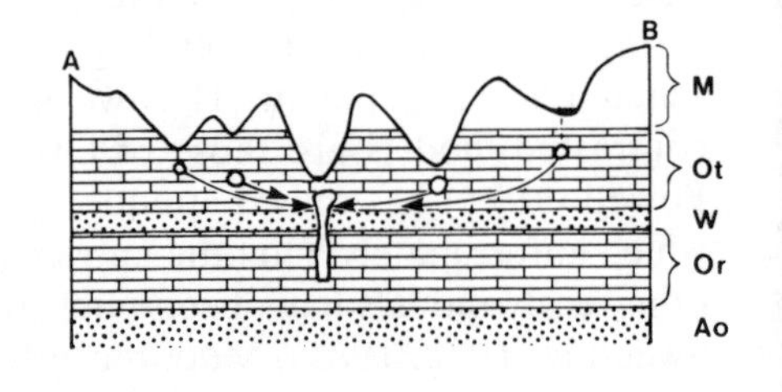

——	Topographic divide
〰	Surface stream
- - -	Course of prior surface stream
⊂〰	Stream rising
●	Depression outlet

M	Mahoenui Group
Ot	Otorohanga limestone
W	Waitomo sandstone
Or	Orahiri limestone
Ao	Aotea sandstone
O	Cave passage
↘	Subterranean drainage line

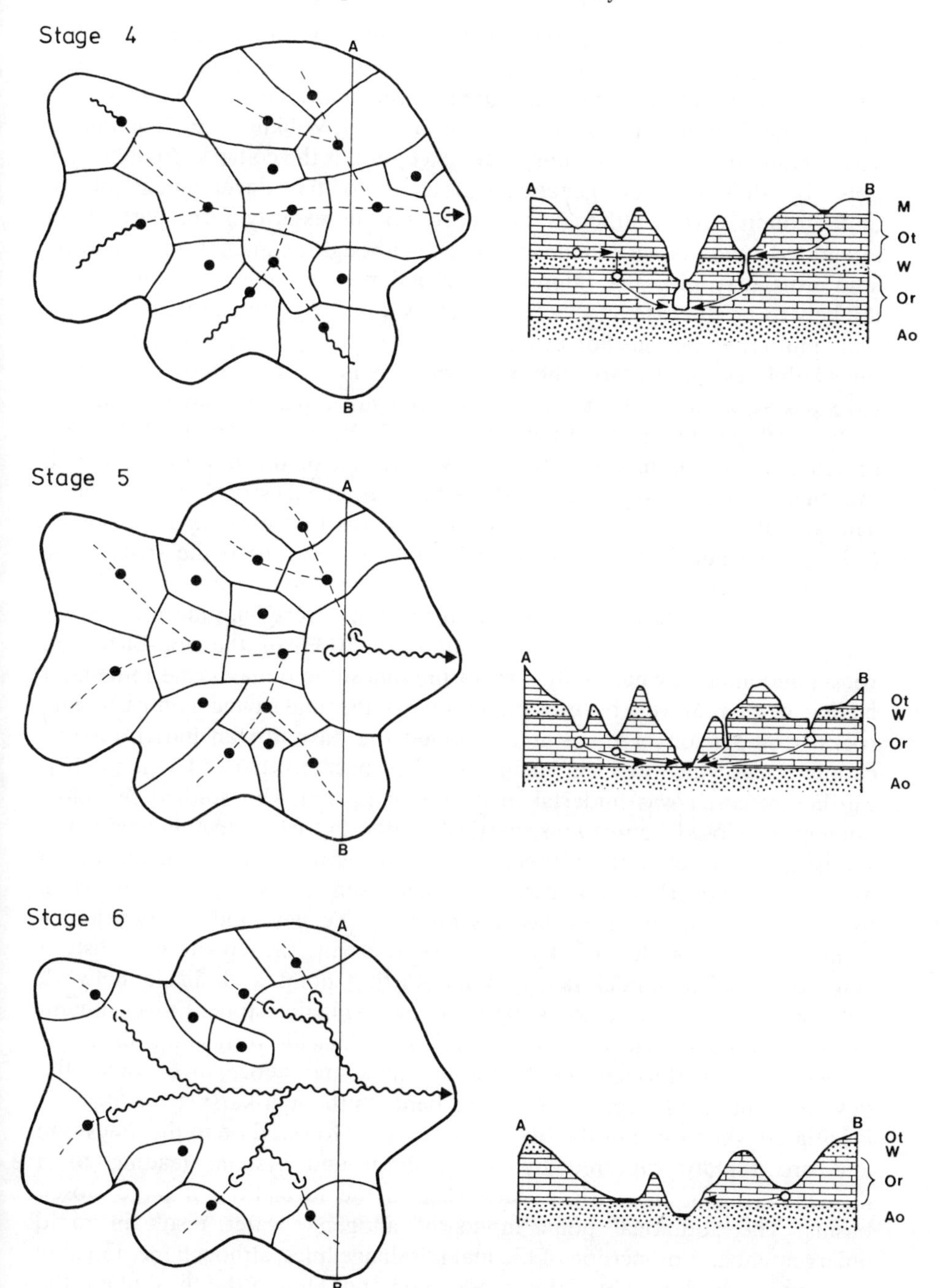

Figure 9.9. Model of karstic evolution in the Waitomo district, New Zealand (Gunn, 1978)

considered in isolation from that of surface landforms. Similarly, most studies of the evolution of closed depressions (dolines) have argued that since solution proceeds from the surface there is little, if any, genetic relationship to underground landforms (Ford and Stanton, 1968; Williams, 1972; Jennings, 1975; Palmquist, 1979). An important exception is the collapse doline which must by definition be genetically related to the development of an underground cavity, although this need not necessarily form part of an integrated cave system (Thomas, 1974). Although endokarst and exokarst landforms may give the impression of independent development, this is unlikely to be the case and there are strong *a priori* grounds for considering their mutual evolution. For example, Bull (1980a, p. 218) has argued that 'many dolines, particularly those in South Wales, are related not only to cave passages up to 180 metres below, but are intricately involved in the early development phases of the cave system'. A recent example of the integrated approach is provided by an investigation of the development, and eventual decay, of polygonal karst in the Waitomo District, New Zealand. The karstification of drainage in the region was first examined by Feeney (1977), who adopted a morphometric approach and used the space–time substitution technique (Abrahams, 1972). Five areas were examined: a 'normal' drainage basin on non-carbonate rocks overlying limestone; areas where erosion has removed 30%, 80% and 100% of the non-carbonate rocks; and an area where only 40% of the limestone remains, the remainder having been removed by erosion to reveal the underlying non-carbonate rocks. Comparison of these areas formed the basis for an initial karstification model which was similar in several respects to that of Cvijic (1918). Further research was undertaken by Gunn (1978) who studied the morphology of closed depressions, vertical shafts and caves together with the spatial patterns of surface topography and subterranean drainage in an area where all overlying non-carbonates had been removed. The results were used to amplify the previous research by Feeney and to develop a comprehensive model of karstic evolution in the Waitomo District (Figure 9.9). The model takes as its starting point a drainage network developed on permeable non-karst rocks overlying limestone. Entrenchment of surface valleys creates hydraulic gradients enhancing vertical movement of agressive water through the limestone mass and hence, over time, the development of subterranean drainage beneath the non-karst cover. Phase 2 is initiated when some of the surface streams are lowered on to the limestone and are rapidly captured by the underground system, leading to a dismembering of the surface network and the formation of large closed basins. The additional point inputs of allogenic water result in rapid enlargement and/or incision of the main drainage lines, although incision may be temporarily slowed by a thin sandstone band. During the third phase the remaining non-carbonate rocks are progressively removed and the large basins are broken up into smaller units by the exposure of limestone and enhanced vertical permeability. New depressions also form on the ridges and interfluves ensuring widespread surface lowering. The depressions also form important sources of point recharge, often via vertical shafts, and the proportion of autogenic inputs grows at the expense of allogenic stream sinks. Subterranean drainage eventually breaches the sandstone band and a period of rapid downcutting ensues. Phase 4 commences with the final removal of the overlying cover rocks, and this is followed by competition for space which ultimately produces a network of drainage divides enclosing depressions of comparable size (a polygonal karst). Lowering of depression floors may be temporarily halted by the sandstone which is frequently breached by shafts. Erosion of the underlying limestone undermines the sandstone, which collapses allowing depression floor lowering to continue. The underground conduits are also incised into the limestone, and phase 5 is marked by their reaching the underlying non-carbonate basement rocks. Further vertical development by solution is not possible, and although the streams may continue to cut down by corrosion this phase marks the beginning of the destruction of the karst system. The large depressions reach the basement first and capture underground flow from smaller, shallower depressions which are eventually incorporated into the larger forms. Lateral widening of depressions and amalgamation into large basins continues into the sixth and final phase, which terminates when all the depression boundaries have been broken down by surface processes and a surface drainage network re-established. This model is essentially one of interstratal karstification and may form a useful basis for evolutionary studies in other regions where limestones are sandwiched between non-karst rocks. However, it should be noted that it describes a sequence of landforms rather than a cycle of landscape evolution, and that different parts of an area may be at different stages of development at any particular point in time. Finally, the research in the Waitomo area has demonstrated that a polygonal karst landscape may form on a relatively thin (100 m) interstratal limestone as well as on the fresh, uplifted limestones with no cover rocks envisaged in the original model of Williams (1972).

9.5.2 The age of karst landforms

In common with other rocks there is a long history of attempts to correlate the evolution of karst landforms with temporal events, and in particular with different climatic regimes during the Quaternary (Warwick, 1971) or even earlier periods (Ford, 1977). In addition to this approach, which may be characterized as 'denudation chronology' the ages of karst landforms have been estimated by backwards extrapolation of solutional erosion rates and by analysis of cave sediments, some of which may also provide absolute dates.

Denudation chronologies

These are the traditional means by which landscape histories are reconstructed. However, severe problems occur as in most areas more recent events have altered or destroyed older features and deposits such that there are more 'gaps' than extant stratigraphic record (Ager, 1973). As a result many studies in both karst and non-karst areas have been based on fragmentary evidence such as the heights of erosion surfaces and river terraces. For example, workers in England (Sweeting, 1950), Belgium (Ek, 1961), Australia (Jennings, 1964) Czechoslovakia (Droppa, 1966) and the USA (Miotke and Palmer, 1972) have attempted to relate the levels of cave passages to river terrace remnants and horizontally extended erosional landforms in order to assess their relative ages. Where no method of assessing the absolute age of landforms and deposits is available, such studies must perforce be of a speculative nature, and a major criticism of denudation chronology is that relatively uniform height distributions have become non-uniform distributions in which the periodicity of erosional flats is similar to the contour interval of the base map (Hodgson *et al*., 1974). Such has apparently been the case in the Morecambe Bay karst, England, where Gale (1981a) was unable to verify the existence of the remnants of 11 erosion surfaces recognized by previous workers who had suggested that they were the results of a long and complex sequence of regional base-level changes.

Another problem with early work is that geomorphic development was generally fitted to the classic scheme of four Pleistocene glaciations, whereas recent work on deep-sea sediments (Shackleton, 1977) and terrestrial deposits (Kukla, 1978) has identified at least 12 glacial/interglacial cycles. The detailed study of the genetic relationship between caves and surface landforms in the Mammoth Cave National Park, USA by Miotke and Palmer (1972) has been criticized in this respect (Hess and Harmon, 1981). Hence it may be concluded that traditional denudation chronology is a relatively poor tool for estimating the age of karst landforms.

Extrapolation of weathering/erosion rates

Weathering rates are frequently assumed to be linear functions of time, but theoretical and empirical evidence, summarized by Colman (1981), fails to support a linear time function for most weathering processes. For example, weathering rates in silicate rocks decrease through time as a result of the inhibiting role of residues in weathering processes. Fortuitously for karst researchers the congruent dissolution of limestone appears to be one of the exceptions to this rule (Coleman, 1981), as control studies of the weathering of limestone tombstones (Goodchild, 1890; Cann, 1974) and carbonate rocks used in buildings (Winkler, 1966) suggest that the solution process proceeds at a constant rate which varies with climate and limestone petrography. However, the backwards extrapolation of present-day solutional erosion rates in order to obtain ages for karst landforms is a hazardous operation both because of the paucity of reliable estimates of modern erosion rates and their spatial and temporal variability (§9.3), and also because of the known changes in climate, and hence of hydrology, during the Pleistocene. Nevertheless, backwards extrapolation can provide order of magnitude estimates such as the age of 1.8 million years which Waltham (1981b) has put forward for a large closed basin in Northern Spain, or Reesman and Godfrey's (1981) assertion that the Central Basin of Tennessee could have developed by chemical means alone during the last 10 million years.

Cave sediment studies

In contrast to the qualitative and speculative nature of earlier work the last decade has seen major advances in the quantitative analysis of cave sediments. This followed the recognition that the relatively stable cave environment and the protection from surface weathering afforded to both cave morphology and sediments makes caves ideal 'museums' for the preservation of evidence for past geomorphic events on the land surface. The value of these museums is increased by the potential length of their record, since 'caves are the longest-lived (least time-limited) elements in our landscapes' (Ford, 1980a, p. 360). Moreover, the advances in sediment analysis have come at the same time as the development of absolute dating techniques for cave deposits, and together they have revolutionized landform chronology since both the landscape history of wide tracts of country such as the Canadian Rockies (Ford *et al*., 1981), and the evolution of specific surface landforms such as Cheddar Gorge (Atkinson *et al*., 1978) may now be reconstructed on the basis of cave studies. Cave sediment studies have also been used to establish the interglacial chronology of the Rocky and Mackenzie Mountains, Canada (Harmon *et al*., 1977); the late Pleistocene sea-level history of Bermuda (Harmon *et al*., 1978); and rates of uplift in north-west South Island, New Zealand (Williams, 1982); and to infer both general palaeoclimates (Harmon *et al*., 1978; Hennig *et al*., 1983) and specific palaeotemperature curves (Hendy and Wilson, 1968; Gascoyne *et al*., 1980). These applications are beyond the scope of the present chapter, but the techniques of cave sediment analysis and dating and their use in estimating the age of karst landforms will be briefly reviewed.

As 'there is no such thing as a unique karst process' (Sweeting, 1981, p. 2) the three main groups of sediments found on the earth's surface—clastic, organic and chemical—also occur in underground cavities and cave entrances. Clastic deposits in caves hardly differ from those on the surface, although the protective nature of the cave environment generally results in better preservation so that most sediments have remained essentially unaltered and undisturbed, since initial deposition (Bull, 1980b). Their value in the reconstruction of palaeoenvironments and detailed histories of cave development has been demonstrated by Bull (1978, 1980b, 1981) and may be further extended in the future by studies of their palaeomagnetic properties (Noel *et al*., 1979). An alternative approach advocated by Gale (1981b) is the use of fluvially transported cave sediments to infer the palaeohydraulics of karst drainage systems. However, the potential of clastic sediment studies can best be realized if they are undertaken in conjunction with absolute dating of chemical sediments. The sedimentary record between absolute dates can then be filled in 'by utilizing the continous sedimentation and spatially invariant properties of the fine-grained sediments that are interbedded with the non-clastic materials' (Bull, 1980b, p. 185).

Chemical and organic cave sediments are usually chemically similar to the corresponding surface forms, although they may differ morphologically (Bogli, 1980, p. 165). White (1976) and Bogli (1980) have provided useful classifications of cave minerals and discussion of the conditions under which they form. However, many of these minerals have little relevance to overall landform evolution; and although 'the identification of certain cave minerals can be useful in elucidating landform genesis, it has only been since the application of isotope studies to calcareous deposits that cave sediment studies have become important' (Bull, 1983, p. 302). The ^{14}C concentration in carbonate deposited on speleothems in limestone caves has long been used to date events occuring within caves (Broecker *et al*., 1960; Hendy, 1970). However, the ^{14}C method has an effective limit of around 40,000 years, which restricts its use in evolutionary studies, and it has now largely been superseded in cave studies by the $^{230}Th/^{234}U$ method which has an effective limit of 350–400 thousand years. Although the use of this method to date terrestrial carbonates was pioneered by Cherdyntsev (1955, 1971—cited by Ford and Drake, 1982), it has come to prominence largely as a result of work at McMaster University, Canada, where over 1000 specimens have been dated (Ford and Drake, 1982). The basic principles of uranium-series disequilibrium dating methods are relatively straightforward (Schwarcz, 1978; Ford and Schwarcz, 1981), and although the laboratory procedures are more complex analyses are now performed on a routine basis in several countries.

At the present-time considerable effort is being devoted to extending the limits of geochronometric dating of cave deposits. To date the most successful technique has been palaeomagnetic analysis of speleothems (Latham *et al*., 1979, 1982) which has produced acceptable dates and allowed deductions to be made on long-term landform evolution and denudation rates in the Canadian Rockies (Ford *et al*., 1981). The thermoluminescence characteristics of calcite may also provide a useful dating method (Wintle, 1978); and although further work on this aspect is necessary, Quinif (1981,

p. 309) has suggested that thermoluminescence may be used to resolve other problems of karst sedimentology including 'definition of lithological units in deposits in caves, origin of the different detrital sediments and palaeogeomorphological reconstructions'. Finally, the potential of Electron Spin Resonance (ESR) dating of cave calcites is currently under investigation at several institutions including the Universities of Bristol (Smart, personal communication) and Auckland (Williams, personal communication). If calibration of ESR against $^{230}Th/^{234}U$ and palaeomagnetic dates proves possible, then this method has the potential to provide dates in excess of one million years with considerable implications for Quaternary research.

The use of geochronometric dates from speleothems to infer ages for karst landforms is based on three main principles:

1. A speleothem date provides a minimum age for the section of cave passage in which it was deposited.
2. Speleothem growth cannot occur in passages which are waterfilled. Hence, speleothem dates from a section of passage which formed under phreatic (waterfilled) conditions provide a minimum age for its drainage. Drainage is a consequence of lowering the 'water table' which, for conduit systems, means a lowering of the elevation of the outlet point. This in turn is usually an indication of valley floor lowering at the outlet point, so that the speleothem date provides a minimum age for valley floor incision. Following incision a second phreatic system may form at a lower level, and if this in turn is drained by later downcutting then a dated sequence of landform evolution may be established (Atkinson *et al.*, 1978). In addition, the difference in elevation between a speleothem sample and the modern valley floor provides a crude indication of the amount of valley deepening since dewatering, and this may be used to calculate the maximum rate of valley incision (Ford *et al.*, 1981). Although potentially extremely useful the use of speleothem dates in this way is somewhat limited by the fact that speleothem growth is often episodic and does not necessarily commence immediately after a cave is dewatered. In addition it is impossible to tell whether the oldest speleothem in a particular passage has been sampled, and the removal of large numbers of samples to increase the chance of finding the oldest is likely to cause conservation problems as well as logistical problems at the preparation and counting stages.
3. Where clastic sediments are overlain by speleothem (stalagmite/flowstone), a basal date will provide a minimum age for their deposition. Similarly, detrital speleothem fragments within or beneath the sediments permit calculation of a maximum date for the initiation of clastic deposition. This may then allow correlation of the floods which deposited the sediments with surface events such as meltwater flow from decaying glaciers or increased rainfall during pluvials (Schwarcz, 1978). However,

the maximum and minimum dates so obtained provide wide limits within which the age of a clastic deposit can vary, since 'a detrital speleothem fragment may be very much older than the sediment in which it is incorporated, whereas a stalagmite or flowstone layer overlying a clastic deposit may be very much younger' (Goede and Harmon, 1983, p. 94). As noted above, sampling may be restricted due to considerations of cave conservation.

The most important point to emerge from consideration of these three principles is that construction of dated landform genesis models for specific regions requires more data than palaeoclimatic investigations which may be based either on detailed study of 1–2 speleothems (Gascoyne *et al.*, 1980) or on grouped data from several caves (Harmon *et al.*, 1977). As a result there exists a growing number of accounts in which palaeoclimatic reconstructions place some temporal constraints on the development of karst landforms (e.g. Harmon *et al.*, 1978; Glazek and Harmon, 1981; and several other short accounts in Beck, 1981), but only five studies of karst landform evolution *per se* had been published at the time of writing, from England (Atkinson *et al.*, 1978; Ford *et al.*, 1983); North America (Gascoyne and Latham, 1981; Hess and Harmon, 1981) and Australia (Goede and Harmon, 1983). Atkinson *et al.* (1978) examined the evolution of karst landforms in north-west Yorkshire which has been covered by ice on at least three occasions during the Pleistocene and in the Mendip Hills which were probably unglaciated. In the Yorkshire study area relict phreatic passages were identified in all major systems at heights of 265–300 m above sea-level, and this was interpreted as indicating the height of the contemporary valley floor. The passages were drained by a major rejuvenation of up to 75 m, and this is ascribed to valley deepening by glacier erosion.

The oldest date from one of these passages is more than 400,000 years, which suggests that the main valley deepening was accomplished by 'an early glaciation before this date' (Atkinson *et al.*, 1978, p. 27). The present author would regard the cold phase at around 440,000 years (stage 12) as the most likely candidate, since evidence from marine cores suggests that more ice accumulated during this stage than in any of the succeeding glacial episodes (Shackleton, 1977). The glacial/interglacial chronologies for England are the subject of considerable debate and continuing investigation (Bowen, 1978; Neale and Flenley, 1981), but the most widely used framework is that of Mitchell *et al.* (1973), who recognized three major glacial episodes in which ice advances deposited identifiable tills, the oldest being the Anglian at around 230–280 thousand years. The Yorkshire speleothem age data have considerable implications for this scheme, and two possibilities present themselves:

1. The stage 12 cold phase represents a major 'pre-Anglian' glaciation. Catt (1981) has provided a summary of the rather limited lithostratigraphic

evidence for such an event and has suggested that the ice cover was more extensive than in any previous or subsequent Quaternary glacial episode.

2. The cold phase between 230–280 thousand years was not a full glacial; that is, there was no major ice advance although periglacial conditions considerably reduced speleothem deposition. This explanation would suggest a correlation of the deposits currently labelled as 'Anglian' with the stage 12 glaciation.

Resolution of this problem will probably have to await the development of techniques capable of dating deposits more than 350,000 years old, but in the present context it may be confidently stated that karstification of drainage in north-west Yorkshire must have commenced at least half a million years ago. Similarly, in the Mendip Hills speleothem dates from G.B. cave indicate that both exokarst and endokarst landforms were well developed prior to 350,000 B.P. (Atkinson *et al*., 1978).

The karst area of England for which the greatest detail on landform evolution is currently available is the Peak District of Derbyshire. There, Beck (1975, 1980) established a relative sequence of developmental stages on the basis of morphological studies of cave passages and their distribution, altitudes, clastic sedimentary fills, and speleothem deposits. Speleothem dating was then used to place the sequence in a chronological framework and to indicate a time span for the formation, filling, abandonment and degradation of some of the major cave levels (Ford *et al*., 1983).

The two most studied karst areas in North America are probably the Southern Canadian Rockies and the Central Kentucky karst. In the former area considerable attention has centred on Castleguard Cave in Banff National Park, both because of its length and particularly because it extends under the Columbia Icefield, the largest surviving glacier in the Rockies (Ford, 1971, 1975, 1980b). At the 8th International Speleological Congress a special symposium was devoted to Castleguard, and eleven papers were presented (Beck, 1981) including one on its antiquity (Gascoyne and Latham, 1981). On the basis of uranium series dates and of two magnetically reversed speleothems, they suggest that most of the presently accessible phreatic passages had been drained and entrenched by vadose streams over 350,000 years ago and that some were drained over 700,000 years ago. In part of the cave eroded speleothems suggest that there was at least one re-invasion and phreatic phase prior to 140,000 years. Since then the accessible cave has been essentially relict, although speleogenesis has continued in a lower drainage system which has not yet been entered despite considerable effort.

The Central Kentucky karst is dominated by the Mammoth–Flint Ridge Cave System, the longest in the world with an integrated network of more than 350 km of passages. The denudation chronology of the cave system and the genetic relationship between cave passages and exokarst landforms were examined by Miotke and Palmer (1972), who identified seven primary cave passage levels and correlated them with terraces in the Green River Valley and with an extensive erosion surface. Absolute ages for speleothems from the various levels were determined by Hess and Harmon (1981). As at Castleguard, speleogenesis probably began in the early Pleistocene as speleothems from the highest cave levels (183–213 m) have magnetically reversed polarity (>700,000 years). Intermediate levels (152–168 m) are more than 350,000 years and passages at about 143 m were probably formed during the Yarmouthian interglacial (180–220 thousand years). No dates are available for the lower levels of the cave, but it is suggested that they developed during the last two glacial episodes. As the marine isotope record shows at least 12 glacial/interglacial cycles during the Pleistocene whereas there are only 7 cave levels, Hess and Harmon suggest the following developmental sequence:

1. *Interglacial*—large phreatic passages form during later stages when the base level (Green River) has stabilized;
2. *Glacial*—passages infilled with sediment and concurrently dissolved upwards at or near base level;
3. *Interglacial*—rapid river entrenchment results in high hydraulic gradients; hence the major passages are re-excavated and narrow vadose canyons cut;
4. *Glacial*—alluviation and sedimentation terminates downcutting and partially infills the canyons;
5. *Interglacial*—during the early stages the sediment from (4) is removed by erosion, and during the later stages there is a long episode of base-level erosion and phreatic conduit development. Hence, at the end of this phase the endokarst landforms will be morphologically similar to those of phase 1.

Although developed for one specific cave system this evolutionary scheme may have wider applicability in karst areas which lay outside but close to the limits of ice advance.

The most detailed geochronometric study of karst landform evolution in the southern hemisphere is that of Goede and Harmon (1983), who used uranium-series dating to obtain estimates of the ages and rates of evolution of Tasmanian caves. In common with other areas of the world the largest cave systems were found to have a long history extending beyond 350,000 years, the limit of the dating method. As expected, the majority of the dates fall in warm periods (isotope stages 1, 3, 5, 7), but there is one anomaly in that 8 of the 43 age determinations are in the range 10–30 thousand years which other evidence suggests was a relatively dry period of maximum cold and glacier advance. This revives the question posed by Gascoyne (1977): 'Does the presence of stalagmites really indicate warm periods?' Goede and Harmon (1983) also estimated the maximum rate of downcutting of one of the cave streams as 106–212 mm per thousand years over a period of 190,000 years, although the maximum rate of valley floor lowering over the last 400,000 years was considered to be somewhat lower at 63 mm per thousand years. Finally, a combination of uranium-series dating and extrapolation of rates of weathering rind formation on dolerite pebbles was used to date an alluvium at 50–75 thousand years. As the morphology of pebbles suggests derivation by frost weathering and transport by short, high-energy streams under cold and wet climatic conditions, Goede and Harmon (1983) suggest that deposition most probably occurred during isotope stage 4 (61–73 thousand years). This event was particularly important in landform evolution as 'it brought about widespread diversion of underground drainage and, at least in some cases, a temporary return to surface drainage' (Goede and Harmon, 1983, p. 97).

In conclusion, it is apparent that cave sediment studies, and particularly absolute dating of speleothems, represent the major growth area of karst research over the last decade. The potential both for estimating the age of karst landforms and long-term rates of solutional erosion (§9.3.2) and also for placing sequences of landform evolution in a time context is considerable. Moreover, they have also been of major significance in Quaternary studies, particularly of palaeoclimates, and it may be argued that they represent the most important contribution of karst researchers to geomorphology as a whole.

9.6 DISCUSSION

Karst landform studies embrace a wide range of topics most of which are related to solute processes. Hence, the present review has perforce been selective both in restricting discussion to carbonate (limestone) karsts and in considering only four major areas of research: solute processes, solutional erosion rates, hydrological processes and landform evolution. In each area it is apparent that although there is a long history of both descriptive work and process-orientated research, there has been a paucity of detailed investigations involving the rigorous application of research methodologies and quantitative techniques. It is therefore appropriate to identify directions for future karst landform studies which have emerged from this review and to consider how greater use of models might provide a better framework for these studies.

9.6.1 Directions for future karst landform studies

Many possible directions exist for future karst studies. For example, Yevjevich (1981) has identified 15 headings and 68 specific topics in the area of physico-environmental research needs. In the present review several areas of potential interest have been considered and it is thought that six of them will provide particularly important contributions to the advancement of karst knowledge over the next decade:

1. Investigation of the role of ground-air carbon dioxide in the limestone solution process (§9.2.3). Although present in the Mendip Hills, England, it has not yet been identified in other karst areas and as a consequence the implications of its existence have not been fully examined. In particular, estimates or measurements of soil carbon dioxide concentrations would no longer be expected to provide an estimate of groundwater carbonate concentrations, and enhanced solution would be expected in the upper layers of bedrock, that is the subcutaneous zone.
2. Detailed hydrochemical studies of net solutional erosion rates and of temporal and spatial (both areally and vertically) variations in solutional erosion rates (§9.3). These are required for a range of climatic zones as a basis for quantitative examination of the influence of climate on erosion rates and landform variability.
3. Development of conservative fluorescent dyes to enhance the value of quantitative water-tracing techniques. Also greater use of pulse-wave experiments and repeated tracing experiments to investigate the form of the phreatic zone (§9.4.4).
4. Study of the subcutaneous (epikarstic) zone (§9.4.5). The existence of this zone in temperate closed depressions has been established (Friederich, 1981; Gunn, 1981a; Williams, 1978, 1983) and its importance as a concentrating mechanism for point recharge demonstrated (Gunn, 1983). In addition Williams (in press) has suggested that the development of solution dolines is a result of spatial variability in subcutaneous solution. However, there is still some doubt as to the overall applicability of these concepts as no evidence of subcutaneous flow has yet been found in tropical depressions, although Williams (1972) inferred its presence in New Guinea.
5. Use and testing of the model of limestone drainage presented in this review (§9.4.5; Figure 9.8). In particular quantitative data from contrasting areas are needed, (a) on the quantities of water stored in the unsaturated zone, in flooded conduits and in flooded fissures, and (b) on the contribution of diffuse and conduit flow systems to karst risings.
6. Establishment of sequences of karst landform evolution involving both endokarst and exokarst landforms and using the geochronological potential of cave sediments (§9.5).

9.6.2 Models in karst

Hydrologists have been the major users of models in karst, and theoretical (§9.4.2), mathematical statistical (§9.4.3) and conceptual (§9.4.5) models have all been employed. Formulation of more accurate predicitive models for outflows from karst systems is regarded as a priority area of research since karst aquifers are increasingly being developed as sources of groundwater (Yevjevich, 1981). Although black-box modelling has achieved some successes, the future models will almost certainly have to possess a greater degree of physical realism which can only be attained by further study of recharge mechanisms and the movement of water through karst rocks. Hence the need for applied predictive models should lead to a greater understanding of hydrological processes.

Karst geomorphologists have made somewhat less use of models, but important developments have taken place in two main directions.

Laboratory models

In natural field situations the solution of limestone and karst landform development proceed relatively slowly and the multitude of interacting variables means that processes are hard to define. Measurements may also be difficult to make and there are many potential hazards such as the wombats, other marsupials, wild horses and frost which have interfered with research on the Cooleman Plain, Australia (Jennings, 1978). The main alternative is to experiment under artificial laboratory conditions 'where at least the conditions can be measured even if they may not replicate natural conditions' (Trudgill, 1976, p. 69). The two main forms of laboratory experiment are scale models in which small blocks of carbonate material are subject to solution by a mineral acid (usually HCl) thereby speeding up the solution process, and analogue models in which plaster of paris (gypsum) and water simulate the karst erosion process. Scale models were used by Kaye (1957) in his study of the effect of solvent motion on limestone solution, and by Mowat (1962) who verified the hypothesis that convex rectangular corners of limestone plates are rounded by solution whereas concave corners retain their angularity. More recently Watts and Trudgill (1979) conducted a more detailed modelling experiment in which dilute hydrochloric acid was dribbled down a limestone block inclined at angles in the range 0–90° in 5° classes. Solvent motion was shown to be the limiting factor affecting limestone solution at all slope angles, and angles of 15° and 40° were identified as respectively marking the commencement of rill flow and its becoming most pronounced. The potential problem of lack of realism as a result of the substitution of hydrochloric for carbonic acid has been reviewed by Trudgill (1976) and Watts and Trudgill (1979), and it is clear that caution is required in interpreting the results of this form of modelling experiment. Greater realism could be obtained by using carbonic acid, perhaps in a slightly more concentrated form than in nature; but these experiments would require greater volumes of solute and would have to run for a much longer period.

Plaster of paris analogue models have been used to investigate the development of scallops (Goodchild and Ford, 1971), solutional openings along bedding planes (Ewers, 1978), rillenkarren (Glew and Ford, 1980) and phreatic roof pendants (Lauritzen, 1981). In all cases it has been claimed that realistic forms have been produced and useful conclusions drawn on the nature of the generating processes. However, the importance of field testing and the potential problems of laboratory models have been emphasized by Dunkerly (1983, p. 202) whose detailed study of nearly 1000 rillenkarren near Chillagoe in north-east Australia provided 'substantial reason for believing that the physical simulations of Glew and Ford may not adequately have modelled the natural process of flute development'.

Mathematical models of karst landform evolution

The earliest, and still the most detailed, mathematical model of exokarst landform evolution is that of Smith *et al*. (1972) who investigated the origins of cockpit (polygonal) karst in Jamaica. Their model takes as its origin a 1° slope, 100 m long, formed on a gently undulating uplifted sea-floor with many shallow internally draining areas. It is argued that if a soil develops it will be thickest at the base of the shallow depressions, tapering to zero thickness at the interfluves. On the basis of a general trend for soil CO_2 concentrations to increase with depth, it is argued that the rate of erosion on the slope will be proportional to soil depth and an equation is derived:

$$y = x(\tan \alpha_i + Dt) + c \tag{9.8}$$

Where y is the height above the base of the slope at x, x is the horizontal distance from the base of the slope, α_i is the initial slope, t is time, and c is a constant of integration; D is given by

$$D = \frac{k_0 - k_{x0}}{x_0}$$

where k_0 is the rate of erosion at the base of the slope, k_{x0} is thae rate of erosion of bare limestone, and x_0 is the horizontal length of the slope.

As $y/x = \tan \alpha$, where α is the slope angle, then Equation (9.8) may be rearranged to obtain the time taken to reach any particular slope angle:

$$t = \frac{(\tan \alpha - \tan \alpha_i)}{D} \tag{9.9}$$

Smith *et al*. (1972) further argue that as slope angle increases so the rate of mechanical erosion will increase and soil cover will gradually be lost. When all the soil is lost the slope will cease to become steeper as the erosion rate will be constant over the whole slope. It is assumed that the critical angle is 30°, and substitution in Equation (9.9) suggests that it will be achieved after about

600,000 years. After this time slope form is retained by constant rate lowering. As most of the Jamaican cockpits have a central shaft its formation was modelled following two assumptions: (a) that in soil-covered depressions water drains laterally to the base where it enters fissures, and (b) that the water is undersaturated when it reaches the slope base. Maximum values for the degree of saturation β may be obtained from:

$$\beta < \bar{k}A/(k_0 + kA) \tag{9.10}$$

where $\bar{k}$ is the mean rate of erosion of the slope, k_0 is the rate of erosion at the base of the slope, and:

$$A = r_1^2/r_2^2 \tag{9.11}$$

where r_1 is the radius of the depression and r_2 the radius of the shaft. Shaft depth d for any time t may then be obtained from:

$$d = [(1 - \beta)\bar{k}A - \beta k_0]t \tag{9.12}$$

(Equations (9.9)–(9.12) are not stated as such by Smith *et al*. (1972) but may be derived from their work.)

From Equations (9.10)–(9.12) it is apparent that shaft depth will tend to increase as the degree of undersaturation $(1- \beta)$ increases, and decrease as radius increases. However, the most important point is that only a small degree of undersaturation is necessary for shaft development. For example, a shaft 20 m in diameter and 38.5 m deep could form in a depression where the basal water was 97.5% saturated over a period of some 620,000 years, and a shaft 2 m in diameter and 57 m deep could form over a similar period even where the basal water is 99.97% saturated. Hence, Smith *et al*. (1972, p. 167) conclude that 'a shaft of some kind will form at the base of the slope as a natural consequence of the presence of the cockpit'. When soil is completely removed from the slope it is envisaged that water will pass into joints and fissures in the bare limestone rather than moving downslope, and that shaft and slope will therefore erode at roughly equal rates. Finally, Smith *et al*. (1972) extend their model to predict equilibrium landforms for degraded karst in which bauxite deposits are present and to predict the length of time required to reach equilibrium.

The models developed by Smith *et al*. are intrinsically useful, and it is both surprising and regrettable that no attempt has been made to test or refine them in other areas. One possible reason may be that the key parameters k_0, k_{x0}, k and β are difficult to quantify and it is not clear how several of the values quoted in the original paper were derived. For example, Smith *et al*. (1972, p. 164) state that 'water flowing on bare rock surfaces . . . will attain a hardness of about 40 mg/l^{-1}, but they use a value of 58 mg/l^{-1} in computing k_{x0}. No indication is given as to how the value of 282 mg/l^{-1} used in computing k_0 was derived, and the figure for mean slope erosion (70 mm per thousand years) is the overall basin erosion rate derived from the mean hardness of spring waters (170 mg/l^{-1}) which is likely to reflect erosion and deposition in the main mass of bedrock as well as slope erosion. More importantly, β appears to have been assigned an arbitrary value less than the maximum predicted by Equation (9.10). The need to quantify this factor can be seen from the fact that an increase of β from 0.975 (as used in Smith *et al.*, Table 2) to 0.98 will result in a decrease of 22 m in the predicted depth of a shaft 10 m in radius, while a decrease in β from 0.9997 to 0.999 will increase the predicted depth of a shaft 1 m in radius by almost 300 m! Future studies should therefore aim to quantify the four key parameters, to consider the reformulation of Equation (9.12) in the light of its sensitivity to very small variations in β, and particularly to consider what changes may be necessary to the model as a whole in order to take account of the work by Williams (in press) which suggests that doline development is a consequence of spatial variability in subcutaneous solution rather than a gradual increase in soil thickness.

An alternative approach to modelling the development of exokarst landforms has been proposed by Brook (1981). The starting point is a surface on a horizontally bedded homogenous limestone crossed by well-defined vertical fractures or narrow fracture zones with great vertical and horizontal persistence. The greater permeability of the fracture zones results in lateral migration of water at the surface and at shallow depth. This in turn produces a focusing of solution erosion, so that ultimately fractures become depressions and the resistant blocks between them hills. Brook (1981) argues that the principal control on landform development is the ratio of vertical solution δ_v to horizontal solution δ_h. Two components are identified: $(\delta_v/\delta_h)_B$ which is a function of bedrock characteristics, particularly mechanical strength; and $(\delta_v/\delta_h)_C$ which is a function of climatic characteristics, of which the most important is probably the frequency of intense rainfall.

Summation of the two components provides an overall ratio for the karst environment $(\delta_v/\delta_h)_{KE}$ which may be approximated as twice the mean depression depth/diameter ratio:

$$(\delta_v/\delta_h)_B + (\delta_v/\delta_h)_C = (\delta_v/\delta_h)_{KE} \sim 2 \times \frac{\text{depth}}{\text{diameter}} \qquad (9.13)$$

If the location of major fracture zones is also known, then the landforms likely to develop can be modelled by summing one sine-wave surface for each fracture set (Figure 9.10(b)), the wavelength being determined by fracture spacing, and amplitude by the $(\delta_v/\delta_h)_{KE}$ ratio (Figure 9.10(a)). As an example, Brook (1981) has generated models for a horizontally bedded limestone crossed by three sets of fractures, two at right-angles, the third intersecting the others at 45°, with different $(\delta_v/\delta_h)_{KE}$ values (Figure 9.10(c)). In this example it is assumed that climate and bedrock are equally important and that the depressions have coalesced. The modelling technique is also used to predict the style of landform evolving under differing $(\delta_v/\delta_h)_B$ and $(\delta_v/\delta_h)_C$ ratios, and it is suggested that shallow doline karst is most likely to develop in weak rocks in areas of low rainfall intensity, and deep cockpit and tower karsts in massive strong limestones situated within areas of frequent high-intensity rainfall.

The model presented by Brook (1981) is much simpler in its formulation than that of Smith *et al*. (1972), and its assumption that water moves laterally towards depression outlets 'at the surface and at shallow depth' (p. 61) is in substantial accord with recent developments in subcutaneous hydrology (Gunn, 1981a; Williams, 1983 and in press). However, the modelling procedure also 'assumes that $(\delta_v/\delta_h)_C$ and $(\delta_v/\delta_h)_B$ have the same range of values, that doline karst develops when $(\delta_v/\delta_h)_{KE}$ is less than 0.002 to 0.039, and cockpit karst when $(\delta_v/\delta_h)_{KE}$ is 0.40 to more than 0.70, and that the only major morphologic difference between doline and cockpit karst is in relative relief' (Brook, 1981, p. 75). These assumptions cannot be sustained in the Waitomo District, New Zealand, where a survey of 80 closed depressions by the writer (Gunn, 1978) yielded a mean depth/diameter ratio of 0.40 and hence, by Equation (9.13), a $(\delta_v/\delta_h)_{KE}$ ratio of 0.80. The Waitomo depressions have a similar plan geometry to tropical cockpits, but their mesh size is much smaller and the hemispherical hills which usually occur between cockpits are absent as the Waitomo depression slopes tend to be concave. Hence, the Waitomo area is quite clearly a well-developed temperate polygonal karst and not a cockpit karst as would be predicted on the basis of Brook's assumptions. It is therefore clear that further developments of the model will require a more substantial database from a wider variety of karsts. In addition:

> 'It will be necessary to determine exact relationships between climate, bedrock, and the $(\delta_v/\delta_h)_{KE}$ ratio, to model stages in karst development, and to model landforms in carbonates where mechanical strength varies, or where fractures are unevenly spaced or differ in extent or in susceptibility to solution.' (Brook, 1981, p. 75)

In marked contrast to the models of exokarst landforms, Dreybrodt (1981c) has developed a mathematical model of cave development which explains karstification from the solution kinetics of mixing corrosion, the saturation length, and the hydrodynamics of water in conduits. The derivation of the model is mathematically complex as Dreybrodt argues from first principles, but three particularly important points emerge from his analysis:

1. Turbulent flow begins at tube diameters of around 1 cm.
2. Once turbulent flow has been established the saturation length increases

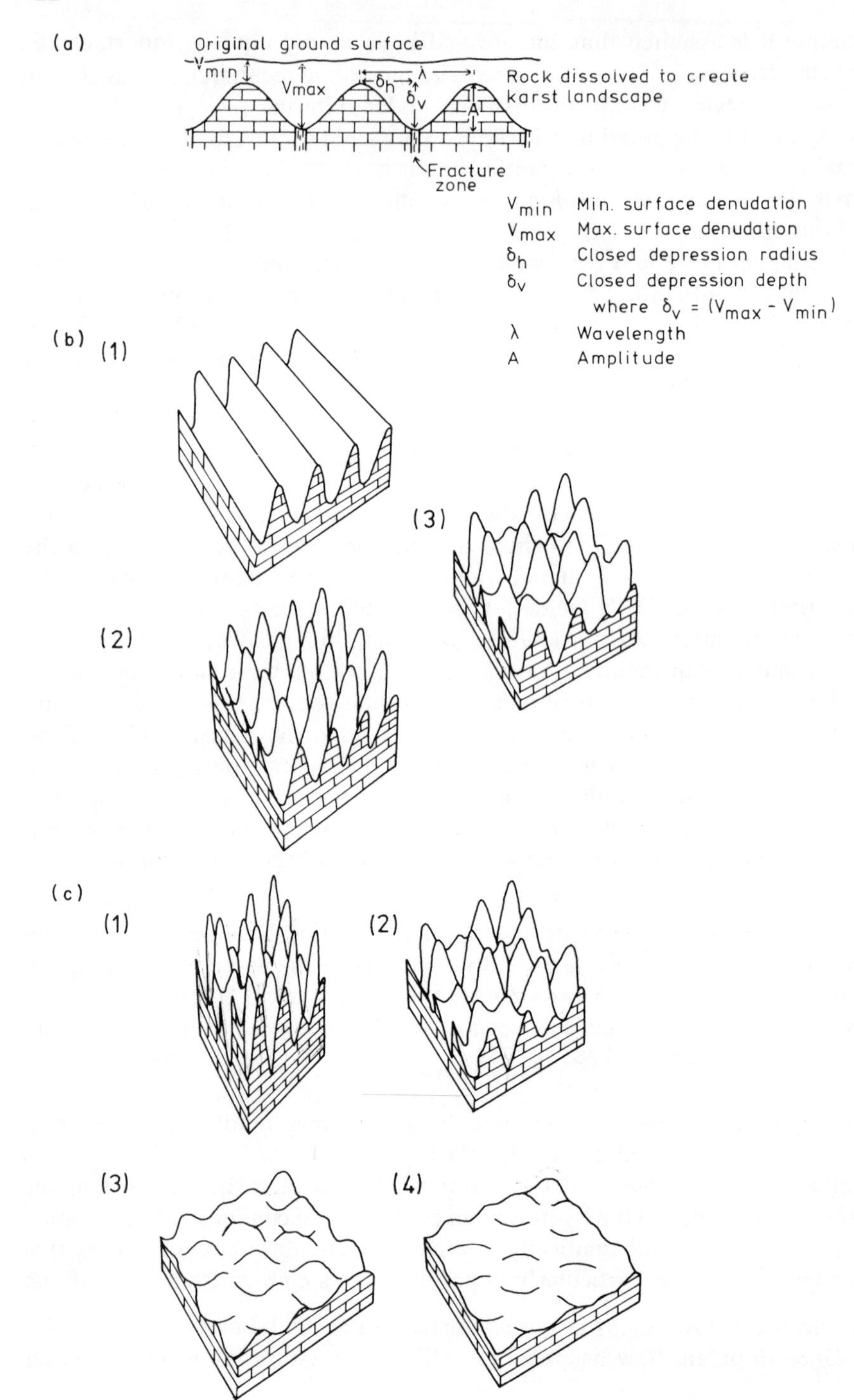
(a)
Original ground surface
V_{min}
V_{max}
δ_h
λ
δ_v
A
Rock dissolved to create karst landscape
Fracture zone
V_{min} Min. surface denudation
V_{max} Max. surface denudation
δ_h Closed depression radius
δ_v Closed depression depth where $\delta_v = (V_{max} - V_{min})$
λ Wavelength
A Amplitude
(b)
(1)
(2)
(3)
(c)
(1)
(2)
(3)
(4)

dramaticaly and dissolution is effective for considerable distances into the limestone.

3. The diameter of phreatic tubes increases linearly in time and may be expressed (in cm per year) by:

$$dR/dt = 5 \times 10^{-3} \exp(-x/x_s) \tag{9.14}$$

where R is the tube radius (cm), t is time (years), x is the distance along the tube (km), and x_s is the saturation length (km).

At hydraulic gradients of 0.01 and tube radii of about 10 cm, x_s is about 35 km. Substitution of this into Equation (9.14) indicates that in the region $x=0$ to $x=2$ an increase of tube radius from 1 cm at the onset of turbulent flow to 50 cm will take approximately 10,000 years. This result is extremely significant as it indicates that there has been sufficient time since the end of the Devensian glaciation for the formation of accessible conduits in the phreatic zone.

On the basis of the mathematical model Dreybrodt suggests that cave development proceeds in three main stages. In stage 1 (Figure 9.11(a)) the limestone is largely impermeable, with surface drainage and minimal percolation of water through tight joints and partings which have widths of the order of 10 μm. The piezometric surface closely follows the ground level, and as saturation lengths are in the order of a few centimetres the water is saturated with respect of $CaCO_3$. Flow is estimated to be of the order of 1 μl/s, but this is sufficient to produce solution channels along joint crossings wherever waters of different chemical compositions mix. Over time (several tens of thousand years) an integrated net of tubes with diameters of ~1 cm develops, and eventually one or more of these tubes reaches the surface. This marks the beginning of stage 2 (Figure 9.11(b)) in which drainage is effected through the integrated tube network. Resistance to flow is much less than in the primary network, so that the piezometric surface falls to a lower level. As the main conduits are still entirely phreatic, passage enlargement and extension of the tube network is envisaged as taking place primarily by mixing corrosion in accordance with Equation (9.14). Passage enlargement increases the storage volume in the phreatic zone so that the piezometric surface drops further, leaving parts of the cave system in the vadose zone (Figure 9.11(c)). The lowering of the piezometric level also increases the hydraulic gradient in

Figure 9.10. Sine-wave models of solution patterns in limestones crossed by equally spaced sets of vertical fractures (Brook, 1981). (a) Two-dimensional model; one fracture set. (b) Three-dimensional models: (1) one fracture set; (2) two fracture sets intersecting at 90°; (3) three fracture sets, two at right-angles, the third intrsetion the others at 45°. (c) Model (b.3) with different $(\delta_v/\delta_h)_{KE}$ ratios: (1) 0.70; (2) 0.57; (3) 0.40; (4) 0.02

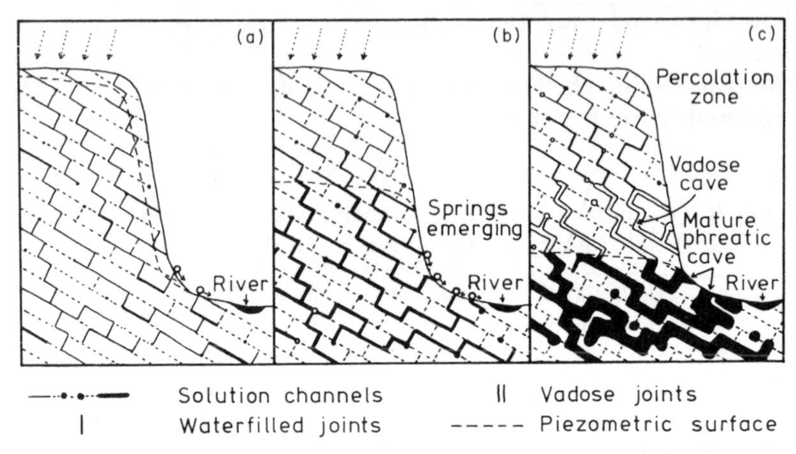

Figure 9.11. Schematic three-stage model for cave development (Dreybrodt, 1981c). See text for details

the vadose zone, and as a result drainage of surface water via the limestone becomes more effective. Although it is not part of Dreybrodt's model, capture of surface streams may be envisaged at this stage, and this would supply increased volumes of aggressive water to the vadose conduits. In the original model two-stage drainage is suggested, with water flowing rapidly through the vadose zone to a phreatic storage area from whence it travels via a phreatic network to a rising or risings. The increased hydraulic gradient during late stage 2 and stage 3 also increase saturation lengths in the vadose zone, and Dreybrodt (1981, p. 235) suggests that as a result 'percolation solution becomes effective in the upper parts of the vadose zone, increasing the permeability of the rock by widening the fissures. In this stage grikes will occur as a feature of mature karst'. Finally, it is notable that both of the main phenomenological theories for cave development—deep phreatic and water table—are possible within the model.

Dreybrodt's model represents one of the most significant advances in karst landform studies for several years as it provides a solid link between solute processes and the development of endokarst landforms. Perhaps the most important omissions are considerations of the effects of (a) allogenic stream inputs which are important in many karsts, and (b) base-level lowering and resultant 'rejuvenation'. Although difficult to incorporate in a geochemical discussion, inclusion of these factors would add greater realism. In addition, the claim that the karstification process is comprehensively modelled cannot be substantiated as the only reference to contemporaneous exokarst landform evolution is the suggestion that grike development occurs towards the end of karstification. Hence there is a need for a geochemist to undertake an equally rigorous 'first principles' investigation of the development of surface landforms, particularly closed depressions. Research on solutional erosion rates (§9.3) and subcutaneous hydrology (§9.4) should provide useful underpinning for this type of study, and geochronological studies 1(§9.5) will aid in testing predicted evolutionary timescales. The ultimate aim should be a linking of exokarst and endokarst models in order to provide a theoretical basis for sequences of landform evolution such as that outlined in §9.5.1.

ACKNOWLEDGEMENTS

Thanks are due to Dr. N. J. Christopher for useful discussions and for comments on earlier drafts; to Catherine Gunn for drawing the diagrams and to Maria Cunningham for typing the text.

REFERENCES

Abrahams, A. D. (1972). Environmental constraints on the substitution of space for time in the study of natural channel networks. *Bull. Geol. Soc. Am.*, **83**, 1523–1530.

Adams, C. S., and Swinnerton, A. C. (1937). Solubility of limestone. *Trans. Am. Geophys. Union*, **18**, 504–508.

Ager, D. V. (1973). *The Nature of the Stratigraphic Record*, Macmillan, London.

Aley, T. J. (1972). Groundwater contamination from sinkhole dumps. *Caves & Karst*, **14**, 17–23.

Ashton, K. (1966). The analysis of flow data from karst drainage systems. *Trans. Cave Res. Grp. G.B.*, **7**(2), 161–204.

Atkinson, T. C. (1968). The earliest stages of underground drainage in limestone—a speculative discussion. *Proc. Brit. Speleo. Ass.*, **6**, 53–70.

Atkinson, T. C. (1977a). Carbon dioxide in the atmosphere of the unsaturated zone: an important control of groundwater hardness in limestones. *Journal of Hydrology*, **35**, 111–123.

Atkinson, T. C. (1977b). Diffuse flow and conduit flow in limestone terrain in the Mendip Hills, Somerset (Great Britain). *Journal of Hydrology*, **35**, 93–110.

Atkinson, T. C., Harmon, R. S., Smart, P. L., and Waltham, A. C. (1978). Palaeoclimatic and geomorphic implications of $^{230}Th/^{234}U$ dates on speleothems from Britain. *Nature*, **272**, (5468), 24–28.

Atkinson, T. C., and Smart, P. L. (1981). Artificial tracer's in hydrogeology. In: *A Survey of British Hydrogeology 1980*, Royal Society, London, pp. 173–190.

Atkinson, T. C., and Smith, D. I. (1974). Rapid groundwater flow in fissures in the chalk: an example from South Hampshire. *Q.J. Eng. Geol.*, **7**(2), 197–205.

Atkinson, T. C., and Smith, D. I. (1976). The erosion of limestones. In: *The Science of Speleology* (T. D. Ford and C. H. D. Cullingford, Eds.), Academic Press, London, pp. 151–177.

Atkinson, T. C., Smith, D. I., Lavis, J. J., and Whitaker, R. J. (1973). Experiments in tracing underground waters in limestones. *Journal of Hydrology*, **19**, 323–349.

Back, W., and Zotl, J. C. (1975). Application of geochemical principles, isotopic methodology, and artificial tracers to karst hydrology. In: *Hydrogeology of Karstic Terrains* (A. Burger and L. Dubertret, Eds.), I.U.G.S. Ser. B, No. 3, I.A.H., pp. 105–121.

Baker, R. C. (1977). Hydrology of karst features in evaporite deposits of the Upper

Permian in Texas. In: *Hydrologic Problems in Karst Regions* (R. R. Dilamarter and S. C. Csallany, Eds.), Western Kentucky Univ. Press, Bowling Green, Kentucky, pp. 333–339.

Basmaci, Y., and Sendlein, L. V. A. (1977). Model analysis of closed systems in karstic aquifers. In: *Hydrologic Problems in Karst Terrains* (R. R. Dilamarter and C. S. Csallany, Eds.), W. Kentucky Univ. Press, Bowling Green, pp. 202–213.

Bathurst, R. G. C. (1972). *Diagenesis of Limestone Sediments*, Elsevier, Amsterdam.

Beck, B. F. (Ed.) (1981). *Proceedings of the 8th International Congress of Speleology*, National Speleol. Soc., Huntsville, Alabama.

Beck, J. S. (1975). The caves of the Foolow-Eyam–Stoney Middleton area, Derbyshire, and their genesis. *Trans. Brit. Cave Res. Assn.*, **2**(1), 1–11.

Beck, J. S. (1980). *Aspects of Speleogenesis in the Carboniferous Limestone of North Derbyshire*, PhD thesis, U. of Leicester, unpublished.

Beede, J. W. (1911). The cycle of subterranean drainage as indicated in the Bloomington, Indiana, Quadrangle. *Indiana Acad. Sci.*, **26**, 81–111.

Birot, P. (1954). Problèms de morphologie karstique. *Ann. Géogr.*, **63**, 160–192.

Birot, P. (1966). Étude d'usure d'un versant calcaire sous un climat tropicale humide. In: L'Evolution des Versants (P. Macar, Ed.), U. of Liège, pp. 69–74.

Bloom, A. L. (1978). *Geomorphology*, Prentice-Hall, New Jersey.

Bögli, A. (1961). Karrentische, ein Beiträg zur Karstmorphologie. *Z. Geomorph.*, **5**(3), 185–193.

Bögli, A. (1964). Mischungskorrosion—ein Beitrag zum Verkarstungsproblem. *Erdkunde*, **18**, 83–92.

Bögli, A. (1980). *Karst Hydrology and Physical Speleology*, Springer-Verlag, Berlin.

Boulton, N. S., and Streltsova, T. D. (1977). Flow to a well in an unconfined fractured aquifer. In: *Hydrologic Problems in Karst Terrains* (R. R. Dilamarter and C. S. Csallany, Eds.), W. Kentucky Univ. Press, Bowling Green, pp. 214–227.

Bowen, D. Q. (1978). *Quarternary Geology*, Pergamon, Oxford.

Bristol Waterworks Company (1972). Pollution of a spring source by silage liquor. *Proc. Water Res. Assoc. Conf. Groundwater Polln.*, **2**, pp. 331–333.

Brodrick, H. (1908). The Marble Arch caves, Co. Fermanagh: main stream series. *Proc. Royal Ir. Acad.*, **27B**, 183–192.

Broecker, W. S., Olsen, E. A., and Orr, R. C. (1960). Radiocarbon measurements and annual rings in cave formations. *Nature*, **185**, 93–94.

Brook, G. A. (1981). An approach to modelling karst landscapes. *S. African Geog. J.*, **63**(1), 60–76.

Brook, G. A., Folkoff, M. E., and Box, E. O. (1983). A world model of soil carbon dioxide. *Earth Surface Processes and Landforms*, **8**, 79–88.

Brown, M. C. (1973). Mass balance and spectral analysis applied to karst hydrologic networks. *Water Resources Research*, **9**(3), 749–752.

Brown, M. C., and Ford, D. C. (1971). Quantitative tracer methods for investigation of karst hydrologic systems. *Trans. Cave Res. Grp.*, **13**(1), 37–51.

Brown, M. C., and Wigley, T. M. L. (1969). Simultaneous tracing and gauging to determine water budgets in inaccessible karst aquifers. *Proc. 5th Int. Speleol. Congr.*, Stuttgart, Hy 3/1–5.

Brown, M. C., Wigley, T. M. L., and Ford, D. C. (1969). Water budget studies in karst aquifers. *Journal of Hydrology*, **9**(1), 113–116.

Bull, P. A. (1978). A study of stream gravels from a cave: Agen Allwedd, South Wales. *Z. Geomorph.*, **22**(3), 275–296.

Bull, P. A. (1980a). The antiquity of caves and dolines in the British Isles. *Z. Geomorph.*, Suppl. Band **36**, 217–232.

Bull, P. A. (1980b). Towards a reconstruction of time-scales and palaeoenvironments from cave sediment studies. In: *Timescales in Geomorphology* (R. A. Cullingford, D. A. Davidson and J. Lewin, Eds.), Wiley, London, pp. 177–187.

Bull, P. A. (1981). Some fine grained sedimentation phenomena in caves. *Earth Surface Processes and Landforms,* **6**,(1), 11–22.

Bull, P. A. (1983). Chemical sedimentation in caves. In: *Chemical Sediments and Geomorphology* (A. S. Goudie and K. Pye, Eds.), Academic Press, London, pp. 301–319.

Burdon, D. J., Eriksson, E., Payne, B. R., Papadimitropoulos, T., and Papakis, N. (1963). The use of tritium in tracing karst groundwater in Greece. In: *Radioisotopes in Hydrology*, Int. Atom Energy Agency Symposium, Tokyo, March 1963, pp. 309–320.

Cann, J. H. (1974). A field investigation into rock weathering and soil forming processes. *J. of Geol. Ed.*, **22**, 226–230.

Catt, J. A. (1981). British pre-Devensian glaciations. In: *The Quaternary in Britain* (J. Neale and J. Flenley, Eds.), Pergamon Press, Oxford, pp. 9–19.

Chilingar, G. V., Bissell, H. J., and Fairbridge, R. W. (1967). *Carbonate Rocks*, Elsevier, Amsterdam.

Christopher, N. S. J. (1980). A preliminary flood pulse study of Russett Well, Derbyshire. *Trans. Brit. Cave Res. Assn.*, **7**(1), 1–12.

Christopher, N. S. J., Trudgill, S. T., Crabtree, R. W., Pickles, A. M., and Culshaw, S. M. (1981). A hydrological study of the Castleton area, Derbyshire. *Trans. Brit. Cave Res. Assn.*, **8**(4), 189–206.

Christopher, N. S. J., and Wilcock, J. (1981a). The classification of karst waters by chemical analysis. *Proc. 8th Int. Speleol. Congr.*, Bowling Green, pp. 526–528.

Christopher, N. S. J., and Wilock, J. (1981b). Geochemical controls on the composition of limestone groundwaters with special reference to Derbyshire. *Trans. Brit. Cave Res. Assn.*, **8**(3), 135–158.

Clayton, K. M. (1966). The origin of landforms on the Malham area. *Field Studies*, **2**, 359–384.

Clayton, K. M. (1980). Geomorphology. In: *Geography Yesterday and Tomorrow* (E. H. Brown, Ed.), R.G.S., London, pp. 168–180.

Colman, S. M. (1981). Rock-weathering rates as functions of time. *Quat. Res.*, **15**, 250–264.

Corbel, J. (1957). *Les karsts du Nord-Ouest de L'Europe*, Inst. Et. Rhod. Mem et Doc., 12.

Corbel, J. (1959). Érosion en terrain calcaire. *Annals de Géorgr.*, **68**, 97–120.

Corbel, J. (1968). Tourbières et morphologie karstique dans la région de Sligo (Irlande). *Rev. de Geog. Alpine (Grenoble)*, **56**, 517–531.

Cowell, D. W., and Ford, D. C. (1980). Hydrochemistry of a dolomite karst: the Bruce Peninsula of Ontario. *Can. J. Earth Sci.*, **17**(4), 520–526.

Crabtree, R. W., and Burt, T. P. (1983). Spatial variation in solutional denudation and soil moisture over a hillslope hollow. *Earth Surface Processes and Landforms*, **8**, 151–160.

Crowther, J. (1983a). Carbon dioxide concentrations in some tropical karst soils, West Malaysia. *Catena*, **10**(1/2), 27–39.

Crowther, J. (1983b). A comparison of the rock tablet and water hardness methods for determining chemical erosion rates on karst surfaces. *Z. Geomorph.*, **27**(1), 55–64.

Curtis, L. F., Courtney, F. M., and Trudgill, S. T. (1976). *Soils in the British Isles*, Longman, London.

Cvijić, J. (1918). L'hydrographie souterraine et l'evolution morphologique du karst. *Revue Georgr. Alpine*, **6**, 375–426.
Davis, W. M. (1930). Origin of limestone caverns. *Bull. Geol. Soc. Am.*, **41**, 475–628.
Day, M. J. (1979). The hydrology of polygonal karst depressions in Northern Jamaica. *Z. Geomorph.*, Suppl. **32**, 25–34.
Derbyshire, E., Gregory, K. J., and Hails, J. R. (1979). *Geomorphological Processes*, Butterworths, London.
Dilamarter, R. R., and Csallany, S. C. (1977). *Hydrologic Problems in Karst Terrains*, Western Kentucky Univ. Press, Bowling Green.
Douglas, I. (1968). Some hydrologic factors in the denudation of limestone terrains. *Z. Geomorph.*, **12**(3), 241–255.
Douglas, I. (1972). The geographical interpretation of river water quality data. *Progress in Geography*, **4**, 1–81.
Douglas, I. (1978). Denudation of silicate rocks in the humid tropics. In: *Landform Evolution in Australasia* (J. C. Davies and M. A. J. Williams, Eds.), A.N.U. Press, Canberra, pp. 216–237.
Drake, J. J. (1980). The effect of soil activity on the chemistry of carbonate groundwaters. *Water Resources Research*, **16**,(2), 381–386.
Drake, J. J. (1983). The effects of geomorphology and seasonality on the chemistry of carbonate groundwater. *Journal of Hydrology*, **61**(1/3), 223–236.
Drake, J. J., and Ford, D. C. (1976). Solutional erosion in the Southern Canadian Rockies. *Canadian Geographer*, **20**(2), 158–170.
Drake, J. J., and Ford, D. C. (1981). Karst solution: a global model for groundwater solute concentrations. *Proc. Jpn. Geomorphol. Union,* **2**(2), 223–230.
Drake, J. J., and Wigley, T. M. L. (1975). The effect of climate on the chemistry of carbonate groundwater. *Water Resources Research*, **11**(6), 958–962.
Dreiss, S. J. (1982). Linear kernel for karst aquifers. *Water Resources Research*, **18**(4), 865–876.
Dreiss, S. J. (1983). Linear unit-response functions as indicators of recharge areas for large karst springs. *Journal of Hydrology*, **61**(1/3), 31–44.
Drew, D. P. (1974). Quantity and rate of limestone solution on the Eastern Mendip Hills, Somerset. *Trans. Brit. Cave Res. Assn.*, **1**(2), 93–100.
Drew, D. P., and Smith, D. I. (1969). Techniques for the tracing of subterranean drainage. *Brit. Geomorph. Res. Grp., Tech. Bull.*, **2**, 1–36.
Dreybrodt, W. (1981a). Kinetics of dissolution of calcite and its application to karstification. *Chem. Geol.*, **31**, 245–269.
Dreybrodt, W. (1981b). The kinetics of calcite precipitation from thin films of calcareous solutions and the growth of speleothems: revisited. *Chem. Geol.*, **32**, 237–245.
Dreybrodt, W. (1981c). Mixing corrosion in $CaCO_3$–CO_2–H_2O systems and its role in the karstification of limestones. *Chem. Geol.*, **32**, 221–236.
Drogue, C., and Guilbot, A. (1977). Representativité d'un bassin témoin en hydrogéologie karstique: application à la modelisation des ecoulements souterrains d'un aquifère de grande extension. *Journal of Hydrology*, **32**, 57–70.
Droppa, A. (1966). The correlation of some horizontal caves with river terraces. *Studies in Speleol.*, **1**(4), 186–192.
Dunkerly, D. L. (1983). Lithology and micro-topography in the Chillagoe karst, Queensland, Australia. *Z. Geomorph.*, **27**(2), 191–204.
Dunne, T., and Leopold, L. B. (1978). *Water in Environmental Planning*, Freeman, San Francisco.
Ek, C. (1961). Conduits souterrains en relation avec les terrasses fluviales. *Annls. Soc. géol. Belg.*, **84**, 313–340.

Engh, L. (1980). Can we determine solutional erosion by a simple formula. *Trans. Brit. Cave Res. Assn.*, **7**(1), 30–32.

Ewers, R. O. (1978). A model for the development of broad scale networks of groundwater flow in steeply dipping carbonate aquifers. *Trans. Brit. Cave Res. Assn.*, **5**(2), 121–125.

Farr, M. (1980). *The Darkness Beckons*, Diadem Books, London.

Feeney, C. M. (1977). *The karstification of drainage: a morphometric approach*. MA thesis, U. of Auckland, unpublished.

Fisk, D. W., and Exley, I. S. (1977). Exploration and environmental investigation of the Peacock Springs Cave System. In: *Hydrologic Problems in Karst Terrains* (R. R. Dilamarter and C. S. Csallany, Eds.), W. Kentucky Univ., Press, Bowling Green, pp. 297–302.

Folk, R. L. (1959). Practical petrographic classification of limestones. *Bull. Amer. Assoc. Petrol. Geol.*, **43**, 1–38.

Ford, D. C. (1971). Alpine karst in the Mt. Castleguard–Columbia Icefield area, Canadian Rocky Mountains. *Arctic and Alpine Research*, **3**(3), 239–252.

Ford, D. C. (1975). Castleguard Cave. *Studies in Speleology*, **2**(7/8), 299–310.

Ford, D. C. (1980a). Threshold and limit effects in karst geomorphology. In: *Thresholds in Geomorphology* (D. R. Coates and J. D. Vitek, Eds.), Allen and Unwin, London, pp. 345–362.

Ford, D. C. (1980b). New discoveries in our greatest cave. *Canadian Geographic*, **100**(4), 12–23.

Ford, D. C., and Drake, J. J. (1982). Spatial and temporal variations in karst solution rates: the structure of variability. In: *Space and Time in Geomorphology* (C. E. Thorn, Ed.), George Allen and Unwin, London, pp. 147–170.

Ford, D. C., and Ewers, R. O. (1978). The development of limestone cave systems in the dimensions of length and depth. *Can. J. Earth Sci.*, **15**, 1783–1798.

Ford, D. C., and Schwarcz, H. P. (1981). Uranium series disequilibrium dating methods. In: *Geomorphological Techniques* (A. Goudie, Ed.), George Allen & Unwin, London, pp. 284–287.

Ford, D. C., and Schwarcz, H. P., Drake, J. J., Gascoyne, M., Harmon, R. S., and Latham, A. F. (1981). Estimates of the age of the existing relief within the Southern Rocky Mountains of Canada. *Arctic and Alpine Research*, **13**(1), 1–10.

Ford, D. C., and Stanton, W. I. (1968). The geomorphology of the south-central Mendip Hills. *Proc. Geol. Ass.*, **79**(4), 401–427.

Ford, T. D. (1976). The geology of caves. In: *The Science of Speleology* (T. D. Ford and C. H. D. Cullingford, Eds.), Academic Press, London, pp. 11–60.

Ford, T. D. (Ed.) (1977). *Limestones and Caves of the Peak District*, Geo. Books, Norwich.

Ford, T. D., Gascoyne, M., and Beck, J. S. (1983). Speleothem dates and Pleistocene chronology in the Peak District, Derbyshire. *Trans. Brit. Cave Res. Assn.*, **10**(2), 103–115.

Foster, I. D. L. (1978). A multivariate model of storm period solute behaviour. *J. of Hydrology*, **39**, 339–353.

Foster, I. D. L. (1979). Intra-catchment variability in solute response: an east Devon example. *Earth Surf. Processes*, **4**, 381–394.

Foster, I. D. L. (1980). Chemical yields in runoff, and denudation in a small arable catchments, East Devon, England. *J. of Hydrology*, **47**, 349–368.

Freeze, R. A., and Cherry, J. A. (1979). *Groundwater*, Prentice-Hall, New York.

Friederich, H. (1981). *The hydrochemistry of recharge in the unsaturated zone, with special reference to the Carboniferous Limestone aquifer of the Mendip Hills*, PhD thesis, U. of Bristol, unpublished.

Froidevaux, R., and Krummenacher, R. (1976). Analyse spectrale des prećipitations et debits mensuels dans un bassin karstique du Jura Français. *Journal of Hydrology*, **29**, 293–313.
Gale, S. J. (1981a). The geomorphology of the Morecambe Bay karst and its implications for landscape chronology. *Z. Geomorph.*, **25**(4), 457–469.
Gale, S. J. (1981b). The palaeohydraulics of karst drainage systems: Fluvial cave-sediment studies. *Proc. 8th Int. Speleol. Congr.*, Bowling Green, pp. 213–216.
Gams, I. (1976). Variations of total hardness of karst waters in relation to discharge. In: *Karst Processes and Relevant Landforms* (I. Gams, Ed.), U. of Ljubljana, pp. 41–59.
Garrels, R. M., and Christ, C. L. (1965). *Solutions, Minerals and Equilibria*, Harper and Row, New York.
Gascoyne, M. (1977). Does the presence of stalagmites really indicate warm periods? New evidence from Yorkshire and Canadian Caves. *Proc. 7th Int. Speleol. Congr.*, Sheffield, pp. 208–210.
Gascoyne, M. (1981). Rates of cave passage entrenchment and valley lowering determined from speleothem age measurements. *Proc. 8th Int. Speleol. Congr.*, Bowling Green, pp. 99–100.
Gascoyne, M., and Latham, A. G. (1981). The antiquity of Castleguard Cave as established by uranium-series dating of speleothems. *Proc. 8th Int. Speleol. Congr.*, Bowling Green, pp. 101–103.
Gascoyne, M., Schwarcz, H. P., and Ford, D. C. (1980). A palaeotemperature record for the mid-Wisconsin in Vancouver Island. *Nature*, **285**, 474–476.
Gerstenhauer, A. (1972). Der einfluss des CO_2–gehaltes der bodenluft auf die kalklösung. *Erdkunde*, **26**, 116–120.
Glazek, J., and Harmon, R. S. (1981). Radiometric dating of Polish cave speleothems: current results. *Proc. 8th Int. Speleol. Congr.*, Bowling Green, pp. 424–427.
Glew, J., and Ford, D. C. (1980). A simulation study of the development of rillenkarren. *Earth Surface Processes*, **5**, 25–36.
Goede, A., and Harmon, R. S. (1983). Radiometric dating of Tasmanian speleothems—evidence of cave evolution and climatic change. *J. Geol. Soc. Australia*, **30**, 89–100.
Goodchild, J. G. (1890). Notes on some observed rates of weathering of limestones. *Geol. Mag.*, **27**, 463–466.
Goodchild, M. F., and Ford, D. C. (1971). Analysis of scallop patterns by simulation under controlled conditions. *J. Geol.*, **79**, 52–62.
Gospodarič, R. and Habič, P. (Eds.) (1976). *Underground Water Tracing*, Inst. for Karst Res., Postojna, Yugoslavia.
Goudie, A. (Ed.) (1981). *Geomorphological Techniques*, Allen and Unwin, London.
Graupe, D., Isailović, D., and Yevjevich, V. (1976). Prediction model for runoff from karstified catchments. In: *Karst Hydrology and Water Resources* (V. Yevjevich, Ed.), Water Resource Publs., Fort Collins, pp. 277–300.
Grund, A. (1914). Der geographische Zyklus in Karst. *Z. Ges. Erdk. Berl.*, **52**, 621–640. (Partly translated in Sweeting (1981), pp. 54–59.)
Gunn, J. (1978). *Karst Hydrology and Solution in the Waitomo District, New Zealand*, PhD thesis, U. of Auckland, unpublished.
Gunn, J. (1981a). Hydrological processes in karst depressions. *Z. Geomorph.*, **25**(3), 313–331.
Gunn, J. (1981b). Limestone solution rates and processes in the Waitomo district, New Zealand. *Earth Surf. Proc. and Landforms*, **6**(5), 427–445.
Gunn, J. (1981c). Prediction of limestone solution rates from rainfall and runoff data: some comments. *Earth Surf. Proc. and Landforms*, **6**(6), 595–597.

Gunn, J. (1982a). Magnitude and frequency properties of dissolved solids transport. *Z. Geomorph.*, **26**(4), 505–511.

Gunn, J. (1982b). Water tracing in Ireland: a review with special reference to the Cuilcagh karst. *Irish Geog.*, **15**, 94–106.

Gunn, J. (1983). Point-recharge of limestone aquifers—a model from New Zealand karst. *J. Hydrol.*, **61**(1/3), 19–29.

Gunn, J. (1984a). A world model of soil carbon dioxide: a discussion. *Earth Surf. Proc. and Landforms*, **9**(1), 83–84.

Gunn, J. (1984b). Applied karst hydrogeomorphology: three case studies from Ireland. *Z. Geomorph.*, Suppl. 51, 1–16.

Gunn, J., and Trudgill, S. T. (1982). Carbon dioxide production and concentrations in the soil atmosphere: a case study from New Zealand volcanic ash soils. *Catena*, **9**(1/2), 81–94.

Gunn, J., and Turnpenny, B. (in press). Stormflow characteristics of three small limestone drainage basins in North Island, New Zealand. In: *New directions in karst* (M. M. Sweeting and K. Paterson, Eds.), GeoBooks, Norwich.

Gustavson, T. C., Simpkins, W. W., Alhades, A., and Hoadley, A. (1982). Evaporite dissolution and development of karst features on the Rolling Plains of the Texas Panhandle. *Earth Surface Proc. and Landforms*, **7**, 545–563.

Halliday, W. R. (1960). Pseudokarst in the United States. *Bull. Nat. Speleo. Soc.*, **22**(2), 109–113.

Harmon, R. S., Ford, D. C., and Schwarcz, H. P. (1977). Interglacial chronology of the Rocky and Mackenzie Mountains based upon ^{230}Th–^{234}U dating of calcite speleothems. *Can. J. Earth Sci.*, **14**, 2543–2552.

Harmon, R. S., Schwarcz, H. P., and Ford, D. C. (1978). Late Pleistocene sea level history of Bermuda. *Quat. Research*, **9**, 205–218.

Harmon, R. S., Thompson, P., Schwarcz, H. P., and Ford D. C. (1978). Late Pleistocene palaeoclimates of North America as inferred from stable isotope studies of speleothems. *Quat. Research*, **9**, 54–78.

Harmon, R. S., White, W. B., Drake, J. J., and Hess, J. W. (1975). Regional hydrochemistry of North American carbonate terrains. *Water Resources Research*, **11**(6), 963–967.

Hatch, F. H., Rastall, R. H., and Greensmith, J. T. (1971). *Petrology of the Sedimentary Rocks*, Murby, London.

Helldén, U. (1973). Limestone solution intensity in a karst area in Lapland, Northern Sweden. *Geog. Ann.*, **54A**(3/4), 185–196.

Helldén, U. (1976). Hydrochemical analyses and karst denudation measurements from Moravian karst in Czechoslovakia. In: *Karst Processes and Relevant Landforms* (I. Gams, Ed.), U. of Ljubljana, Yugoslavia, pp. 81–95.

Hendy, C. H. (1970). The use of C14 in the study of cave processes. In: *Radiocarbon variations and absolute chronology* (I. U. Olsson, Ed.), Wiley, London, pp. 419–442.

Hendy, C. H., and Wilson, A. R. (1968). Palaeoclimatic data from speleothems. *Nature*, **219**, 48–51.

Hennig, G. J., Grün, R., and Brunnacker, K. (1983). Speleothems, travertines, and palaeoclimates. *Quat. Research*, **20**, 1–29.

Herak, M., and Stringfield, V. T. (1972). *Karst*, Elsevier, Amsterdam.

Hess, J. W., and Harmon, R. S. (1981). Geochronology of speleothems from the Flint Ridge-Mammoth Cave System, Kentucky, U.S.A. *Proc. 8th Int. Speleol. Congr.*, pp. 433–435.

High, C. J., and Hanna, F. K. (1970). A method for the direct measurement of erosion on rock surfaces. *Brit. Geomorph. Res. Grp., Tech. Bull*, **5**.

Hodgson, J. M., Rayner, J. H., and Catt, J. A. (1974). The geomorphological significance of clay-with-flints on the South Downs. *Trans. Inst. Brit. Geogr.*, **61**, 119–129.

Jakucs, L. (1977). *Morphogenetics of Karst Regions*, Adam Hilger, Bristol.

Jennings, J. N. (1964). Bungonia caves and rejuvenation. *Helictite*, **3**(4), 79–84.

Jennings, J. N. (1967). Some karst areas of Australia. In: *Landform Studies from Australia and New Guinea* (J. N. Jennings and J. A. Mabbutt, Eds.), Cambridge University Press, pp. 256–292.

Jennings, J. N. (1971). *Karst*, A.N.U. Press, Canberra.

Jennings, J. N. (1972). Observations at the Blue Waterholes, March 1965–April 1969, and limestone solution on Cooleman Plain, N.S.W. *Helictite*, **10**, 3–46.

Jennings, J. N. (1975). Doline morphometry as a morphogenetic tool: New Zealand examples. *N.Z. Geographer*, **31**, 6–28.

Jennings, J. N. (1978). Limestone solution on bare karst and covered karst compared. *Trans. Brit. Cave Res. Assn.*, **5**(4), 215–220.

Jennings, J. N. (1982). Principles and problems in reconstructing karst history. *Helictite*, **20**,(2), 37–52.

Jennings, J. N., and Sweeting, M. M. (1963). The limestone ranges of the Fitzroy Basin, Western Australia. *Bonn. geogr. Abh.*, **32**, 321–360.

Katzer, F. von (1909). Karst und Karsthydrographie. *Zur Kunde der Balkanhalbinsel*, **8**, 94 pp.

Kaufman, W. J., and Orlob, G. T. (1956). Measuring groundwater movement with radioactive and chemical tracers. *J. Am. Water Works Assoc.*, **48**, 559–572.

Kaye, C. A. (1957). The effect of solvent motion on limestone solution. *J. Geol.*, **65**, 35–46.

Kirkby, M. J. (Ed.) (1978). *Hillslope Hydrology,* Wiley, London.

Knežević, B. (1976). Optimization of parameters of discharge hydrograph models. In: *Karst Hydrology and Water Resources* (V. Yevjevich, Ed.), Water Resource Publs., Fort Collins, pp. 259–275.

Knisel, W. G. (1972). Response of karst aquifers to recharge. Colorado State Univ. Hydrology Papers, No. 60.

Kukla, G. (1978). The classical European glacial stages: correlation with deep-sea sediments. *Trans. Nebraska Acad. Sci.*, **11**, 57–93.

Láng, S. (1977). Relationship between world-wide karstic denudation (corrosion) and precipitation. *Proc. 7th Int. Speleol. Congr.*, Sheffield, pp. 282–283.

Latham, A. G., Schwarcz, H. P., Ford, D. C., and Pearce, G. W. (1979). Palaeomagnetism of stalagmite deposits. *Nature*, **280**, 383–385.

Latham, A. G., Schwarcz, H. P., Ford, D. C., and Pearce, G. W. (1982). The palaeomagnetism and U-Th dating of three Canadian speleothems: evidence for the westward drift, 5.4–2.1 Ka BP. *Can. J. Earth Sci.*, **19**, 1985–1995.

Lauritzen, S-E. (1981). Simulation of rock pendants—small scale experiments on plaster models. *Proc. 8th Int. Speleol. Congr.*, Bowling Green, pp.407–409.

Le Grand, H. E. (1952). Solution depressions in diorite in North Carolina. *Am. J. Sci.*, **250**, 566–585.

Leighton, M. W., and Pendexter, C. (1962). Carbonate rock types. *Mem. Am. Assn. Petrol. Geol.*, **1**, 33–61.

Löffler, E. (1974). Piping and pseudokarst features in the tropical lowlands of New Guinea. *Erdkunde*, **28**, 13–18.

Löffler, E. (1978). Karst features and igneous rocks in Papua New Guinea. In: *Landform Evolution in Australasia* (J. L. Davies and M. A. J. Williams, Eds.), A.N.U. Press, Canberra, pp. 238–249.

Mangin, A. (1974–75). Contribution à l'étude hydrodynamique des aquiferes karstiques. *Ann. Spéléol.*, **29**,(3), 283–332; **29**(4), 495–601; **30**(1), 21–124.

Marker, M. E. (1970). Some problems of a karst area in the Eastern Transvaal, South Africa. *Trans. I.B.G.*, **50**, 73–85.

Martel, E. A. (1910). La théorie de la 'Grundwasser' et les eaux souterrains du karst. *La Géogr.*, **21**, 126–130.

Martel, E. A. (1921). *Nouveau Traité des Eaux Souterrains*, Paris.

Maurin, V., and Zötl, J. (1959). Die Untersuchung der Zusammenhänge unterirdischer Wässer mit besonder Berücksichtigung der Karstverhältnisse. *Steir. Beitr. Hydrogeologie,* **10/11**, 5–184.

Miotke, F-D. (1972). Die messung des CO_2–gehaltes der bodenluft mit dem Dräger-Gerät und die beschleunigte kalklösung durch höhere fließgeschwindigkeiten. *Z. Geomorph.*, **16**(1), 93–102.

Miotke, F-D. (1974). Carbon dioxide and the soil atmosphere. *Abh. Karst -u. Hohlenkunde*, Reihe **A**, Heft 9, 49 pp.

Miotke, F-D., and Palmer, A. N. (1972). *Genetic Relationship between Caves and Landforms in the Mammoth Cave National Park Area*, Bohler, Würzburg.

Mitchell, G. F., Penny, L. F., Shotton, F. W., and West, R. G. (1973). *A Correlation of Quaternary Deposits in the British Isles*, Geol. Soc. London, Spec. Rept. No. 4.

Mowat, G. D. (1962). Progressive changes of shape by solution in the laboratory. *Cave Notes*, **4**, 45–49.

Nakayama, F. S. (1968). Calcium activity, complex and ion-pair in saturated $CaCO_3$ solutions. *Soil Science*, **106**(6), 429–434.

Neale, J., and Flenley, J. (1981). *The Quaternary in Britain*, Pergamon Press, Oxford.

Nicod, J. (1976). Karsts des gypses et des évaporites associś. *Annales de Geographie*, **85**, 513–554.

Nöel, M., Homonko, P., and Bull, P. A. (1979). The palaeomagnetism of sediments from Agen Allwedd. *Trans. Brit. Cave Res. Assn.*, **6**(2), 85–92.

Özis, U., and Keloglu, N. (1976). Some features of mathematical analysis of karst runoff. In: *Karst Hydrology and Water Resources* (V. Yevjevich, Ed.), Water Resource Publs., Fort Collins, pp. 221–235.

Palmer, H. S. (1927). Lapiés in Hawaiian basalts. *Geog. Rev.*, **17**, 627–631.

Palmquist, R. (1979). Geologic controls on doline characteristics in mantled karst. *Z. Geomorph., N.F.*, Suppl. Band **32**, 90–106.

Parker, G. G. Shown, L. M., and Ratzlaff, K. W. (1964). Officer's Cave, a pseudokarst feature in altered tuff and volcanic ash of the John Day formation in eastern Oregon. *Bull. Geol. Soc. Am.*, **75**, 393–402.

Peterson, J. A. (1982). Limestone pedestals and denudation estimates from Mt. Jaya, Irian Jaya. *Australian Geogr.*, **15**, 170–173.

Pfeiffer, D., and Hahn, J. (1972). Karst of Germany. In: *Karst* (M. Herak and V. T. Stringfield, Eds.), Elsevier, Amsterdam, pp. 189–224.

Picknett, R. G., Bray, L. F., and Stenner, R. D. (1976). The chemistry of cave waters. In: *The Science of Speleology* (T. D. Ford and C. H. D. Cullingford, Eds.), Academic Press, London, pp. 213–266.

Pirs, M., Udovc, H., and Toplisek, M. (1976). Tracing with Cr-51, In: *Underground Water Tracing* (R. Gospodaric and P. Habic, Eds.), Inst. Karst Research, Ljubljana, pp. 192–195.

Plummer, L. N., Wigley, T. M. L., and Parkhurst, D. L. (1978). The kinetics of calcite dissolution in CO_2-water systems at 5 to 60°C and 0.0 to 1.0 atm CO_2. *Am. J. Sci.*, **278**, 179.

Plummer, L. N., Wigley, T. M. L., and Parkhurst, D. L. (1979). Critical review of the kinetics of calcite dissolution and precipitation. In: *Chemical Modelling in Aqueous Systems* (E. A. Jeune, Ed.). Am. Chem. Soc., Symp. Ser. 93, pp. 537–576.

Powell, R. L. (1975). Theories of the development of karst topography. In: *Theories of Landform Development* (W. L. Melhorn and R. C. Flemal, Eds.), S.U.N.Y., Binghamton, pp. 217–242.

Quinif, Y. (1981). Thermoluminescence: a method for sedimentological studies in caves. *Proc. 8th Int. Speleol. Congr.*, Bowling Green, pp. 309–313.

Quinlan, J. F. (1982). Groundwater basin delineation with dye-tracing potentiometric surface mapping, and cave mapping, Mammoth Cave Region, Kentucky, U.S.A. *Beitraege zur Geologie der Schweiz-Hydrologie*, **28**(1), 177–190.

Ramljak, P., Filip, A., Milanovic, P., and Arandjelovic, D. (1976). Establishing karst underground connections and responses by using tracers. In: *Karst Hydrology and Water Resources* (V. Yevjevich, Ed.), Water Resources Publications, Colorado, pp. 237–257.

Reed, J. C., Bryant, B., and Hack, J. T. (1963). Origin of some intermittent ponds on quartzite ridges in western North Carolina. *Bull. Geol. Soc. Am.*, **74**, 1183–1188.

Reesman, A. L., and Godfrey, A. E. (1981). Development of the Central Basin of Tennessee by chemical denudation. *Z. Geomorph., N.F.*, **25**(4), 437–456.

Reynolds, D. L. (1961). Lapies and solution pits in olivine-dolerite sills at Slieve Gullion, Northern Ireland. *J. Geol.*, **69**, 110–117.

Roglic, J. (1960). Das Verhältnis der Flusserosion zum Karstprozess. *Z. Geomorph.*, **4**(2), 116–128.

Russell, E. W. (1973). *Soil Conditions and Plant Growth* (10th edn.), Longmans, London.

Schwarcz, H. P. (1978). Dating methods of Pleistocene deposits and their problems. II: Uranium series disequilibrium dating. *Geoscience Canada*, **5**(4), 184–187.

Shackleton, N. J. (1977). Oxygen isotope stratigraphy of the middle Pleistocene. In: *British Quaternary Studies* (F. W. Shotton, Ed.), Clarendon Press, Oxford, pp. 1–16.

Smart, P. L. (1976). Catchment delimination in karst areas by the use of quantitative tracer methods. *Proc. 3rd Int. Symp. Underground Water Tracing*, Bled, Juoslavia, pp. 291–298.

Smart, P. L. (1982). A review of the toxicity of 12 fluorescent dyes used for water tracing. *Beitr. Zur. Geol. der Schweiz-Hydrologie*, **28**(1), S1O1–112.

Smart, P. L., Finlayson, B. L., Rylands, W. D., and Ball, C. M. (1976). The relation of fluorescence to dissolved organic carbon in surface waters. *Water Research*, **10**, 805–811.

Smart, P. L., and Hodge, P. (1980). Determination of the character of the Longwood Sinks to Cheddar Resurgence conduit using an artificial pulse wave. *Trans. Brit. Cave Res. Assn.*, **7**(4), 208–211.

Smart, P. L. and Laidlaw, I. M. S. (1977). An evaluation of some fluorescent dyes for water tracing. *Water Resources Research*, **13**(1), 15–33.

Smith, D. I. (1977). Applied geomorphology and hydrology of karst regions. In: *Applied Geomorphology* (J. Hails, Ed.), Elsevier, Amsterdam, pp. 85–117.

Smith, D. I., and Atkinson, T. C. (1976). Process, landforms and climate in limestone regions. In: *Geomorphology and Climate* (E. Derbyshire, Ed.), Wiley, London, pp. 367–409.

Smith, D. I., and Atkinson, T. C. (1977). Underground flow in cavernous limestones with special reference to the Malham area. *Field Studies*, **4**, 597–616.

Smith, D. I., Atkinson, T. C., and Drew, D. P. (1976). The hydrology of limestone terrains. In: *The Science of Speleology* (T. D. Ford and C. H. D. Cullingford, Eds.), Academic Press, London, pp. 179–212.

Smith, D. I., Drew, D. P., and Atkinson, T. C. (1972). Hypotheses of karst landform development in Jamaica. *Trans. Cave Res. Grp.*, **14**(2), 159–173.

Smith, D. I., and Mead, D. G. (1962). The solution of limestone with special reference to Mendip. *Proc. Univ. Bristol Spelaeol. Soc.*, **9**(3), 188–211.
Stanton, W. I., and Smart, P. L. (1981). Repeated dye traces of underground streams in the Mendip Hills, Somerset. *Proc. Univ. Bristol Spelaeol. Soc.*, **16**(1), 47–58.
Sweeting, M. M. (1950). Erosion cycles and limestone caverns in the Ingleborough district. *Geog. J.*, **115**, 63–78.
Sweeting, M. M. (1966). The weathering of limestones. In: *Essays in Geomorphology* (G. H. Dury, Ed.), Heinemann, London, pp. 177–210.
Sweeting, M.M. (1972). *Karst Landforms*, Macmillan, London.
Sweeting, M. M. (1976). Some comments on the lithological basis of karst landform variations. *Proc. 6th Int. Speleol. Congr.*, Olomouc, 1973, pp. 319–329.
Sweeting, M. M. (1979). Karst morphology and limestone petrology. *Prog. in Phys. Geog.*, **3**,(1), 102–110.
Sweeting, M. M. (1980). Karst and climate. *Z. Geomorph., N.F.*, Suppl. Band **36**, 203–216.
Sweeting, M. M. (Ed.) (1981). *Karst Geomorphology*, Hutchinson Ross, Pennsylvania.
Swinnerton, A. C. (1932). Origin of limestone caverns. *Bull. Geol. Soc. Am.*, **43**, 662–693.
Tate, T. (1879). The source of the River Aire. *Proc. Yorks. Geol. Polytech. Soc.*, **7**, 177–187.
Thomas, T. M. (1974). The South Wales interstratal karst. *Trans. Brit. Cave Res. Assn.*, **1**(3), 131–152.
Thompson, G. M., Hayes, J. M., and Davis, S. N. (1974). Flurocarbon tracers in hydrology. *Geophysical Research Letters*, **1**(4), 177–180.
Thrailkill, J. V. (1968). Chemical and hydrologic factors in the excavation of limestone caves. *Bull. Geol. Soc. Am.*, **79**, 19–45.
Trudgill, S. T. (1975). Measurement of erosional weight-loss of rock tablets. *Brit. Geomorph. Res. Grp. Tech. Bull.*, **17**, 13–19.
Trudgill, S. T. (1976). Rock weathering and climate: quantitative and experimental aspects. In: *Geomorphology and Climate* (E. Derbyshire, Ed.), pp. 59–99.
Trudgill, S. T. (1977). Problems in the estimation of short-term variations in limestone erosion processes. *Earth Surface Processes*, **2**(2/3), 251–256.
Trudgill, S. T., High, C. J., and Hanna, F. K. (1981). Improvements to the Micro-Erosion Meter. *B.G.R.G. Tech. Bull.*, **29**, 3–18.
Tschang, H. L. (1961). The pseudokarren and exfoliation forms of granite on Pulai Ubin, Singapore. *Z. Geomorph., N.F.*, **5**, 253–259.
Verstraten, J. M. (1977). Chemical erosion in a forested watershed in the Oesling, Luxembourg. *Earth Surface Processes*, **2**(2/3), 175–184.
Wall, J. R. D., and Wilford, G. E. (1966). A comparison of small scale solution features on microgranodiorite and limestone in West Sarawak, Malaysia. *Z. Geomorph., N.F.*, **10**, 461–468.
Walling, D. E. (1978). Reliability considerations in the evaluation and analysis of river loads. *Z. Geomorph.*, Suppl. Band **29**, 29–42.
Walling, D. E. (1979). Hydrological Processes. In: *Man and Environmental Processes* (K. J. Gregory and D. E. Walling, Eds.), Butterworths, London, pp. 57–81.
Walling, D. E., and Webb, B. W. (1975). Spatial variation of river water quality: a survey of the River Exe. *Trans. Inst. Brit. Georgr.*, **65**, 155–172.
Walling, D. E., and Webb, B. W. (1978). Mapping solute loadings in an area of Devon, England. *Earth Surf. Processes*, **3**(1), 85–99.
Waltham, A. C. (1981a). Origin and development of limestone caves. *Prog. in Phys. Geog.*, **5**(2), 242–256.

Waltham, A. C. (1981b). The karstic evolution of the Matienzo depression, Spain. *Z. Geomorph., N. F.*, **25**(3), 300–312.

Waltham, A. C., Smart, P. L., Friederich, H., Eavis, A. J., and Atkinson, T. C. (1983). The caves of Gunung Sewu, Java. *Trans. Brit. Cave Res. Assn.*, **10**(2), 55–96.

Ward, R. C. (1975). *Principles of Hydrology* (2nd edn.), McGraw-Hill, London.

Warwick, G. T. (1971). Caves and the ice age. *Trans. Cave Res. Grp., G.B.*, **13**(2), 123–130.

Watts, S. T., and Trudgill, S. T. (1979). An investigation into the relationship between solvent motion and the solutional erosion of an inclined limestone surface. *Trans. Brit. Cave Res. Assn.*, **6**(1), 18–29.

Waylen, M. J. (1979). Chemical weathering in a drainage basin underlain by old red sandstone. *Earth Surface Processes*, **4**(2), 167–178.

Webb, B. W., and Walling, D. E. (1980). Stream solute studies and geomorphological research: some examples from the Exe Basin, Devon, U.K. *Z. Geomorph.*, Suppl. Band **36**, 245–263.

Webb, B. W., and Walling, D. E. (1982). The magnitude and frequency characteristics of fluvial transport in a Devon drainage basin and some geomorphological implications. *Catena*, **9**(1/2), 9–23.

Weeks, E. P., and Thompson, G. M. (1982). Use of atmospheric flurocarbons F-11 and F-12 to determine the diffusion parameters of the unsaturated zone in the Southern High Plains of Texas. *Water Resources Research*, **18**(5), 1365–78.

White, K. E. (1974). Use of radioactive tracers to study mixing and residence-time distributions in systems exhibiting three-dimensional dispersion. *Proc. First European Conf. on Mixing and Centrifugal Separation*, Cambridge, A6:57–76.

White, K. E. (1977). Tracer method for the determination of groundwater residence time distributions. *Proc. Reading Conf. on Groundwater Quality*, Water Research Centre, pp 246–273.

White, W. B. (1969). Conceptual models for carbonate aquifers. *Ground Water*, **7**(3), 15–21.

White, W. B. (1976). Cave minerals and speleothems. In: *The Science of Speleology* (T. D. Ford and C. H. D. Cullingford, Eds.), Academic Press, London, pp. 267–327.

White, W. B. (1977). Conceptual models for carbonate aquifers: revisited. In: *Hydrologic Problems in Karst Terrains* (R.. R. Dilamarter and C. S. Csallany, Eds.), W. Kentucky Univ. Press, Bowling Green, pp. 176–187.

White, W. B., Jefferson, G. L., and Haman, J. F. (1966). Quartzite karst in southeastern Venezuela. *Int. J. Speleol.*, **2**, 300–314.

White, W. B. and Longyear, J. (1962). Some limitations on speleogenetic speculation imposed by the hydraulics of groundwater flow in limestones. *Nittany Grotto Newsletter*, **10**(9), 155–167.

Wigley, T. M. L., Drake, J. J., Quinlan, J. F., and Ford, D. C. (1973). Geomorphology and geochemistry of a gypsum karst near Canal Flats, British Columbia. *Can. J. Earth Sci.*, **10**(2), 113–129.

Wigley, T. M. L., and Plummer, L. N. (1976). Mixing of carbonate waters. *Geochim et Cosmochim Acta*, **40**, 989–995.

Wilcock, J. D. (1968). Some developments in pulse-train analysis. *Trans. Cave Res. Grp., G. B.*, **10**(2), 73–98.

Williams, P. W. (1963). An initial estimate of the speed of limestone solution in County Clare. *Irish Geog.*, **4**, 432–441.

Williams, P. W. (1966). Limestone pavements. *Trans. I.B.G.*, **40**, 155–172.

Williams, P. W. (1968). An evaluation of the rate and distribution of limestone solution and deposition in the River Fergus basin, Western Ireland. Aust. Nat. Univ., Dept. Geog., Pub. G/5, pp. 1–40.

Williams, P. W. (1972). Morphometric analysis of polygonal karst in New Guinea. *Bull. Geol. Soc. Am.*, **83**, 761–796.
Williams, P. W. (1977). Hydrology of the Waikoropupu Springs: a major tidal karst resurgence in northwest Nelson (New Zealand). *Journal of Hydrology*, **35**, 73–92.
Williams, P. W. (1978). Interpretations of Australasian karsts. In: *Landform Evolution in Australasia* (J. L. Davies and M. A. J. Williams, Eds.), A.N.U. Press, Canberra, pp. 259–285.
Williams, P. W. (1982). Speleothem dates, Quaternary terraces and uplift rates in New Zealand. *Nature*, **289**, 257–260.
Williams, P. W. (1983). The role of the subcutaneous zone in karst hydrology. *Journal of Hydrology*, **61**(1/3), 45–67.
Williams, P. W. (in press). Subcutaneous hydrology and the development of dolines. In: *New Directions in Karst* (M. M. Sweeting and K. Paterson, Eds.), GeoBooks, Norwich.
Williams, P. W., and Dowling, R. K. (1979). Solution of marble in the karst of the Pikikiruna Range, Northwest Nelson, New Zealand. *Earth Surface Processes*, **4**, 15–36.
Wimpenny, J. W. T., Cotton, N., and Statham, M. (1972). Microbes as tracers of water movement. *Water Research*, **6**, 731–739.
Winkler, E. M. (1966). Important agents of weathering for building and monumental stone. *Eng. Geol.*, **1**, 381–400.
Wintle, A. G. (1978). A thermoluminescence study of some Quarternary calcite: potential and problems. *Can. J. Earth Sci.*, **15**, 1977–1986.
Wolman, M. G., and Miller, J. P. (1960). Magnitude and frequency of forces in geomorphic processes. *J. Geol.*, **68**, 54–74.
Wood, C. (1976). Caves in rocks of volcanic origin. In: *The Science of Speleology* (T. D. Ford and C. H. D. Cullingford, Eds.), Academic Press, London, pp. 127–150.
Yevjevich, V. (Ed.) (1976). *Karst Hydrology and Water Resources* (2 Vols.), Water Research Publs., Fort Collins.
Yevjevich, V. (Ed.) (1981). *Karst Water Research Needs*, Water Resource Publs., Fort Collins.

Solute Processes
Edited by S. T. Trudgill

CHAPTER 10

Mathematical models for solutional development of landforms

M. J. Kirkby

School of Geography
University of Leeds

Solute processes within soil profiles and down hillslopes may be seen from previous chapters to show considerable complexity. No existing models can explain more than the broadest patterns revealed by field and laboratory studies. Hydrological and chemical processes interact to produce solutional effects at all scales. These two groups of processes dominate almost all the research described above, and attempts at modelling must marry hydrology and chemistry in a way which retains adequate reality without unnecessary complication. In this chapter a series of models are described which relate to particular aspects of solution and its effects. They range in chemical complexity from assumptions of constant solutional loss to approximations for the equilibrium thermodynamics of the soil; and in hydrological complexity from assumptions of constant downward percolation to detailed routing of flows into the soil and downslope. Throughout the intention will be to describe the simpler models in some detail, and to sketch in the concepts and potential applications of more complex models but without going into comparable detail.

A basic division between the chemical bases of alternative types of model depends on whether waters are assumed to reach some kind of equilibrium with soil solids or not. The latter, kinetic approach has many theoretical advantages, but its greater complexity has severely limited their application. Simple slope profile models may be made on each set of assumptions and it will be seen that the role of solution in forming convexities differs appreciably between equilibrium and kinetic assumptions. Investigation of the soil profile, initially at least, requires some consideration of chemical processes in detail, though it is finally argued below that a simple linear model for solution is appropriate under many circumstances. Models for soil profile development are then based on this linear model. The soil profile approach can be further simplified to provide a fuller basis for slope profile modelling, with an explicit representation of regolith development which may be traced back to a chemical description of the soil weathering profile. Finally some rather crude hydrological models will be used to explain the gross features of observed differences in solution rates with climate and lithology. This chapter follows the sequence dscribed above in an attempt to describe the present, very incomplete state of the art of solute process and form modelling.

10.1 Equilibrium and kinetic approaches

If water is brought into contact with soil solids then it will tend towards a quasi-equilibrium in which the solute content of the water is in balance with the existing soil composition. This quasi-equilibrium is closely approached within a matter of hours or days (Bricker, 1968), depending on the rate of the solution reactions. On a very much longer timescale, commonly to the measured in thousands or millions of years, a true equilibrium will be reached in which the soil material also reaches equilibrium with the water (Garrels and Christ, 1965). This true equilibrium is not generally of concern in considering the movement of ions within the soil, and the term 'equilibrium' is applied below exclusively to the quasi-equilibrium described above. In reaching it, the composition of the soil solids is changed to such a small extent that it may be ignored in calculating the concentration of solutes in the water. Where water flows through the soil slowly enough to continually reach this equilibrium, the rate of leaching loss from the soil may be calculated as the product of equilibrium concentration and flow discharge. In this way the gradual loss of soil substance may be followed as the soil solids very gradually approach their long-term equilibrium; at a rate which is controlled by the quasi-equilibium with the water and its flow rate, rather than by the reaction rates of the soil solids themselves. Only under exceedingly static groundwater conditions could reaction rates control the soil evolution, and even then diffusion of solutes in the water is commonly the rate-determining process.

It is therefore argued that the long-term kinetics of the soil solids are determined by the removal of solutes in soil water. A second, less clear aspect of the soil reaction rates concerns the extent to which the short-term equilibrium is fully reached for the concentration of solutes in the water. The times required to approach equilibrium depend on both the rates of the solubility reactions and on the degree of contact between the soil and the water. For finely divided soil materials shaken in water, a time of approximately 100 hours appears to be sufficient (Bricker *et al.*, 1968) to allow soils to approach within $\pm 10\%$ of their equilibrium concentrations, even though some of the aluminium ion reactions are still incomplete. This value will be used as a guide below, but may need to be revised to suit particular data sets. For periods shorter than 100 hours, or where water is in poor contact with the soil, the assumption of equilibriation is thus likely to provide an inadequate basis for modelling.

Within the soil, contact between water and soil solids is intricate in detail. Within the soil matrix, water flow is very slow, particularly if there is an appreciable clay content. The intra-pedal water within an individual soil aggregate is therefore expected to remain close to equilibrium. Water flowing past an aggregate within inter-pedal voids is in poor contact with the soil solids, and must rely partly on ionic diffusion to gain solutes from the more highly concentrated solutes of the intra-pedal water. The larger the soil voids, the less effective will be solute pick-up within them, although the total contact times for flow within the soil may still be long. Perhaps the least effective contact between soil and water is for overland flow, which in some cases may be assumed to flow downslope without any appreciable solute gain. A simple approach for modelling purposes is to partition hillslope flow into components which respectively reach equilibrium and pick up no solutes; but such a model can be no more than an approximation.

The advantage of using an equilibrium model is that the process is both simpler and better understood. In the simplest case of a single ion, the equilibrium solute concentration is constant for each material at given conditions of temperature and pressure. The saturated solubilities are well-established for most relevant substances. For a mineral assemblage the position is somewhat more complicated, as will be seen below (§10.3), but still reasonably manageable. For a kinetic model (e.g. Lerman, 1979), the constant concentration is replaced by an inverse exponential approach to saturation, with the rate of increase of concentration given by a kinetic constant k, so that:

$$dc/dt = k(c_* - c) \tag{10.1}$$

where c is solute concentration, c_* is equilibrium concentration, and t is the elapsed time. The resulting expression for concentration starting from initially pure water is then:

$$c = c_*[1 - \exp(-kt)]$$

This expression is not only a little harder to deal with, containing time explicity and the additional kinetic constant; but the kinetic constants are a good deal less well-established then the equilibrium constants for the same reactions, especially for soils containing several ions. In some cases an experimental value for k may be used, often a single value for an entire mineral assemblage, but the approximations involved are generally as great as those in partitioning the flow between equilibriated and uncontaminated water. Partitioning has therefore been preferred as the basis for most of the models presented below, but it is recognized that considerable development is needed for improved models based on kinetic considerations.

10.1.1 Water paths through soils and down hillsides

For both equilibrium and kinetic models it is helpful to examine the various routeways over and through the soil and to assess the residence time of water in contact both with individual soil peds and with hillslope materials as a whole. For the slope as a whole, it is convenient to separate overland and sub-surface flow, even though there is some exchange between them through infiltration and seepage downslope. At a local scale within the soil, a similar division may be recognized between intra- and inter-pedal flow, again with some exchange between them.

Flow velocities vary very considerably, and a much fuller discussion of the evidence and principles may be found in, for example, Dunne (1978). To generalize, however, the relative velocities (in metres per hour) commonly take the following ranges:

Overland flow:	100–1000
Saturated sub-surface flow:	1–10
Unsaturated sub-surface flow:	0.01–0.1

These rates give residence times for a 300 m slope of respectively about 3 hours, 2 weeks and three years. In other words, overland flow will generally be expected to pick up little solute load, saturated sub-surface flow should approach equilibrium, and unsaturated flow be fully equilibriated. Unsaturated flow, however, is rarely thought to be an effective mode of downslope flow, but is primarily of importance during the infiltration process. The relevant residence time is that required for infiltration to a level of saturated flow, either as groundwater or within the soil, if one exists. Periods of the order of 3–24 hours are thought appropriate. The critical process is therefore that of saturated sub-surface flow, if it occurs. The range of velocities is such that for slow flows equilibrium may reasonably be assumed, while if flow is relatively rapid a kinetic approach is to be preferred. A complication to this simple view arises through mixing of old and new slopewater in a storm, which alters the degree of equilibriation achieved.

On the more detailed scale within the soil, percolation of rainfall has been modelled (Beven and Germann, 1981; Trudgill *et al*. 1983) as a process of infiltration into soil peds until their infiltration capacity is exceeded. The additional ponded rainfall then overflows into soil cracks and other voids, where it travels at higher velocities than within the peds. Along the walls of the voids infiltration to sub-surface peds may continue laterally if they are not already saturated. Within peds, particularly if they are fine-textured, flow is slow and water has ample residence time to equilibriate fully with the soil solids, but little escapes to percolate downwards under gravity: instead most is ultimately lost by evapotranspiration.

Water flowing in the void spaces is in contact with the soil for a much shorter period so that it has little opportunity to equilibriate directly. Diffusion is, however, a highly effective process in transferring solutes from the interior of the ped into the relatively clean void water. Within a saturated ped with no water movement, equilibrium between ion diffusion and solute uptake is achieved when:

$$\frac{d}{dx}\left(-D\frac{dc}{dx}\right) = k(c* - c) \qquad (10.2)$$

where D is the rate of ionic diffusion, c is the solute concentration and x is the distance from the ped centre. The resulting concentration profile in equilibrium with a zero concentration at the ped margin ($x = x_0$) is then:

$$c = c* \{1 - \cosh[x\sqrt{(k/D)}]/\cosh[x_0\sqrt{(k/D)}]\}$$

and the outflow at the ped margin is:

$$q_0 = c_0\sqrt{(kD)}\tanh[x_0\sqrt{(k/D)}] \simeq c_0 k x_0$$

with the approximation valid for small peds ($x < \sqrt{(D/k)}$).

For a square network of inter-pedal cracks, the total outflow per unit area is:

$$2q_0/x_0 \simeq 2kc_0$$

Thus for a diffusivity of 0.01 m^2 per year and an equilibriation time of 100 hours, the approximation is valid for peds of up to about 20 mm diameter, and the outflow over a 1 m depth of saturated ped walls is equivalent to 10 mm per hour of equilibriated water. In other words it is likely that a substantial fraction even of stormwater will approach equilibrium in most cases. Exceptions are likely to occur for highly permeable soils and in the largest continuous voids where some effective solute bypassing may occur. It is in these cases that is it thought to be most important to develop a kinetic model.

10.2 EVOLUTION OF SLOPE PROFILES UNDER SOLUTIONAL DENUDATION

The most naive models of solution treat the detailed chemical processes as a 'black box', which gives an overall rate. The rate may be constant or depend in some way on position in the landscape. This type of approach may be used to model soil profile development but is perhaps most useful to follow the evolution of slope profiles as a whole. Hillslope models with 'black box' processes are well-established in the literature (e.g. Ahnert, 1976), although solution processes are not usually included in them.

At the simplest, solution may be considered to give constant denudation all over the landscape, at a rate which is implicitly assumed to depend mainly on rainfall and lithology. The basis for this model is an assumption that rainfall (or net rainfall) reaches solute equilibrium where it falls (Carson and Kirkby, 1972, chapter 9), and subsequently passively transports its dissolved load. Soils are thus assumed areally constant; or else percolation is assumed to penetrate to and equilibriate with a uniform bedrock beneath the soil. Violations of this set of assumptions are widespread, both because of areal variations in soil parent materials and because of percolation to variable depths. Nevertheless the constant denudation model provides a basic model for solution evolution of the landscape.

Constant solutional denudation is a 'weathering-limited' process description, and most plausible black-box slope models also imply weathering-limited removal, although denudation may be assumed to vary in rate with topographic or soil factors. Soil factors may be included if soil depth is also modelled downslope. The simplest way to budget for soil depth is to augment it by a weathering rate (related in some way to the solution rate) and decrease it by the amount of mechanical denudation loss. Weathering and mechanical processes may then be related to soil depth, as has been done by Ahnert (1964) and Armstrong (1976). Another type of topographic factor which may be included in a simple slope model is some allowance for overland flow of chemically clean water. Even without detailed budgeting, it is reasonable to assume that chemical denudation falls to zero on slope-base plains of sufficiently low gradient where all flow is likely to be overland. This kind of assumption is needed to allow a constant denudation model to be compatible with low rates of slope-base downcutting.

10.2.1 Continuity and process relationships

Models for slope evolution are inevitably constrained by the continuity equation for movement of sediment and solutes:

$$(\partial/\partial x)(S + V) = -\partial z/\partial t \tag{10.3}$$

where S is the mechanical sediment transport in volumetric units, V is the chemical solute transport in volumetric units, z is the elevation measured as rock-equivalent height, x is the distance downslope from the divide, and t is elapsed time.

This equation states that if more material is removed from a slope section than is brought into it, then the balance must be made up by net erosion within the section, and vice versa for deposition. To produce a usable slope model, some process laws are needed for the variation of transporting capacities with topographic position. Some asumption must also be made about the relationship between transporting capacity and actual transport rates: for instance removal may be assumed to be weathering-limited or transport-limited. Slope evolution is also constrained by boundary conditions, of which the simplest are a fixed divide (at $x = 0$) and a fixed slope length, at the base of which removal is specified in terms of transport rate or basal elevation, one of which is considered to remain constant or to vary in a specified way over time, thus representing the behaviour of a river etc. at the foot of the slope.

The simplest possible model for solution is one of constant lowering, at a rate D determined by climate and bedrock:

$$-\mathrm{d}z/\mathrm{d}t = D \tag{10.4}$$

This is the basic weathering-limited model. As has been noted above, it implies that solute equilibrium is reached locally, and that the water involved takes no further part in solute reactions. A second, equally simple equilibrium model assumes that all the hillslope water reaches equilibrium with the soil at each consecutive point downslope. In that case removal is transport-limited, but is in practice only different from Equation (10.4) when the slope profile crosses lithological boundaries, when it gives abrupt changes in solute transport:

$$V = W = Dx \tag{10.5}$$

where W is the solute transporting capacity.

A kinetic approach requires only a slight increase in complexity. It may be seen that the actual rate of solute transport may be written as:

$$V = Pcx \tag{10.6}$$

where P is the net addition to downslope flow per unit distance (roughly the net precipitation).

If the actual solute concentration c is replaced by the equilibrium concentration c_*, then the capacity solute transport W is similarly obtained. Substitution into Equation (10.1) gives:

$$\partial c/\partial t = k(c_* - c) = k[c_* - V/(Px)] \tag{10.7}$$

Substitution into Equation (10.3) then leads to:

$$-\partial z/\partial t = \partial V/\partial x = \partial V/\partial t$$

$$u = (Px/u)(\partial c/\partial t) = (Pkxc_* - kV)/u \tag{10.8}$$

where u is the downslope velocity of water flow in the soil.

In comparing this expression with the weathering-limited removal cases of Equation (10.5), the potential rate of solute denudation D is given by the first term on the right-hand side, namely:

$$D = (Pkc_*)\,(x/u) \tag{10.9}$$

The coefficient of V in the final term of Equation (10.8) may also be identified as the reciprocal of a characteristic 'travel distance' $h = u/k$, which is the distance required for clean water to reach the capacity transport rate if denudation proceeds at rate D. In Equation (10.8) the terms which vary with topography, for a fixed parent material (which mainly influences k and c_*) are the distance x and the flow velocity u. The latter is most plausibly assumed as proportional to gradient, at least provided that soil properties remain constant. Thus the potential denudation rate is expected to remain roughly constant over profile convexities and to increase rapidly in concavities, while the 'travel distance' is inversely proportional to gradient. The effect of this model is transitional between the weathering-limited and transport-limited cases, and provides both some gradient dependence and also a transitional zone at sharp lithological boundaries. For all three of the cases discussed above, it will be found realistic to assume that solutional loss falls to zero once basal gradients are sufficiently low, on the assumption that downslope flow is then largely overland with negligible solute pick-up. The implication of these models for slope profile form are followed below in §10.2.2 (equlibrium models) and §10.2.3 (kinetic models).

10.2.2 Evolution assuming solute equilibria

For the weathering-limited and transport-limited cases, the associated slope forms may be seen by assuming that downcutting is occurring at a rate T which may either be considered constant or else to vary in some manner downslope. For example, on most 'mature' slopes the local rate of denudation may be expected to decrease steadily downslope in some manner; and over time as denudation proceeds. In the transport-limited case described above, the solute transport is taken as equal to the local denudation rate D times distance from the divide x, so that the average denudation rate from the divide is always equal to the local rate. In the weathering-limited case the solute transport is assumed to increase with distance downslope at a rate given by local lithology. The average denudation from the divide is then a weighted average of all local values upslope.

In both the weathering-limited and transport-limited cases, the residual mechanical transport rate is, by subtraction:

$$S = \int_0^x (T - D)\,\mathrm{d}x = (\bar{T} - \bar{D})x \tag{10.10}$$

where T and D refer to local values of total and solutional denudation; and $\bar{T}$

and $\bar{D}$ refer to values averaged from the divide down to the point considered. If solution is accompanied by mechanical transport processes which are governed by a law of the form:

$$S = f(x)\, g \tag{10.11}$$

where f is a distance function representing the influence of flow processes and g is the local slope gradient, then the slope gradient required to produce the specified rate of overall denudation is:

$$g = (\bar{T} - \bar{D})\, x / f(x) \tag{10.12}$$

This expression provides a basis for obtaining the general form of slopes subject to solutional denudation in the presence of other, mechanical processes and were average total denudation exceeds the average solutional denudation. Where the former condition is not met, then denudation proceeds at the local solutional rate everywhere, with change of form only where lithology is areally variable. Where the latter condition is not met it may be assumed that gradients fall to zero, and all flow is overland, so that solution falls to match the total denudation rate. The simplest case where these restrictions do not apply is where T and D are constant everywhere, corresponding to constant downcutting imposed over uniform lithology. Equation (10.12) then shows that gradients are reduced in constant proportion from those which would apply in the absence of solution. Thus, for example, creep or solifluction will continue to evolve convex slopes, but the slope gradients and the landforms will be more subdued.

In the case of mature slopes, with total denudation rates declining downslope, and in the extreme case of a fixed basal elevation reaching zero at the slope foot, then average rates of total denudation will also decrease downslope, though never to zero. Even if the solute denudation remains constant downslope, its effect will be more marked than in the previous case, especially downslope where it will introduce marked profile concavities. Figure 10.1 illustrates this behaviour for slopes with soil creep, with a range of constant solution rates. The total rate of local denudation is, for simplicity, assumed to decline linearly downslope (to zero if the slope is long enough). It may be seen, and readily verified from Equation (10.12), that for zero solution rates the slope is convex throughout, reaching a point of inflexion at the slope base. As the solution rate is increased from zero, the lower part of the slope becomes increasingly concave, and for high solution rates features a relatively sharp break in slope to a basal plain. This type of profile appears to illustrate at least some of the features of tower karst and some tropical slope profiles. For the higher rates of denudation, the exact nature of the mechanical processes becomes increasingly irrelevant to the general form of the profile which is dominated by the solutional concavity. It should, however, be stressed that the mechanical process plays a vital role in

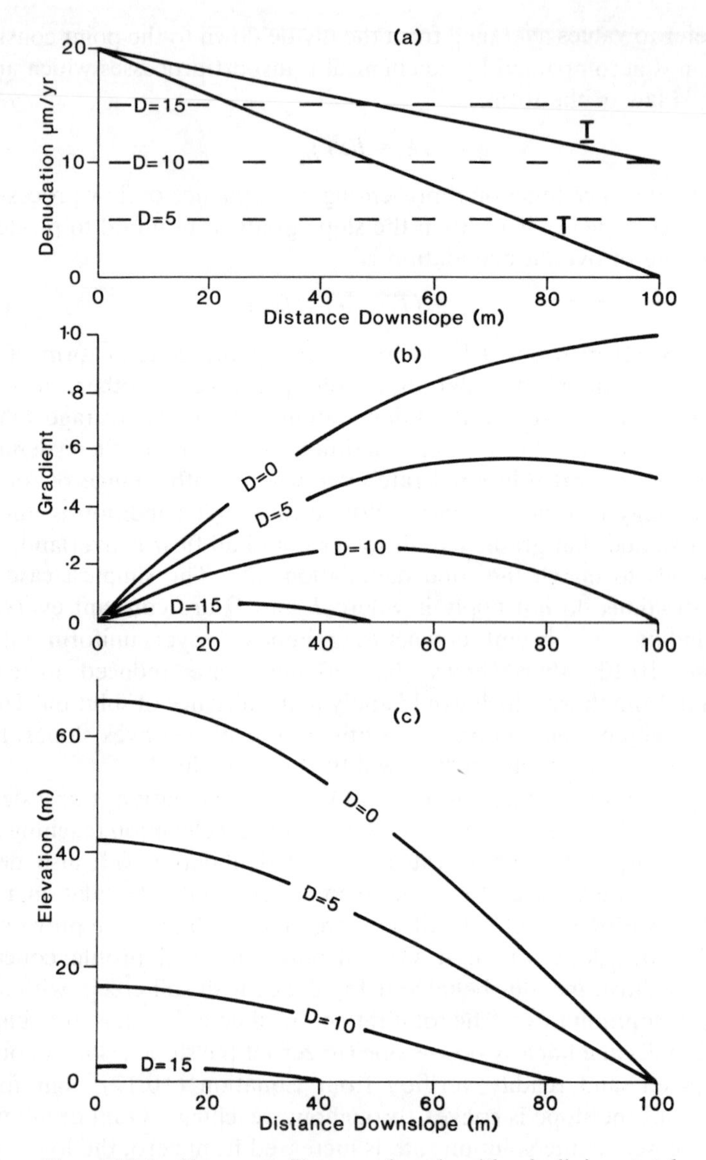

Figure 10.1. Slope profiles produced with declining total denudation rate *T* downslope, and for various solute denudation rates (constant downslope), in the presence of creep processes only: (a) assumed denudation rates; (b) resulting gradients; (c) resulting slope profiles

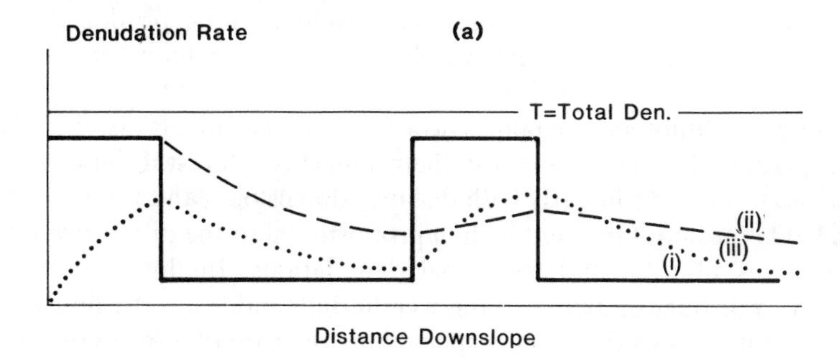

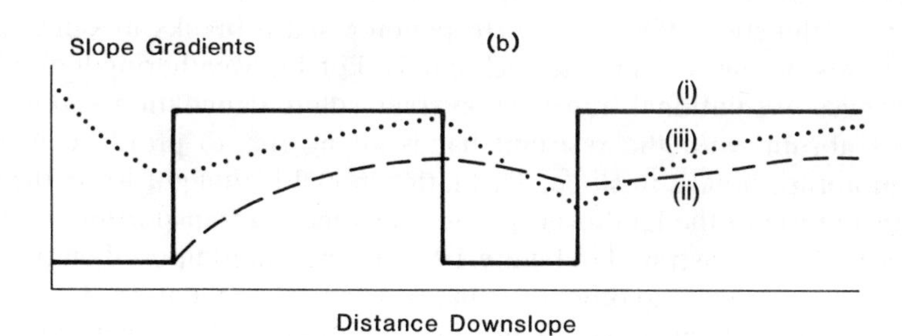

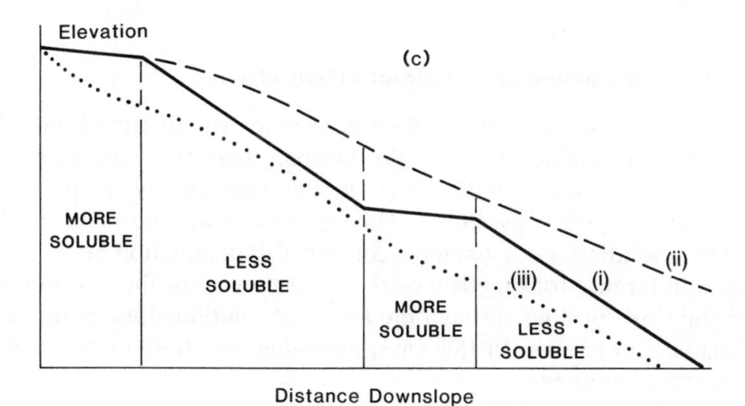

Figure 10.2. Influence of downslope difference in solutional denudation on slope gradients: (a) assumed variation in denudation rates with lithology (T for total; D for solution; (b) slope gradients; (c) slope profiles. (i) = transport-limited case; (ii) = weathering-limited case; (iii) = kinetic case with h_0 = 10 m. Total slope length = 60 m

producing these landforms, so that they could only be produced under a continuous regolith cover which would act as the medium for mechanical transport.

The effect of lithological changes down the slope profile is also readily seen from Equation (10.12). In this case their impact is clearest if the process rate is assumed to increase linearly with distance downslope, and total denudation rate is held constant. Gradient is then proportional to the difference between average rates of total and solutional denudations. In this case it will be necessary to distinguish between the weathering- and transport-limited cases. Figure 10.2 illustrates the forms expected for the case of a series of restricted outcrops of a rock less soluble than that on either side of it. In the transport-limited case where abrupt changes in transport rate occur at each lithological junction, there are corresponding sharp breaks in slope, giving steep facets on the less soluble rock bands. For the weathering-limited case the changes in solutional transport, average solute denudation and gradient are less abrupt, with the resistant bands giving rise to profile convexities without abrupt breaks in slope. This latter model is thought to be the more realistic in term of the landforms produced beneath a transporting soil cover. The local effects illustrated in Figure 10.2 may be superimposed on the effect of other process laws and other distributions of total denudation. Thus for soil creep the local changes would be superimposed on the convexity of the overall profile.

10.2.3 Evolution assuming significant kinetic effects

If it is assumed that rainwater is unable to reach solute equilibrium before flowing an appreciable distance downslope, then the kinetic model of Equations (10.8) and (10.9) above is the simplest available basis for forecasting slope profile evoltuion. The expressions are, however, much less amenable to solution, even assuming constancy of denudation rates. One very important difference for kinetic models is that they allow the development of meaningful slope profiles through the action of solution alone in the absence of mechanical processes. In this case, assuming that flow rates are directly proportional to gradient:

$$g = (D_0 - T)\, x \,/\, (h_0 T) \tag{10.13}$$

where $D_0 = Pc_*$ is the potential rate of solute denudation, and h_0 is the travel distance u/k for unit slope gradient. For constant denudation rates downslope, this gives a uniformly convex slope, with steeper slopes for more soluble rocks if total denudation is held constant. If total denudation declines downslope, then the slope profiles show increasing convexity downslope, truncated at the base with a very sharp break in slope to a basal plain of zero denudation (Figure 10.3). These profiles look even more like tower karst

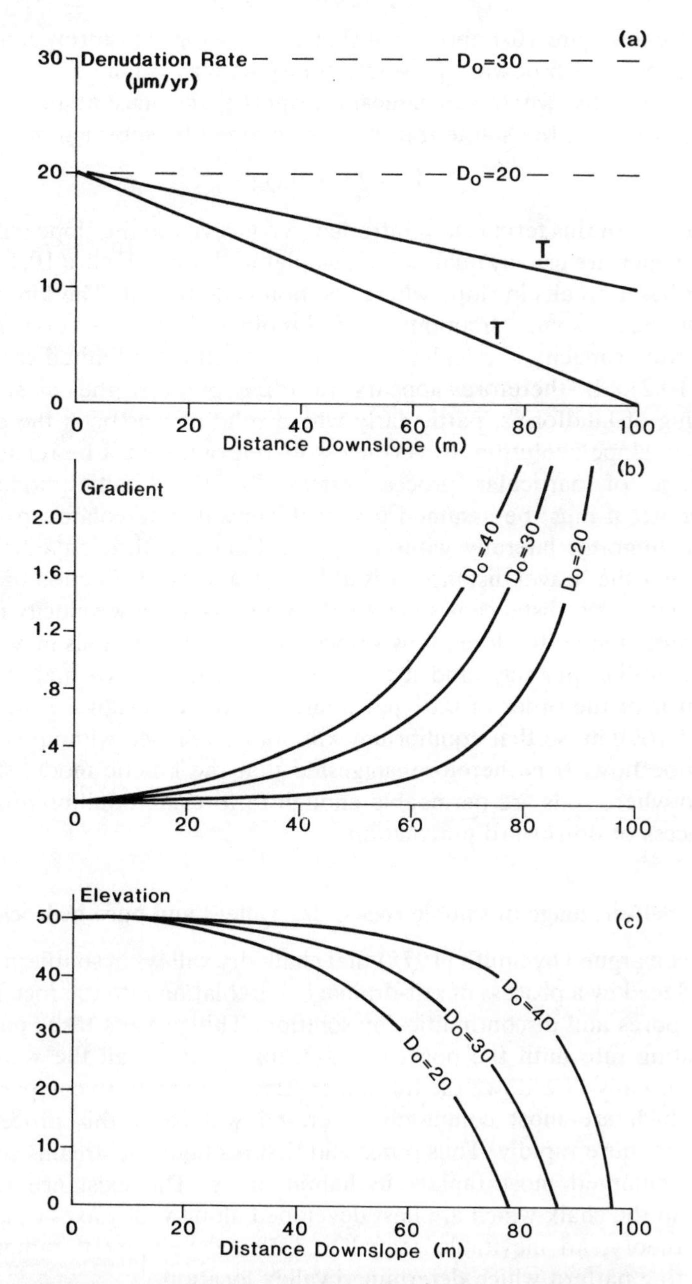

Figure 10.3. (a) Assumed denudation rates, (b) resulting gradients and (c) elevations for a kinetic solution model

than those in Figure 10.1 above, but they still rely for their formation on a soil cover, within which downslope solute transport may occur.

More generally, where mechanical transport takes place at a rate given by Equation (10.11), the solute transport is obtained by subtraction as:

$$V = Tx - f(x)g \qquad (10.14)$$

The effect of this term is to reintroduce a concavity at the slope base, giving profiles which are usually qualitatively similar to those of Figure 10.1, but with sharper basal breaks in slope where solution is dominant. The kinetic model also provides some smoothing of lithological breaks downslope, if transporting capacity is calculated as for the weathering-limited case ((iii) in Figure 10.2). It therefore appears to offer perhaps the most realistic modelling of landforms, particularly where solution has been the dominant process in slope evolution. Nevertheless its relevance must be related to our knowledge of particular process rates. For the kinetic model to be appropriate, it must be assumed first that rainwater percolates to a level at which it migrates laterally without approaching chemical equilibrium, and second that the 'travel distance' h is at least a significant fraction of the total slope length. This distance is given by the ratio u/k of flow velocity to kinetic uptake rate. On a 30° slope, flow velocities in permeable soils may be of the order of 400 m per day, and it has been suggested above that the kinetic constant is of the order of 0.01 per hour. The travel distance is then of the order of 1600 m, so that equilibrium will not be reached without substantial downslope flow. It is therefore suggested that the kinetic model should be adopted where soils are permeable enough to forestall equilibriation during the process of downward percolation.

10.2.4 Self-drainage in soluble rocks: dry valleys and bare bedrock slopes

It has been argued by Smith (1975) that chalk dry valleys in southern England are produced by a process of self-drainage. Percolation into the rock is able to enlarge pores and discontinuities by solution. This process takes place at an accelerating rate until the pores etc. are able to drain all the water which would normally flow down the hillside or stream channel. In comparing sites, those which are most commonly saturated will carry this process to its conclusion more rapidly. Thus pores and fissures beneath streams commonly become enlarged most rapidly in humid areas. The existance of fissure systems in the chalk which are best developed along valley axes is, according to this theory, an inevitable sequel to valley development rather than a pre-existing pattern which determined valley location.

In a humid environment, stream valleys commonly develop, even on limestones except where large open fissure systems pre-date drainage. In more arid areas, any percolation loss to groundwater reduces the already small surface runoff, so that complete underground diversion may occur at an earlier stage in landscape development, and surface valley networks may therefore be poorly integrated or completely absent. It is also possible, at least in principle, for self-drainage to occur at many points on the hillsides, as well as in valley bottoms. This stage will normally only be reached after valley axes are already draining underground. If self-drainage becomes widespread, then soils which previously covered hillslopes and acted as the medium for lateral water flow may be carried down through open fissures with percolating water. This process of self-drainage thus provides one mechanism, inherent in the continuing solution process, by which bare rock slopes may develop in limestones with a form which is broadly convex when viewed from a distance, but intricately dissected in detail. It is difficult to see how such a form might alternatively develop through mechanical stripping of the soil by overland flow once the connected surface is broken.

The effect of the self-drainage process may be illustrated by considering the flow along a tubular pore of radius a. For laminar flow under a hydraulic head at (sine) gradient s:

$$Q = \pi \rho g s a^4 / 8\eta \quad (10.15)$$

where ρ is the water density and η is its viscosity. If this flow is assumed to reach solute equilibrium, and enlarges the pore uniformly, over a total height drop of h (i.e. with a total tube length of h/s), then the pore radius at any time is given by:

$$1/a^2 = 1/a_0^2 - \rho g s^2 c_* t / 8h\eta \quad (10.6)$$

where a_0 is the initial pore radius (at $t = 0$). It may be seen from the form of this expression that the diameter increases slowly at first and then faster and faster, apparently reaching infinite dimensions at a time:

$$t_c = 8\eta h / (\rho g c_* a_0^2 s^2) \quad (10.17)$$

Because the final stages of enlargement are so rapid, this time is a useful estimate of the time needed for total drainage into the fissure. If reasonable values are substitued into Equation (10.17)—including a concentration $c_* = 100$ mg/1; $h = 100$ m fall; $s = 0.05$ (30°)—then for initial pore diameters, of 1–10 μm, the time required for indefinite enlargement is approximately 4×10^3 to 4×10^5 years. For an initial overall porosity of 1%, the initial percolation rate for pores of this diameter would be at 6–600 mm per year, so that realistic enlargement times may lie within this range of values.

This model is a great oversimplification of the process of pore enlargement, particularly since pores will tend to enlarge most rapidly at their upstream ends, giving funnel-shaped forms. Enlargement of the pores down-flow will be much slower than predicted above, increasing the time needed for effective self-drainage. Another complication is that, as a pore enlarges, flow will fill it less often because self-drainage is already partially effective, and times are therefore extended for completion of the process. Turbulence has also been ignored so that the model is only appropriate for small pores. Nonetheless the forecast of Equation (10.17) above is a useful guide to orders of magnitude in comparing different sites and lithologies. It is, for example, clear that initial pore size has the strongest influence on the process. The effect of climate and of different sites within a catchment may be seen by interpreting the time t_c as the period over which flow was able to enter the pore owing to saturation of its entrance. Thus valley-bottom sites will reach this critical time period more quickly than hillslope sites. Similarly self-drainage is likely to be completed more quickly in a humid than in a semi-arid climate, though valley development may by then be less complete. The influences of lithology (c_*) and available hydraulic routes (h and s) may also be seen.

It is thought that the self-drainage model may go some way to explain not only the development of limestone dry valleys in temperature areas, but also some of the features of cockpit karst. It is argued that where the time required for self-drainage is short compared with the time required for valley development, then valleys will not mature, but be interrupted by self-drainage leaving a surface without an integrated valley network. Self-drainage might also occur on hillslopes at some stage after the initial development of convex hilltops by solution (perhaps with other processes). In this way the original soil cover could be carried away by water percolating into the self-drainage fissures to leave a fretted bedrock surface which inherits a broad overall convexity.

10.3 SIMPLIFIED THERMODYNAMIC VIEW OF EQUILIBRIUM SOLUTION BY LEACHING WATER

Where solute equilibrium may reasonably be assumed, then knowledge of chemical equilibria for solutions may be applied to provide estimates for the concentration of solutes in soilwater for a given soil composition. Strictly this is a quasi-equilibrium (§10.1) since the soil solids are very gradually changing in composition as a result of the dissolution process, but the degree of approximation involved is negligible for the rather weak soil solution relative to other errors. Even where equilibrium cannot be assumed, then a knowledge of the equilibrium state is one necessary component of any kinetic model (e.g. Equation (10.1) above). It is unimportant whether equilibrium is achieved by direct contact between water and soil or with the assistance of ionic diffusion within peds (§10.1.1), provided that the end point is one of equilibrium. Although the equilibrium thermodynamics is sufficiently well known (Garrels and Christ, 1965; Berner, 1971) and parameterized to give a fairly exact value to the equilibrium concentration for any soil composition, a number of approximations are made here to simplify the problem. It is considered that only unimportant errors are generally introduced by these procedures.

The main approximating assumption made is to assume that the soil consists of a physical mixture of its constituent oxides, replacing its full mineralogy (Kirkby, 1976a). The free energy values used must then be adjusted somewhat from standard tabulated values (e.g. Robie and Waldbaum, 1968) to correct for the errors introduced by this assumption, but the corrections remain almost constant for a given weathering sequence of bedrock to the soil weathered from it. This approximation not only simplifies the calculations but, more important, avoids the necessity of recalculating the entire soil mineralogy at each stage: if required a 'normative' mineralogy can be produced separately for each stage of weathering. Other assumptions made below to simplify the procedure are:

1. the chemical 'activity' of each oxide in the solid soil is equal to the proportion of its molecules in the soil (the mole fraction);
2. the chemical activity of each solute is equal to its concentration in the water measured in moles per litre;
3. there is no interaction between solutes in the solution to influence their activities.

The first of these assumptions is most strictly true for soils with a single constituent, and the others at very low solute concentrations. None of these approximations is necessary to the procedures described, and may be abandoned if desired where unusually large errors might arise.

10.3.1 Approximating soils as mixtures of oxides

Expressions of thermodynamic equilibrium may be expressed in terms of the Gibbs free energy or solubility product, which depends on the reaction involved and the ambient temperature and pressure. The equilibrium reactions may be written in chemical balance equations in which the number of atoms and ionic charges are equal on both sides of the equation. The symbol $\rightleftharpoons$ indicates that the reaction might go in either direction. Thus for the solution of solid quartz (SiO_2) to form one of its products, the negatively charged ion $H_3SiO_4^{-aq}$, the balance equation is:

$$SiO_s^{=} + 2H_sO^{l} \rightleftharpoons H_3SiO_4^{-aq} + H^{+aq} \qquad (10.18)$$

where the subscripts indicate the number of atoms of an element in each molecule and the superscripts indicate the ionic charge (± and the number of free electrons if more than one) and the state; whether solid (s), in aqueous solution (aq) as an ion, as a liquid (l) or a gas (g).

The statement of solubilities is expressed in terms of the activities of each item in the balance equation above, which are indicated by writing square brackets round the molecular formula. Since activities are multiplicative, the equilibrium equation is usually expressed in terms of logarithms (to base 10). The activity of the hydrogen ion is usually re-expressed in terms of the pH (which is defined as minus the logarithm of the hydrogen ion, i.e. $pH = -\log[CH^{+aq}]$). The reaction above then gives, for equilibrium solution:

$$\log[H_3SiO_4{}^{-aq}] = \log[SiO_2{}^{0ws}] - 13.71 + pH \qquad (10.19)$$

It may be noted that a term does not appear in this expression for the activity of the water. This is because its activity, as the only liquid present, is equal to one, so that the logarithm is zero. Similar equations may be written for each of the ions commonly present in solution in significant quantities. Many oxides will have more than one ion, including silica which also has the uncharged ion $H_4SiO_4{}^{0aq}$. The degree of pH dependence will be directly related to the sign and value of the hydrogen-ion term (if any) in the balance equation. Thus, for example, there is no pH dependence for an uncharged ion. Combining the solubilities for all the ions associated with a particular oxide gives a total solubility curve for the oxide. Thus, in the case of silica the uncharged ion gives a constant solubility up to pH = 10, above which the negatively charged ion gives a rapid increase of solubility with pH.

Similar expressions can be written down for all of the major oxides found in soil. The thermodynamic background is described and many more examples given in standard texts on the subject (e.g. Garrels and Christ, 1965). Free-energy values for relevant minerals are tabulated in, for example, Berner (1971) and Robie and Waldbaum (1968). Calculations for correction factors to allow for the use of oxides in place of full mineral assemblages are illustrated in Kirkby (1976a). The units for solubilities in equilibrium equations are in moles per litre. To convert moles to grams the solubility for each ion must be multiplied by its molecular weight (e.g. for $H_3SiO_4^-$; 3×1 (atomic wt of hydrogen) + 28 (silicon) + 4×16 (for oxygen) = 95 grams to 1 mole). Most rock minerals (with the exception of some aluminium ions) provide cations (negatively charged) which must, to reach an overall balance of charges in the water, be offset be an equal number of positively charged anions. The original source of these anions is largely atmospheric, via carbon and nitrogen fixed by plants, fungi etc. in the soil. Similar equations must also be written for solute equilibrium with the soil atmosphere, essentially of nitrogen, oxygen and carbon dioxide, the last of which is produced by humus decomposition and plant respiration. Some complications arise in writing equations for iron and nitrogen, since these elements and their ions commonly appear with different valencies related not only to the pH but also to the oxidation–reduction potential (*Eh*) of the soil. These problems are not discussed further here but can be dealt with without losing the model structure presented.

The procedure described above gives a series of equations for the equilibrium solubility of each ion for a given soil composition (which provides the activities of the solid phase) and for a given composition of the soil gas (for the anion equations). The equations are, however, not uniquely soluble because many of them show dependence on pH, which is itself a property of the soil. The final equation which determines pH is obtained as a total ion balance. Thus the total number of negative charges is proportional to the sum of the activities of all singly-charged cations, plus twice the sum of the activities of all the doubly-charged cations, plus three times the sum of activities of triple-charged cations, and so on. This total must balance the sum of all the positive charges from the anion activities. Thus, for example, an appropriate equation might be:

$$\begin{aligned}&[H^+] + [Al(OH)_2^+] + 3[Al^{3+}] + [K^+] + 2[Ca^{++}] + 2[Mg^{++}] \\ &\quad = [HCO_3^-] + [OH^-] + \approx[CO_3^{--}] + [Al(OH)_4^-] + [CH_3SiO_4^-] \qquad (10.20)\end{aligned}$$

All of the terms in this equation may be re-expressed in terms of the known soil and gas compositions and the unknown pH using the individual equations of which Equation (10.19) is an example. This substitution then provides an equation in positive and negative powers of $[H^+]$ which at first site appears formidable, but which is well-behaved, with a single solution which an iterative or graphical solution procedure will readily find. The value of pH is thus obtained, which is not only a summary variable of interest for itself but can also be substituted back into the individual equilibrium equations to give the total solute concentration and its ionic composition. A worked example of this procedure is given as an appendix in Kirkby (1976a).

The procedure described in this section is seen to be routine and unambiguous, even if complex in details and relying on a body of chemical knowledge which may be unfamiliar to some readers. It is thus well suited to computational methods which may additionally be designed to eliminate some of the simplifying assumptions of this presentation. The overall effect of this model is to show that the composition of the soil and its gas uniquely determines the equilibrium solute composition and concentration. It will be seen below that the gas composition is largely influenced by the processes of plant growth and decay, while the soil composition is slowly being influenced by the loss of solutes to the water. The latter process is taken further in the next section in terms of the soil evolution at a point as weathering proceeds, on the assumption that water remains in contact with the soil long enough to equilibriate. The additional complications of a kinetic approach, together with the lesser knowledge of kinetic constants, provides a strong encouragement to the modeller to work within the equilibrium assumption wherever possible! It has been argued above that this assumption is commonly a reasonable one, but caution should be exercised in applying an equilibrium model (or any other!) in inappropriate circumstances.

10.3.2 Evolution of soil at a site as leaching proceeds

In a non-spatial context the rate of loss of each soil oxide as leaching water passes through it is equal to the solute concentration (assumed as at equilibrium) multiplied by the total rate of flow of clean water (expressed as litres of water passing through each milligram of soil in unit time). At the end of a brief time increment, the total loss for each soil oxide can be subtracted from the total to give a new soil composition with slightly changed proportions. New solute concentrations may be calculated and the process repeated. Thus the process of solute equilibriation produces a progressive and gradual evolution of soil composition over time. At any stage, the total proportion of soil remaining from an original weight of parent material can be followed to give a useful index of the overall course of weathering. This may be compared with a by-volume analysis of field soils in which a similar proportion is calculated on the assumption that some soil constituent (usually Al_2O_3 or TiO_2) is completely insoluble (so that the proportion obtained is likely to be a slight overestimate).

The resulting evolution may be expressed in a number of ways. One useful way to present the results is as a triangular diagram, showing the change in composition over time (expressed in litres per mg) in proportions of bases, silica and sesqui-oxides (of iron and aluminium). An example of evolution for fixed soil gas environments and different parent materials is shown in Figure 10.4. The soil gas is represented by the partial pressures (equal to thermodynamic activities at atmospheric pressure) of CO_2, which commonly takes values over the range 3×10^{-5} atmospheres in free air to a maximum of about 0.1 atmospheres in some soil air for environments of active organic breakdown. The values and courses shown broadly match those of real soils, although with rather higher base solubilities than are effective in reality. The plotted points indicate 'time' spans of 0.04 l/mg of water. To turn this value into years, the rate of percolation must be included. Thus 500 mm of rainfall passing through a 1 m depth of soil each year is equivalent to 5×10^{-7} l/mg per year: the plotted points at intervals of 0.04 l/mg are then equivalent to time intervals of about 80,000 years. This type of diagram illustrates the relative change in soil composition over time, together with comparative rates of weathering. For example, it is apparent that the equilibrium model adopted shows the high rate of weathering of carbonate rocks in comparison with igneous rocks in the presence of organic matter (0.1 atmosphere partial pressure), and the much smaller differential where organic matter is absent, as in arid or arctic regions.

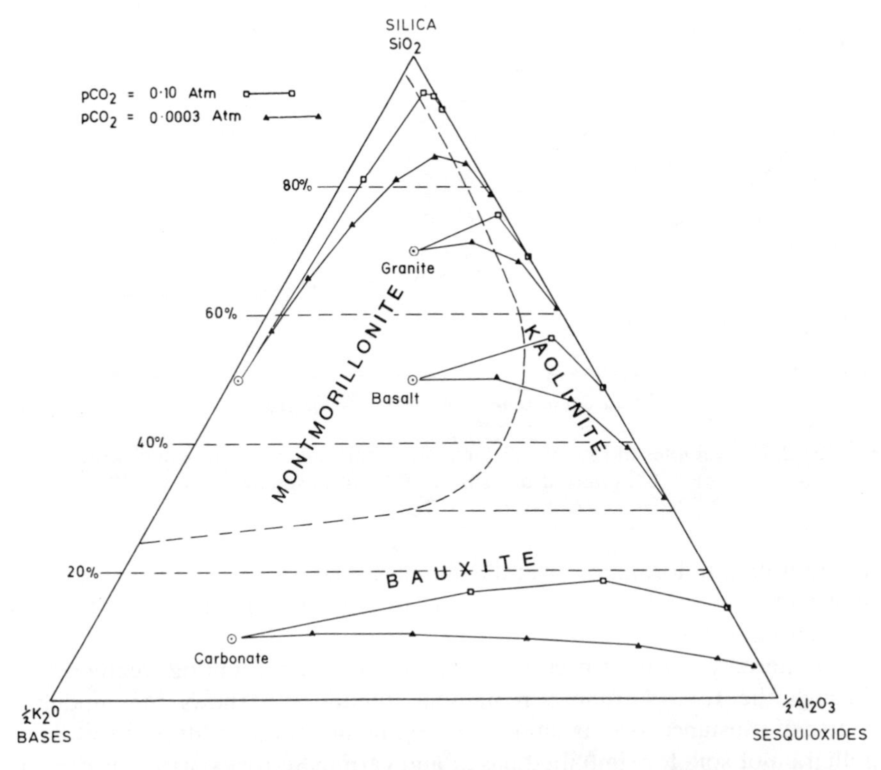

Figure 10.4. The course of weathering for simple soils. Plotted points show composition at intervals of 0.04 litres of water per mg of soil. The broken lines roughly define the regions of dominant clay mineral production

The fields assigned to the clay minerals montmorillonite, kaolinite and bauxite are obtained by writing approximate equilibrium relationships between them. Where the course of weathering passes through each field it indicates the predominant clay mineral type which will be formed. Thus basalts are indicated as producing montmorillonites in the early stages of weathering and subsequently kaolinite and then bauxites. This sequence compares well with the differences in Hawaiian soils reported by Sherman (1952) and related to climate. The interpretation given here is that although the soils are of similar ages, the different rainfall regimes produce different degrees of weathering when measured in litres of water per milligram of soil. In other words the montmorillonite-rich soils of the more arid areas are interpreted as showing a lesser degree of weathering rather than primarily a different course of weathering. (In fact the suggestion of Figure 10.4 is that the course of arid weathering (low CO_2) tends to be slightly less favourable to

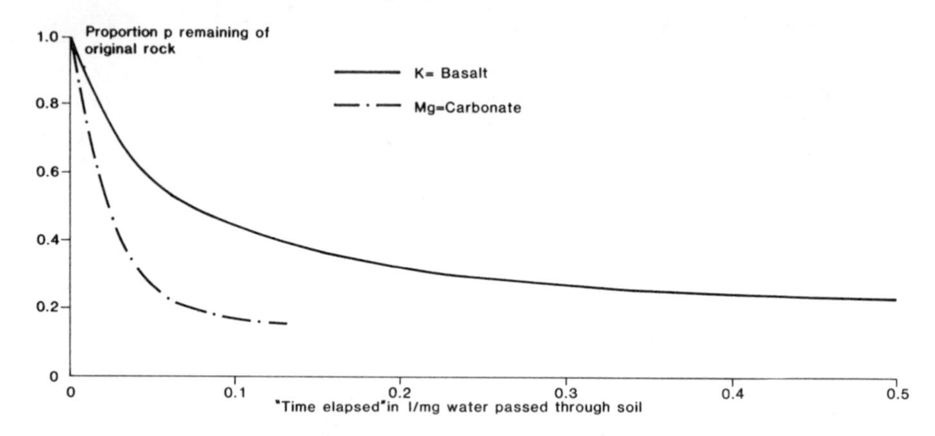

Figure 10.5. Examples of loss of substance for a carbonate and an igneous rock for a CO_2 partial pressure of 0.03 atmospheres

montmorillonite formation overall.) Similarly the predominance of kaolinite in granite weathering is readily seen throughout the major part of its weathering history.

An alternative way of presenting changes in the soil during weathering is through the total substance remaining. Figure 10.5 shows the modelled course of substance loss, again on a time scale measured in litres of water per milligram of soil, for simplified basalt and carbonate rocks, with low organic activity levels. The insoluble residue from the carbonate rock is shown to reach low levels (<0.2) in periods of 200,000 years (for the percolation rates assumed above), while the basalt still retains almost one half of its original substance. Either of these formulations may, as has been noted above, be replaced by a full composition in terms of oxides, but these summary curves are thought to give a clearer view of the overall progress of weathering. Comparison between these curves and the behaviour of real weathering profiles shows good general agreement, which can be made excellent for particular cases by selection of appropriately modified free-energy values (to allow for the mineralogy as noted above). In other words the observed course of weathering lends support to the assumption made at the outset; namely that solute equilibrium could be assumed without gross error.

10.3.3 Approximate linear solution model

Figure 10.6 shows the equilibrium solute concentration for sample parent materials in terms of p, the proportion of substance remaining in the solid soil. Expressing the process of weathering in this form has the advantage that the time element is not explicitly included, as it was in Figures 10.4 and 10.5. The

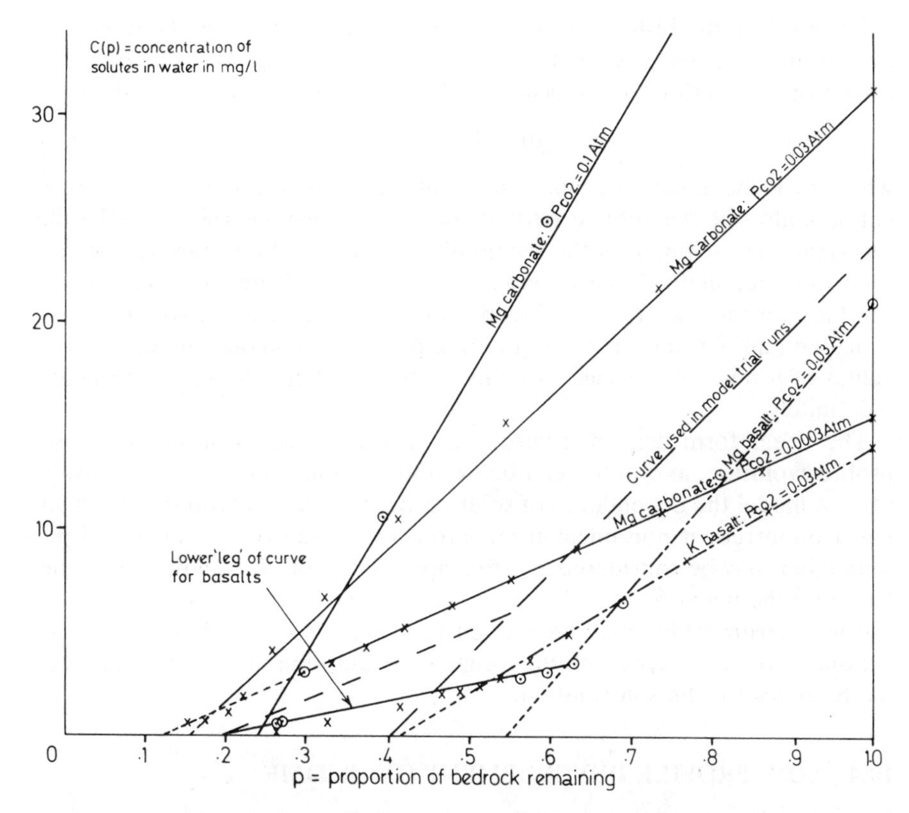

Figure 10.6. Relationship between total solute concentration $c(p)$ and total proportion remaining p for a number of simulated weathering sequences

form shown by the relationship, for a variety of parent materials of simple form, consists of a series of essentially straight-line segments. Given the complex interplay of soil components with changing pH as weathering proceeds, this simplicity is remarkable. A straight-line segment can be readily compared with the behaviour of an idealized simple material, a proportion, initially q_0 of which is simply soluble and the remainder (initially $1 - q_0$) totally insoluble. At some subsequent stage when the total substance remaining is p, the soluble part is reduced to $p - (1 - q_0)$, and the solute concentration is linearly reduced in proportion if PCO_2 is unchanged. Comparing this behaviour with the more complex actual soil model, the intercept of each straight-line segment with the horizontal axis gives an apparent 'insoluble residue' and the slope of the segment an apparent average solubility for the remainder. The points at which the course of weathering changes from one segment to the next appear to be related to the complete removal of successive components of the soil.

For much of the history of soil development by weathering, the right-hand straight-line segment is all that is needed for adequate forecasting, so that the course of weathering can be described by the single linear relationship:

$$c(p) = k_*(p - p_*) \tag{10.21}$$

where p_* is the apparent proportion of insoluble residue and k_* is the apparent solubility of the soluble part. It has been shown (Kirkby, 1985b) that this right-hand segment of the relationship between solute concentration and proportion remaining has a second significance, in that the apparent insoluble residue p_* is the lowest value of proportion remaining near the soil surface for which an equilibrium for the weathering profile as a whole can exist. Lower values of p near the surface can only be found where the soil is thickening indefinitely.

The linear form described here is a great convenience in modelling soil profile evolution, as will be seen below. Other forms, may, however, also be used to model the dependence of solute concentration on proportion remaining if preferred: if non-linear forms are chosen then the linear model presented her may be considered as a first approximation. It is, however, argued here that the linear form in fact gives a good approximation to the course of soil weathering which may be applied to provide soil profile models which continue to fully represent the original assumption of equilibrium thermodynamics for the soil solution.

10.4 SOIL PROFILE DEVELOPMENT OVER TIME

In this section development of the soil profile is considered in some detail (Kirkby, 1985b), using the assumption of equilibrium for the soil solution, in the form summarized by Equation (10.21), which expresses total solute concentration in terms of the 'proportion remaining' p in the soil relative to its parent material. Individual ions are not separately modelled, although it has been argued above that chemical composition may be reconstructed from the current value of p and for a given parent material. The model described here is disaggregated into three components which are largely independent and which operate on very different timescales: a weathering component, an inorganic nutrient-cycling component, and an organic soil component. The total composition of the soil is then seen as an additive combination of these three.

Several processes are incorporated in the model, while some others have, at this stage, been neglected. The most important process of weathering is thought to be leaching by percolating water, which is modelled as reaching equilibrium with the soil solids at a concentration given by Equation (10.21). Weathering may also proceed, however, in the absence of water movement, as has already been discussed above (§10.1.1) in the context of the removal of solutes from the interior of individual peds. Ionic diffusion is thus seen as a second significant weathering process, which is particularly important at the base of the soil where parent material is impermeable. In the upper part of the soil profile mechanical mixing of the soil occurs, primarily through the action of soil meso-fauna such as earthworms or termites. This mixing acts as a second diffusion process for the transfer up or down the profile of all soil components. It is thus a very important influence on the weathering profile, but is equally important for nutrient cycling and for organic soil development. Nutrient uptake and carbon fixation by plants are significant processes for the inorganic nutrient-cycling profile and the organic profile respectively, and are of course closely linked by plant metabolic requirements. Leaf fall (including root decay etc.) is the source of material for organic soil; and decomposition releases inorganic minerals and carbon to complete the inorganic and organic nutrient cycles. These processes may all be explicitly modelled, and examples of plausible relationships are outlined below (§10.4.1).

Processes which have not been incorporated into the model outlined here include complexing and lessivage, so that serious errors may arise where these processes are significant, as perhaps in the formation of podsols. Both processes have the effect of mobilizing aluminium and iron at the expense of silica. In an extreme case, the curves shown in Figure 10.4 above might converge on the silica rather than the sesqui-oxide corner of the diagram instead of eventually veering away from it as modelled for even the most acidic parent materials. In its simplest form a complex is an additional ion produced from the combination of two existing simple ions. There may thus be $MgCO_3^{0aq}$ complex ions in addition to the simple Mg^{++aq} and CO_3^{--aq} ions in solution. The complex ions increase the solubility of Mg in this example. The increase in solubility is greatest where the simple ions have high ionic charges, particulary for Al^{3+} and Fe^{3+} in soils, and the most soluble complexes appear to be those with organic anions. As for equilibrium concentrations of other ions, the concentration of complex ions is greatest where their constituents are abundant, that is where pH is low (giving high concentrations of Al^{3+} and Fe^{3+}) and organic content is high. Ideal conditions are therefore found in acid 'mor' humus, for example. The complex ions are commonly redeposited further down the soil profile where the chemical environment is less favourable, giving the hardpans characteristic of podsol soils.

Water percolating through the soil is able to carry some suspended material (lessivage). Appreciable amounts are only carried where pores are large relative to the particles carried ($>10\times$). Clay particles can thus be carried through the textural voids of sands and through cracks and other structural voids. One result of this process is the deposition of clay skins as a lining to structural voids. Another result is that clays are washed out from very sandy horizons, making them into even purer sands. This process removes aluminium (in the clays) at the expense of silica (in the sands). It thus to some extent parallels the operation of organic complex removal, in removing aluminium preferentially. Both of these processes partially reverse the normal direction of weathering by mobilizing aluminium and carrying it back down the soil profile. It has not been found possible to model these processes adequately to date, so that caution should be used in comparing the model described with soils where substantial cheluviation or lessivage are thought to occur.

10.4.1 Modelling some major soil processes

In order to model the process of leaching, some assumptions must be made about the percolation process. The simplest possible assumption is that the net rainfall percolates straight through the soil and down into the parent material. While this may be correct in some cases, a more general procedure is to assume that some flow is diverted laterally at various depths. In this way leaching may occur from the upper soil horizons even above a totally impermeable parent material. For the simplest case of a steadily convex slope profile with a soil transmissibility $T(z)$ integrated over depths greater than z, the maximum flow which can be carried below depth z is $gT(z)$, where g is the hydraulic gradient which will be assumed equal to the slope gradient. To support a percolation into the soil at depth z of $F(z)$, spread over the whole slope above:

$$F(z) = F_0 + T(z)g/x \tag{10.22}$$

where f_0 is the percolation rate into the parent material and x is the distance from the divide. For a convex slope profile the ratio g/x is constant, and the transmissibility may rationally be related to the total soil deficit below depth z, defined as:

$$w(z) = \int_z^\infty (1 - p)\mathrm{d}z \tag{10.23}$$

In this expression the deficit w has the dimension of depth and is, in effect, a generalization of soil depth which makes no assumption about the exact base of the soil. A simple relationship to use is the linear one, so that:

$$F(z) = F_0 + w(z)/t_0 \tag{10.24}$$

where t_0 is a constant with the dimensions of time.

The actual percolation modelled at depth z is then the value derived from Equation (10.23) or the total net rainfall, whichever is less. This approach normally takes no account of seasonal or storm variations in percolation rate because it is thought that the weathering process is too slow to respond to such short-term changes. They may, however, be incorporated in the form of a distribution of net rainfalls.

At any depth the quantity of solutes carried downward in percolating water at depth z is $F(z)c(p)$. Taking account of diversions as sub-surface flow, represented by the down-profile reduction in $F(z)$, the loss of substance produced by leaching may be expressed as:

$$dp/dt = -F(z)dc(p)/dz \tag{10.25}$$

For the weathering process, $c(p)$ may be substituted from Equation (10.21) for the linear case. Equation (10.25) is also valid, however, for inorganic nutrient cycling. In this case the concentration $c(p)$ should be interpreted as the marginal concentration and suitably weighted for the chemical composition of the plant nutrient stream. The use of marginal concentrations is correct in the context of a nutrient profile which is added to the weathering profile. For the model runs presented, marginal solute concentration has been taken as directly proportional to the marginal soil solid concentration of inorganic nutrients. For the organic soil, it has been assumed that leaching is negligible in importance compared to other processes.

Ionic diffusion cannot on its own export solutes from the profile but may be significant in redistributing them. For a diffusivity constant D_I the rate of downward transport is given by:

$$-D_I dc(p)/dz \tag{10.26}$$

For the weathering profile solute concentration is again substituted from Equation (10.21). The diffusivity D_I varies only slightly for different ions, and also with the soil porosity, but a constant value of 0.01 m^2 per year has been adopted here. For the inorganic nutrient profile and for the organic profile, ionic diffusion is generally negligible as it is dwarfed by faunal mixing.

The intensity of mixing by soil meso-fauna is related to the base status of the soil, which indirectly provides nutrition for its inhabitants. Adequate soil aeration is also necessary, so that activity generally declines with depth and is negligible in waterlogged soils. Mechanical mixing transports material from zones of high to zones of low concentration, and so acts as a diffusion process with a transport rate:

$$-D_O dp/dz \tag{10.27}$$

where D_O is the appropriate diffusivity. Estimates from a variety of direct and indirect sources suggest that diffusivities range from about 10^{-4} to 10^{-2} m^2 per year, the extremes corresponding roughly with mor and mull humus conditions. These rates are similar to or lower than ionic diffusivities, but mixing acts on the whole body of the soil, whereas ionic diffusion acts only on the solute concentrations and therefore acts at a rate which is at least 100 times slower. Organic mixing declines with depth in the soil, and for present purposes has been assumed to fall off exponentially so that mixing at depth z

is given as:

$$D_O \exp(-z/z_2) \quad (10.28)$$

for a characteristic depth z_2 which is thought to be about 0.5 m.

The process of nutrient uptake may be modelled via the transpiration stream together with an assumed constant concentration of nutrients in it. A vegetation cover is established and reaches some sort of equilibrium in a time which is short compared with that required for soil development. The process of evapotranspiration is therefore modelled as having a fixed distribution with depth, related to the equilibrium distribution of plant roots with depth. The concentration of nutrients has been assumed to be constant, controlled by the vegetation cover alone, even though this a somewhat simplified view of plant physiology. Root densities decline with depth, and an exponential decay of evapotranspiration with depth has been adopted for modelling purposes, with a scale depth z_0.

Leaf fall may again be considered to reach equilibrium quickly enough for transient states to be ignored over the time spans required to soil development. It may therefore be taken as a constant rate of input to the organic soil. Decomposition of organic soil material in fact occurs at a wide variety of rates depending on the chemical composition of the plant residues. An average rate has been adopted here for modelling purposes, but its use underestimates the penetration of organic matter into the soil because there is a consistent decline in decomposition rate with depth (as more resistant components have time to penetrate farther). An annual average proportion of 20% of the organic soil decomposing each year, and releasing its inorganic nutrients, is though appropriate for an area with a mean annual temperature of 10°C, and temperature is thought to be the primary control where soil aeration is adequate.

The final process which needs to be considered for the development of a soil profile is mechanical denudation of the soil surface by erosional processes. There is in fact some interchange of material downslope, but the only simple way of incorporating erosion into an individual soil profile is as a net denudation. Rates are normally negligible for the time spans required for the development of organic and nutrient cycling profiles, but may be highly significant for the development of weathering profiles. In the model structure used here, changes in bulk density are not explicitly modelled, and the soil is, conceptually at any rate, considered to occupy the same volume as the parent material from which it was derived. This provides a simple coordinate framework for modelling and identifying changes in a particular piece of soil over time. A correction must, however, be used to convert the denudation rate to an equivalent depth of soil lost, since the surface soil is reduced to a proportional substance p_s. If the rate of denudation is T, the appropriate loss of soil depth is therefore T/p_s. All of these processes have been incorporated into a family of soil profile models, within the mass-balance framework described below.

10.4.2 Mass balance and weathering equations for the soil profile

Any net loss of soil material must be associated with an increase in the rate of solute transport to remove the lost material. Thus for the weathering profile:

$$\partial p/\partial t = -\partial V/\partial z \tag{10.29}$$

where V is the total rate of downward transport of soil materials due to all relevant processes combined. Similar mass-balance equations apply to the nutrient cycling and organic soil profiles. Neglecting processes which are thought to be unimportant in each case, the three profiles are then governed by equations obtained by substituting for V as:

$$\partial p/\partial t = (\partial/\partial z)[D_{\mathrm{I}}\partial c(p)/\partial z + D_{\mathrm{O}}(z)\partial p/\partial z] - F(z)\partial c(p)/\partial z + (T/p_{\mathrm{s}})\partial p/\partial z \tag{10.30}$$

$$\partial q/\partial t = (\partial/\partial z)[D_{\mathrm{O}}(z)\partial q/\partial z] - F(z)k\,\partial q/\partial z + \beta r - (mE/z_0)\exp(-z/z_0) \tag{10.31}$$

$$\partial r/\mathrm{d}t = (\partial/\partial z)[D_{\mathrm{O}}(z)\mathrm{d}r/\mathrm{d}z] - \beta r \tag{10.32}$$

where q is the marginal quantity of inorganic nutrients cycling, k is their marginal solubility in the soil water, β is the proportional rate of decomposition, r is the quantity of organic soil at depth z, m is the rate of nutrient uptake by the vegetation, and E is the annual rate of actual evapotranspiration.

In Equation (10.30) the use of the linear form of Equation (10.21) may be seen to simplify the form of the equation after substitution, although numerical solutions may also be obtained with other expressions for solute concentration. The inclusion of the final term in Equation (10.30) represents a selection of coordinate axes in which depth is measured from the current soil surface, as this is continually modified by mechanical denudation.

These equations may be solved numerically using a computer in all cases. Analytical solutions may be found in only the very simplest cases, usually including equilibrium cases. To obtain a solution additional statements must be made about behaviour at the boundaries of the soil profile. Thus at great depth p tends to 1.0 (parent material) while q and r tend to zero. At the soil surface there is no transfer of material into the profile, except for the organic profile which receives leaf fall at a specified rate. The implications of these boundary conditions and some solutions are discussed below for the three profiles separately.

10.4.3 Some simple cases of equilibrium weathering profiles

An equilibrium profile is one which is no longer changing over time. Thus dp/dt is zero at all depths and p_s is constant. If $F(z)$ and T are also assumed constant over depth and time respectively, then Equation (10.30) may be integrated to give:

$$p = 1 - \{Fk_*(1 - p_*)/[T + Fk_*(1 - p_*)]\} \exp(-Az) \\ \{[D_I k_* + D_O \exp(-z/z_2)]/(D_I k_* + D_O)\}^{-Az2} \qquad (10.34)$$

where $A = (T - Fk_* P_*)/(D_I k_*)$. In this expression, the linear equilibrium solution form (Equation (10.21)) has been substituted, and organic mixing is assumed to decline exponentially with depth. It may be seen that constants have been evaluated to meet the boundary conditions of $p = 1$ at depth and that the surface proportion at equilibrium is given by:

$$p_s = T/[T + Fk_*(1 - p_*) \qquad (10.35)$$

which is equal to the proportion of mechanical to total (mechanical plus chemical) denudation. Examples of this distribution are shown in Figure 10.7. Where there is any organic mixing, it can be seen to dominate the upper part of the profile; giving a zone of almost constant composition equal to the surface composition. Organic mixing rates fall to those for ionic mixing at depth:

$$z = z_2 \ln(D_O/D_I) \qquad (10.36)$$

Below this level there is a rather sharp weathering front in which the proportion p increases rapidly towards its bedrock value of 1.0. Equilibrium profiles of this sort only occur where the appropriate value of p_s is greater than the apparent insoluble residue p_*. Otherwise the weathering profile does not reach an equilibrium but instead thickens indefinitely. It should be noted that these profiles are emphatically not deep-weathering profiles: the depth of the constant composition zone is relatively limited and does not thicken with time once equilibrium has been reached. It should also be noted that the constant composition zone is not a saprolite but instead a zone of substantial physical mixing by the soil fauna.

An equilibrium weathering profile can only be formed in the presence of appreciable amounts of mechanical denudation, comparable to or greater than the solutional loss. Equilibrium profiles are thus associated with sloping rather than with level sites. It will be seen below that the time required to approach equilibrium is least where mechanical denudation is highest, so that the typical equilibrium profile is more likely to be an immature mountain soil rather than a deep lowland soil, especially where post-glacial time spans are all that are available for soil development.

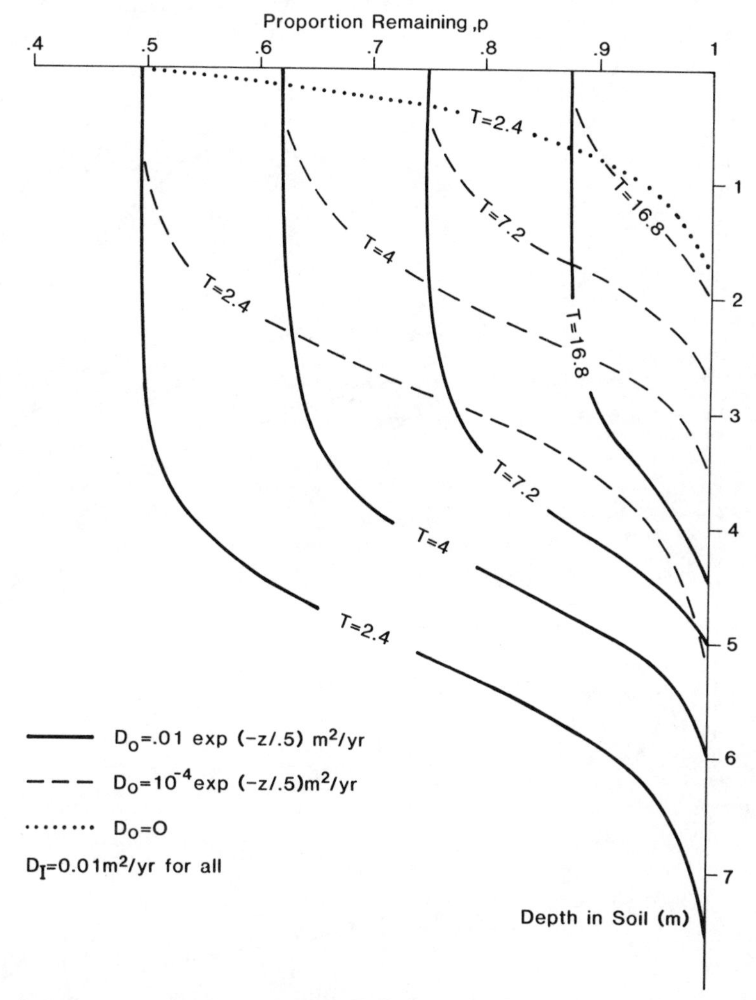

Figure 10.7. Examples of equilibrium weathering profiles for a range of values for organic mixing and mechanical denudation. $F = 0.01$ m per year; $p_* = 0.04$; $k_* = 4 \times 10^{-5}$

10.4.4 Evolutionary soil weathering profiles

Numerical solutions, usually via computer programs, are appropriate for all cases involving the evolution of weathering profiles over time. They are also necessary to incorporate variations in percolation rate with depth, for example. Figure 10.8 illustrates several sequences of profile evolution over time, including a falling percolation rate; and for a number of values of mechanical denudation. It may be seen that for long times the general form of

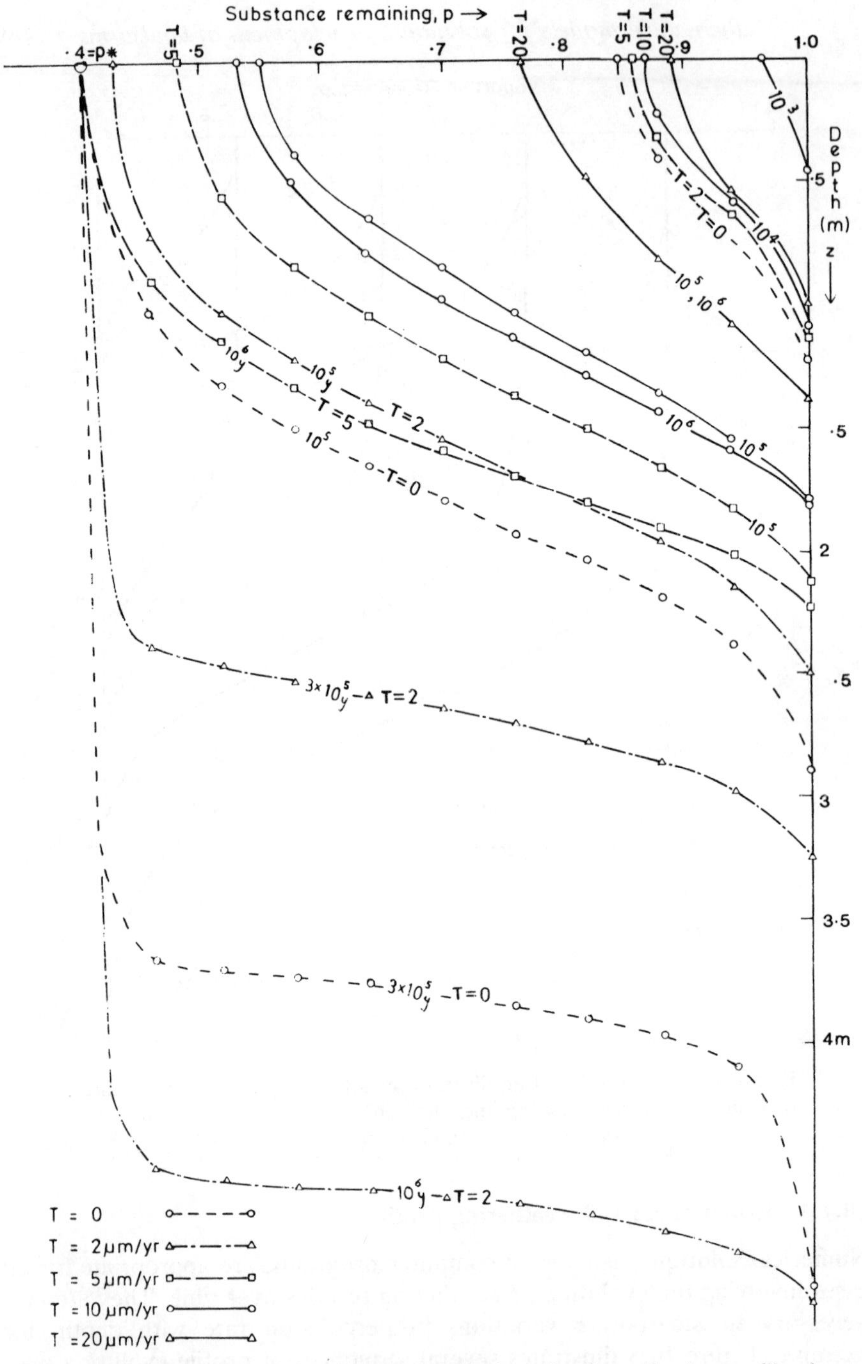

Figure 10.8. Examples of the simulated evolution of soil weathering profiles for a range of rates of mechanical denudation T. $D_0 = 10^{-4}$ m^2/per year; other values as in Figure 10.7

these profiles is very similar to those shown in Figure 10.7 (allowing for the varying percolation rate) at equilibrium. In general it may be seen that the zone of constant composition is a little less marked, with variations in the leaching rate controlling the final removal of solutes and hence influencing the form of the profile.

For low rates of mechanical stripping (less than about 4 mm per 1000 years for the example shown) equilibrium would never be achieved. Instead the soil tends to thicken indefinitely, developing a zone of very constant compostion at proportions p close to p_*. This zone extends considerably below the influence of organic mixing (about 2 m) and is a zone of saprolite or volumetric weathering where the skeleton of the parent material is undisturbed by mixing. As time goes on this zone gets progressively deeper without limit. This overall pattern is summarized more succinctly in §10.5.2.

An important point to note in examining the evolution of the weathering profile in Figure 10.8 or from other points of view, is that there is only a very slight crossover between the curves obtained for different geomorphic environments (defined by T). It follows that the area to the right of the curve, defined above as the total soil 'deficit' at $z = 0$ by Equation (10.23), is related to the surface proportion p_s at any stage with only a small degree of ambiguity. In other words, for a given climatic and lithological environment, it seems reasonable to establish a kind of rating curve connecting soil deficit w to p_s, and this rating curve may be used as a still further simplified means of modelling the variation of weathering profiles downslope and its evolution over time, while at the same time retaining a link to the equilibrium concept, to the soil composition and to the form of each individual soil profile. This approach is pursued in §10.5.

10.4.5 Organic soil profiles

For soil development purposes, the most relevant organic soil profile is that for equilibrium, and other solutions are ignored here. For the simple case where D_O is constant, then equation (10.32) may be solved immediately to give:

$$r = (mEz_1/D_O)\exp(-z/z_1) \qquad (10.37)$$

where mE is the total leaf fall, which is, in equilibrium, equal to total uptake, and z_1 is a scale depth for organic matter, equal to $\sqrt{(D_O/b)}$. Equation (10.37) describes an exponential decay of organic matter content with depth. For the values quoted above ($D_O = 10^{-4}$–10^{-2} m^2 per year and $b = 0.2$ per year) the scale depths for this decay lie between 2 and 20 centimetres. Comparison of these depths with organic soil profiles is, in fact, the simplest way of estimating an appropriate value for diffusivity of mixing D_O. Where the rate of organic mixing declines with depth, the organic content at the surface is the same, but the decline with depth is more rapid.

10.4.6 Nutrient cycling profiles and 'B' horizon formation

For equilibrium, Equation (10.31) above shows that if F and D_O are taken as constant the nutrient cycling profile takes the form:

$$q = mE/(Fk + D_O/z_0)\exp(-z/z_0) - mE/(Fk + D_O/z_1)\exp(-z/z_1) \quad (10.38)$$

where the terms are as previously defined. This difference between two exponential decay terms has the same sign as $(z_0 - z_1)$: that is it is all-positive if the roots extract water from beneath the organic soil, and all-negative if the roots only extract water from within the organic soil. The all-positive case appears to correspond to the typical humid soil, in which it is a good strategy for plants to intercept nutrients leached from the roots. The all-negative case arguably relates to arid areas and base-rich rocks where plants may grow shallow roots to minimize their uptake of bases and encourage leaching losses. The curves rise to a maximum (if positive) before declining gradually to zero at depth. If, as is usual, the organic mixing term D_O is much larger than the leaching term Fk, then the maximum is very close to the surface. The decay with depth is roughly scaled by the larger of z_0 or z_1. Thus in the 'humid' case the final inorganic profile associated with nutrient cycling shows a base-rich zone extending to the bottom of the root zone; and the 'arid' case shows a corresponding base-deficient zone.

In simulations of evolution over time, it is apparent that the times required to reach the equilibrium nutrient cycling profile are much longer than for the organic soil, commonly requiring at least several hundred years, and in some cases several thousand. It may readily be seen that the nature of the transient nutrient profiles is completely different for the humid and arid cases. In both cases the initial removal of nutrients to supply a living biomass cover and an organic soil leads to a depletion of the inorganic nutrients. This depletion builds up until vegetation and organic soil reach equilibrium. Subsequently this depletion zone is carried downwards into the soil and dispersed, as the nutrient equilibrium is approached, from the surface downwards. For the 'humid' case this gives a complete reversal from initial depletion to accumulation at equilibrium; whereas for the 'arid' case the initial depletion merely speeds the attainment of the final equilibrium, also of depletion.

Figure 10.9 illustrates two examples of simulations based on Equation (10.31) for the 'humid' case. In (a) the conditions roughly match that of a mull humus with high rates of organic mixing penetrating down to a scale depth of 0.5 m; with roots also penetrating to 0.5 m and the organic soil to about 0.2 m. In this case the deficit zone builds up and becomes more marked and deeper for about 1000 years before declining in intensity. In

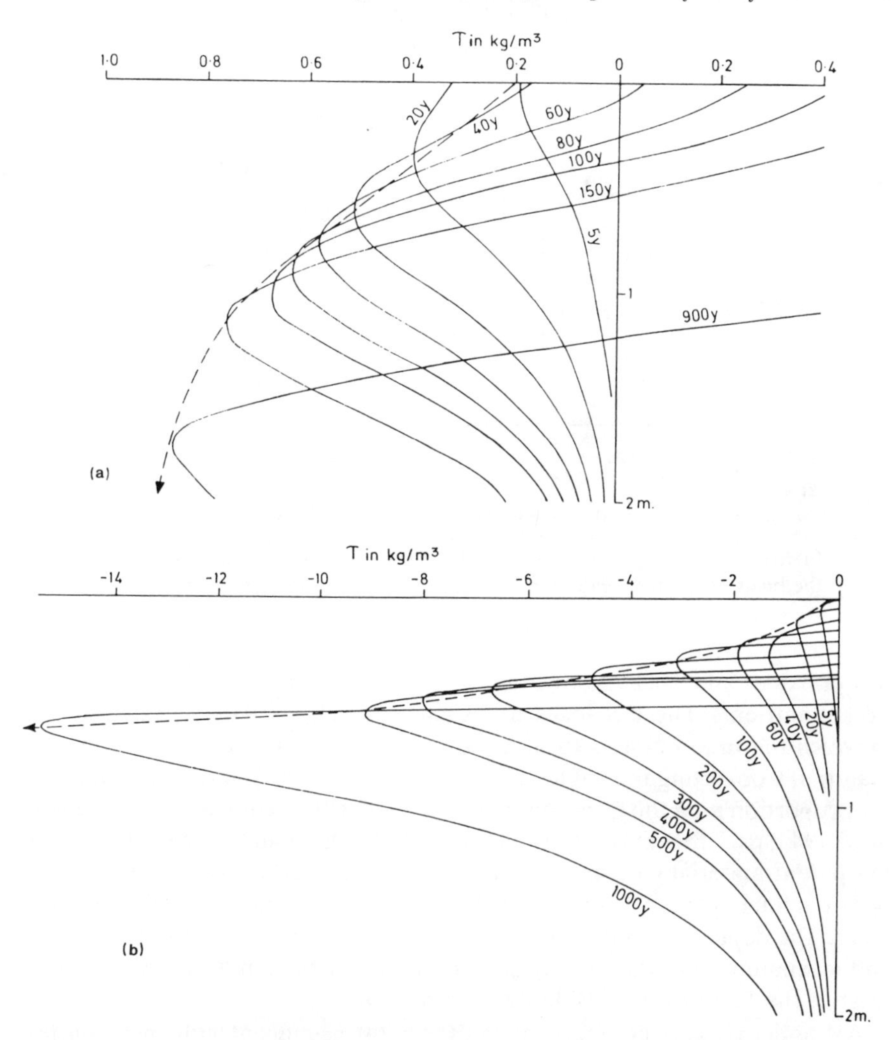

Figure 10.9. Examples of the simulated evolution of nutrient-cycling inorganic soil profiles showing 'B' horizon features: (a) $D_O = 0.01 \exp(-z/0.5)$; (b) $D_O = 0.0001 \exp(-z/0.2)$

Figure 10.9(b) conditions are like those of a mor humus, with lower intensities and depth (0.2 m) of organic mixing, organic soil only a few centimeters deep and roots still extending to 0.5 m. In this case the organic mixing process is too shallow to disperse the deficit zone, which therefore continues to build up for several thousand years.

The deficit zone shows depletion of nutrients used by plants, so that it shows strong depletion of some bases and only very week depletion of

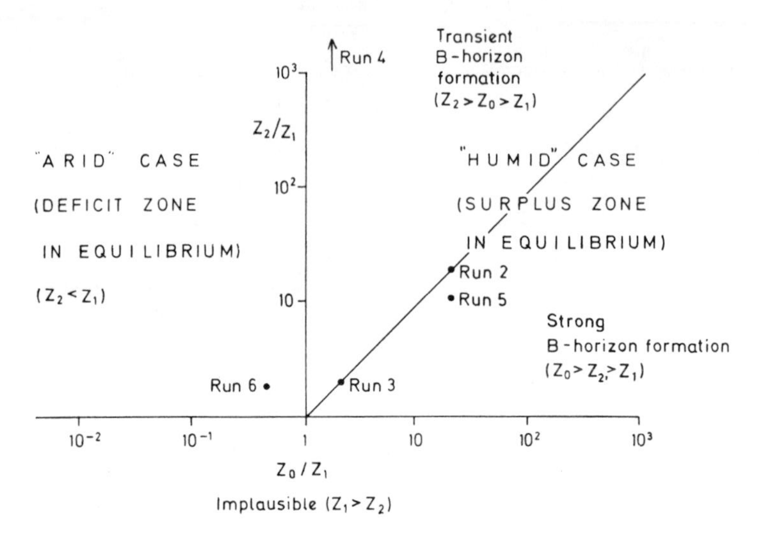

Figure 10.10. Suggested basis for classifying soil 'B' horizon types on the basis of relative scale depths: z_0 for rooting; z_1 for organic soil; z_2 for organic mixing

sesqui-oxides and silica. Above the depletion zone there is strong accumulation of nutrients. These changes are usually superimposed on a weathered soil in which the ratio of bases to sesqui-oxides is well below that for the parent material. Converting the values shown in Figure 10.9(b) to the same units as for proportion p on the basis that bedrock ($p = 1.0$) is equivalent to a density of 2500 kg m^{-3}, the ratios of sesqui-oxides to bases could rise from, say, 1.0 in the parent material to 7 in the weathered soil and 10 in the depletion zone before falling again to 1.4 in the surface zone of organic and nutrient enrichment. The depletion zone is therefore interpreted as a zone of low base status and maximum proportional clay content. It is thus thought to exhibit some at least of the features of a 'B' horizon in the soil.

As model parameters are changed the most significant influences on the extent and type of the 'B' horizon described above appear to be the relative depths of the three scales depths; for rooting (z_0), organic soil (z_1) and organic mixing (z_2). Three cases are distinguished in Figure 10.10. First it is noted that organic soil is never deeper than organic mixing so that only cases for which $z_2 > z_1$ are relevant. The distinction between 'arid' and 'humid' soil types is determined by whether z_0 or z_1 is the greater. If the latter, then both the initial response to plant growth and the final equilibrium show a nutrient-deficit zone developing. There is a slight accumulation zone which in base-saturated conditions might correspond with the formation of concretions, but this zone is not at all marked and rather transient (*c.* 100 years). Within the humid soil domain two further cases may be distinguished, according to whether z_2 or z_0 is the greater. If the former then organic mixing extends beyond the root uptake zone and disperses the initial deficit within a few hundred to 1000 years. If the rooting depth is deeper than organic mixing then the deficit becomes more concentrated and disperses only very slowly over periods of several thousand years. This distinction of three classes of nutrient profile is seen as an important one which reflects some of the major differences between soils on a global scale.

10.5 REGOLITH DEVELOPMENT ON SLOPE PROFILES AND SOIL CATENAS

The methods and results of the soil profile models described above may be applied to the evolution of whole slope profiles and to the soils on them. At a rather high level of complexity, it is possible to amalgamate slope and soil profile models, and although this may be done in principle, a much simpler approach is preferred, and presented here. This depends on the relationship noted in §10.4.4 between total soil deficit w and the surface proportion remaining in the weathering horizon p_s.

First it is noted that the time spans for hillslope evolution are at least as long as those for the development of a weathering profile in the soil, so that the long-term slope evolution should be related to the weathering profile rather than to the organic soil ('A' horizon) or nutrient-cycling profile (?'B' horizon). Secondly the almost single-valued relationship demonstrated in Figure 10.8 (and in other simulations of the weathering profile) between total soil deficit and surface proportion allows an empirical rating curve to be used as a basis for regolith modelling on a slope profile. It will be shown below (§10.5.4) that such a relationship allows solutions to be obtained for the mass-balance and process equations. This approach will also be used in an aspatial sense to summarize the course of soil development at a site for a specified erosional environment and history.

The soil deficit basis for a slope model outlined in the paragraph above retains a fair degree of simplicity as will be seen. At the same time it also retains a rational link to the detailed solute equilibrium models for soil in its profile context. This is certainly an advance on the very simple models of §10.2 in which solution loss was held constant or controlled by a single kinetic constant. It is interesting to compare this approach with that of Ahnert (1964) in which weathering rate is treated as an empirical function of soil depth.

10.5.1 Soil deficit 'rating curve': weathering as a function of soil depth

Close to bedrock the tail of the weathering profile for both the equilibrium case (Figure 10.7 and Equation (10.34)) and during evolutionary development (Figure 10.8) has an inverse exponential saturation form, behaving like:

$$1 - p = (1 - p_k) \exp(-z/z_k) \tag{10.39}$$

for constant p_k, z_k. If successive weathering profiles are considered to be formed by the downward displacement of this profile, then for any surface value p_s the soil deficit is obtained by integrating Equation (10.39) as:

$$w = \int_z^\infty (1 - p_k) z_k \exp(-z/z_k)\, dz = (1 - p_s) z_k \tag{10.40}$$

That is to say that the exponential tail on the weathering profile implies a linear dependence of soil deficit w on the proportion removed $(1 - p_s)$ for very slightly weathered soils. Analysis of evolutionary profile data such as that shown in Figure 10.8 shows that the linear relationship breaks down appreciably for p_s less than about 0.7, and that the deficit then increases more rapidly, becoming very large as p_s falls towards the apparent insoluble residue p_*. Three examples are shown in Figure 10.11. A simple functional form for the overall dependence is:

$$u_s = u_* w/(w + z_k) \tag{10.41}$$

where $u = (1 - p)/p$ with subscripts $*$ and s as before to represent the apparent insoluble residue and the surface value respectively and z_k is the same constant as in Equation (10.40). Values of z_k are thought to be about 1–2 m, in line with the curves of Figure 10.11. The expression (10.41) will be used in the simulations presented, although other forms can readily be substituted.

The soil profile models presented above may also be analysed to show variations in total solution loss with soil deficit. It has been hypothesized, largely on the basis of Gilbert's (1877) work, that the form of the relationship between solutional loss and soil depth (or deficit as defined here) is one of increase for shallow depths and then a gradual decline. For a bare bedrock surface, it is argued that all rainfall runs off immediately without reaching solute equilibrium. Solution loss is therefore low. As soil thickens, the water is retained longer, and hence is expected to dissolve more soil material, until some optimum is reached, probably for a rather shallow soil. For very deep soils it is argued that soilwater, although well equilibriated with the soil, is also very stagnant, so that little solutional *loss* occurs. Thus increases in depth tend to reduce the solutional loss.

The above argument may be compared with the simulations described, one of which is illustrated in Figure 10.12. The expected rise for shallow soil depths is well shown, and is plainly inherent in the general form of the percolation model described in §10.4.1. A subsequent gentle fall in solution rate is shown in the figure, but only occurs for profiles with rather restricted percolation (large t_0 in Equation (10.24)). In other simulations this fall is less

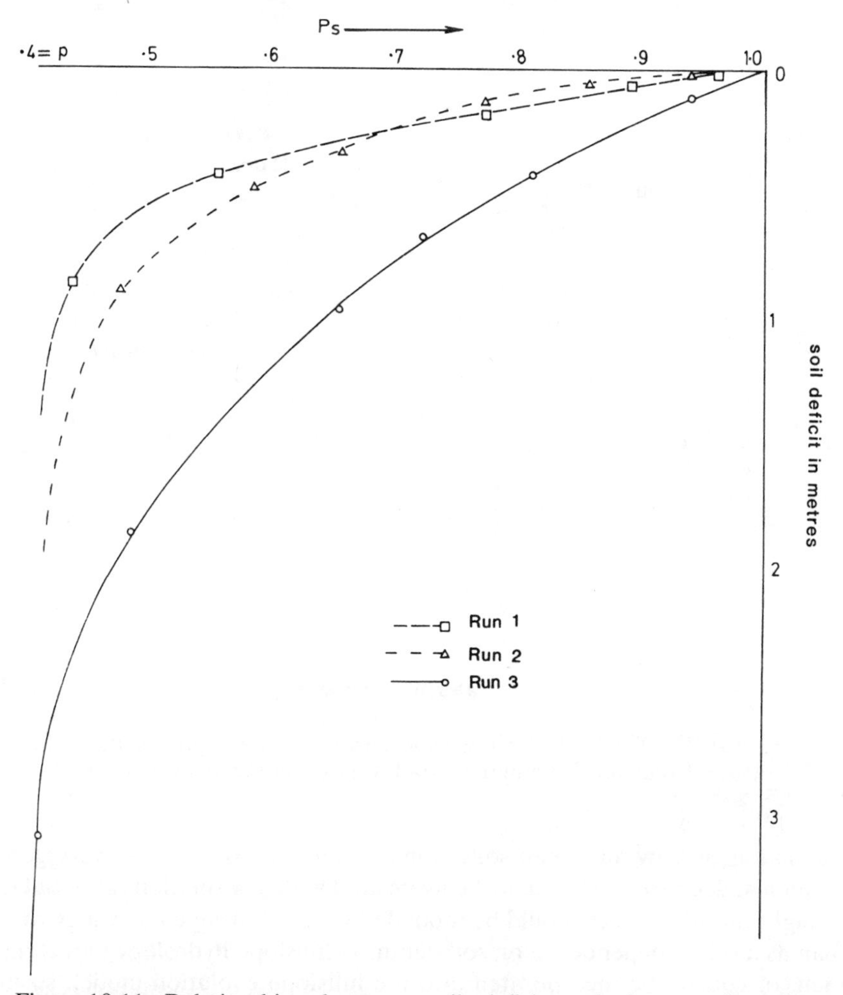

Figure 10.11. Relationships between soil deficit and degree of surface weathering obtained in simulations of soil weathering profiles like those shown in Figure 10.8

marked or absent. Where a reduction in solution rate was found in the simulations, it occurred because sub-surface flow was concentrated just above the weathering front. As weathering deepened the soil profile this front became sharper, so that more of the flow was diverted within highly weathered soil, and so carried away fewer solutes. In other words the reduction in solute removal appeared to be less marked and occurred for different reasons from those in Gilbert's argument. The stagnation hypothesis is nevertheless an important one, but it is argued here that it is not inherent in soil depth *per se* but depends on the overall hillslope geometry. Since both

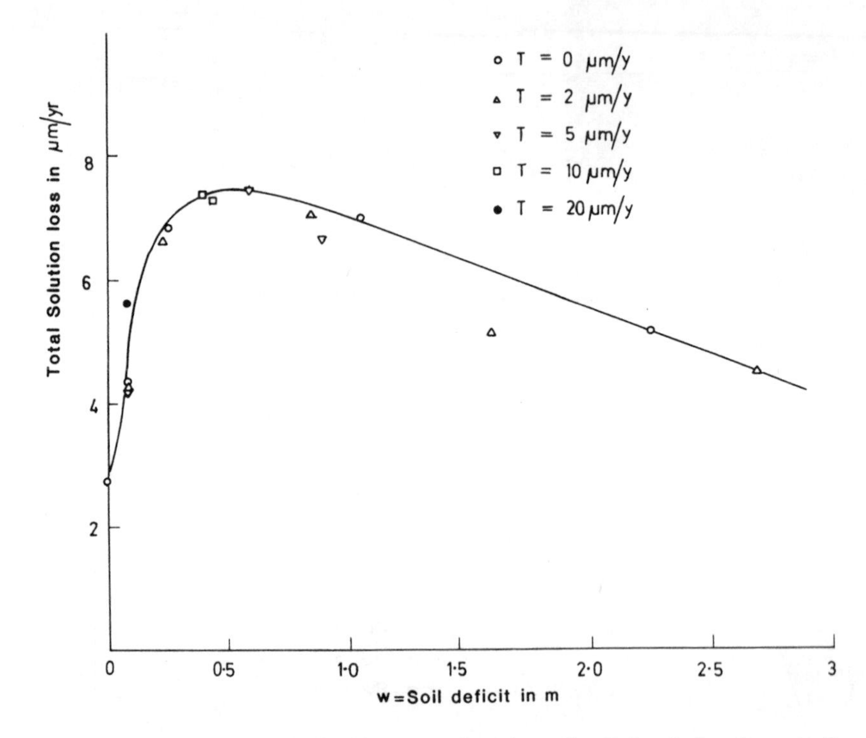

Figure 10.12. Total solution loss as a function of soil depth for the set of simulated weathering profiles shown in Figure 10.8

stagnation of flow and deep soils commonly occur together on very gentle gradients, deep soils will usually be associated with low solution rates, but it is thought that this effect should be modelled via the hillslope hydrology rather than as a direct dependence on soil depth. A hillslope hydrology model must then, of course, be incorporated into the hillslope evolution model, so that there is some increase in complexity to trade against any improvement in physical reality.

10.5.2 Soil profile development in a given erosional environment

The deficit rating curve discussed above may be used out of the context of the slope profile as a whole provided that the erosional environment may be specified independently, as rates of net mechanical and solutional denudation for the site in question. A mass balance for the soil deficit w can be written as:

$$\partial w/\partial t = (\partial/\partial x)[V - S(1 - p_s)/p_s] = (\partial/\partial x)[V - Su_s] \qquad (10.42)$$

where S and V are respectively the rates of net mechanical and chemical

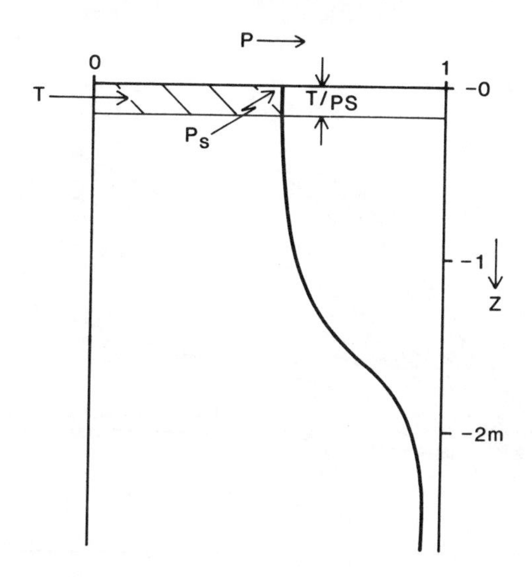

Figure 10.13. Definition sketch for obtaining a mass balance for soil deficits

transport. For present purposes this expression will be considered in an aspatial form, by assuming that p_s is varying only slowly downslope: the differential on the right-hand side may then be removed, and S and V replaced by the mechanical and chemical denudation rates, denoted by T and U. In writing down this expression it is clear that all solute denudation adds directly to the soil deficit w. Mechanical denudation indirectly reduces the deficit by truncating the profile. A loss of T corresponds, as has been seen above, to a loss of T/p_s of soil depth; and the loss of deficit is then obtained by multiplying by $(1 - p_s)$ to give the expression above. As it stands this expression is insoluble except when there is no mechanical denudation (when $w = Ut$) and in the equilibrium case, for which:

$$p_s = T/(T + U) \quad \text{or} \quad u_s = U/T \tag{10.43}$$

This result is in fact merely a re-expression of Equation (10.35), although in a simpler context here. To proceed farther the rating curve of Equation (10.41) or an alternative form is needed to allow a solution.

Substituting Equation (10.41) in (10.42) gives:

$$\mathrm{d}w/\mathrm{d}t = U - Tu_* w/(w + z_k) \tag{10.44}$$

This can be integrated directly for a constant erosional environment (T, U) to give an expression for the increase of soil thickness over time, which if the soil

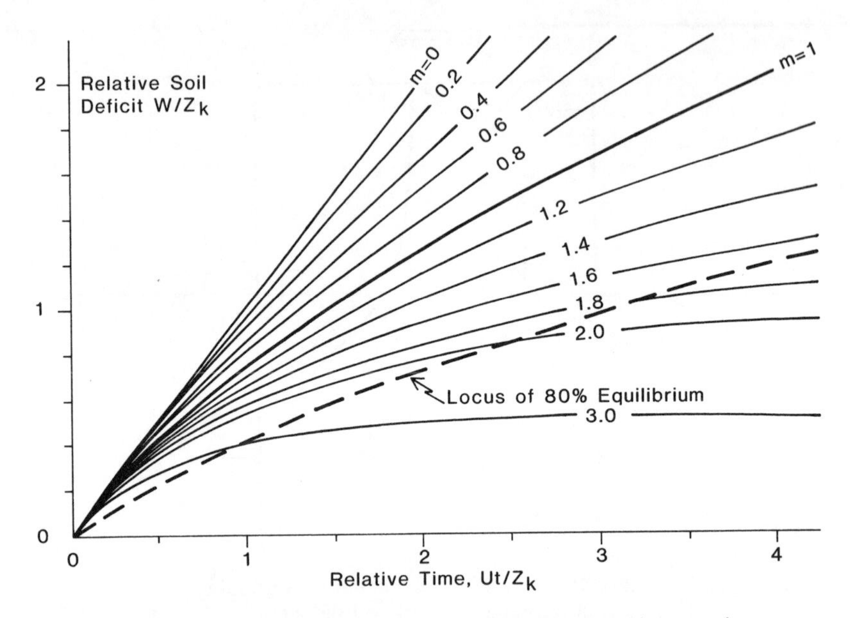

Figure 10.14. Simulated growth of soil deficit (i.e. depth) over time as a function of the ratio of mechanical to chemical denudation

grows from zero at time zero, takes the form:

$$w/z_k - [m/(1 - m)] \ln[1 + (1 - m)w/z_k] = (1 - m)Ut/z_k \quad (10.45)$$

where m is the ratio Tu_*/U. This ratio is critical to the long-term course of soil development. Where it is greater than 1.0—that is where mechanical denudation is high enough—then the soil tends towards an equilibrium. Where $m < 1.0$, however, the soil does not reach an equilibrium but instead thickens indefinitely. This relationship is illustrated in Figure 10.14 for a fixed rate of chemical denudation, as the rate of mechanical denudation is changed. Using a value of 5 μm/per year solute loss, 50% apparent insoluble residue ($p_* = 0.5$ or $u_* = 1.0$) and the scale depth $z_k = 1$ m, it may be seen that soils show negligible divergence for the first 50,000 years as mechanical denudation varies over a normal range of rates. Thus little divergence is to be expected between post-glacial soils except on the most severe gradients.

For mechanical stripping at 10 μm per year the soil takes about half a million years to approach (80%) equilibrium. Over this time span the total landscape denudation is 7.5 m which, although appreciable, is only a small fraction of the relief normally available in the landscape associated with such denudation rates (*c*. 150 m according to Ahnert (1970)). Soil evolution is therefore argued to be rapid relative to evolution of the landscape as a whole, and very substantial soil evolution can occur without appreciable change in the erosional environment if climatic and other conditions are held constant. Where erosional environments change, for example through climatic change, they may be followed numerically using Equation (10.44) but this topic is not pursued further here.

10.5.3 Soil catenas in equilibrium with slope processes

For mature slopes which have been developing over periods long enough for equilibrium soil weathering profiles to have formed, downslope differences in the geomorphic environment may be expected to produce consistent downslope differences in weathering profile. Such weathering catenas take a long time to develop, commonly at least 10^5 years and sometimes much longer, and equilibrium will be reached soonest, as has been seen above, in areas and parts of the slope where the ratio of mechanical to chemical denudation is highest. At this point it is preferable to return to the full form of Equation (10.46) also removes an apparent anomaly in (10.43) where denudation rates are locally negative (aggradation). In Equation (10.46) it

$$p_s = S/(S + V) \text{ or } u_s = V/S \tag{10.46}$$

In this expression the soil is seen to respond to the whole sequence of denudation rates upslope, so that Equation (10.43) may also be corrected by replacing local rates of denudation by their averages down from the divide. Equation (10.46) also removes an apparent anomaly in (10.43) where denudation rates are locally negative (aggradation). In Equation (10.46) it may be seen that the equilibrium p_s always lies between zero and one because S and V must necessarily be positive, indicating downslope transport only.

Figure 10.15 illustrates the form of weathering catena expected for a mature humid region, making reasonable assumptions about overall slope form and denudation. (a) represents a valley-side slope well away from base level where the slope-base river is still actively down-cutting. It is assumed firstly that total denudation $(T + U)$ is proportional to elevation above base level, so that the divide is being lowered fastest and the stream least (curve (i)). Secondly it is assumed that, for a humid area chemical denudation U is roughly constant (curve (ii)). By summing values downslope, estimates may be obtained for total solute transport (V in curve (iii)) and mechanical transport (S in curve (iv)). According to the level of solute denudation, there may or may not be a zone of mechanical aggradation at the slope base (and decreasing mechanical transport). It may be seen that in all cases the equilibrium surface proportion p_s declines steadily downslope; at first gradually and then somewhat faster, especially if there is mechanical aggradation. Soils therefore become more weathered downslope in a humid environment which approximates to these assumptions.

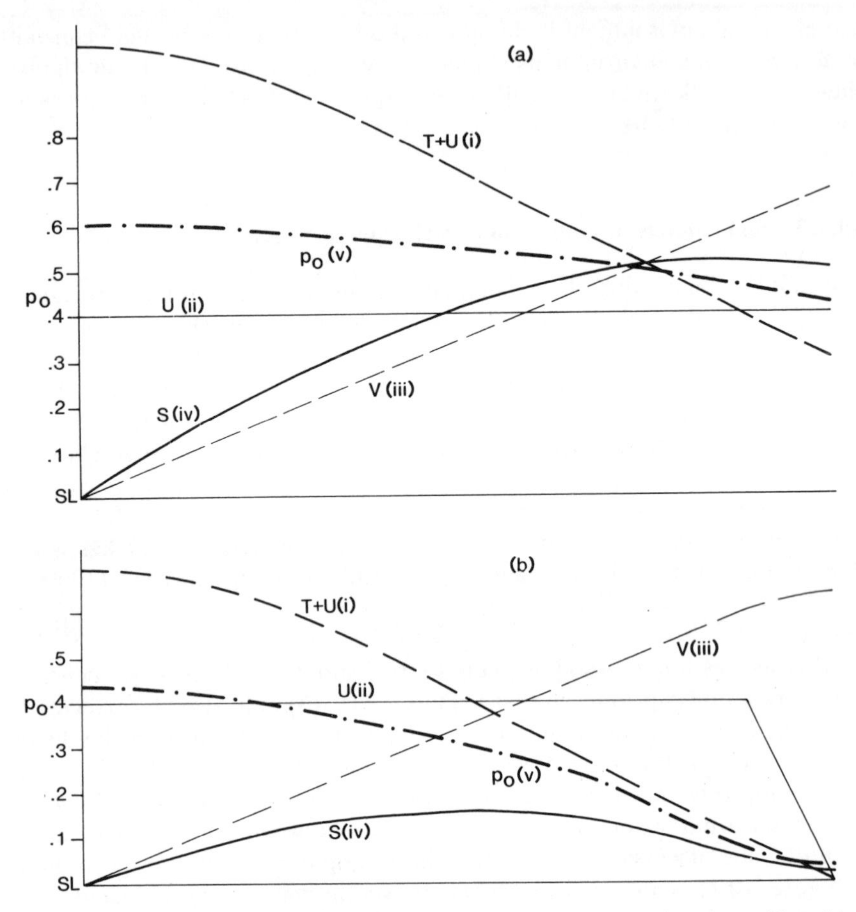

Figure 10.15. Expected general form of weathering catenas in areas of humid climate: (a) well above base level; (b) close to base level. p_o refers to the surface value of p in equilibrium

Referring back to Figure 10.14 it may be seen that the times taken to reach equilibrium become longer downslope, so that the equilibrium catena develops soonest near the divide, and gradually extends downslope. In Figure 10.15(b) the same assumptions are applied to a site where the river is close to base level and therefore no longer actively downcutting. Solutional denudation is assumed to be at substantially the same rate, except for a sharp decline close to the slope base where soil saturation is likely to impede sub-surface drainage. Mechanical denudation is scaled down in proportion to the lower overall elevations, falling to zero at the river. In this case it is seen that soils are everywhere somewhat more weathered, and that the catenary differences become more marked except in the saturated zone near the river.

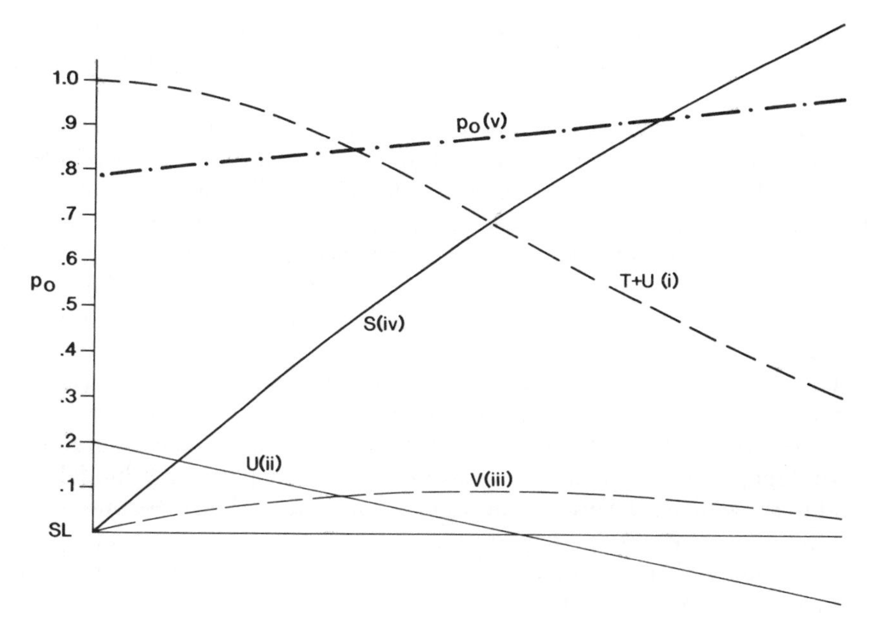

Figure 10.16. Expected form of weathering catenas in arid and semi-arid areas

The overall pattern within a catchment is thus seen to be of equilibrium catenas becoming established first near upstream divides and gradually spreading downslope and down-valley. After a sufficiently long period for equilibrium weathering profiles to be established everywhere, the degree of weathering will follow a similar pattern; from least on upstream divides to most in downslope valley bottoms. It will be apparent from Figure 10.14 that some equlibria will be unattainable so that some sites will instead show indefinite thickening of the soil profile. These sites, if they occur, will tend to be in valley bottoms and close to base level.

For a semi-arid to arid area, different assumptions should be made about mature weathering catenas. There mature slopes may still be assumed to show total denudation roughly proportional to elevation, but chemical denudation tends to be low, and to decrease downslope as upslope runoff is lost to evaporation with consequent redeposition of solutes (this argument is expanded in §10.6). The curves for solute and mechanical transport are therefore as shown in Figure 10.16. Mechanical transport increases steadily downslope, though at a decreasing rate, while solute transport rises to a low peak before falling to an even lower slope-base value. The equilibrium soil is thus thin (low p_s) and becomes still thinner downslope, with much shorter times needed to reach equilibrium than in the humid case. Equilibrium weathering profiles will therefore tend to develop first at the base of slope profiles.

Although equilibrium weathering profiles are unlikely to have been attained in youthfully dissected areas, it is instructive to apply the arguments above to them and to other cases. For example, in a recently dissected area, the rate of mechanical denudation tends to increase downslope since incision tends to proceed by headward erosion. In this case, for a humid region with little areal variation in chemical denudation rates, equilibrium soils tend to be less weathered downslope. The incision strips off pre-existing soil, so that equilibrium is quickly approached. Thus recently formed valleys will commonly show less mature soils along their axes than the side slopes into which they are cut; even if this trend is ultimately reversed as maturity is reached and relative incision stops.

The catenas described above are all due to weathering, and take very long time spans to achieve. Other catenary effects develop over much briefer time spans of decades to thousands of years, in response to consistent differences downslope, most commonly of hydrology. For example, many humid soils tend to show higher levels of saturation at the base of slopes, because of the relatively large collecting area and the low gradient to carry soilwater away. As a result differences in oxidation conditions which can, for example, influence the mobility and form of iron compounds, and modify the mesofaunal population and so the organic soil profile.

10.5.4 Mass-balance and process equations for the soil deficit

The relevant mass-balance equations for continuity of total material and soil deficit have already been presented above as Equations (10.3) and (10.42) above. Their joint solution requires a series of process equations to specify the rates of mechanical and chemical material transport, S and V respectively, a description of the soil profile form which gives the proportion of surface weathering p_s in terms of soil deficit or other variables. Initial values of elevation and soil deficit are needed, together with boundary conditions for behaviour at the divide ($x = 0$) and at the slope base.

Process equations for mechanical transport processes are not discussed here. Suggestions may be found in Ahnert (1976) or Kirkby (1976b, 1984) among other sources. Process rates may be expressed in terms of topographic factors alone, or in terms of an underlying hydrological sub-model, in which case process rates are expressed in terms of gradient and appropriated flow partitions such as overland flow. In the context of a soil-linked model it is possible to introduce terms which curtail mechanical sediment transport in the presence of a thin soil (Kirkby, 1985a), or in other cases merely cut off transport when soil thins to zero (Armstrong, 1976).

Equations (10.5) and (10.8) above show alternative simple formulations for chemical transport rates over constant lithology. The first expresses a constant rate of solute pick-up per unit discharge and implies that solute equilibrium is being reached for all sub-surface flow. The second equation (10.8) instead implies a kinetic pick-up process in which water flows a significant distance downslope (i.e. tens of metres) before reaching equilibrium. Both of these expressions may readily be adapted to variable lithologies by changing the equilibrium concentrations. In the model described below the equilibrium model has been used, although its limitations are recognized. The process law of Equation (10.5) has been used in the form:

$$dV/dx = Pc_* \tag{10.47}$$

where P is the net addition to downslope flow per unit distance, and c_* is the equilibrium concentration for the underlying parent material as it changes downslope. These definitions are as before, and P may either be taken as constant or derived from a hydrological model for sub-surface flow.

Boundary conditions at a divide are straightforward, with $S = V = 0$. At the slope base, either elevation or sediment transport is held constant or specified as a given function of time. The same approach is not, however, readily applicable to the soil deficit. It is not thought appropriate to set soil deficit to a fixed value or zero at the river since weathering may proceed below the river bed where its downcutting is constrained. The least restrictive choice is suggested as $dw/dx = 0$, which is a requirement of symmetry along the valley axis without further constraint, and this has been adopted in the examples presented below.

10.5.5 An integrated slope and regolith model and its implications

Under conditions of constant downcutting, equilibrium solutions may be found for the mass-balance equations. In the simplest case, which corresponds roughly to humid conditions, both mechanical and chemical denudation are constant downslope, leading to a slope profile which is closely related to the mechanical process laws. Thus for transport-limited mechanical processes Equation (10.12) remains appropriate. The only effect of the soil deficit equation is to show that this slope profile, whatever its form, will have a constant soil composition and depth down its length, with p_s given by Equation (10.43). It may be seen that, for a given *total* denudation rate T the slope gradients will be more gentle and the soil more weathered as the chemical solution rate is increased.

More interesting results are shown where basal denudation is constrained or prevented. In that case, as may be exprected from the analyses of soil catenas above, the soil tends to thicken over time and downslope, approaching a steadily changing equilibrium with current slope processes in the upper parts of the slope and steadily thickening near the slope base. Deep

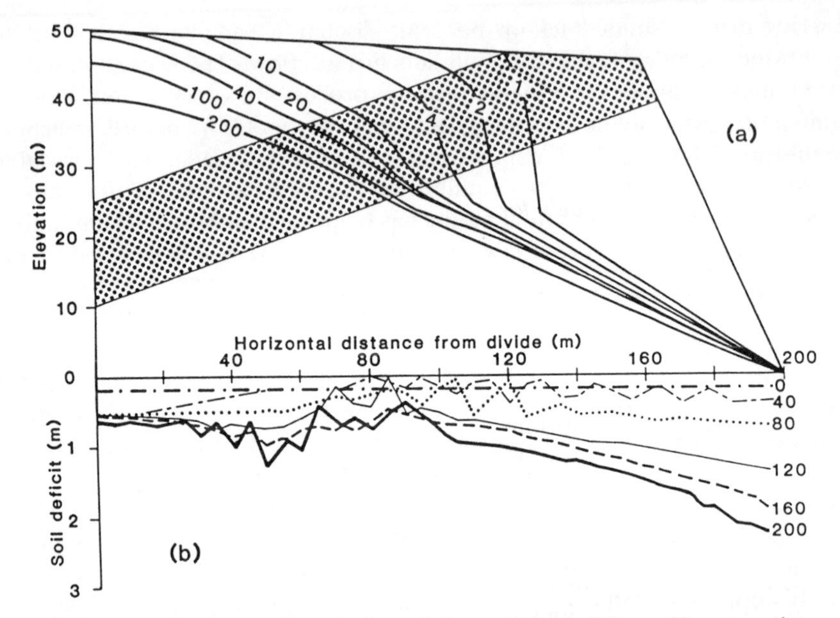

Figure 10.17. Simulated evolution of slope and regolith profiles over time for a slope with a band which is less soluble and more resistant to landsliding: (a) slope profile; (b) soil deficits

soils are seen to be associated with the slower mechanical processes, and are not normally able to develop where landsliding or soil wash are active. Figure 10.17 shows a more complex example with a band of differing lithology. Although the resolution of the model does not locate breaks in slope very accurately, it may be seen that the hard band has a substantial influence on slope form and soil thickness. The resistant band shown in the figure is in fact one of higher resistance to landsliding and lower solubility, but in a range of model runs it was clear that geotechnical properties generally had a much more marked influence than differences in solubility.

In the very long term lithological units of greater solubility will develop gentler gradients, as is shown in Figure 10.2. Nevertheless geotechnical properties may for a long time mask this effect. Limestones are in fact among the common relief-forming units in many areas, especially those where the relief is of Quaternary age. The evidence of simulation modelling is that their elevation owes very much to their geotechnical resistance to landsliding, and much less to the effect of their permeability in suppressing overland flow and other erosional processes.

The models presented in this chapter for slope and soil evolution under solutional processes generally require very long time spans for significant change. Particularly in a country like Great Britain with a short period of soil development, often only post-glacial, the development of weathering catenas or weathering profiles is commonly rudimetary, so that direct field evidence for long-term development is lacking. In the short term, patterns of solutional denudation generally reflect local differences in the soil, however these may have been derived.

10.6 VARIATIONS IN SOLUTION RATES WITH HYDROLOGICAL FACTORS

So far the rate of solutional transport and denudation has been expressed in terms of sub-surface flow, and it has been assumed that, at least for humid temperate climates, flow increases linearly downslope so that, for uniform lithology, solutional denudation is more or less constant over the landscape. Here the hydrological content of this assumption is looked at more carefully at a range of time and space scales, to demonstrate some of the ways in which hydrology influences solution rates and distributions. The assumption of solute equilibrium will, however, be maintained; so that corrections are needed where kinetic assumptions should be made, generally leading to a relative increase in solution rates downslope.

10.6.1 Variations in solution rate downslope

On the scale of a hillslope profile, the average hydrology may initially be analysed on the basis of annually or seasonally averaged conditions. For a point on the slope at distance x from the divide, the rate Q of increase of steady-state discharge per unit width is equal to the difference between mean rainfall intensity i and the evapotranspiration loss e, which is itself related to its potential rate e_P and to the soil moisture deficit d. For an area between two adjacent flow lines w apart (which varies as the flow converges or diverges):

$$d(wQ)/dx = w(i - e) = w[i - e_P f(d)] \qquad (10.48)$$

where $f(d)$ is the ratio of potential to actual evapotranspiration, decreasing in some manner as the deficit d increases.

To obtain explicit forms for the change in deficit downslope, an explicit flow law must be used to give discharge in terms of deficit and gradient, $f(d)$ must be specified and the slope geometry is needed to give the downslope variation in gradient and flow-strip width w. For a convex ridge (w constant, and gradient and therefore discharge increasing linearly downslope) the deficit is constant downslope. Discharge and solute transport area then increase linearly, so that the solutional denudation is constant. Since divides are normally convex, they may therefore be expected to show roughly constant solutional loss.

For other slope configurations the form of the evaporation term becomes important. For an arid slope, with a linear flow law [$Q = Ks(d_0 - d)$] and evaporation rate $[f(d) = d/d_0]$ flow on a slope of uniform gradient s rises along an inverse exponential saturation curve towards a steady value, over a distance scaled by Ksd_0/e_P which is the distance travelled downslope by runoff before it evaporates (which will commonly be of the same order of magnitude as the slope length). The solute transport, which is proportional to the flow, similarly rises rapidly at first and then more and more slowly, so that solute denudation falls exponentially. On a sufficiently marked slope concavity the flow will actually decline, slowing flow velocities and allowing increasing evaporative losses. Solute deposition then occurs, a result that has been used in §10.5.3, and is much more general than the particular expressions used above, requiring only that average rainfall intensities are less than potential evapotranspiration rates. In the concavities deficits are then reduced, allowing lusher vegetation but encouraging chemical deposition. The rate of this deposition may be estimated from models based on Equation (10.21), though (10.51) may be preferred. The effect of storm rainfalls on this background pattern is for some increase in overland flow production from the concavities, although this is generally slight since deficits are commonly large compared to storm rainfalls.

In a humid climate ($i > e_P$) the variation in rates of $E - T$ is slight, and the form of the flow can be adequately approximated by assuming a constant rate. The steady-state flow at any point can then be written in the form:

$$Q = (i - e)a \tag{10.49}$$

where a, the area drained per unit contour length along the flow strip is given by:

$$a = \int_0^x w \, dx / w$$

If discharge is proportional to gradient for a given deficit, then discharge everywhere increases downslope; while deficit is constant on convexities, decreases on straight slopes and decreases rapidly on concavities and even more so in hollows (areas of flow convergence). If concavities or hollows are sufficiently marked they may form perennial seepage zones in which discharge is too great to be contained entirely within the soil, forcing some saturation overland flow to the surface. In storms, areas of low deficit are likely to become saturated, so that they generate overland flow even though upslope areas may generate none. It has been argued above the overland flow is unlikely to reach chemical equilibrium, so that the water available for chemical denudation is only a part of the total rainfall. Hollows and slope-base concavities are therefore likely to be areas of reduced chemical denudation in a humid climate. When overland flow recombines with sub-surface flow in the streams, the mixed solute concentration will in consequence be somewhat less than the equilibrium value. This dilution effect will of course be greatest in times of flood, and is one component of the variation in concentration over time which is discussed further below.

10.6.2 Influence of evapotranspiration on concentration

Water entering the soil is assumed to reach solute equilibrium with successive layers of soil, while at the same time there are losses of percolation water to evapotranspiration and sub-surface lateral flow. Evapotranspired water carries nutrients away with it, but these are ultimately returned by leaf fall and decomposition, so that in the long term the lost water leads to no net loss of solutes. The percolating water is thus enriched with solutes and there is some consequent tendency to redeposition at the base of the rooting zone. This effect is most evident in the case of semi-arid chernozems and similar soils where the 'B' horizon is of humid type with $z_2 > z_0 > z_1$ (Figure 10.10), and the rate of organic mixing is low whereas the solubility of the bases is high (as may be seen from Equation (10.38). In this section a variant of this model is used which allows much higher solute concentrations, particularly in arid areas.

The model used here is built on the assumption that all water entering the soil comes into solute equilibrium with it; and that subsequent losses by evaporation then concentrate the solutes in the remaining water. For each mineral or oxide constituent, the process of concentration is allowed to proceed until saturation occurs beyond which point the constituent in question is redeposited in the soil. The maximum concentration of solutes is thus given by the sum of the individual saturated (as opposed to equilibrium) concentrations for each consituent. This model has been proposed by Carson and Kirkby (1972, Chapter 9) and shown to give useful results, even though its theoretical basis appears to be somewhat naive. Where redeposition is appreciable, however, it is in some respects preferable to a solubility model based on Equation (10.21) because redeposition of bases generally occurs in mineral forms more soluble than those in the parent material. The result of this process is to raise potential concentrations to the saturated levels proposed here. In assessing the initial equilibrium it should also be remembered that the relevant soil material is that within the root zone rather than the parent material, since percolation only penetrates to the root zone before evapotranspiration is lost.

The extent of possible concentration by the evaporative process depends on the constitution of the parent soil. The maximum ratio is:

$$m = \Sigma k/\bar{k} = 1/\bar{p} \qquad (10.50)$$

where the sum is made over all constituents, and p is the mean constituent proportion, weighted by their solubilities. Thus for a parent material dominated by a single constituent like a limestone, there is little increase in concentration possible; while a rock with many constituents, like most igneous rocks, permits a much greater proportionate increase, up to a higher maximum value. An expression which shows empirically the correct form for the increase in concentration is:

$$k = \bar{k}(m - r)/[1 + (m - 2)r] \quad (10.51)$$

where k is the final solute concentration, and r is the proportion of total percolation which ultimately runs off (after subtracting evapotranspiration).

10.6.3 Implications for differences in solutional denudation with climate and lithology

The relevant form of the relationship, exemplified by Equation (10.51), is a 1 : 1 rise in k at low evaporation values as r falls below 1.0, and a value of m for high evaporation rates (r close to zero). Figure 10.18 shows the implications as rainfall changes for a given *potential* evapotranspiration, assuming that all rainfall initially percolates into the soil. (r has been estimated as $x^4(x - 1)/x(x^4 - 1)$, where x is the ratio of rainfall to potential $E - T$.) Values broadly match those described empirically (e.g. Langbein and Dawdy, 1964). It may be seen that for hot desert areas, igneous rocks with their high value of m and low values of k may actually be more soluble than limestones, while cold areas are among those in which limestones are dissolved many times faster than igneous rocks (even if not necessarily at high absolute rates).

Figure 10.19(a) similarly shows the pattern of total denudation as rainfall changes while keeping $E - T$ constant (i.e. temperature roughly constant). A rapid increase is shown at low rainfalls, with the most rapid for igneous rocks etc, with high m values. At high rainfalls the total denudation rate continues to increase; though more slowly as solute concentrations fall. In Figure 10.19(b) denudation is expressed in terms of potential evapotranspiration keeping rainfall constant, roughly showing the effect of changing temperature alone. Once more a general increase is seen, though its form differs showing a definite tendency to level off at high temperatures where available rainfall becomes the controlling influence.

10.6.4 Distribution of solute concentrations over time

Equation (10.51) is equally relevant to variations in solute concentrations over time within a catchment. For a seasonal time span, differences in water balance reflect directly on the proportion of runoff, and so on the seasonal solute levels. As for broader climatic effects, rocks with diverse chemical composition are expected to be more responsive in this respect than chemically simpler rocks like limestones. Changes in water balance are also

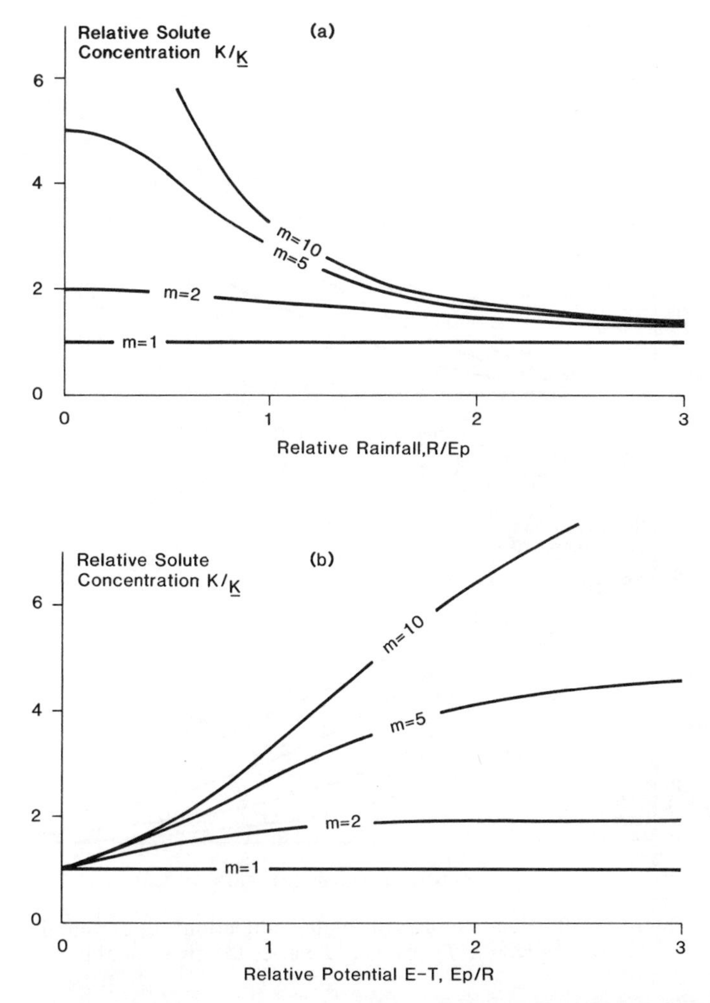

Figure 10.18. Solute concentrations relative to equilibrium concentration for a range of values of m (= saturated/equilibrium concentration): (a) as rainfall varies for given potential *E-T* (or temperature); (b) as *E-T* varies for given rainfall

relevant on the timescale of a single storm. For any unit of rainfall which may be traced down a hillslope, Equation (10.51) remains valid with the modification that any overland flow or direct channel precipitation may be considered to be chemically only as clean as the input precipitation. Thus the concentration given by Equation (10.51) should be multiplied by the ratio $r'/(1 + r')$, where r' is the proportion of rainfall entering the soil to provide

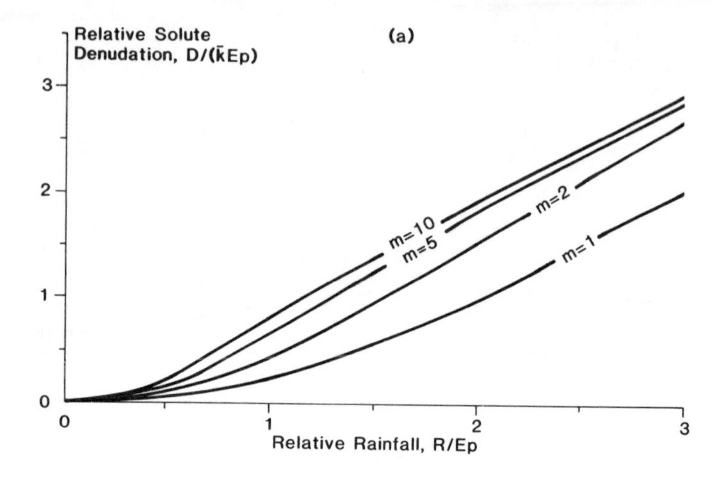

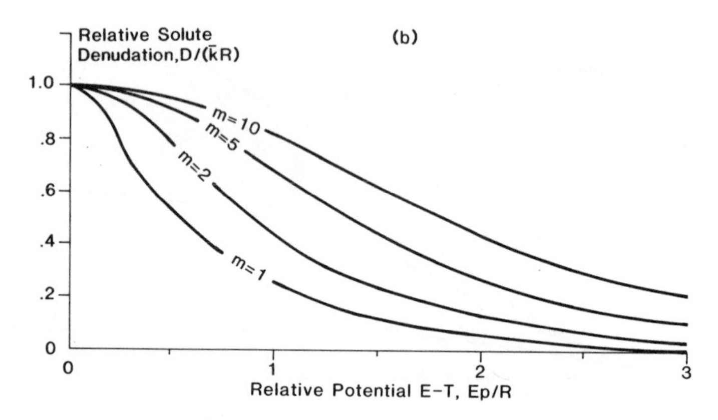

Figure 10.19. Relative rates of solute denudation: (a) as rainfall varies for given *E-T*; (b) as *E-T* varies for given rainfall

sub-surface flow, and r in Equation (10.51) is interpreted as the proportion of this sub-surface flow which finally reaches the stream after subtracting evapotranspiration losses.

The multiplying factor above is the normal dilution effect, which explains at least part of the common reduction in stream solute concentration during storms. In many humid areas, however, the reduction in evaporative losses during storms may be equally important, and is still effective where there is no overland flow. A complete model of concentration during the course of a storm is difficult to construct, since water of many ages is pushed out from the slope during a storm. Thus the initial effect of new rainfall may be to push out water from deep within the soil which combines maximal evaporation losses with equilibrium in a relatively base-rich zone of the soil. Concentration may in these circumstances rise briefly to begin with. Subsequently dilution by overland flow and channel precipitation occurs. At the same time the addition of increasingly young sub-surface flow, which has lost less by evaporation and thus has a lower concentration than older slope water, also lowers the solute concentration appreciably in multi-component rocks (although insignificantly for limestones) following Equation (10.51). On the falling limb of a flood hydrograph, this latter influence is likely to be far the more important since overland flow and channel precipitation are generally negligible soon after storm rainfall ceases, whereas the recovery in solute concentration levels and the decline in stream discharge are spread over a much longer period.

10.7 CONCLUSION: THE VALUE AND LIMITATIONS OF SOLUTION MODELS

The models presented in this chapter briefly review a range of possible approaches varying in complexity and reality. Not all the approaches are mutually consistent, most notably the forecasts implicit in Equations (10.21) and (10.51). Although more detailed models of soil geochemistry are available, the intention here has been to focus on the scale of whole soil profiles and whole slopes. The advantage of this approach is its relationship to other aspects and problems of geomorphology: its chief disadvantages are that it frequently requires very long time spans for appreciable change and that its variables are often not readily comparable with measurement data which must necessarily relate to short periods at single sites.

As with other models for slope or soil evolution, simple formulations tend to imply constancy of conditions over unrealistically long time spans. This constancy is, however, only inherent in the models if they have no explicit climatic inputs, and this difficulty has been avoided in at least some of the approaches outlined here. Within the limitations implied, models of solutional slope development seem to offer some progress in understanding the forms of tower karst and other extreme solutional forms, and to suggest that the morphological expression of solution is muted in landscapes where it is not the dominant process. Some explicit modelling of the solution process is also thought to be important to reach any useful understanding of soil catenas developed by weathering. The proposed formulation of soil processes is recognized as incomplete, but is believed to be almost the first positive step for the time and space scales of the landscape as a whole. It opens up the possibility of models with either a better empirical basis or a stronger geochemical content. Perhaps the most useful models for short-term forecasting of solute concentration and denudation are those presented in §10.6, where the slope hydrology may be examined at a range of time/space scales to show worldwide variations in solution rate with climate or lithology, downslope differences in solutional removal in the short term or variations of concentration within single storms. Hydrology has been seen as one of the strongest influences of solutional removal in all of the models presented, and much progress can still be made in solution modelling by adopting very simplified views of the geochemistry in combination with existing levels of hydrological understanding. Such models naturally include the additive mixing of solute loads from waters following different flow paths, but can very fruitfully go far beyond that simple and necessary principle.

REFERENCES

Ahnert, F. (1964). Quantitative models of slope development as a function of waste-cover thickness. *Abstract of Papers; 20th IGU Congress*, London.

Ahnert, F. (1970). Functional relationships between denudation, relief and uplift in large, mid-latitude basins. *Am. J. Science*, **268**, 243–263.

Ahnert, F. (Ed.) (1976). Quantitative slope models. *Zeitschrift für geomorphologie*, Supp. Band **25**, 168 pp.

Armstrong, A. C. (1976). A three-dimensional simulation of slope forms. In: Ahnert (ed.), 1976. *qv*, pp. 20–28.

Berner, R. A. (1971). *Principles of chemical sedimentology*. McGraw Hill, New York, 240 pp.

Beven, K. J., and Germann, P. (1981). Waterflow in soil macropores: I and II. *J. Soil Sci.*, **32**, 1029.

Bricker, O. P. (1968). Cations and silica in natural waters: control by silicate minerals. Int. Assoc. for Sci. Hydrology, Publ. 78, pp. 110–119.

Bricker, O. P., Godfrey, A. E., and Cleaves, E. T. (1968). Mineral water interactions during the chemical weathering of silicates. In: *Trace Inorganics in Water*, Advances in Chemistry Series 73, pp. 128–142.

Carson, M. A., and Kirkby, M. J. (1972). *Hillslope Form and Process*, Cambridge University Press, 475 pp.

Dunne, T. (1980). Field studies of hillslope flow processes. In: *Hillslope Hydrology* (M. J. Kirkby, Ed.), pp. 227–293.

Garrels, R. M., and Christ, C. L. (1965). *Solutions, Minerals and Equilibria*, Harper, NY, 450 pp.

Gilbert, G. K. (1877). *The Geology of the Henry Mountains*, US Geol. and Geog. Survey, Washington.

Kirkby, M. J. (1976a). Soil development models as a component of slope models. *Earth Surface Process*, **2**, 203–30.

Kirkby, M. J. (1976b). Hydrological slope models—the influence of climate. In: *Geomorphology and climate* (E. Derbyshire, Ed.), John Wiley, Chichester, pp. 247–267.

Kirkby, M. J. (1984). Modelling cliff development in South Wales: Savigear reviewed. *Zeitschrift für Geomorphologie*. **28**(4), 405–426.

Kirkby, M. J. (1985a). A model for the evolution of regolith-mantled slopes. In: *Models in Geomorphology* (M. Woldenberg, Ed.), Allen and Unwin, p. 213–217.

Kirkby, M. J. (1985b). A basis for soil profile modelling in a geomorphic context. *J. Soil Science*, **36**, 97–121.

Langbein, W. B., and Dandy, D. R. (1964). Occurrence of dissolved solids in surface waters in the United States *USGS Professional Paper*, 501D, 115–117.

Lerman, A. (1979). *Geochemical Processes: Water and Sediment Environments*, John Wiley, 481 pp.

Robie, R. A., and Waldbaum, D. R. (1968). Thermodynamic properties of minerals and related substances at 298.15K (15°C) and one atmosphere (1.013 bars) pressure, and at higher temperatures. *US Geol. Survey, Bulletin 1259*.

Sherman, G. D. (1952). The genesis and morphology of the alumina-rich laterite clays. In: *Problems of Clay and Lateritic Genesis*, Am. Inst. of Mining & Metallurgical Engineers, NY, pp. 154–161.

Smith, D. I. (1975). The problems of limestone dry valleys—implications of recent work in limestone hydrology. In: *Processes in Physical and Human Geograph* (R. F. Peel, M. D. Chisholm and P. Haggett, Eds.), Heinemann, pp. 130–147.

Trudgill, S. T., Pickles, A. M., and Smettem, K. J. (1983). Soil water residence time and solute uptake, 2: Dye tracing and preferential flow predictions. *J. Hydrology*, **62**.

Solute Processes
Edited by S. T. Trudgill

CHAPTER 11

Solute processes and landforms: an assessment

S. T. Trudgill

Department of Geography,
University of Sheffield

The review of rates of surface processes on slopes by Saunders and Young (1983) shows that there is considerable knowledge of rates of chemical denudation on a variety of lithologies (their Table IV). In 1974 Young wrote that 'measurement of solution loss on slopes is the greatest gap in current knowledge of surface processes'. Some ten years later, this would still appear to be substantially the case since the 1983 review is a presentation based mostly on calculations from dissolved river loads. Thus the rates are for large areas and there is little indication of the distribution of rates in the landscape. The rates are thus of very limited application in the study of landform evolution: they may be used at a general level for comparison of one area to another, but they are of little use in indicating which parts of the landscape are being eroded more than others, e.g. slope crest, slope foot, mid-slope, etc. They may be used at a general level for comparison of one area with another, but the role of, say, climate in giving rise to areal differentiation may often be difficult to assess, since many of the variables change from catchment to catchment and therefore there is little in the way of control situation.

Clearly, more data are needed on the areal differentiation of erosion rates within catchments and with specific reference to different parts of landforms. Ahnert (1980) lends support to this view by writing 'all geomorphological work deals . . . with the shaping of landforms'; this, 'in turn is the result of the spatially and temporally varying denudational/depositional mass balance at the surface of the earth . . . measurements should therefore aim at identifying the mass balance and its variation from point to point in space'. Data such as Young's 1978 observation on a sandstone slope in the north of England are of interest within this context. Here, he showed that, by the movement of buried rods, the annual vertical dissolution loss was 0.31 mm while horizontal movement was less, at 0.25 mm; but greater knowledge of rate distribution on slopes is needed, such as are discussed in Chapters 6, 8 and 10 of this book.

However, it must be stressed again that such measurements as are carried out at the present day will refer to the present-day evolution of a slope and will be of predictive value, rather than of value in explaining the present form of the slope—unless the slope is in equilibrium with present-day processes. This would appear to be unusual, except for very rapidly eroding slopes on unconsolidated material. In a review of slopes and slope processes, Mosley (1982) writes that 'it seems incontrovertible that much of the land surface (that is, many slopes), is thousands of millions of years old, and not in dynamic equilibrium with present-day processes' (as discussed by Young, 1969). Mosley's review covers the points that 'much of the landscape in Europe may reflect evolution during Pleistocene or Tertiary times; even in the southern Alps of New Zealand, a landscape widely regarded as very dynamic, the general occurrence of glacially oversteepened valley walls, periglacial slope deposits, and inactive screes and blockfields attest to the age of the landscape'. The general principle of the antiquity of the landscape is illustrated by the example of the rates of cave and landform development in the Yorkshire Dales limestone area in the UK described by Gascoyne *et al.* (1983). They obtained uranium-series ages for speleothems collected from caves present at different levels, the sequences of ages enabling them to evaluate rates of downcutting of base levels to which caves formerly flowed—speleothem initiation being assumed to have occurred subsequent to the abandonment of caves as water drained to new, lower levels formed by valley downcutting. Mean downcutting rates quoted are 2–5 cm per 1000 years, though this mean rate masks the episodic nature of downcutting; maximum entrenchment in glacial times would have been 5–20 cm per 1000 years, with up to 24 m lowering per glacial–interglacial cycle. Taking the present depths of the Yorkshire Dales, it is clear that the upper beds of limestone were incised to form the main valleys between 1 and 2 million years ago. This emphasizes the limitations of current process measurements in providing insight into landform evolution. This is especially the case since Gascoyne *et al.* drew attention to the fact that their mean rates of downcutting (2–5 cm per 1000 years) is an order of magnitude different from the rates measured at the present day in cave stream beds using a micro-erosion meter (Trudgill *et al.*, 1981) as quoted by High and Hanna (1970) of 0.4–0.5 mm per year. Coward (1975) quotes rates equivalent to 76 cm per 1000 years. Gascoyne *et al.* write 'These disparate findings emphasize the essential difference between long-term average rates . . . and site-specific direct measurements. The former may not measure the true intensity of erosion because they average in periods of aggradation (no channel deepening) that probably occurs, the latter simply should not be extrapolated to the time spans required to generate landforms of the magnitude considered in this paper'. This emphasizes the point of Trudgill *et al*. (1981) and Trudgill, (1977a) that short-term micro-erosion meter data are best reported in mm per year while long-term data should be reported in units per 1000 years.

In many ways the paper of Gascoyne represents a milestone in geomorphology because it provides an actual dated sequence of landform evolution such as has not been provided before. This is a major development in karst geomorphology and also a development which stands out in the rest of geomorphology. Such work on the antiquity of limestone landforms has been made in parallel with the sedimentological work of Bull (1980) who also stresses the antiquity of caves and related doline features, with some cave remains showing deposits reflecting a variety of climatic changes. Bull discusses the sensitivity of cave development to changes in climate, a topic discussed in general terms by Brunsden (1980), who writes of variation in landscape sensitivity to external change. Brunsden differentiates 'sensitive' and 'insensitive' areas in a similar fashion to the 'labile' and 'non-labile' areas of Trudgill (1976). Sensitive areas are prone to change and react quickly to external change (e.g. river channels and beaches), while insensitive areas are slowly responding to process change. Process changes in the Pleistocene in temperate areas have been described in general terms by Brown (1980) (Table 11.1) showing how the relative importance of chemical weathering is likely to have changed. Brunsden (1980), however, stresses that we cannot accept a simple glacial model, with long periods of interglacial stability:

Table 11.1. Past weathering environments (Brown, 1980)

Dates at start of time	Time	Climate	MW/CW[a]
10,000	Holocene	Temperate	cw
10,800	Loch Lomond Readvance	Glacial	MWcw
12,000	Interstadial	Temperature	cw
118,000	Devensian Glaciation	Clacial	MWcw
128,000	Ipswichian Interglacial	Temperate	CW
6–700,000	Earlier Glaciations	Glacial and Interglacial	MWcw
1–6M	Pre-glacial Pleistocene	Cold/Temperate	cw
65M	Tertiary	Wet, sub-tropical, cool	CW
240M	Mesozoic	Tropical	CW

[a]M = mechanical; C = chemical; W = weathering; capitals = dominant process

'instead we are forced to recognize that climatic controls are almost continuously varying with time.' He suggests that 'small scale, labile and sensitive systems can adjust to new environmental conditions in 10^2–10^3 years but that insensitive, hard rock systems, large integrated river basins, interfluves and plateaux may require relaxation periods in excess of 10^4–10^5 years' (where a relaxation period implies the adjustment of a form to a change in process). In the latter, insensitive areas, adjustment may be so minimal that, for example, the complex history of the Pleistocene may not be recognizable.

What role, then, do the measurements of present-day solute processes and rates have? Certainly, they help to satisfy a perceived need for experimental work, as promoted by Ahnert (1980) and Slaymaker (1980); but their role in landform studies is far from clear (Douglas, 1980). Clearly, their promotion gives researchers and students alike a sense of involvement in active geomorphology and the measurements of rates and processes enables scientific method to be used, with experimentation to test hypotheses and to assess the importance of variables under control conditions. However, the review of Slaymaker (1980) on field experiments reveals that only a small proportion (13%) of so-called 'experiments' are experiments *sensu stricto*—most are measurements or monitoring. In addition, of some 100 projects reported, only a few are related to solutes—though from other sources, such as the journals *Earth Surface Processes and Landforms* and *Zeitschrift fur Geomorphologie*, it is clear that there is a steady proportion of papers on solute monitoring and solute processes. Such papers as those by Waylen (1979) and Douglas (1964) are, however, geomorphologically related rather than focusing on solute concentrations in runoff. Reiterating Ahnert's (1980) point that 'all geomorphological work deals in the end with the shaping of landforms' it is perhaps Webb and Walling (1980) who make the most apposite comment: 'Consideration of stream solute research . . . has demonstrated the need for careful design in solute monitoring programmes if further advances in geomorphological applications are to be achieved'—particularly in refining maps of chemical denudation, presented by the authors on a kilometre grid basis, to the distribution of rates over landforms. In this context. Ahnert (1980) suggest that 'perhaps the buried tablet technique of Trudgill (1977) will provide the technique to measure at least relative rates of weathering at different depths below the surface and at different points in an area.'

Certainly, the technique has been used with profit, as discussed in Chapters 6 and 8, but the method is not without its problems. It is essential that the errors due to cleaning of the tablet for reweighing after retrieval are less than the weight losses. The errors are, however, often variable, and only data from tablets left in the soil for 'long' periods of time ('long' being as yet not closely evaluated), or where erosional loss is rapid, can be reliably reported. Several authors have reported on the use of this technique and some have pointed out variance between data gained by this technique and other methods. This variance may be because:

1. short-term methods should not be compared with long-term data;
2. data have been derived from different sites which should not be compared (e.g. catchment rates *v*. soil profile rates);
3. the errors of the technique mean that absolute rates should not be calculated from the data.

It is probable that only *relative* rates should be reported, and where erosional weight losses are substantially greater than any error from the retrieval and cleaning technique. Thus, relative weight losses, say, downslope, could be reliable compared precisely the data needed if the distribution of solution erosion losses in the landscape are to be estimated.

Day *et al*. (1980) reported on data for a Welsh hillslope on Silurian mudstone, though details of soil type were not given. Weight losses were in general greater on the upper slope (mean 11.0 mg in 2 years) than at the mid-slope (9.2 mg), and these were greater than the on lower slope (3.7–4.4 for 2 sites). In addition, in a tropical environment in Malaysia, rock discs exposed at the surface weathered more slowly (about 16–22 mg loss in two years) than those at depth (22 cm, about 19–38; 140 cm, about 25–33).

Crowther (1983) reported on limestone rock tablet weight loss in a Malaysian tropical environment, stressing the qualified nature of the data gained—confidence limits of 5% were only attainable where losses exceed 40 mg. Crowther compared tablet weight loss with erosion rates calculated from runoff solute load in micro-catchments. He concluded that the latter provides the only reliable means of estimating erosion rates where the time period of observation is less than several months. Such an approach may mean that short-term data may have a seasonal bias, and it is probable that both methods are best used on as long a term basis as possible. Crowther's tablet rates were reported in millimetres per 1000 years and ranged from 0.94–0.97 on bare rock to 1.10–1.45 under shallow, organic rich soils and 1.58–2.26 under deeper, mineral soils at 60 cm. Similarly, Trudgill (1977b) reported on losses under shallow organic soils as being greater than under subaerial conditions (Table 11.2). In Australia, however, Jennings reported a reverse trend, with higher rates under subaerial conditions. Here, it is likely that under highly evaporative conditions, soilwater runoff, and therefore solute losses, might be less under soils more retentive of moisture which then evaporates, leaving weathering products behind in the soil.

In general, however, under pluvial conditions, there would appear to be a correlation between erosion rate and soil type—and especially soil pH. This would appear to be the case in limestone districts (Trudgill, 1985; Table 11.3), but there is no reason to suppose that this would not also be the case on

Table 11.2. Limestone tablet weight loss in milligrams per year, Cockpit Country, Jamaica (Trudgill, 1977b)

Bare surface	0.000–0.010
Mineral soils	0.050–0.053
Organic soils	0.055–0.065

siliceous rocks: Saunders and Young (1983) report rates of solution on siliceous rocks to be significant—about half those occurring on limestone.

It should be emphasized that care has to be taken with the emplacement position and interpretation of results of tablet experiments. Part of the good relationship between pH data and erosional weight loss is due to the fact that tablets placed up in the upper parts of soil profiles are almost bound to give high results since these are the most acid positions. They do not necessarily reflect true rates of slope development if there is little weatherable mineral matter at that position. Data from the soil–bedrock or regolith interface are the only ones of interest unless there is weatherable bedrock mineral matter in the profile. Thus, on a limestone bedrock, a quartz-based drift soil means that data should be derived from limestone tablets at the drift–bedrock interface; on a carbonate soil, data should be derived from limestone tablets in the zone where detectable carbonate exists in the profile. In addition, the use of fresh tablets may give misleadingly high rates, and it is possible that the use of natural rock fragments might be preferred, though this may increase the problems of retrieval, reweighing and standardization. Nevertheless, since adequate data do exist in the broad relationship between soil pH and erosion rates (see Chapter 8), it is probable that pH is the most easily measured surrogate variable for an indication of the distribution rates on slopes. Clearly, water flow is also important; but the acid zones of soil tend also to be the leached, well-drained ones where weathering products may also be removed—with the exception of poorly drained acid peat bogs. Otherwise waterlogged areas also tend to be the less acid sites, either because they are

Table 11.3. Tablet weight loss under soils (% per year) (Trudgill, 1985)

Acid soil (pH = 4–6)	Alkaline soil (pH = 7–8)
$CaCO_3$ 0.01%	$CaCO_3$ 1–10%
0.04, 0.05, 0.04, 0.05, 0.02, 0.04, 0.03, 0.11, 0.06, 0.04, 0.08, 0.36	0.004, 0.002, 0.007, 0.004, 0.002, 0.002, 0.005, 0.004, 0.002, 0.002, 0.002, 0.002, 0.002, 0.003, 0.003, 0.009, 0.005

drainage-receiving sites or because, on flat plateau sites, leaching is weak, so that, in general, pH can in fact be taken as the most useful variable to measure, together with a more general assessment of soil type.

It can be concluded, then, that if the distribution of present erosion rates in the landscape is of interest, then rather than measuring solute losses in runoff and redistributing these losses over a catchment area, and rather than time-consuming measurements of rock tablet weight loss, then the possible association of soil types and solutional erosion rates may provide a profitable line of enquiry—especially if soil maps for an area exist. This is logical, since soil type provides an integrated indication of the weathering environment. Further work could thus usefully attempt to clarify the erosion regimes which exist under contrasting soil types developed on differing slope positions as a basic step in the understanding of the spatial distribution of erosion rates—and hence the understanding of current erosional trends, whether the soil be developed *in situ* or from allotochthonous material.

Since most soils on slopes are wetter and less acid at the base of the slope, and more acid upslope, many slopes are liable to retreat by slope decline, as discussed in Chapter 8, where chemical weathering and solute transport form the dominant erosional processes. Some possible situations are illustrated in Figure 11.1.

If the task of a geomorphologist is to provide explanations of landforms, it remains to ask what constitutes an explanation of a landform. It is clear that there are two separate exercises involved here. Firstly, there is the question of how the landform got like it is; and secondly, how the landform is developing at the moment. Current process data can provide many of the answers to the latter question, especially if the spatial distributions of erosion rates are known. The answer to the second question requires more inference and dating of sequences. Certainly, it is important to separate these two topics in any endeavour. It is also clear that much solute data has helped with process models and with inference about process and form; while endeavour concerning dated sequences is vital to answer the historical questions, it is evident that, in terms of current processes, rates and landform development, more spatially distributed models are necessary, together with information on the relative importance of solutional and non-solutional processes. This book focuses on solute processes, reflecting recent trends in geomorphology; future books may be able to focus more on solutional processes and landforms.

In conclusion, it will be appropriate to consider the nature of current information and the requirements of future modelling. Firstly, compartment models of solute sources will be examined; and secondly, the topic of the spatial distribution of solutional erosion will be discussed.

Compartment models are important since they focus on the significance of individual solute sources within the ecosystem. This helps to indicate which are the more important processes which should be emphasized in modelling

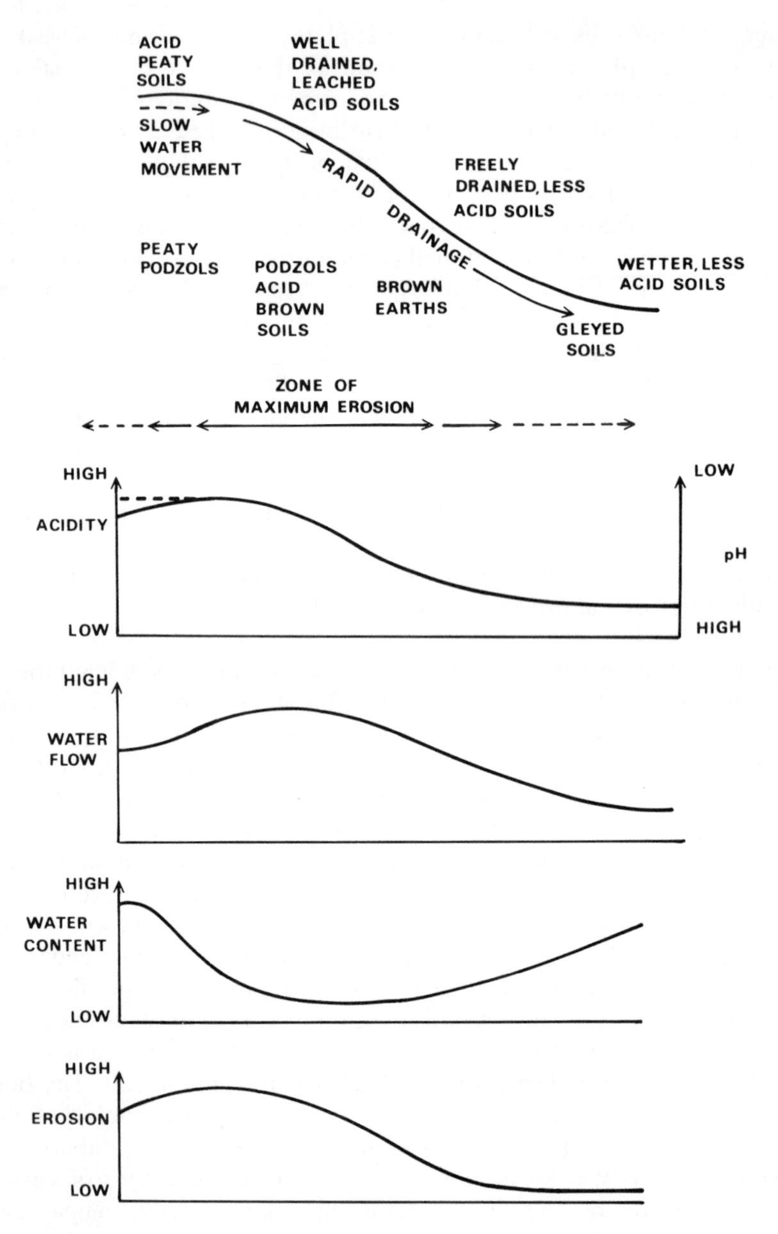

Figure 11.1. Models of soil changes and implications for erosion on hillslopes

and field research—and which are the less important aspects of solute sources which can be played down in initial modelling. Clearly, if biological sources dominate solute output from a catchment ecosystem, this needs to be taken into account when interpreting stream solute data: it can be suggested that the geomorphological effects of the solutes present in the streams may be minimal. Conversely, if soil or bedrock sources dominate stream output, stream solute data can be more directly interpreted in terms of a geomorphological effect.

One of the difficulties in this type of endeavour is, however, gaining appropriate data. All that is often possible is an input–output analysis. Thus, it may be possible to establish that, for example, ecosystem output in streams is greater than atmospheric inputs, and thereby infer that the balance is made up by rock weathering. But for any one measurement of ionic concentration, in streamwater output, such mathematics do not reveal the source or pathway unless isotopic labelling has been employed. For example, Likens *et al.* (1977) concluded that bulk precipitation and gaseous input are the major sources for sulphur, nitrogen, chloride and phosphorus, while weathering is a major source for calcium, magnesium, potassium and sodium. An aggrading northern hardwood forest studied showed gains of nitrogen, sulphur, phosphorus and chloride and losses for silicon, calcium, sodium, aluminium, magnesium and potassium—the latter losses being made up by weathering of primary minerals. Their data on losses and gains from a number of study areas are shown in Table 11.4. Here stream outputs are compared with atmospheric inputs. It is clear that in many ecosystems, losses in streams are greater than atmospheric inputs, with a net ecosystem loss, presumably from weathering sources, including most data for calcium, magnesium, sodium and potassium and possibly sulphur, but with inputs greater than outputs for nitrogen and phosphorus. Even a detailed flux analysis for, say, calcium (their Figure 30, summarized in Table 11.5) reveals various options on pathways and the separation of the biotically and abiotically controlled solute sources is not clear. A similar point was also made by Trudgill (1977c) when discussing Ovington's (1962) data on calcium cycling (Trudgill, 1977c, pp. 99–101). Mass-balance data do not necessarily reveal sources or pathways and several interpretations of the data are possible. In particular, the partitioning of leaching losses to streams to sources from litter decomposition and from weathering is a crucial one in a geomorphological context. A direct weathering-to-leaching output is attributable to a direct geomorphological effect; a litter-to-leaching output may be only indirectly attributable to a geomorphological effect as the solute may have been derived from weathering at the root–mineral interface or from a relatively closed cycle of nutrients, such as that described by Jordan and Herrera (1981) for oligotrophic soil tropical ecosystems. Clearly, the identification of the most important solute pools is important in both interpretation of processes and in

Table 11.4. Input–output budgets (kg ha yr^{-1}) for study ecosystems (Likens *et al.*, 1977)

	inputs	Stream outputs	Loss (−) or Gain (+) to system −	+	*n*
Calcium					
1	2.2–14.9	6.9–148	0.1–167		9
2	2–6.7	4.5–57.7	54	0.7	10
3	9	53.8	44.8		1
4	0.8–21.8	4.7–43.1	3.9–21.3		2
Magnesium					
1	0.6–9	3.1–77.1	3–75		9
2	0.6–3	2.4–10.8	1.5–10.1		9
3	4.4	8.7	4.3		1
4	2.0–4.9	3.0–15.0	1–10.1		2
Sodium					
1	1.6–50	4.5–18.9	0.6–21		9
2	0.7–19.1	3.7–38.4	2.1–27		9
3	25.5–27.2	43.7–45.2	16.5–19.7		2
4	57.2	64.5	1.5–7.3		1
Potassium					
1	0.8–5	2.4–13	1.5–8		8
2	0.8–2.5	1–4.8	2.3	0.7	9
3	1.6–3.1	2.6–9	1.0–5.9		2
4	18.2	20.8	2.6		1
Nitrogen					
1	2–20.7	1–5.6		1–16.7	6
2	1.1–14.5	0.4–2.3		0.5–12.3	8
3	8.2	3		5.2	1
4	5.6	4.7		0.9	1
Phosphorus					
1	0.07–0.54	0.02–0.26	0.13	0.017–0.52	6
2	0.1–0.39	0.015–0.51	0.22	0.27	8
3	0.6	0.4		0.2	1
4	0.2	0.1		0.1	1
Sulphur					
1	7–21	7–38	17	7.5	5
2	1.4–11.5	3.2–28	19	0.9	6

1 = temperate deciduous; 2 = temperate coniferous and evergreen; 3 = temperate bog; 4 = tropical evergreen.

Table 11.5. Partitioning of calcium (Likens *et al.*, 1977)

	($kg\ ha^{-1}$)
Above-ground living biomass	383
Below-ground living biomass	101
Forest floor litter	370
Soil available	510
Mineral bound	9600 soil + 64,000 rock
Fluxes ($kg\ ha\ yr^{-1}$)	
Input	2.2
Uptake	62.2
Litter + mineralization	42.4
Weathering	21.1
Leaching	13.7
Export	13.9

terms of management. However, separation of the denudational component in stream output remains somewhat conjectural unless very detailed compartment models can be quantified, preferably with isotopic labelling. Many workers have concluded that biotic regulation of solute losses from ecosystems is very strong (e.g. Gosz *et al.*, 1973), especially on nutrient-poor soils where biotic nutrient conservation mechanisms are well-developed (Jordon and Herrera, 1981). It may be suggested that in more eutrophic soils, solute losses and denudation may be more closely related, with biological cycling involving relatively few of the nutrients available in the whole system. The converse would then be true for oligotrophic soils, where nutrients are more likely to be conserved by biotic regulation and solute losses would be more related to leakages from nutrient cycles, rather than to denudation.

Modelling the sources of solute outputs is, however, only one endeavour in solute geomorphology. The second endeavour is modelling the location of solutional erosion in relation to soil properties and topography, as mentioned earlier in this chapter. Prediction of the location of greatest erosion in terms of soil acidity and water status is liable to prove quite fruitful. Indeed, Furley (1971) and Whitfield and Furley (1971) use soil acidity and soil type to predict the location of maximum erosion on a hillslope. They show that the point of maximum erosion is often focused in mid-slope where the factors involved combine to maximize erosional processes. As discussed earlier (Figure 11.1), upslope areas may be acid but downslope water movement is limited; water movement increases downslope, increasing erosion until chemical equilibriation is approximated to. These sorts of models need much wider testing, and the distribution of solutional erosion rates under soils is liable to be a fruitful field for the future. So many slopes do approximate to the kind of pattern, in general, shown in Figure 11.1, that it should be possible to produce quite powerful predictive models of the distribution of solutional erosion in the landscape from a study of the distribution of soils and soil properties. Furley (1971) present data, for example, on pH and gradient, showing how steeper slopes are less acid; although referring to chalk, there is liable to be generality in this statement because of the fundamental downslope movement processes involved. Complicating factors arise on non-uniform parent materials, however. Often there is some cover on the plateau top, colluvial material collects at the slope foot and bedrock is more exposed in steeper, mid-slope sections. Thus, local departures will be seen from any general models. It can, however, be concluded that the ways ahead should usefully include more detailed compartment models—useful not only in geomorphology, but also in environmental management—and also closer field investigation and modelling of the relationship between solutional erosion and soil types on hillslopes.

REFERENCES

Ahnert, F. (1980). A note on measurements and experiments in geomorphology. *Zeitschrift für Geomorphologie*, Suppl. Band **35**, 1–10.

Brown, E. H. (1980). Historical geomorphology—principles and practice. *Zeitschrift für Geomorphologie*, Suppl. Band **36**, 9–15.

Brunsden, D. (1980). Applicable models of long term landform evolution. *Zeitschrift für Geomorphologie,* Suppl. Band **36**, 16–26.

Bull, P. A. (1980). The antiquity of caves and dolines in the British Isles. *Zeitschrift für Geomorphologie*, Suppl. Band **36**, 217–232.

Coward, J. M. H. (1975). quoted in Gascoyne, *qv*.

Crowther, J. (1983). A comparison of the rock tablet and water hardness methods for determining chemical erosion rates on Karst surfaces. *Zeitschrift für Geomorphologie*, **27**, 55–64.

Day, M. J., Leigh, C., and Young, V. & A. (1980). Weathering of rock discs in temperate and tropical soils. *Zeitschrift für Geomorphologie*, Suppl. Band **35**, 11–15.

Douglas, I. (1964). Intensity and periodicity in denudation processes with special reference to the removal of material in solution by rivers. *Zeitschrift für Geomorphologie*, **8**, 453–473.

Douglas, I. (1980). Climatic geomorphology. Present-day processes and landform evolution. Problems of interpretation. *Zeitschrift für Geomorphologie*, Suppl. Band **36**, 27–47.

Furley, P. A. (1971). Relationships between slope form and soil properties developed over chalk parent materials. In: *Slopes Form and Process* (Brunsden, Ed.), Institute of British Geographers, Special Publication No. 3.

Gascoyne, M., Ford, D. C., and Schwarcz, H. P. (1983). Rates of cave and landform development in the Yorkshire Dales from Speleotherm age data. *Earth Surface Processes and Landforms*, **89**, 557–568.

Gosz, J. R., Likens, G. E., and Bormann, F. H. (1973). Nutrient release from decomposing leaf and branch litter in the Hubbard Brook Forest, New Hampshire. *Ecological Monographs*, **43**, 173–191.

High, C., and Hanna, F. K. (1970). A method for the direct measurement of erosion on rock surfaces. British Geomorphological Research Group, Technical Bulletin No. 5.

Jennings, J. N. (1977). Limestone tablet experiments at Cooleman Plain, New South Wales, Australia, and their implications. *Abhandlugen zur Karst Hohlenkunde*, **15**, 526–538.

Jordan, C. F., and R. Herrera (1981). Tropical rain forests: are nutrients really critical? *American Naturalist*, **117**, 167–180.

Likens, G. E., *et al*. (1977). *Biogeochemistry of a Forested Ecosystem*, Springer-Verlag.

Mosley, M. P. (1982). Slopes and slope processes. *Progress in Physical Geography*, **6**, 115–121.

Saunders, I., and Young, A. (1983). Rates of surface processes on slopes, slope retreat and denudation. *Earth Surface Processes and Landforms*, **8**, 473–501.

Slaymaker, O. (1980). Geomorphic field experiments. Inventory and prospect *Zeitschrift für Geomorphologie*, Suppl. Band **35**, 183–194.

Trudgill, S. T. (1976). Rock weathering and climate: quantitative and experimental aspects. *Geomorphology and Climate* (E. Derbyshire, Ed.), London, Wiley, pp. 59–99.

Trudgill, S. T. (1977a). Problem in estimation of short-term variations in limestone erosion processes. *Earth Surface Processes*, **2**, 251–256.

Trudgill, S. T. (1977b). The role of a soil cover in limestone weathering, Cockpit Country, Jamaica. *Proceedings of the International Speleological Congress*, Sheffield, 1977, British Cave Research Association.

Trudgill (1977c). *Soil and Vegetation Systems*, Oxford University Press.

Trudgill, S. T. (1985). Limestone weathering under a soil cover and the evolution of limestone pavements, Malham district, N. Yorkshire, U.K. In: *New Directions in Karst* (M. M. Sweeting and K. Paterson, Eds.), Geo Books, Norwich.

Trudgill, S. T., High, C., and Hanna, F. K. (1981). Improvements to the micro-erosion meter (MEM). *British Geomorphological Research Group, Technical Bulletin*, **29**, 3–17.

Waylen, K. J. (1979). Chemical weathering in a drainage basin underlain by old red sandstone. *Earth Surface Processes*, **4**, 167–178.

Webb, B. W., and Walling, D. E. (1980). Stream solute studies and geomorphological research: some examples from the Exe Basin, Devon, U.K. *Zeitschrift für Geomorphologie*, Suppl. Band **36**, 245–263.

Whitfield, W. A. D., and Furley, P. A. (1981). The relationship between soil patterns and slope form in the Ettrick Association, south-east Scotland. In: *Slope form and process* (D. Brunsden, Ed.), Institute of British Geographers, Special Publication No. 3.

Young, A. (1969). Present rate of land erosion. *Nature*, **224**, 851–852.

Young, A. (1974). The rate of slope retreat. In: *Progress in Geomorphology* (E. H. Brown and R. S. Waters, Eds.), Institute of British Geographics, Special Publication No. 7, pp. 65–78.

Index